From Eudoxus to Einstein

A History of Mathematical Astronomy

Since man first looked towards the heavens, a great deal of effort has been put into trying to predict and explain the motions of the Sun, Moon, and planets. Developments in man's understanding have been closely linked to progress in the mathematical sciences. Whole new areas of mathematics, such as trigonometry, were developed to aid astronomical calculations, and on numerous occasions throughout history breakthroughs in astronomy have been possible only because of progress in mathematics. This book describes the theories of planetary motion that have been developed through the ages, beginning with the homocentric spheres of Eudoxus and ending with Einstein's general theory of relativity. It emphasizes the interaction between progress in astronomy and in mathematics, showing how the two have been linked inextricably since Babylonian times. This valuable text is accessible to a wide audience, from amateur astronomers to professional historians of astronomy.

CHRISTOPHER LINTON is head of the Mathematical Sciences Department and Reader in Applied Mathematics at Loughborough University. He is a member of the London Mathematical Society, the British Society for the History of Mathematics, and the British Society for the History of Science.

T0276021

From Eudoxus to Einstein

A History of Mathematical Astronomy

C. M. LINTON

Department of Mathematical Sciences, Loughborough University

CAMBRIDGE UNIVERSITY PRESS
Cambridge, New York, Melbourne, Madrid, Cape Town, Singapore, São Paulo

Cambridge University Press
The Edinburgh Building, Cambridge CB2 8RU, UK

Published in the United States of America by Cambridge University Press, New York

www.cambridge.org
Information on this title: www.cambridge.org/9780521827508

First published 2004
This digitally printed version 2007

A catalogue record for this publication is available from the British Library

ISBN 978-0-521-82750-8 hardback
ISBN 978-0-521-04571-1 paperback

To Matthew and Heleni

Contents

Preface

'It is extremely hard these days to write mathematical books, especially astronomical ones.' Thus begins Kepler's *New Astronomy*, published in 1609. While I would not attempt to claim that the problems I have faced are comparable with those Kepler had to tackle, he was after all working at the cutting edge of both the mathematics and astronomy of his day; the peculiar difficulties that face authors of books containing mathematics, but the major thrusts of which lie elsewhere, do not appear to have changed much over the passage of time. A balancing act is required: how does one include sufficient technical detail accurately to describe the procedures involved and not end up losing sight of the main focus? In the context of a historical survey there is another question: how faithful should one be to the form in which the mathematics was originally written? Here, historians and mathematicians writing about history tend to take different approaches. As a member of the latter category I have taken the view in this book that it is more important to understand what it is that was accomplished than precisely how it was achieved.

The mathematical details that are given hopefully serve to provide a more comprehensive description of the development of astronomical theories than is usually found in general histories of astronomy. Sometimes the mathematics is described using the methods available at the time, but on other occasions modern mathematical language has been used to make the discussion easier to follow for the modern reader. I have tried to ensure that anachronisms are labelled clearly as such and that the resulting mixture of old and new is both informative and not misleading.

There is an enormous number of books written about the history of astronomy, a vast subject spanning 4000 years of human history. The story of man's gradual appreciation of the nature of the heavens and the development of techniques for predicting the future positions of celestial bodies is fascinating. It can be appreciated on many levels, and histories of astronomy abound which are designed to appeal to the general public as well as to expert astronomers. Having a continuous history from 2000 BC to the present day, it is inevitable that astronomical thought has interacted with progress in many other branches of science (e.g. mathematics and physics) as well as areas such as philosophy and theology. Most astronomy today is concerned with objects which lie outside our Solar System but, until the nineteenth century, astronomy was the study of the motion of the Sun, Moon and planets measured against the background of 'fixed' stars. It is my aim in this book to describe the theories of planetary motion that have been developed, beginning with the homocentric spheres of Eudoxus and ending with Einstein's general theory of relativity, with particular regard to the interaction between progress in astronomy and in mathematics.

Since Babylonian times, astronomy and mathematics have been linked inextricably. The needs of astronomy have provided the impetus for research into many areas of mathematics, and whole new branches of mathematics (e. g. trigonometry) were developed to aid astronomical calculations. Conversely, on numerous occasions throughout history breakthroughs in astronomy have been possible only because of progress in mathematics. This two-way process pervades science:

> Mathematics is an indispensable medium by which and within which science expresses, formulates, continues, and communicates itself. And just as the language of true literacy not only specifies and expresses thoughts and processes of thinking but also creates them in turn, so does mathematics not only specify, clarify and make rigorously workable concepts and laws of science which perhaps, partially at least, could be put forward within it; but at certain crucial instances it is an indispensable constituent of their creation and emergence as well.
>
> (Bochner (1966), p. 256.)

However, most books on the history of astronomy provide only a cursory treatment of the underlying mathematics on the assumption that such topics would put people off, and in most books on the history of mathematics the interaction with astronomical thought is discussed only briefly. For example, in the preface to *The Cambridge Concise History of Astronomy* (1998), Michael Hoskin, describing the post-Newtonian celestial mechanicians, writes: 'But while their conclusions were of the keenest interest to astronomers, they were not themselves astronomers but mathematicians working in the service of astronomy, and so we can disregard the details of their calculations with a clear conscience.' All this is entirely understandable: the history of astronomy *can* be told in a non-mathematical way, and there is far more to the history of mathematics than just those parts which are relevant to astronomy.

At the other extreme, there are books that investigate episodes in the history of mathematical astronomy in great detail – O. Pedersen's *A Survey of the*

Almagest or N. M. Swerdlow and O. Neugebauer's *Mathematical Astronomy in Copernicus's De revolutionibus*, for example – but a huge amount of technical astronomical knowledge is required before these works become accessible.

This state of affairs can be frustrating for mathematicians, or people interested in mathematics, who want to explore the history of astronomy but who are not already well-versed in technical astronomy. For example, it is difficult to find out about the mathematical problems that Kepler struggled with in his *New Astronomy* without access to technical journals, and yet a proper appreciation of Kepler's contributions to astronomy surely requires a knowledge of the mathematical difficulties which he had to overcome. It is my aim in this book to provide a history of theoretical astronomy that to some extent fills this gap. The book is not simply a study of the interactions between mathematical and astronomical developments – its main aim is to tell the story of how we have come to understand the motions of the Sun, Moon and planets – but it does so recognizing that the mathematics that challenged astronomers throughout preceding millennia forms an integral part of the tale.

The first mathematical model of the Universe was that of Eudoxus in the fourth century BC, and this provides (more or less) the starting point. Eudoxus' model represented the motions of the celestial bodies in a qualitative way, but was not satisfactory when it came to quantitative prediction. Over the centuries, through the work of men such as Ptolemy, Ibn al-Shāṭir, Copernicus, Tycho Brahe, Kepler, and Newton, models of the heavens came to reproduce the results of observations with greater and greater accuracy. During the eighteenth and nineteenth centuries, astronomers believed that the fundamental principles on which Newtonian dynamics rested provided the final word on celestial mechanics (a phrase coined by Laplace) though mathematical progress would be required to enable people to extract more information from Newton's theory of universal gravitation.

Discoveries of new celestial bodies led to new challenges, and the sophistication of the mathematical theories designed to meet them increased. The discovery of Neptune in 1846 brought the power of theoretical astronomy to the attention of a wide audience and cemented its position as the ultimate exact science. And yet, within 70 years, its omnipotence had been challenged successfully – twice. First, new techniques for analysing problems in mechanics led Poincaré to the conclusion that there was a theoretical limit to the predictive power of the differential equations of celestial mechanics; here were the beginnings of 'chaos theory'. Second, and finally as far as this book is concerned, Einstein showed in 1915 that the whole of Newtonian mechanics is an approximation valid only at speeds much less than that of light. The general theory of relativity provided the explanation for the one planetary phenomenon that had refused to succumb to the power of Newtonian theory: the anomalous advance of the perihelion of Mercury. Since 1915, theory and observation can be said to be in agreement within the limits of observational accuracy. This provides a suitable endpoint.

The book is intended for anyone who is interested in the history of astronomy and who is not afraid of mathematics. Large parts of the book can be read by someone with only a rudimentary mathematical knowledge, but there are parts where the level of mathematics required is that taught at undergraduate level. No prior knowledge of astronomy is required, however, and so the book will be suitable particularly for undergraduates reading a mathematics-based degree programme, and who are taking a course in the history of either astronomy or mathematics. It was through teaching a course on the history of mathematics at Loughborough University that I first saw the need for, and decided to write, this book.

For those wishing to explore specific topics in more detail, extensive references are provided via footnotes. These include both works of scholarship and popular accounts. For reasons of practical necessity I have restricted myself to sources written in English – which of course, rules out the original works of most of the people discussed in the book – but every effort has been made not to propagate errors and misconceptions from the secondary literature. Some will remain, and some errors will have been committed that I cannot blame anyone for but myself. For these I apologize in advance.

I would like to express my gratitude to colleagues and friends for their encouragement and, in particular, to Peter Shiu for reading a preliminary draft and making numerous helpful suggestions. Finally, I would like to thank Joanna for her patience.

1

Introduction

Basic astronomical phenomena

A great deal of human effort has been expended over the past 4000 years or so in trying to predict and explain the motions of the Sun, Moon, planets and stars. Since Babylonian times, this quest has relied heavily on mathematics, and the developments in man's understanding of the heavens have been inextricably linked to progress in the mathematical sciences. As far as the Sun, Moon and planets are concerned, attempts at an explanation of their motion using mathematical techniques began in ancient Greece and the first mathematical model of the heavens was constructed by Eudoxus in the fourth century BC. The final piece of the celestial jigsaw was supplied by Einstein's theory of general relativity in the twentieth century and, since 1915, all major phenomena associated with planetary motion have possessed theoretical explanations. This should not be taken to imply that we know everything about the future positions of the bodies in our Solar System. Indeed, the researches of Poincaré in the late nineteenth century have led us to a much clearer understanding of the limitations of theoretical predictions. Before embarking on the fascinating story that describes the endeavours of men such as Ptolemy, Copernicus, Kepler, Newton, and Laplace, we will begin by familiarizing ourselves with the various heavenly phenomena that people have for so long sought to explain.

For a variety of reasons, early astronomers thought that the Earth was stationary and that the heavenly bodies moved around it. On the face of it this is an extremely natural assumption to make, the evidence to the contrary is far from obvious. The fact that the natural interpretation of the situation is wrong is one of the reasons why astronomy has such an absorbing history. Progress in man's understanding of the nature of the Universe has not been a gradual refinement of simple and intuitive ideas, but a struggle to replace the seemingly obvious by, what to many, was patently absurd. Nowadays, we accept that the idea of a

1

stationary Earth at the centre of the Universe is wrong and that the Earth, Sun, planets, and stars are all in motion relative to each other, but in order to understand early approaches to astronomy it is often helpful to throw away our modern notions and to try and picture the Universe as the ancients would have done.

The most obvious of the objects visible in the sky is the Sun, and systematic observations of its motion across the sky were made by the Babylonian and Egyptian civilizations using a *gnomon*, which was simply a primitive sundial consisting of a stick placed vertically on a horizontal surface.[1] During the course of a day both the length and direction of the *gnomon*'s shadow vary; when the Sun is high in the sky the shadow is short and although the minimum length of the shadow varies from day to day, the direction of the Sun at the time when the shadow is shortest is always the same. For an observer in the northern hemisphere (and we will assume throughout that the observer is in the northern hemisphere) this direction defines due north. Each day, the Sun rises in the eastern part of the sky, travels across the sky reaching its highest point in the south in the middle of the day, and then sets in the western part of the sky. However, the amount of time for which the Sun is visible and its daily path across the sky are not constant, e.g. the points on the horizon at which the Sun rises and sets undergo variations. Observations over a sufficient period of time would show that these variations repeat, and that the period with which they do so is related to the weather (in most parts of the world); in this way one is led to the concept of a year with its seasonal changes in climate.

In Egypt, where the seasons are not particularly noticeable, this period was recognized to be the same as that associated with the flooding of the Nile and, hence, crucial to people's lives. The Egyptians noticed that for part of the year Sirius, the brightest star in the sky, was invisible as it was too close to the Sun, but that the floods began soon after the time when the star rose in the eastern sky just before dawn, and they measured the period between these so-called heliacal risings at a little over 365 days. We call this period the **tropical year**, and it forms the basis of our calendar. For calendrical purposes, the Egyptians used a year of precisely 365 days and the simplicity of such a system was beneficial in the extreme to astronomers; indeed, it was used by Copernicus as late as the sixteenth century. The division of a day into 24 h is also of ancient origin, though originally these were of unequal length and depended on one's location and the time of year in such a way that there were always 12 h of daylight and 12 h of darkness.[2]

[1] The early use of shadow sticks is described in Fermor (1997).
[2] For the purposes of everyday life, the change to 24 equal hours did not take place until the fourteenth century with the advent of the mechanical clock, but hours of equal length were used

The next most obvious heavenly body is the Moon, and this is seen to undergo not only variations in its track across the sky but also in its form, changing within a period of about $29\frac{1}{2}$ days from thin crescent to circular disc (**full moon**) back to thin crescent again and then disappearing for two or three nights (**new moon**). These changes in form are known as the 'phases of the Moon' and are associated with another characteristic length of time, known as a **lunation** or **synodic month**, that formed the bases of many ancient calendars. When the Moon is full, it is opposed diametrically to the Sun, and we say that the Sun and the Moon are in **opposition**. When the Sun and the Moon are in the same direction – which happens at the new moon – they are said to be in **conjunction**. This same terminology is used for any pair of heavenly bodies. Oppositions and conjunctions are known collectively as **syzygies**.

If observers look toward the sky on a clear night, they see a large number of stars that appear to lie on a spherical surface with themselves at the centre; this imaginary surface is called the **celestial sphere**. The stars appear to move over the surface of this sphere but the distance between them remains constant, and it was for this reason that many ancient astronomers regarded the celestial sphere as a real entity with the stars attached physically to it. This physically real celestial sphere had its centre at the centre of the Earth, but it is more convenient to take the imaginary celestial sphere (illustrated in Figure 1.1) as having its centre at the observer O. The point on the celestial sphere directly above the observer is called the **zenith**, Z.

Although we now know that the stars are not all the same distance from the Earth, it is not at all hard to imagine when looking at them that they are equidistant from us, though we have no immediately simple way of determining just how far away they are. Ignoring the question of distance, the celestial sphere gives us an easy way of describing the direction of a star: it is the same as the direction of the line that points from an observer to the point on the celestial sphere where the star appears to lie. We can then talk about the **apparent distance** between two stars as the angle between the lines pointing towards each of them.

Some careful observation will show that the stars move as if they were attached to the celestial sphere and it was rotating about an axis that intersects the sphere at a point in the northern sky. This point, which nowadays is close to the star Polaris, is called the **north pole**, P_N, of the celestial sphere and we can easily imagine that there is a **south pole**, P_S, in the part of the sky invisible to us.

from ancient times in the context of scientific discussions. A division of the day into equal temporal units was used in China from the second century BC. See Dohrn-van Rossum (1996) for a complete history of the hour.

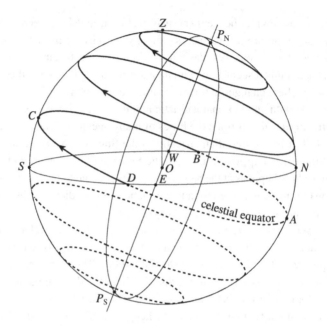

Fig. 1.1. The celestial sphere.

The stars move around the north pole in an anticlockwise direction with some stars always above the **horizon** ($NWSE$), so-called **circumpolar** stars (or, as some ancient observers termed them, 'those that know no weariness'),[3] and some dipping below the horizon as indicated by the dashed lines in Figure 1.1. We can imagine easily that there are other stars that remain invisible below the horizon. The circle $ABCD$, which lies in a plane through O midway between the poles, is called the **celestial equator**. If we were at the north pole of the Earth, then the north pole of the celestial sphere would correspond to the zenith and all visible stars would be circumpolar, whereas if we were on the equator, the north pole of the celestial sphere would be on the horizon and there would not be any circumpolar stars. For a general point in the northern hemisphere, the angle between OZ and OP_N is $90°$ minus the latitude of the observer. The celestial sphere completes 1 revolution in about 1 day, and this rotation is called the 'daily' or **diurnal motion**. The Moon also participates in this daily rotation

---

[3] It is these circumpolar stars that are most suggestive of the spherical nature of the heavens. Ptolemy, in AD second century stated explicitly that these stars were instrumental in leading astronomers to the concept of a celestial sphere (PTOLEMY *Almagest*, Book I, 3 (see Toomer (1984)).

as does the Sun, though since the stars and the Sun are not seen at the same time, this is harder to appreciate.

Provided we have an accurate means of measuring time, we can observe that the stars actually complete a revolution about the pole in about 23 h 56 min, so that they return to the same place at the same time in 1 year.[4] For an ancient observer, this observation was more difficult. However, by observing stars rising above or setting below the horizon at about sunrise or sunset, one notices that the stars are moving faster than, and therefore gradually changing their position with respect to the Sun, returning to the same position after 1 year. If we regard the stars as fixed on the celestial sphere, then the Sun must move relative to the stars in the direction opposite to the diurnal motion (i.e. from west to east), completing one circuit of the celestial sphere in a year. There are other features of the motion of the Sun that need to be explained. For example, the points on the horizon at which the Sun rises and sets vary from day to day, as does the midday height of the Sun and, indeed, the length of time the Sun is above the horizon. A great deal of careful observation led ancient astronomers to the conclusion that the complex motion of the Sun was built up from two much simpler motions: the first was the daily rotation of the celestial sphere and the second was a much slower **annual motion** that took place on an oblique great circle ($AQCR$ in Figure 1.2). This circle is called the **ecliptic**, the reason being that for eclipses to occur, the Moon must be on or near it, and the angle at which it cuts the celestial equator is called the **obliquity** of the ecliptic, ε. Nowadays, the value of ε is about $23° 27'$, but the obliquity actually decreases very slowly with time. In 3000 BC, it was about $24° 2'$.[5]

In order to describe the position of a body on the celestial sphere, we need some sort of coordinates. One possibility, introduced by Babylonian astronomers, is to measure angles relative to the ecliptic. The **ecliptic longitude** measures the distance around the ecliptic, whereas the **ecliptic latitude** measures the distance north or south of this line. Hence, by definition, the ecliptic latitude of the Sun is $0°$. A common alternative is to use a system based on the celestial equator, and here the distance around the equator is known as the **right ascension**, whereas the distance north or south of the equator is the **declination**.

From Figure 1.1, it can be seen that when the Sun lies on the celestial equator the lengths of the day and night are equal (since the arcs DCB and BAD are the same length) and, hence, the points A and C in Figure 1.2 are known as the **equinoctial points**, and the times when the Sun is at these points are the **equinoxes**. The time at which the Sun is at A (i.e. when it crosses the celestial

---

[4] Twenty-four hours divided by 365.25 (the number of days in 1 year) is just under 4 min.
[5] Thurston (1994).

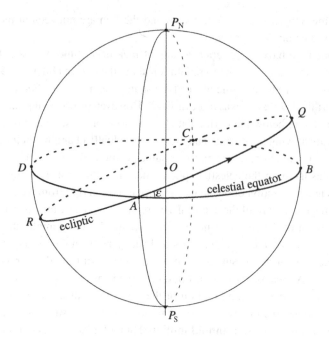

Fig. 1.2. The annual motion of the Sun.

equator from south to north) is the **spring** or **vernal equinox**, whereas the Sun
is at C at the **autumnal equinox**. Ancient astronomers were aware of the fact
that the Sun does not move around the ecliptic at a uniform rate – the autumnal
equinox is about 186 days after the vernal equinox, but only 179 days pass
before the autumnal equinox is reached again.

The Sun is at point Q – its most northerly extreme – at the **summer solstice**,
and at its most southerly extreme, R, at the **winter solstice**. The points Q and
R are known as the **solstitial points**. Observations over a short period of time
suggest that the equinoctial and solstitial points remain fixed with respect to
the fixed stars, but actually they do not; the equinoctial points rotate around
the celestial equator with a period of about 26 000 years. Surprisingly, perhaps,
this phenomenon, called the 'precession of the equinoxes', was recognized as
early as the second century BC. The fact that the equinoxes precess means that
the time taken for the Sun to return to the same position on the ecliptic (the
tropical year) is different from the time taken to return to a fixed star. This latter
period is known as the **sidereal year**. The modern values for these periods
are approximately 365.242 days for the tropical year and 365.256 days for the
sidereal year. When we use the term 'year' without qualification we mean the
tropical year, though in most situations the difference is unimportant.

Table 1.1. *The periods of the moon.*

Month	Length in days
synodic	29.531
sidereal	27.322
draconitic	27.212
anomalistic	27.555

Observations of the Moon show that its monthly path round the celestial sphere is also a great circle that is very close to the ecliptic but inclined slightly to it. The orbit of the Moon crosses that of the Sun at two points called 'the **nodes** of the orbit' and the straight line joining these points is called the **nodal line**. At one of its nodes, the Moon is moving from south of the ecliptic to north of it, and we call this the **ascending node**, the other point of crossing being the **descending node**. The period of the rotation of the Moon around the celestial sphere is known as the **sidereal month**, which is, of course, different from the synodic month because that measures the motion of the Moon with respect to the Sun, which itself is moving around the ecliptic. There are two other important periods associated with the Moon which were recognized in ancient times. First, there is the **draconitic month**, which is the period between successive ascending (or descending) nodes,[6] or equivalently the time it takes for the Moon to return to the same distance from the ecliptic (i.e. to the same ecliptic latitude) while travelling in the same direction. Second, the speed with which the Moon moves across the sky relative to the stars is variable (varying between about 12 and 15° per day) and the period of time it takes for the Moon to return to the same speed is called the **anomalistic month**. The modern mean values for the various periods of the Moon are shown in Table 1.1, though the actual values for any given month may differ from these by as much as 7 h.

Perhaps the most dramatic of celestial events which are easily observable are **eclipses**, and these are of two types: lunar and solar (see Figure 1.3).[7] In a **lunar eclipse** the Earth passes between the Sun and the Moon and casts its shadow over the Moon's surface, whereas in a **solar eclipse** it is the shadow

---

[6] In medieval times, the part of the Moon's orbit south of the ecliptic was known as the 'dragon' (which devoured the Moon during eclipses) and from this we get the terminology 'dragon's head' for the ascending node and 'dragon's tail' for the descending node. An example of this usage can be found in Chaucer's *Treatise on the Astrolabe* (1391), one of the oldest surviving scientific works written in English. The periods between successive nodes has, over time, been termed the dracontic, draconic and draconitic month, the words deriving from the Greek for 'dragon'.

[7] Eclipses can be categorized further depending on the precise arrangement of the three bodies (see, for example, Payne-Gaposchkin (1961)).

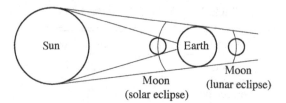

Fig. 1.3. The nature of eclipses.

of Moon that passes over the Earth. Thus, a lunar eclipse takes place when the Moon is in opposition to the Sun (i.e. when it is full) and simultaneously at (or near) one of its nodes. Similarly, a solar eclipse occurs when the passage of the Moon corresponds, through one of its nodes, to a conjunction with the Sun. Now, if the nodal line of the orbit of the Moon had a fixed orientation with respect to the stars, one of its nodes would lie directly between the Earth and the Sun every half year. Actually, the nodal line rotates (a fact known to the ancient Greeks) and as a result a node lies on the Earth–Sun line once every 173.3 days. A solar eclipse will occur if there is a new moon close enough to this time. It turns out that the shadow of the Moon will touch the Earth if the new moon appears within about $18\frac{3}{4}$ days either side of the alignment of a node and, thus, there is a $37\frac{1}{2}$-day eclipse season every 173 days during which a solar eclipse may be visible.

A similar analysis applies to lunar eclipses, but since the shadow cone of the Earth narrows as one moves further from the Sun, the full moon needs to appear nearer to the point of alignment than the $37\frac{1}{2}$-day window that exists for solar eclipses. As a result, lunar eclipses actually are less frequent than the solar variety. In any one calendar year, there are between two and five solar eclipses, but there can be no more than three lunar eclipses per year and there might not be any. The reason solar eclipses are seen so much more rarely than lunar ones is, of course, due to the fact that the area of the Earth covered by the shadow of the Moon in a solar eclipse is very small, and so a particular solar eclipse is visible from only a small part of the surface of the Earth. On the other hand, the shadow of the Earth can block out the light of the Sun from the whole of the surface of the Moon, and such a lunar eclipse will be visible from anywhere on the Earth from which the Moon normally would be visible.

Eclipses of both the Sun and Moon were observed by ancient astronomers and would no doubt have aroused great interest. Over a period of time it would have become clear that eclipses of the Moon occur only when the Moon is full, and that eclipses of the Sun occur only at new moon. The fact that solar eclipses are caused by the Moon passing in front of the Sun would not have been hard to appreciate, but the fact that eclipses of the Moon are caused by the shadow of

Table 1.2. *The zodiacal and synodic periods*
of the planets.

	Zodiacal period (years)	Synodic period (days)
Mercury	1	115
Venus	1	584
Mars	1.88	780
Jupiter	11.86	399
Saturn	29.46	378

the Earth passing over it requires a rather deeper understanding of the situation and was probably not realized until much later. The reasons behind both types of eclipse were, however, fully understood by the time of the ancient Greeks.

Ancient astronomers were also aware that five of the star-like objects in the sky changed their position relative to the other stars. These five objects – now named after the Roman gods Mercury, Venus, Mars, Jupiter and Saturn – are the **planets**, from the Greek for 'wanderer'. Careful observations of these objects reveal that, like the Sun, as well as participating in the daily rotation of the heavens, they, too, move around the celestial sphere though with differing periods, and also that while they move predominantly in the same direction as the Sun – from west to east – they sometimes switch back and, for a time, move from east to west in so-called **retrograde** motions. The periods of retrograde motion are linked to the motion of the Sun – the centre of the retrograde motion for Mars, Jupiter and Saturn always occurring when the planet is in opposition to the Sun – whereas for Mercury and Venus this phenomenon occurs at conjunction. The planets also remain close to the ecliptic, the maximum deviation for any of them being 8°, and thus all the wandering heavenly bodies can be found within a strip on the celestial sphere 16° across centred on the ecliptic. This strip, therefore, is very important, and is known as the **zodiac** and was divided by the Babylonians into twelve equal parts: the signs of the zodiac. The average time it takes for a planet to complete 1 revolution around the ecliptic is its **zodiacal period**, and the average period between successive occurrences of retrograde motion is known as the **synodic period** of the planet; for the five planets visible with the naked eye these are given in Table 1.2.

The fact that the Sun, Moon and planets are nearer the Earth than the stars may have been suggested by **occultations**, the temporary disappearance of one heavenly body behind another. The most obvious example is the Moon, which sometimes eclipses the Sun and also is easily observed to pass in front of the planets and stars. Thus, all ancient astronomers agreed that the Moon was the closest of the heavenly bodies. This view was, of course, supported by the Moon's enormous size compared to the other planets and by the fact that various details could be made out on its surface. However, there is no easy way of determining how far away each celestial body is, and so ancient astronomers took the view that their distance probably was related to the speed with which they traversed the celestial sphere. This was consistent with the fact that the Moon is the swiftest of the heavenly bodies.

The Sun's journey around the ecliptic takes, by definition, 1 year, and it was found that Mars returned to the same place among the stars on average after about 2 years, Jupiter after 12 years and Saturn after $29\frac{1}{2}$ years. These planets, being slower than the Sun, were therefore thought to be further away and, hence, higher in the sky, and were termed the **superior** planets. On the other hand, Venus and Mercury both complete 1 revolution about the Earth in the same average time (1 year) and there was considerable disagreement about the correct ordering of these two planets. Eventually, a consensus was reached that they were closer to the Earth than the Sun, and they became known as the **inferior** planets. The observed behaviour of the superior and the inferior planets is different in another important way: Mercury and Venus are never seen far from the Sun, the maximum differences in longitude being about 29 and 47°, respectively, and so they are only ever visible near dawn or sunset, whereas the difference in longitude between the Sun and Mars, Jupiter or Saturn can take any value and they can be visible at any time of night.

The Sun, Moon and five planets, together with the fixed stars, were the only heavenly bodies recognized in antiquity, and this situation did not change until Galileo pointed a telescope at the night sky in 1609–10 and discovered the moons of Jupiter. No further planets were discovered until the eighteenth century, and comets were not considered as celestial bodies until the pioneering observations of Tycho Brahe in the late sixteenth century (prior to this they were thought of as atmospheric phenomena).

The problem for astronomers, then, was to explain the phenomena that have been described above. In the beginning, attempts were limited to qualitative explanation, but the models that were developed could not produce accurate quantitative predictions for astronomical events. The quantitative problem is much, much harder, and it exercised many of the greatest minds over a period of more than 2000 years, from Babylonian times to the twentieth century.

Babylonian astronomy

The heavenly phenomena were of great importance to the Babylonians, as they were perceived as omens and just about every possible astronomical event had some significance. For example, when it came to the retrograde motion of the planets it was not simply the retrograde motion itself, but also where it took place with respect to the stars, that was important:

> When Mars comes out of the constellation Scorpius, turns and reenters Scorpius, its interpretation is this: ... do not neglect your guard: the king should not go outdoors on an evil day.[8]

In order to improve their abilities to predict such phenomena, the Babylonians developed a tradition of observing and recording celestial events; examples of Babylonian astronomical observations exist which date back as far as 1600 BC. These are not particularly accurate, but show that the practice of observing and then recording the results has a long history in Babylonian culture. Scribes systematically began documenting celestial phenomena (e.g. eclipses) in about the eighth century BC, and over the next 700 or 800 years produced a large quantity of data that would later prove invaluable to Greek astronomers.[9] In order to carry out their work, astrologers needed tables of the future positions of heavenly bodies (which we now call 'ephemerides') and this desire was the driving force behind the production of such tables for over 2000 years.

The Babylonian number system was sexagesimal (base 60) and positional.[10] The reason for the use of 60 is unclear, but it may well have been because 60 is divisible exactly by lots of small integers, so many calculations can be done without the use of fractions. Nevertheless, because of the positional nature of the system, the treatment of fractions was vastly superior to the methods employed in Egypt and Greece, and as a result it was used by virtually all Western astronomers up until the sixteenth century, when decimal fractions began to take over. The superiority of the positional system of numeration helps to explain why quantitative astronomy reached a much greater level of sophistication in Babylonia than in other contemporary civilizations. The influence of the sexagesimal number system is still apparent in our measurement of time and angles, with 1 h being divided into 60 min, and so on. The Babylonians are

---

[8] This is just one of many such examples described in Swerdlow (1998a).

[9] The earliest Babylonian observations that the later Greek astronomers were able to use were made during the reign of Nabonassar (747–33 BC) and thus this period became a key reference point in the Greek system of reckoning time. Babylonian astronomers are often referred to as Chaldeans, a tribe from the southern part of the area known as Babylonia that established a new dynasty in 625 BC. Our knowledge of Babylonian astronomy was transformed during the twentieth century with the deciphering of hundreds of astronomical documents. The way this new information was absorbed into mainstream history of science is described in Rochberg (2002).

[10] We will follow standard modern practice of writing sexagesimal numbers with a semicolon representing the sexagesimal point. Thus, $1, 24; 20 = 60 + 24 + \frac{20}{60} = 84\frac{1}{3}$.

also responsible for dividing the circle up into 360 equal parts, which we call 'degrees', though this would not appear to have been a direct consequence of their sexagesimal system, but seems to have been due to the length of the year being about 365 days, and so in 1 day the Sun moves about 1° with respect to the stars.

The Babylonians began to turn their observational records into a mathematical theory around 500 BC. It was also around this time that they created the concept of the zodiac, dividing the strip straddling the ecliptic into twelve equal subdivisions of 30° each, so as to aid computations. From their data they realized that there were fixed periods of time associated with many astronomical events and, since knowledge of these events had great benefits in terms of prognostication, they spent considerable time and effort determining these periods accurately. Typically, they would develop relationships in which m intervals of one type were equated with n different intervals. For example, they discovered the approximate equivalence between 19 years and 235 synodic months; a relationship that formed the basis of their luni-solar calendar and is now known as the 'Metonic cycle' after the Athenian who tried unsuccessfully to incorporate it into the Greek calendar (since $235 = 19 \times 12 + 7$, years consisted of 12 months, with 7 intercalary months every 19 years). Another example is the Venus cycle in which the planet completes five synodic periods in 8 years ($8 \times 365 = 5 \times 584$).

The Babylonians developed sophisticated mathematical techniques that enabled them to develop an accurate predictive astronomy.[11] Most Babylonian astronomy was concerned with the motion of the Moon (e.g. determining when the new moon would first become visible, since this was the basis of their calendar), but they also tabulated planetary oppositions and conjunctions. Our knowledge of their astronomy comes largely from ephemerides, which list the positions of the Moon or planets over a number of years at regularly spaced intervals. The values listed are the results of calculations rather than of observations, so they reveal the underlying mathematical techniques that the Babylonians used to model the heavenly phenomena.

Fundamental to the phenomena the Babylonians were interested in was the motion of the Sun, and from their ephemerides it has been possible to deduce how they modelled the Sun's non-uniform motion. Interestingly, there were two methods, both of which were in use during the whole period for which ephemerides have been found (the last three centuries BC). The simpler version, known as 'System A', implied a motion for the Sun of 30° each mean

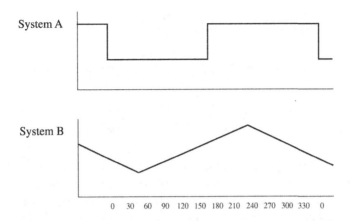

Fig. 1.4. Variations in the solar motion in Babylonian astronomy.

synodic month for just over half the year and then 28; 7, 30° for the remainder. Calculations reveal that this system corresponds to the relation

$$1 \text{ year} = 12; 22, 8 \text{ synodic months.}$$

In 'System B', rather than have the Sun's motion change abruptly at two points during the year, it was made to vary continuously, oscillating between the maximum and minimum values of 30; 1, 59° and 28; 10, 39, 40° per mean synodic month. In order to represent this, the Babylonians used alternate increasing and decreasing arithmetic progressions that are best illustrated as a zigzag function. The relation between the length of the year and that of the synodic month was the same as in System A. Both systems are illustrated in Figure 1.4, in which the numbers denote the ecliptic longitude of the Sun measured from the vernal equinox. Ephemerides also exist that concern the daily motion of the Moon in which its speed is treated in a similar way by means of a zigzag function.

The planetary phenomena are more complex, but the mathematical techniques with which the Babylonians attempted to model them were essentially the same as the ones they used to represent the Sun's motion.[12] For the case of a superior planet, there were five characteristic phenomena that concerned Babylonian astronomers, and these are shown schematically in Figure 1.5. When the planet is in conjunction, it cannot be seen because of the Sun's glare but as it moves around the ecliptic relative to the Sun there comes a point, Γ, where it first becomes visible just before dawn (its heliacal rising). Similarly, the planet

---

[12] Detailed descriptions of the technicalities of Babylonian astronomy can be found in, for example, Aaboe (1958, 1964, 1980), Swerdlow (1998a, 1999), Steele (2000, 2003).

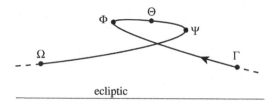

Fig. 1.5. The characteristic phenomena for a superior planet in Babylonian astronomy.

eventually will disappear as it approaches conjunction again, and its last appearance just after sunset (its heliacal setting) is labelled Ω. In-between, the planet will undergo a period of retrograde motion around its opposition, Θ, and the two positions at which its motion along the ecliptic is stationary, Φ and Ψ, make up the points of interest. The main problem of Babylonian planetary astronomy was, given the longitude of one of these phenomena for a particular planet, to predict the longitude of the next occurrence of the same phenomenon. It was these differences in longitude that the Babylonians attempted to model with their zigzag (System B) or step (System A) functions.

As well as developing fairly sophisticated techniques with which to model the periodic phenomena they observed in the heavens, the Babylonians achieved a high level of skill in numerical techniques. What they do not appear to have done, however, was seek to explain what went on in the heavens in terms of a geometrical or physical model – their approach was entirely arithmetical. The desire to construct a model of the Universe, a cosmology, which corresponds to the observed celestial phenomena has its origins in the works of ancient Greek philosophers. However without the accurate quantitative astronomy that the Babylonians created, and the vast amount of data they catalogued, Greek astronomers would not have been able to achieve what they did.

Early Greek astronomy

Early cosmologies were based heavily on man's earthly experiences and owed little to accurate astronomical observation. They fulfilled a well-documented psychological need, providing a stage for the drama of daily life, for the actions of the gods and also supplying meaning to man's existence. Before the ancient Greek civilization, cosmologies were constructed that made no attempt to explain any but the most rudimentary of astronomical phenomena. For example, the Egyptians explained the motion of the Sun by having the Sun god Ra take a daily trip through the air and then each night make a passage through the water. The Moon, on the other hand, was attacked on the fifteenth day of each month by a sow and after 2 weeks of agony the Moon died and was reborn.

Gradually, as the Greek civilization progressed, cosmologies became more sophisticated, and the idea that the way to find out about the nature of the Universe is through accurate astronomical observation, now a well-established tenet of Western thought, developed. One of the earliest examples we have showing Greek interest in the heavens comes from the writings of Homer in which the Earth is a flat circular disc surrounded by the Ocean river and covered by the vault of heaven. The Homeric epics show that in the eighth century BC the Greek understanding of the natural world was very primitive, and their awareness of astronomical phenomena was considerably less than that of their Babylonian contemporaries. Further evidence of early Greek astronomical knowledge can be obtained from Hesiod's poem 'Works and Days', written about a century after Homer's time, which among other things, contains what might be described as an agricultural calendar. Hesiod described how the seasons are related to the heliacal risings and settings of certain stars and, in contrast to Homer, connected astronomical events to the lives of ordinary people.

Up until about 300 BC, Greek astronomy was almost entirely qualitative. Indeed, there is no evidence that the accurate prediction of heavenly phenomena was even thought of as a desirable goal. This situation changed when the Greeks came into contact with the quantitative methods of Babylonian astronomers during the expansion of their empire under Alexander the Great. Before then, however, celestial phenomena were the subject of a great deal of philosophical debate that provided the basis on which later astronomers could build. Four major schools of philosophical thought existed during the 300 or so years prior to the construction of the first mathematical model of the Universe by Eudoxus.[13] There were the Ionians, a group founded by Thales of Miletus (in Asia Minor, modern-day Turkey) in about 600 BC. During the sixth century BC the region of Asia Minor went through a considerable upheaval due to the expansion of the Persian empire and many philosophers travelled to other parts of the Greek empire. One member of the Ionian School, Xenophanes of Colophon, migrated to southern Italy and eventually settled in Elea, which became an important philosophical centre. Perhaps the most famous of the early Greek philosophical schools was that set up by Pythagoras (who is said to have been taught by Thales), again in southern Italy, and finally there was the celebrated school centred around Plato's Academy in Athens. It should always be borne in mind

---

[13] For a modern introduction to ancient Greek philosophy, see Kenny (1998).

that it is difficult to be certain about the specific theories espoused by these early thinkers, since most of our knowledge comes from remarks of later (and not always reliable) authors.

The Ionian philosophers began the process of determining the nature of things but did not progress very far towards a rational description of the Universe. We know very little about the people who made up this School and none of their writings have survived. What little we do know is due to Greek historians and commentators. As far as astronomy was concerned, the philosophers of the Ionian School thought the Earth was flat, and they had a very poor understanding of the nature of the Sun and Moon.

According to Aristotle, the founder of Ionian natural philosophy was Thales, whose ideas (cultivated during travels around Egypt, the Mediterranean and Near East) were built around the fundamental belief that water is the essence of all things. The significance of this is the suggestion that there is an underlying unity to physical phenomena and, hence, that nature is not quite as haphazard as our senses would have us believe. In terms of the structure of the Universe, however, Thales' ideas were about as primitive as those found in Homer. Later authors have ascribed knowledge of the sphericity of the Earth and of the causes of eclipses to him, though it is extremely doubtful that Thales actually possessed this knowledge. He is also credited with predicting a solar eclipse in 585 BC though this is probably false also.[14] The sayings which are attributed to him show us that over 2500 years ago Greek intellectuals were attempting to understand concepts such as space and time:

> Of all things that are, the most ancient is God, for he is uncreated; the most
> beautiful is the universe, for it is God's workmanship; the greatest is Space, for it
> contains everything; the swiftest is Mind, for it speeds everywhere; the strongest is
> necessity, for it masters all; and the wisest Time, for it brings everything to light.[15]

Thales also holds a significant place in the history of mathematics. He is reputed to have been the first person to *prove* geometrical theorems and a number of such proofs are attributed to him, though their nature is unclear. Certainly, it is true that the use of logical proofs in mathematics was developed over the three centuries before Euclid wrote the *Elements* in about 300 BC, and that later Greek writers credited Thales with inventing this deductive method, but the precise nature of Thales' contribution is uncertain.

Perhaps the most important of the Ionian philosophers was Anaximander, a pupil of Thales. Almost all we know of his doctrines comes ultimately from Theophrastus' *Physical Opinions*. In Anaximander's system, the essence of all

[14] See Neugebauer (1969), though an explanation of how Thales might have achieved this feat is given by Hartner (1969). See also Aaboe (1972).
[15] Quoted from Heath (1932).

things was not water or any other physical quality, but the infinite, and there are similarities between Anaximander's 'infinite medium' and the concept of the ether, that played a significant role in science right up to the twentieth century.[16]

He described the Sun, Moon and stars as hollow wheels full of fire, each of which had a small hole in it through which the fire was visible. These narrow openings could close, thus accounting for phenomena such as eclipses or the phases of the Moon. The Earth, about which the wheels rotate, was a flat disc that he thought of as the top surface of a cylinder, suspended freely in space, the height of which was for some reason taken to be one-third of the diameter of the disc.[17] Anaximander also speculated on the sizes and distances of the Sun and Moon, though with no apparent basis for his conjectures:

> the sun is a circle twenty-eight times the size of the earth; it is like a chariot-wheel, the rim of which is hollow and full of fire, and lets the fire shine out at a certain point in it through an opening like the nozzle of a pair of bellows: such is the sun.[18]

The significance of Anaximander's ideas lies much less in his conclusions than in the questions he was trying to answer: how big are the stars?, how far away are they? Perhaps most significant of all, in his cosmology Anaximander replaced the actions of gods with mechanisms familiar to those on Earth.

Anaximander's ideas were extended by Anaximenes, who was about 20 or 30 years younger than Anaximander, though his cosmology remained very primitive. Perhaps wishing to make philosophy more tangible, Anaximenes made air the essence of things:

> Anaximenes of Miletus, son of Eurystratus, who had been an associate of Anaximander, said, like him, that the underlying substance was one and infinite. He did not, however, say it was indeterminate, like Anaximander, but determinate; for he said it was air.[19]

Anaximenes, like Thales before him, believed that all things were alive and that, just as air sustains human life, a different form of air sustains the life of all things in the Universe. He believed that the flat disc of the Earth was carried by the air, as were the flat discs of the Sun and Moon, while the stars were fixed

---

[16] Anaximander used the Greek word *apeiron* and his meaning was somewhat ambiguous; see Toulmin and Goodfield (1965), p. 66. Bochner (1966) describes *apeiron* as 'some kind of stuff (perhaps so tenuous as to be little more than "spatiality") of which the universe was composed'.
[17] According to Heath (1932), p. xxiii, Anaximander said that the wheels of the Sun and the Moon 'lie obliquely', indicating that he was in some sense aware of the ecliptic and its obliquity to the celestial equator. However, Eudemus, in his lost *History of Mathematics* written in the fourth century BC, attributes this discovery to Oenopides in the fifth century BC (Heath (1913), Chapter XIV). There is no evidence that Oenopides made any measurements of the obliquity, but he did calculate the length of the year as $365\frac{22}{59}$ days.
[18] AËTIUS *De placitis*, Book II, Volume 20, 1. Translation from Heath (1932).
[19] SIMPLICIUS *Physics*. Translation from Heath (1932).

to the crystalline celestial vault. The heat of the Sun came from the speed of its motion, the stars providing no heat due to their great distance.

The Ionian philosophers asked many interesting questions, most importantly: what is the nature of things?, what is the Sun?, what is the Earth? While their attempts at answers may seem slightly comical, their curiosity was followed up by the greatest minds of antiquity, with impressive results. Modern science grew out of the pioneering spirit of people such as Thales, Anaximander, and Anaximenes.

The most important of the Eleatic philosophers was Parmenides, who was probably a pupil of Xenophanes, who left Ionia for southern Italy in 545 BC, during the time of the Persian conquest of Asia Minor. Xenophanes' most profound thoughts concerned the philosophy of religion, and he believed that certainty of knowledge was unattainable, but he also had some interesting, but still very primitive, ideas about the nature of heavenly bodies. He wrote that the Sun was born again each time it rose and that eclipses occurred due to the light of the Sun being put out. He also believed that the Moon shone by its own light. In Xenophanes' philosophy, the basic element was not water or air, but earth, and he claimed that the world was evolved from a mixture of earth and water and that the earth gradually would be dissolved again by moisture. This process would then repeat in cycles.

Aristotle's pupil Theophrastus credited Parmenides with the significant discovery that the Earth is spherical, but others credit the Pythagoreans with this observation.

> Further we are told that Pythagoras was the first to call the heaven the universe, and the earth spherical, though according to Theophrastus it was Parmenides, and according to Zeno it was Hesiod.[20]

It is, of course, not unlikely that the idea of a spherical Earth occurred independently to a number of people, observant Greek navigators, for example. Parmenides conceived of a Universe consisting of a number of concentric layers centred on the spherical Earth. Here we see for the first time the idea of a system of geocentric spheres that was to play such an important role in the development of future astronomical theories.

The Ionian and Eleatic thinkers represent the origins of Western philosophy and, although their thoughts on the nature of the Cosmos helped stimulate debate, their direct influence on the history of astronomy was slight. The Pythagoreans, on the other hand, had a much more direct bearing on future astronomical thought. Pythagoras – a mystical figure who was born in the first

---

[20] DIOGENES LAËRTIUS *Lives of Philosophers*, Book VIII, 48. Translation from Heath (1932).

half of the sixth century BC – left Asia Minor and founded a School in southern Italy which over a period of about 200 years produced a wealth of influential ideas in astronomy and mathematics. Almost nothing is known of Pythagoras himself and, since the group he founded was very secretive, much of their output is attributed simply to the Pythagoreans as a whole rather than to any one individual. According to Aristotle:

> ... those who are called Pythagoreans, taking up mathematical things, were first to promote these, and having been reared on them, they supposed that the sources of them were the sources of all things.[21]

Thus, it was mathematics that held the key to understanding the Universe. To the Pythagoreans, mathematics could be broken down into four distinct branches, which in the Middle Ages became known as the *quadrivium*: arithmetic, geometry, music and astronomy. Arithmetic and geometry represent pure mathematics, and music and astronomy are applied mathematics, musical harmony being an application of arithmetic while astronomy is an application of geometry. Like the earlier Ionian philosophers, the Pythagoreans sought to understand the essence of things, but they found their answer in numbers (positive integers). To them, numbers were the basis of all physical phenomena. The Pythagoreans discovered that the notes obtained by plucking two strings sounded harmonious if the lengths of the strings were in ratios of simple whole numbers. For example, a string twice the length produces a note an octave higher, while if the ratio is 3 : 2 we get a fifth.

We learn from Aristotle how the notions of harmony and number were also applied to the study of the heavens:

> ... they assumed that the elements of numbers were the elements of all things, and that the whole heaven was a harmony and a number. And as many things among numbers and harmonies as had analogies to the attributes and parts of the heavens and to the whole cosmic array, they collected and fit together.[22]

Ratios of whole numbers could be found among the motions of the heavenly bodies and so the Pythagoreans believed that they emitted harmonious sounds – this is the origin of the phrase 'harmony of the spheres'. The idea that a mathematical structure underlies the Universe originated with the Pythagoreans, and has been extremely influential ever since. It represents one of the cornerstones of modern science.

In terms of astronomical phenomena, the Pythagorean School is generally credited with knowledge of the sphericity of the Earth, and also with some of

---

[21] ARISTOTLE *Metaphysics*, Book *A*5, 985b. Translation from Aristotle (1999).
[22] ARISTOTLE *Metaphysics*, Book *A*5, 986a. Translation from Aristotle (1999).

the consequences of this, e.g. the existence of regions of the Earth's surface with days and nights each lasting half a year. As we have mentioned, most discoveries of the Pythagoreans are not attributed to any one person, but in the field of astronomy one particular idea has come down to us with the name of its originator attached. Philolaus of Tarentum developed a doctrine in which the centre of the Universe is a fire called *Hestia* – the goddess of the sacred fireplace of homes and public buildings. The spherical Earth, like the other planets, described a circle about this fire with the uninhabited side always turned toward it, completing 1 revolution each day. Outside the Earth were the spheres of the Moon, Sun, Venus, Mercury, Mars, Jupiter, Saturn and the fixed stars. He thought of the Sun as a transparent globe receiving its light and heat from the central fire and from the outside of the heavens. To balance his system he invented a counter-Earth which is always positioned on the opposite side of the central fire, bringing the total number of spheres of heavenly bodies to the sacred number ten. Philolaus' hypothesis had to be abandoned when travellers expanded the horizons of the known world beyond Gibraltar in the west and to India in the east, but still saw no signs of the central fire or counter-Earth.

The ideas that the Earth was not the centre of the Universe and that it was in continuous motion were very bold ones, contrary to all prevailing views of the time and, indeed, for the next 2000 years. It would appear that these views received little support from anyone outside the Pythagorean brotherhood. On the other hand, the idea that the universe was made up of bodies moving in uniform circular motions around a fixed centre was to dominate astronomical thought until the work of Kepler in the seventeenth century. In AD first century Geminus wrote:

> It is presupposed in all astronomy that the sun, the moon, and the five planets move in circular orbits with uniform speed in a direction opposite to that of the universe. For the Pythagoreans, who were the first to apply themselves to investigations of this kind, supposed the motions of the sun, the moon and the five planets to be circular and uniform ... For which reason, they put forward the question as to how the phenomena might be explained by means of circular and uniform motions.[23]

After the defeat of the Persians in 479 BC, Athens became an extremely wealthy city, attracting a number of intellectuals. One of the first to move there, and to bring with him the knowledge of Ionian natural philosophy, was Anaxagoras of Clazomenae, who became a friend of the city's ruler Pericles,

---

[23] Quoted from Goldstein and Bowen (1983). In older books, and also in the *Dictionary of Scientific Biography*, Geminus is said to have lived in the first century BC, but this is now thought to be an error caused by a confusion between ancient calendars (see Neugebauer (1975), pp. 579–81).

and devoted his life to investigations of the heavens. He proclaimed Mind as the moving principle in the Universe. Many people before him had speculated on the origin of the light from the Moon, but Anaxagoras was the first to state clearly that the cause was the reflection of light from the Sun, which he thought of as a vast mass of incandescent metal, and he understood that lunar eclipses occur when the Earth blocks the illumination of the Sun. He was the first to think of the seven wandering heavenly bodies in the order Sun, Moon, followed by the five other planets, an arrangement adopted by a number of later astronomers.

When Pericles' popularity began to dwindle, Anaxagoras, his protégé, became vulnerable and his speculations about the sizes of the Sun and Moon (that the Moon was as large as the Peloponnese – an area of mainland Greece – and the Sun was larger!), and those on the causes of Earthly phenomena and the nature of the heavens (he maintained that they were of the same general nature), led to him being convicted of impiety. He died in exile.

During the fourth century BC a number of centres for advanced learning were set up in and around Athens, by far the most influential being the Academy set up by Plato, who was a pupil of the philosopher Socrates. While not a mathematician himself, Plato was influenced strongly by Pythagorean doctrines. He believed that abstract mathematical concepts had a real, independent, existence, and that true knowledge could be obtained only through study of these so-called Platonic forms. The natural world was to be understood via unchanging mathematical laws, untarnished by the imperfections of our senses. His Academy became a major centre for the study of mathematics and philosophy; almost all the noteworthy mathematical achievements of the fourth century BC are due to his friends and pupils. After 300 BC, the seat of mathematical learning shifted to Alexandria, but the Academy remained pre-eminent in philosophy until finally it was closed in AD 529 as a result of its support for pagan learning.

Much of what we know about Plato's teaching comes from a series of works in the form of dialogues between Socrates and various historical and fictitious people. As far as astronomy is concerned, two dialogues are particularly important: the *Republic* and the *Timaeus*. We can gauge some of Plato's opinions about astronomy from the following passage in the *Republic* where Socrates says:

> Thus we must pursue astronomy in the same way as geometry, dealing with its fundamental questions. But what is seen in the Heavens must be ignored if we truly want to have our share in astronomy ... Although celestial phenomena must be regarded as the most beautiful and perfect of that which exists in the visible world (since they are formed of something visible), we must, nevertheless, consider them

as far inferior to the true, that is to the motions ... really existing behind them. This can be seen by reason and thought, but not perceived with the eyes.[24]

Plato's ideas on the structure of the Universe are found in his cosmological myth, the *Timaeus*, Timaeus being the main character in the dialogue and through whom Plato expresses Pythagorean ideas. In Plato's universe, the Earth is fixed at the centre and around it are the spheres of the Moon, Sun, Venus, Mercury, Mars, Jupiter, Saturn and the fixed stars.[25] Plato's model explained only the crudest of observed phenomena, but it laid the foundations for the construction of the first mathematical model of the Universe by Plato's collaborator, Eudoxus. The cosmology that Plato described in the *Timaeus* had a great influence on Western thought in the Middle Ages, in part because this was the only one of his works that was available in Latin before the twelfth century.

In the *Timaeus*, Plato described his teachings on the creation of the Universe and the composition of nature. Like most of the physicists of his time, he followed the Sicilian philosopher Empedocles and distinguished four elements of matter: fire, earth, water, and air. The idea that all things are made from these four elements was a synthesis of the various views of the early Ionian philosophers, and may well have been derived from a natural misinterpretation of the action of fire. When something burns, it appears to go from a complex state to a simpler one; thus, it is resolved into its constituent parts. When green wood is burned, for example, the *fire* clearly is visible, the smoke disappears into *air*, *water* boils off from the ends and the ashes clearly are the nature of *earth*.[26]

A member of Plato's Academy who made considerable progress in mathematics and who, indirectly, influenced astronomy, was Theaetetus; the Platonic dialogue bearing his name was a commemorative tribute by Plato to his friend. Among other things, Theaetetus is credited with the original constructions of the octahedron and icosahedron, two of the regular, or Platonic, solids (polyhedra which have the property that all the faces and all the vertices are identical). The other three (the tetrahedron, cube and dodecahedron) were discovered by the Pythagoreans.[27]

One significant fact about the regular solids that led both Plato and, later, Kepler to develop a physical theory around them, is that there are only five

---

[24] Quoted from Pedersen (1993).
[25] Detailed descriptions of Plato's cosmic scheme can be found in, for example, Knorr (1990), Pecker (2001), pp. 55–70.
[26] This example is taken from Dampier (1965), p. 21.
[27] See Waterhouse (1972), in which the historical problems associated with tracing the early history of these shapes are discussed in some detail.

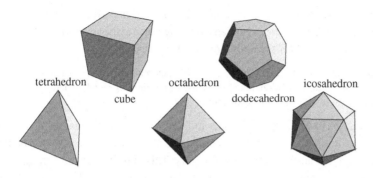

Fig. 1.6. The five Platonic solids.

of them (illustrated in Figure 1.6). This is easy to prove.[28] We use the fact that the sum of the interior angles of the faces which meet at the vertex of a regular polyhedron must be less than 360°. This condition is fulfilled only by three, four, or five equilateral triangles (since the interior angles of an equilateral triangle are 60° and 6 × 60 = 360), corresponding to a tetrahedron, octahedron and icosahedron, respectively, three squares (the interior angles of a square are 90° and 4 × 90 = 360), corresponding to a cube, and three pentagons (the interior angles of a pentagon are 108° and 4 × 108 > 360), corresponding to a dodecahedron. Three hexagons is already too many (the interior angles of a hexagon are 120° and 3 × 120 = 360).

Plato was searching constantly for a parallel between the hierarchy of material things and that of mathematical objects, and this led him to consider these solids. The fact that there are only five of them (unlike the regular polygons of which there is an infinite number) made them special. Fire, on account of the shape of a flame, was compared with the tetrahedron, and water, the bulkiest of the elements, corresponded to the icosahedron. Air, having an intermediate density, corresponded to the octahedron, which has a number of triangular faces lying between those of the tetrahedron and icosahedron. Earth was likened to a cube since it is the most immobile of bodies and should be represented by the most stable figure. What about the dodecahedron? Plato got round this by saying that the god had used it for the whole Universe, a statement that did not satisfy many of Plato's disciples; later the dodecahedron became associated with the ether.

As will become evident, many of the ideas about the Universe promoted by people who were the product of Plato's Academy were extremely influential.

---

[28] The proof given here (which is probably due to Theaetetus) is given in the concluding proposition of Euclid's *Elements*, written in about 300 BC.

But some, which in retrospect we can see to have been correct, were not. One of Plato's pupils, Heraclides, assumed a rotating Earth at the centre of the Universe in order to account for the daily motion of the heavens.

> Heraclides of Pontus and Ecphantus the Pythagorean move the earth, not however in the sense of translation, but in the sense of rotation, like a wheel fixed on an axis, from west to east, about its own centre.[29]

Here, we have an example of a modern view, but one that attracted few followers among Heraclides' contemporaries and which was not considered seriously again until the fourteenth century. This may at first seem strange, but, as with other similar examples, the perceived advantages of Heraclides' theory were far outweighed by the obvious disadvantage of requiring the Earth to spin round, something that is quite alien to everyday experience.[30]

---

[29] AËTIUS *De placitis* Book III, Volume 13, 3. Translation from Heath (1932). Virtually nothing is known about Ecphantus. Since all his known opinions correspond to those of Heraclides, Heath (1913) thinks it likely that he was in fact a character used by Heraclides in dialogues.

[30] It is often stated that Heraclides believed Mercury and Venus rotated around the Sun rather than the Earth, but Neugebauer (1972) puts this down to a mistranslation of a sentence from Chalcidius' commentary on Plato's *Timaeus*.

2

Spheres and circles

Eudoxus' system of concentric spheres

The Babylonians were concerned with predicting the time at which a particular phenomenon (e.g. a planetary opposition) would occur, since it was the date that was ominous. Their astronomy was thus concerned with the analysis of discrete processes. On the other hand, as we shall see, the Greeks in their astronomy focused on predicting where a celestial body would be at a given time, and they were thus concerned with modelling a continuous process, which naturally leads to the use of geometrical methods. The first person to produce a geometrical model of celestial motions was Eudoxus, who was born in Cnidus on the western coast of Asia Minor in about 400 BC. According to Diogenes Laërtius' *Lives of Philosophers* (written in AD third century) he was taught geometry by Archytas of Tarentum, one of the leading Pythagorean philosophers, and studied with Plato, who was about 30 years older than him. None of Eudoxus' works have survived. Most of the information we have about his system of concentric spheres comes from a brief contemporary account in Aristotle's *Metaphysics* and a more substantial description due to Aristotle's influential commentator Simplicius (AD sixth century).[1]

---

[1] Simplicius based his work on that of the philosopher Sosigenes (AD second century – not to be confused with the Sosigenes who helped Julius Caesar reform the calendar in the first century BC), and ultimately on the lost history of astronomy by Eudemus, who lived only a generation after Eudoxus. It is important to recognize that Simplicius was writing 900 years after the event and he may well have been putting his own interpretation on Eudoxus' work. By itself, the information supplied by Aristotle and Simplicius does not provide a clear picture of Eudoxus' model, but in a classic paper in 1877, the Italian Giovanni Schiaparelli produced a reconstruction of the Eudoxan system which has remained the accepted version ever since. Recently, however, Yavetz (1998, 2001) has argued that Schiaparelli's interpretation is not the only one that can be put on the original sources which is consistent with what we know about Greek astronomy of the time, and that there is no reason to believe one rather than the other. Here we will stick to the standard interpretation.

Eudoxus had a considerable influence in many areas of mathematics. In fact, he was the inspiration behind two of the most profound mathematical advances of the fourth century BC. The first of these was the theory of proportion, which formed the basis of much of Euclid's *Elements* (see p. 36), and the second was the method of exhaustion, which was used extensively in the same work and represents the beginnings of the subject now known as the 'integral calculus'. Both of these contributions were of fundamental significance for the development of mathematics, but it would appear that, as far as other scholars from antiquity were concerned, his main claims to fame were his astronomical discoveries,[2] one of which was an accurate estimate for the length of the tropical year, namely 365 days 6 h.

Eudoxus' theory based on concentric spheres marks the beginning of a long development in Greek mathematical astronomy, culminating in the work of Ptolemy about 500 years later.[3] Usually, it is claimed that Eudoxus' theory, described in his treatise *On Speeds*, was a response to a challenge of Plato. Simplicius tells us that:

> Eudoxus of Cnidus was the first of the Greeks to concern himself with hypotheses of this sort, Plato having . . . set it as a problem to all earnest students of this subject to find what are the uniform and ordered movements by the assumption of which the phenomena in relation to the movements of the planets can be saved.[4]

In other words, what uniform circular motions could be used to represent the motions of the planets? Quite how much of the impetus for a geometrical theory of the heavens was due to Plato is open to question, however.[5]

The discovery of the spherical nature of the Earth, together with Pythagorean ideas about uniform circular motions, may well have suggested to Eudoxus that the motion of the stars and planets could be modelled by means of a system of concentric spheres with the Earth at their common centre. He supposed that each of the Sun, Moon and planets occupied its own sphere which was attached via its poles to a larger sphere that rotated about other poles. Third and fourth spheres could be added, each rotating at different rates and about different axes, according to mathematical and observational needs. Apart from the sphere of the fixed stars, Eudoxus used three spheres each for the Sun and Moon and four spheres for each of the planets, making twenty-seven in all. In each case, one

---

[2] Lasserre (1964), p. 124.

[3] It has been claimed (Landels (1983)) that Eudoxus' model of the Universe represents the first ever example of the technique of mathematical modelling, but while the most obvious mathematical phenomena are modelled quantitatively by Eudoxus' theory, the more subtle effects are only reproduced qualitatively, if at all.

[4] SIMPLICIUS on *De caelo*. Translation from Heath (1932).

[5] See, for example, Goldstein (1980), (1983), Knorr (1990).

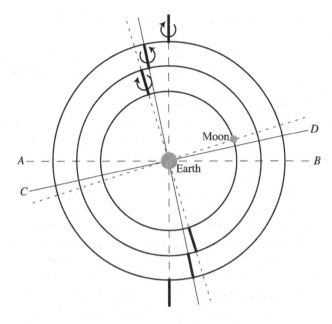

Fig. 2.1. Eudoxus' scheme for the motion of the Moon, according to Simplicius.

of the spheres performed exactly the same function: it rotated once every 24 h and accounted for the daily rotation of the heavens. Thus, the motions of the Sun, Moon and planets with respect to the fixed stars actually were modelled using only nineteen uniform circular motions. It is not clear whether Eudoxus thought of the spheres as being real physical entities, or whether he regarded them as purely mathematical constructions. Indeed, he may not have thought that the distinction was significant.[6]

In Eudoxus' theory for the Moon (see Figure 2.1, in which AB represents the celestial equator and CD the ecliptic), the outermost of the three spheres completes 1 revolution from east to west in 1 day, thus accounting for the diurnal rotation of the Moon. In order to account for the motion of the Moon around the ecliptic, Eudoxus introduced a second sphere attached to the first through an axis passing through the poles of the ecliptic and rotating from west to east. Eudoxus also knew that the Moon did not follow exactly the ecliptic, but deviated a little above and below it, and accounted for this by including a

---

[6] Wright (1973) has argued that the very nature of the system makes it probable that Eudoxus did think of his spheres as having a physical reality.

third sphere, the axis of rotation of which makes a small angle with that of the second sphere.

It is not known what speeds Eudoxus chose for his second and third spheres, or what inclination he gave the third sphere relative to the second; Simplicius merely states that the rotation of the third sphere was slow. It seems likely that Eudoxus would have noted that the Moon's variation in latitude amounted to something like $\pm 5°$ and so would have used a value close to 5° for the inclination of the third sphere to the second. For the inclination of the second sphere to the first he would have used his value for the obliquity of the ecliptic and there is reason to believe that this was 1/15 of a circle, or 24°. This is the angle subtended at the centre by a regular fifteen-sided polygon, the construction of which is given in Euclid's *Elements*, Book IV, Proposition 16, and which, according to Proclus, was included because of its usefulness in astronomy.[7]

Eudoxus' theory for the Sun was very similar to his lunar theory. The axis of the innermost sphere was inclined at a very small angle to the second, thus giving the Sun a deviation from the ecliptic. The ecliptic seems to have been defined vaguely as some great circle passing through the zodiac rather than as the path of the Sun, the definition in terms of the motion of the Sun not being formulated until the second century BC by Hipparchus. The solar theory implied that the Sun moves with constant speed relative to the fixed stars, from which it follows that the seasons are all of equal length – a statement known to be false, since in 432 BC two astronomers from Athens – Meton and Euctemon – measured the times between summer solstice and autumn equinox, autumn equinox and winter solstice, winter solstice and spring equinox, and spring equinox and summer solstice, as 90, 90, 92 and 93 days, respectively. Eudoxus' theory also assumed that the heavenly bodies were each at a fixed distance from the Earth, and so was incapable of modelling the observed changes in the apparent diameters of the Sun and Moon. This is most significant for the Moon,

---

[7] As to the speeds of the second and third spheres, there is a certain amount of disagreement. According to Dicks (1970), the period of rotation of the second sphere would have been the synodic month and the slow rotation of the third sphere in the opposite direction would not have affected the speed of the Moon very much, but would have accounted for the Moon's movement in latitude and the fact that its greatest deviations occur at points which shift steadily westwards. However, many have taken the view that Simplicius got it wrong and that it was the second sphere that rotated slowly, with a period of over 18 years, and the innermost sphere rotated once in a period of a little over 27 days. This provides a much more accurate representation of the Moon's deviations in latitude and also models the so-called *Saros* period of 223 synodic months ($\approx$ 18 years), which was used by the Babylonians to investigate the periodic recurrence of various lunar phenomena, including eclipses. This view, which originated in the nineteenth century and has been repeated many times since, is, according to Dicks, an example of a misleading interpretation of early astronomical thought by attributing to it a sophistication inconsistent with contemporary knowledge. The fact that Eudoxus' system of concentric spheres is poor quantitatively in virtually all other respects lends credence to Dicks' opinion, but others (e.g. Thoren (1971)) do not share his view.

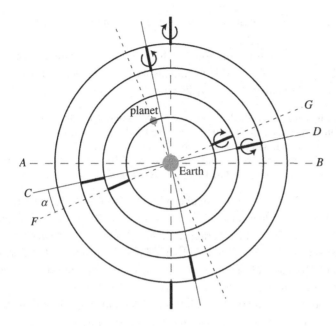

Fig. 2.2. Eudoxus' scheme for the motion of the planets.

the greatest apparent diameter of which is 14 per cent larger than its smallest. Simplicius was well aware that, even judged by the astronomical knowledge of Eudoxus' day, the system of concentric spheres had serious deficiencies.

> Nevertheless the theories of Eudoxus and his followers fail to save the phenomena, and not only those which were first noticed at a later date, but even those which were before known and actually accepted by the authors themselves.[8]

Perhaps the main fascination with Eudoxus' theory is with his method for accounting for the retrograde motion of the planets, which he did by adding an extra sphere (see Figure 2.2).[9] Just as in the case of the Sun and the Moon, the first and second sphere together account for the diurnal rotation and the regular motion around the ecliptic in the appropriate zodiacal period. The axis of the third sphere was in the plane of the equator of the second sphere (i.e. in the ecliptic plane) and the axis of the fourth sphere was inclined at a small angle α to that of the third. Crucially, the third and fourth spheres rotated in

---

[8] SIMPLICIUS on *De caelo*. Translation from Heath (1932).
[9] Since Simplicius lived nearly 1000 years after Eudoxus, it is quite possible that the phenomena he attempted to describe using Eudoxus' scheme, such as retrograde motion, were not those that Eudoxus originally had in mind (see Goldstein (1997)). However, Yavetz (1998) claims that, given the nature of the Eudoxan model, the most likely explanation is that retrograde motion was the object of the design.

opposite directions but at the same rate of 1 revolution per synodic period of the planet.

The motion of the third and fourth spheres together combine to generate a figure-of-eight-type curve, which Eudoxus termed a *hippopede*, named after the device used to tether a horse by its feet.

> ... the fourth sphere, which turns about the poles of the inclined circle carrying the planet and rotates in the opposite sense to the third, i.e. form east to west, but in the same period, will prevent any considerable divergence (on the part of the planet) from the zodiac circle, and will cause the planet to describe about this same zodiac circle the curve called by Eudoxus the *hippopede* ...[10]

When superimposed on the regular motion induced by the outermost two spheres, the hippopede has the effect of producing small deviations in latitude and occasional periods of retrograde motion, as required. This curve was only of limited success in modelling the observed planetary motion, but is an early example of the study of curves in space and its construction demonstrates a great deal of mathematical skill.[11]

With modern mathematical techniques, visualization of Eudoxus' hippopede is much simpler than it was for the ancients. If we introduce orthogonal unit vectors $\{i_1, j_1, k_1\}$, where k_1 is along the axis of rotation of the fourth sphere (FG in Figure 2.2) and i_1 and j_1 lie in its equatorial plane and rotate with it, then the position of the planet is given, for all times, by the vector i_1. Next, we introduce a second set of vectors $\{i_2, j_2, k_2\}$ which are the same as the first except that the vectors i_2 and j_2 do not rotate with the sphere. Hence, the vectors i_2, j_2 are related to i_1, j_1 via a 2×2 rotation matrix, and if the angular rotation rate of the fourth sphere is $-\Omega$ we have:

$$\begin{pmatrix} i_2 \\ j_2 \\ k_2 \end{pmatrix} = \begin{pmatrix} \cos \Omega t & \sin \Omega t & 0 \\ -\sin \Omega t & \cos \Omega t & 0 \\ 0 & 0 & 1 \end{pmatrix} \begin{pmatrix} i_1 \\ j_1 \\ k_1 \end{pmatrix} = A_1 \begin{pmatrix} i_1 \\ j_1 \\ k_1 \end{pmatrix},$$

for example. Finally, we introduce an orthogonal set of vectors fixed in the third sphere with k_3 along its axis of rotation (CD in Figure 2.2) and, hence, in the ecliptic plane, which makes an angle α with k_1 and k_2. Since the angular speed

---

[10] SIMPLICIUS on *De caelo*. Translation from Heath (1932).

[11] The question as to how Eudoxus devised his construction of the hippopede has been the subject of much research and speculation (see, for example, Neugebauer (1953) and Riddell (1979)). Aaboe (1974) claims that Eudoxus' model could have been meant only as a qualitative description of the planets, because if one enters in the appropriate periods for Venus or Mars there is simply no way of producing retrograde motion. Yavetz (1998) disputes this for the case of Mars.

Fig. 2.3. An example of a planetary path in Eudoxus' theory.

of this sphere is Ω, we have:

$$\begin{pmatrix} \mathbf{i}_3 \\ \mathbf{j}_3 \\ \mathbf{k}_3 \end{pmatrix} = \begin{pmatrix} \cos\alpha\cos\Omega t & -\sin\Omega t & \sin\alpha\cos\Omega t \\ \cos\alpha\sin\Omega t & \cos\Omega t & \sin\alpha\sin\Omega t \\ -\sin\alpha & 0 & \cos\alpha \end{pmatrix} \begin{pmatrix} \mathbf{i}_2 \\ \mathbf{j}_2 \\ \mathbf{k}_2 \end{pmatrix} = A_2 \begin{pmatrix} \mathbf{i}_2 \\ \mathbf{j}_2 \\ \mathbf{k}_2 \end{pmatrix},$$

for example. It follows that the position of the planet in terms of $\{\mathbf{i}_3, \mathbf{j}_3, \mathbf{k}_3\}$ is:

$$A_2 A_1 \begin{pmatrix} 1 \\ 0 \\ 0 \end{pmatrix} = \begin{pmatrix} \cos\alpha\cos^2\Omega t + \sin^2\Omega t \\ (\cos\alpha - 1)\cos\Omega t \sin\Omega t \\ -\sin\alpha\cos\Omega t \end{pmatrix}.$$

This, then, is the parametric equation of the hippopede. For the purpose of illustrating the type of planetary paths that can result from Eudoxus' model, we can superimpose a uniform motion in the $\mathbf{k}_3$ direction, corresponding to the motion around the ecliptic, and obtain various curves like that shown in Figure 2.3, in which the angle α was taken as $5°$. The curve shows how Eudoxus' construction leads to paths that deviate by a small amount above and below the ecliptic, and are predominantly in one direction, but with regular periods of retrograde motion.

As a quantitative predictive tool, Eudoxus' system of concentric spheres would not have been much use and, since it was developed before the Greeks came into contact with the accurate arithmetic astronomy of the Babylonians, it is reasonable to think that quantitative prediction was not its purpose. However, as the goals of Greek astronomy evolved, other astronomers attempted to modify the scheme so that it better represented the observed phenomena. Callippus of Cyzicus, who was a pupil of a pupil of Eudoxus, added seven more spheres to Eudoxus' scheme.

> Callippus set down the same arrangement of spheres as did Eudoxus, and gave the same number as he did for Jupiter and Saturn, but for the sun and moon he thought there were two spheres still to be added if one were going to account for the appearances, and one for each of the remaining planets.[12]

---

[12] ARISTOTLE *Metaphysics*, Book Λ8, 1073b. Translation from Aristotle (1999).

The two extra spheres for the Sun were introduced to accommodate the variation in the length of the seasons – summer, autumn, winter and spring – which were measured by Callippus, around 330 BC, as 92, 89, 90 and 94 days, respectively, each being accurate to within 1 day.

   The details of Eudoxus' theory are not known with any certainty, but we do know that the scheme exerted a profound influence over the development of astronomical thought. It was not as good at predicting the future positions of heavenly bodies as the arithmetical schemes of the Babylonians, but it was far more influential. This was because it demonstrated the power of geometrical techniques, in that superpositions of simple uniform rotations could be used to model extremely complex behaviour, and because it (as modified by Callippus) was adopted by the giant of Greek scientific philosophy – Aristotle – whose teachings dominated intellectual thought for the next 2000 years. As a scientific theory, Eudoxus' system of concentric spheres is best described as *ad hoc*. It explains the phenomena only in as much as they are built into the model. It does not predict or explain any independent result, and is untestable because the model's intrinsic parameters simply can be modified whenever observations fail to fit in with predictions.[13]

Aristotle

Aristotle was a pupil of Plato, but the two men differed significantly in their approach to the understanding of the natural world. Plato was an idealist who focused on mathematics as the underlying reality, whereas Aristotle took a more pragmatic approach and emphasized physical aspects (e.g. cause and effect); he searched for reasons why things were as they were. Aristotle's modification of Eudoxus' system (he described his cosmology in the work *On the Heavens*, or *De caelo* in Latin) was not undertaken from the point of view of a mathematical astronomer, but as part of his bold attempt to unify all the separate branches of natural philosophy. As mentioned above, it is unclear whether Eudoxus regarded the spheres in his system as mathematical constructions or material entities, but in Aristotle's version the spheres were regarded as real physical objects, motion being imparted to the planets by the divine and eternal 'ether' (a word derived from the Greek for 'runs always') which filled the spheres. This insistence on the physical reality of the spheres led to the addition of many more spheres to

---

[13] The status of Eudoxus' homocentric spheres as a scientific theory is discussed and compared with later models in O'Neil (1969).

Callippus' model, twenty-two in all, so as to undo the motion of each planet before beginning the set of spheres associated with the next planet.

Aristotle wrote on many other subjects (e.g. physics, chemistry and biology) and attempted, with a considerable degree of success, to organize all knowledge into a unified whole. As a result, it became very difficult to criticize one aspect of Aristotle's teachings without criticizing it all. This led to a degree of stability and is one of the reasons why Aristotelian cosmology had such a pervasive influence. Central to Aristotle's teaching was the idea of a stationary Earth at the centre of the Universe:

> ... the natural movement, both of parts of the earth and of the earth as a whole, is toward the centre of the universe – this is why the earth now actually lies at the centre ... It happens that the same point is the centre both of the earth and of the universe. ... It is clear from these considerations that the earth neither moves nor lies away from the centre.[14]

These conclusions followed from Aristotle's terrestrial physics in which natural motion was toward or away from the centre of the universe, depending on the 'heaviness' of the material concerned. When thought of in terms of the physics of the time, ideas like a moving Earth were indeed ridiculous, even if they could be used accurately to model celestial phenomena. A new terrestrial physics would be needed before astronomers could make the leap to a planetary Earth.

Aristotle made a complete distinction between the sublunary world (in which he included comets and the Milky Way!) and heavenly objects, beginning with the Moon and extending to the stars. The physics of these regions were totally different. In the former (described in his *Meteorology*) everything was explained in terms of the four elements earth, water, air, and fire. The central Earth was surrounded by water (the oceans) which itself was surrounded by a shell of air (the atmosphere). Between the atmosphere and the Moon there was a shell of fire. These regions represented the natural habitat of each element and thus explained why fire travelled upwards but a stone fell toward the earth. Within the sublunary world everything, including life itself, was imperfect and changing continually.

On the other hand, the celestial regions were ordered perfectly and never changed. This supralunary world was formed out of a fifth element, the ether (or in medieval terminology 'quintessence') and uniform circular motions were considered natural for heavenly bodies, straight line motion being natural

---

[14] ARISTOTLE *On the Heavens*, II. Translation from Heath (1932).

on Earth. Aristotle believed that all the heavenly bodies where spherical; such a shape was in his view 'appropriate' for heavenly bodies.[15] The belief that the Moon was spherical is attributed to the Pythagoreans, who argued that the shape of the phases of the Moon are those that would be observed if the Moon were spherical and illuminated on one side by the Sun. Aristotle argued that if one heavenly body was spherical, they all must be. He also supported the idea that the Earth was a sphere based on a number of observations. Among these were the facts that during a lunar eclipse the shadow cast by the Earth on the Moon is consistent with a spherical Earth, and that when one travels north or south the stars which are visible above the horizon change, indicating that the angle between the observer's horizon and the celestial equator has changed. A spherical Earth is consistent also with Aristotle's terrestrial physics in which the matter making up the Earth is subject to a constant tendency to move toward its centre.

Aristotle was aware that the apparent daily motion of the stars could equally well be accounted for by a rotation of the Earth about its own axis, as suggested by his contemporary Heraclides, but he rejected this idea on the ground that circular motion was a property of the heavenly bodies only. Thus, for Aristotle the diurnal rotation of the stars was real and he needed to determine its cause. Plato had thought of the heavenly bodies as gods who supply their own movement, but in Aristotle's philosophy every motion required some driving force. To set the celestial mechanism in motion, Aristotle invented the concept of a supreme Prime Mover. The Prime Mover did not actually do anything to move the sphere of the stars, since that would imply motion and, hence, a new cause; instead, the sphere moved due to the desire for perfection that He aroused in it. Aristotle used the motion of the sphere of the stars to prove that the Universe is finite, since, he argued, a line from the Earth of infinite length could not possibly complete 1 revolution in a finite time, whereas the fixed stars rotated once every 24 h. He backed this up by pointing out that a body with a centre cannot be infinite and yet the Universe has a centre, namely the centre of the Earth.

Another Aristotelian principle, closely related to the finiteness of the Universe and very significant in later developments, was the impossibility of a vacuum. This principle, which was supported by a number of simple experiments, followed from Aristotle's theory of motion in which the speed of a body was, other things being equal, inversely proportional to the resistance

---

[15] The status of the Moon itself in Aristotle's writings is somewhat ambiguous. It was common in Greek intellectual culture to ascribe an Earthly nature to the Moon and when Aristotle refers to the relation between the Moon and the sublunary world such characteristics are often present. When discussing celestial motion, however, the Moon is a prefect, unchanging sphere (see Montgomery (1999)).

of the medium through which it was moving. In a vacuum there would be no resistance and it would thus take no time at all to travel between two points. Aristotle argued that space and matter could not be considered independently: it was impossible to have one without the other. There was no matter outside the sphere of the fixed stars and, hence, no space; there was nothing. Some Greek philosophers (the Stoics) agreed with Aristotle that there could be no vacuum within the Cosmos, but believed that outside the sphere of the fixed stars there was an infinite void. There were those who took a contrary view however, most notably the atomists, led by Leucippus and Democritus, who believed that the Universe consisted of atoms moving about in an otherwise infinite void.

Aristotle's cosmology certainly was ingenious but, from a technical point of view, turned out to be unsatisfactory as there were a number of observable phenomena it could not hope to reproduce or explain. Within a century of Aristotle's death, mathematicians who had come into contact with the quantitative predictive astronomy of the Babylonians were developing geometrical theories that violated many of Aristotle's fundamental physical principles in order to account for subtle irregularities in the positions of the heavenly bodies. The influence of Aristotle's cosmology did not diminish, however. Following the demise of the Greek civilization, a large quantity of Aristotle's work (unlike that of Plato) was translated and became available to scholars in the Islamic world and then in Europe during the Middle Ages. For more than two millennia Aristotle's conceptually simple view of the Universe reigned supreme, though many of those who adhered to it were unaware of its deficiencies.

Most people believed that the heavenly bodies had great influence over things that happened on Earth, and the study of these influences through astrology was undertaken widely by professional astronomers. Some influences were obvious (e.g. the Sun's position on the ecliptic is tied to the seasons) and it was natural to generalize and assume similar influences for all the stars and planets. Aristotelian cosmology helped to reinforce these ideas, since it provided a plausible mechanism by which the motion of heavenly bodies could affect terrestrial phenomena. The constant friction between the celestial spheres, and ultimately between the sphere of the Moon and the sublunary world, was the primary mechanism for all change on Earth. Much of the desire for accurate planetary tables came from those who wished to use them to cast horoscopes, and these were often relied upon when making important decisions. As a result, astrology was a major driving force behind technical astronomy. Also, since astrology fits in well with an Earth-centred cosmology, but loses much of its ideological force when the Earth is relegated to a planet, it helped to shackle astronomers to geocentrism.

Developments in mathematics

Eudoxus' concentric spheres encouraged the study of spherical astronomy
(sphaerics), the first known works on which were *On the Moving Sphere* and *On
Risings and Settings* by Autolycus of Pitane, written between 330 and 300 BC.
Autolycus' works, which contain a variety of simple geometrical theorems,
seem to have been superseded quickly by Euclid's *Phaenomena*, believed to
have been written about 300 BC.[16] From then on, sphaerics became a standard
branch of Greek mathematics. Euclid's *Phaenomena* is a geometrical treatment
of various problems relating to the risings and settings of stars and, in particular,
addresses at length the problem of how to determine the duration of daylight
on a given day at a given location.

The *Phaenomena* is, however, only a minor work of Euclid, who was responsible for the most successful textbook ever written: the *Elements*, also written
in about 300 BC, and used as the primary source for teaching geometry to
schoolchildren until the 1970s.[17] Of Euclid's life we know virtually nothing,
but it is assumed that he studied in Athens and then, after the death of Alexander
the Great in 323 BC, migrated to Alexandria where he established a school,
known as the Museum.

The *Elements* was not the first textbook of elementary Greek mathematics,
but clearly it was so superior to the others that it alone has survived. It consists of thirteen books dealing with the basic properties of rectilinear figures
and circles, the theory of proportion, the theory of numbers, the classification
of incommensurables, and solid geometry. The style of the book, consisting
entirely of definitions, axioms, theorems and proofs, determined the nature of
all subsequent Greek mathematical works and, indeed, much mathematics that
is written today still bears the hallmark of this early Greek paradigm.

The contents of the *Elements* show that in the three centuries since the beginnings of deductive mathematics in Ionia, the Greeks had made great progress in
mathematics. Two things are particularly important for our discussion of mathematical astronomy. First, Greek mathematics was almost entirely geometry.
The algebraic methods that we find indispensable today were not introduced
until AD ninth century, and the idea that algebraic equations could be used to
represent curves and surfaces did not become fully understood until the seventeenth century. Second, the only curves that appear in the *Elements* are the
circle and the straight line, and these two geometrical objects dominated Greek
mathematical thought for 1000 years.

---

[16] A translation of the *Phaenomena* is given in Berggren and Thomas (1996).
[17] The *Elements* has appeared in more editions than any other book apart from the Bible. The
most famous English translation is that of Sir Thomas Heath (Euclid (1956)).

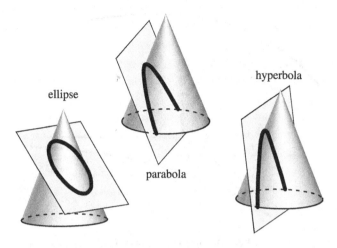

Fig. 2.4. Conic sections.

Other curves were discovered slowly however, and the most important of these were the conic sections – the ellipse, parabola and hyperbola – credited to Menaechmus, a pupil of Eudoxus, in about 350 BC. About 130 years later, Apollonius wrote his treatise on conic sections that is one of the masterpieces of Greek mathematics. Apollonius obtained these curves as the intersections of a plane and a cone, with the three curves being the result of using a different angle of intersection, as shown in Figure 2.4. If the intersecting plane exactly is parallel to one of the generating lines of the cone, we obtain a parabola; otherwise, we get an ellipse or hyperbola, as shown.

Menaechmus, however, used a different construction in which the intersecting plane is always perpendicular to one of the generating lines of the cone, and the different curves are obtained by varying the vertex angle of the cone. It is not known for certain what led to the discovery of the conic sections. One possibility is that they were stumbled upon in the course of attempting to solve the three famous construction problems of antiquity (squaring the circle, duplicating the cube and trisecting the angle). However, it has been suggested[18] that Menaechmus' curious construction – which insists on the perpendicularity of the intersecting plane to the generating lines of the cone – suggests an astronomical heritage. As the Sun moves in a circular arc across the sky, the surface described by the ray that just clips the top of a *gnomon* clearly is conical, and the path mapped out by the end of the *gnomon*'s shadow can thus be thought of as the intersection between a plane (the ground) and this cone. Hence, it will be a

---

[18] Neugebauer (1948).

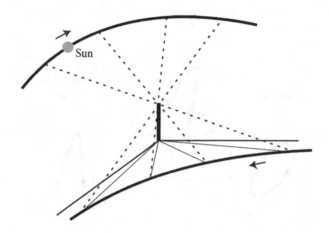

Fig. 2.5. The shadow of a sundial.

conic section and, provided the Sun dips below the horizon during a revolution, it will be a hyperbola (see Figure 2.5). If at some point during the motion of the Sun it is directly overhead, then the *gnomon* will be parallel to one of the generating lines of the cone and perpendicular to the intersecting plane.

The conic sections were recognized by the Greeks as being important geometrical objects, but they were not thought to have any physical significance. The only physically significant curves were the circle – circular motion being natural for heavenly bodies – and the straight line – rectilinear motion being natural for earthly bodies. This mode of thinking did not change until the seventeenth century when Galileo discovered that projectiles follow parabolic paths and Kepler showed that planets have elliptical orbits.

Aristarchus

The predominant Greek view of the Universe had the Earth at the centre. This was Eudoxus' starting point and was fundamental to the whole of Aristotelian physics. An alternative cosmology in which the Sun, Earth and the other celestial bodies revolved around a central fire had been suggested by the Pythagorean Philolaus, and another notable exception to the prevailing viewpoint was that espoused by Aristarchus, from the island of Samos in Asia Minor.[19] Aristarchus

---

[19] Very little is known about Aristarchus. According to Hipparchus he observed the time of the summer solstice in 281 BC.

is credited by Archimedes as having postulated that the Earth is not at the centre of the Universe, but that it orbits the Sun. It was Copernicus, in the sixteenth century, who first developed a serious cosmology based on this principle, and so Aristarchus is often referred to as the Copernicus of antiquity. However, the attribution of a fully fledged heliocentric theory to Aristarchus is based actually on extremely scant evidence. Much is made nowadays of the following passage from Archimedes:

> But Aristarchus of Samos brought out a book consisting of certain hypotheses, in which the premises lead to the conclusion that the universe is many times greater than that now so called. His hypotheses are that the fixed stars and the sun remain motionless, that the earth revolves about the sun in the circumference of a circle, the sun lying in the middle of the orbit, and that the sphere of the fixed stars, situated about the same centre as the sun, is so great that the circle in which he supposes the earth to revolve bears such a proportion to the distance of the fixed stars as the centre of the sphere bears to its surface.[20]

Archimedes' work, of which the above quotation forms a part, is a demonstration of his skill in using the Greek numeration system to manipulate very large numbers. Ostensibly, Archimedes was attempting to calculate the number of grains of sand that the Universe could hold and he concluded that this had to be less than the unimaginably large number, 10^{63}. Clearly, this required a knowledge of the size of the Universe, and Archimedes tried deliberately to overestimate this. He criticized Aristarchus' heliocentric hypothesis because, in order for it to reproduce the observed phenomena; the stars would have to be infinitely far away, contradicting Archimedes' hypothesis that the Universe is finite. The problem is that of stellar parallax: if the Earth is orbiting the Sun, then the observed longitude at which a fixed star is found would vary as the Earth moved around its orbit (see Figure 2.6). Such changes were, however, not observed.

There are many theories, all very speculative, of why the heliocentric theory did not catch on and was superseded so completely by geocentric astronomy. Perhaps the most plausible is the simple fact that the geometrical skill of the Greeks allowed them to devise ingenious constructions that modelled accurately the motions of all the heavenly bodies without having to take the drastic step of removing the Earth from its privileged position at the centre of the Universe.

---

[20] ARCHIMEDES *Psammites* (*Sand Reckoner*). Translation from Heath (1932). Aristarchus' heliocentric theory is discussed at length in Heath (1913), but Wall (1975) is of the opinion that Heath has made rather too many assumptions and that there is no evidence that Aristarchus ever wrote a treatise on his heliocentric hypothesis. Indeed, it is very likely that all known references to Aristarchus' theory are derived from this one remark of Archimedes.

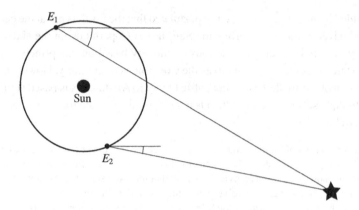

Fig. 2.6. Stellar parallax. E_1 and E_2 are two points on the orbit of the Earth.

The earliest complete astronomical treatise that has survived from ancient Greece is Aristarchus' *On the Sizes and Distances of the Sun and Moon.*[21] The organization of the work, starting with axioms and then making logical deductions from them, is similar to that of the great contemporary works in geometry such as Euclid's *Elements*. The treatise is based on six hypotheses:

(i) The Moon receives its light from the Sun.
(ii) The Earth is positioned at the centre of the sphere in which the Moon moves.
(iii) When the Moon appears to us halved, the great circle which divides the dark and bright portions of the Moon is in the direction of our eye.
(iv) When the Moon appears to us halved, its angular distance from the Sun is 87°. (This value is considerably in error. The true value is about 89° 50′.)
(v) The breadth of the Earth's shadow is that of two Moons. (This figure presumably was based on the length of lunar eclipses. The correct figure is nearer three lunar diameters.)
(vi) The Moon has an angular diameter of 2°. (About 4 times too large.[22])

[21] A translation is given in Heath (1913).
[22] Further on in the work Aristarchus assumed that the angular diameter of the Sun is the same as that of the Moon, i.e. 2°. Archimedes claimed that Aristarchus later discovered the much more accurate value of 1/2° for the angular diameter of the Sun, in agreement with his own observations (see Shapiro (1975)). It is likely that Aristarchus' figure of 2° for the Moon's angular diameter was not based on measurements at all, but simply assumed for the purposes of his demonstration (see van Helden (1985), p. 8).

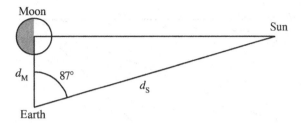

Fig. 2.7. Aristarchus' method for determining the relative distances from the Earth of the Sun and Moon.

Using the first four of these hypotheses, which are illustrated in Figure 2.7, Aristarchus concluded that the ratio of the distance from the Earth to the Sun, d_S, to that of the moon, d_M, satisfies the inequality

$$18 < d_S/d_M < 20.$$

The derivation of the value of d_S/d_M from Figure 2.7 would now be a simple exercise in trigonometry ($d_M/d_S = \cos 87° = \sin 3°$) but this had not yet been invented. The reason for the inequality that results from Aristarchus' calculations has nothing to do with an estimate of the accuracy of the experimental data, but instead represents the accuracy with which Aristarchus could determine what we would now write as $\sin 3°$. Here, we will demonstrate his argument for showing that $d_S/d_M > 18$, which is particularly elegant. In Figure 2.8, E, S and M represent the Earth, Sun and Moon with $\angle ESM = 3°$. The line EA is a continuation of ME such that $|AM| = |MS|$, and SC bisects the angle MSA. The line BC is perpendicular to AS so that $|BC| = |CM|$.

Aristarchus began by noting that

$$\frac{\angle CSM}{\angle ESM} = \frac{22\frac{1}{2}}{3} = \frac{15}{2},$$

from which it follows, using the equivalent of the fact that $\tan \alpha / \tan \beta > \alpha/\beta$ if $\beta < \alpha < 90°$ (a result which was well known in Aristarchus' time; a proof can be found in Euclid's *Optics*), that

$$\frac{|CM|}{|EM|} > \frac{15}{2}. \tag{2.1}$$

Next, we observe that

$$\frac{|CA|^2}{|CM|^2} = \frac{|CA|^2}{|CB|^2} = 2 > \frac{49}{25},$$

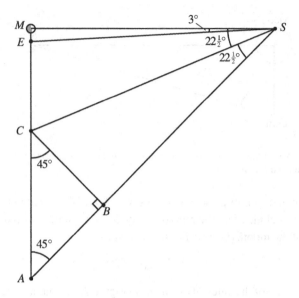

Fig. 2.8. Aristarchus' method for showing that $d_S/d_M > 18$.

and, hence, that $|CA|/|CM| > 7/5$, which in turn implies that

$$\frac{|AM|}{|CM|} = \frac{|CA| + |CM|}{|CM|} > \frac{12}{5}. \qquad (2.2)$$

(Note that here Aristarchus has used the well-known Pythagorean approximation to $\sqrt{2}$, namely 7/5.) Putting Eqns (2.1) and (2.2) together we obtain

$$\frac{|AM|}{|EM|} = \frac{|AM|}{|CM|} \cdot \frac{|CM|}{|EM|} > \frac{12}{5} \cdot \frac{15}{2} = 18.$$

But $|ES| > |MS| = |AM|$, and so we arrive finally at the desired result

$$\frac{d_S}{d_M} = \frac{|ES|}{|EM|} > 18.$$

Having determined the relative distances of the Sun and Moon, Aristarchus immediately could deduce the relative sizes of the Sun and Moon since (from the observed 'fact' that the Moon exactly eclipses the Sun) he assumed that they have the same angular diameter. Hence, the radii of the Sun and Moon, r_S and r_M, for example, satisfy

$$18 < r_S/r_M < 20.$$

Aristarchus' geometry based on the bisected Moon is flawless, but the inaccuracies in his observations mean that his results are not. The correct value

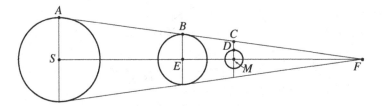

Fig. 2.9. A simplified illustration of Aristarchus' method for determining the relative sizes of the Sun and Moon.

of r_S/r_M is about 400. However, the conclusion that the distance to the Sun is about 19 times greater than the distance to the Moon appears frequently in astronomical writings over the course of the 2000 years following Aristarchus.

Aristarchus then went on to determine the sizes of the Sun and Moon relative to the Earth. His method was fairly complicated, but a simplified version is given below and illustrated in Figure 2.9.[23] The centres of the Sun, Earth and Moon are S, E and M, respectively, and the lines SA, EB and MC are all perpendicular to SEM and intersect the circles representing the Sun, Moon and Earth at A, B and D, respectively. Then ABF is a good approximation to the tangent to the Sun and Earth and, hence, $|MC|$ is close to the radius of the shadow of the Earth at the distance of the Moon, which is $2r_M$ according to Aristarchus' fifth hypothesis.

Using the similarity of the triangles ASF, BEF and CMF we can deduce that

$$r_S d_M + 2r_M d_S = r_E(d_S + d_M),$$

and if we substitute into this equation Aristarchus' mean value for the relative sizes of the Sun and Moon (i.e. $d_S = 19d_M$, $r_S = 19r_M$) we obtain[24]

$$\frac{r_S}{r_E} = \frac{20}{3} \quad \text{and} \quad \frac{r_M}{r_E} = \frac{20}{57}.$$

The correct value for r_S/r_E is about 109. Aristarchus used his results to show that the volumes of the Sun and Earth are in the ratio of between 254 : 1 and

---

[23] Based on that given in Boyer (1989). This type of eclipse diagram played a leading role in distance determinations until the seventeenth century.

[24] Aristarchus actually carried out his geometrical calculations maintaining the consequences of the inequalities for the relative sizes of Sun and Moon and obtained the estimates

$$\frac{19}{3} < \frac{r_S}{r_E} < \frac{43}{6} \quad \text{and} \quad \frac{43}{108} < \frac{r_M}{r_E} < \frac{19}{60}.$$

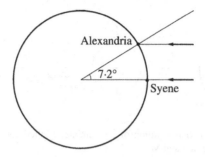

Fig. 2.10. Eratosthenes' method for determining the size of the Earth.

368 : 1. It may be that the fact that the Sun was thus such a massive object compared to the Earth was one of the reasons Aristarchus was led to believe that the Sun was at rest with the Earth orbiting round it.

While the relative sizes of the Sun, Moon and Earth were understood poorly by the ancient Greeks, the actual size of the Earth was determined with some accuracy. In *On the Heavens* Aristotle credits 'the mathematicians' with the figure of 400 000 *stade*s for the circumference of the Earth, but left no indication of how, or by whom, this figure was determined. A *stade* was a unit of distance the value of which varied from one place to another so it is hard to judge the accuracy of Aristotle's figure, but it would appear to be a significant overestimate.[25]

This value was improved significantly in the following century by Eratosthenes, the director of the Library at Alexandria (see Figure 2.10). He found that at noon on the summer solstice the Sun was directly overhead at Syene (present-day Aswan) while at the same time in Alexandria, which he assumed to be about 5000 *stade*s due north, the Sun was $7\frac{1}{5}°$ (i.e. one-fiftieth of a circle) from the zenith.[26] Eratosthenes concluded that the circumference of the Earth is $5000 \times 50 = 250\,000$ *stade*s. This was later changed to 252 000, possibly in order to have a round figure of 700 *stade*s per degree. Depending on the interpretation of a *stade*, this could be anything from about 5 per cent too small to 25 per cent too big.[27]

---

[25] According to Diller (1949), 1 *stade* could be anything between about 150 and 200 m. Thus, 400 000 *stade*s is somewhere between 60 000 and 80 000 km. The Earth's circumference is actually about 40 000 km.

[26] Actually the difference in latitude between Aswan and Alexandria is slightly less than Eratosthenes' value, and there is a difference in longitude of approximately 3°.

[27] Eratosthenes' original work on the measurement of the Earth is lost and our knowledge of his technique comes from a number of later sources, notably Cleomedes, a popular writer on astronomical matters. There has been much debate about the precise details of Eratosthenes'

Eratosthenes is credited[28] also with measuring the obliquity of the ecliptic, arriving at a value of $22/83$ of a right angle,[29] or about $23° 51'$, which is within 1 per cent of the true value for that time.

Eccentric circles and epicycles

Some 150 years after Eudoxus, Apollonius of Perga (who spent most of his life in Alexandria and is now most famous for his work on conic sections) devised a new solution to Plato's challenge of saving the phenomena using uniform circular motions. No astronomical works of Apollonius survive; we know of his writing only through the later work of Ptolemy. It was well known that Eudoxus' scheme of homocentric spheres failed to account for certain readily observable phenomena (e.g. the varying brightness of the planets) and Apollonius suggested a scheme based on eccentric circles and epicycles in order to get round this difficulty.[30]

He developed two equivalent models for the motion of the Sun (illustrated in Figure 2.11) that were designed to account for the fact that the time between the vernal equinox and summer solstice (measured by Callippus as 94 days) is longer than that between the summer solstice and the autumnal equinox (92 days). In the first scheme, the annual motion of the Sun S was around a circle (the dashed circle in the diagram) the centre of which was not the Earth E, but a point D called the 'eccentre', displaced away from the Earth, the distance $|DE|$ being known as the 'eccentricity'. Alternatively, we can consider the Sun positioned on a small circle (the epicycle) the centre C of which rotates uniformly about the Earth around another circle (the deferent). The epicycle rotates in the opposite sense to that with which C rotates around the deferent, but with the same angular speed, so that $EDSC$ remains a parallelogram. The path traced out by the Sun S will again be the dashed circle. The point C, which rotates around the Earth at a uniform rate at the same average speed as the Sun, represents the mean position of the Sun.

method; see Dutka (1993) and the references cited therein; also, Fischer (1975), Rawlins (1982), Goldstein (1984). Posidonius used a similar method a century later, involving Alexandria and Rhodes, to obtain a similar estimate for the Earth's size. On the relative merits of Posidonius' and Eratosthenes' contributions (see, for example, Taisbak (1974)).

[28] By Ptolemy, *Almagest*, Book I, 12 (see Toomer (1984)).

[29] An explanation for this curious ratio is given in Fowler (1987), pp. 51–2.

[30] Apollonius was probably not the first person to develop an epicycle theory; indeed, van der Waerden (1974, 1982) provides evidence that the Pythagoreans were in possession of a primitive epicycle theory. However, as far as we know, Apollonius was the first to show how epicycles could be used to model all the main features of planetary motion, including retrograde motion.

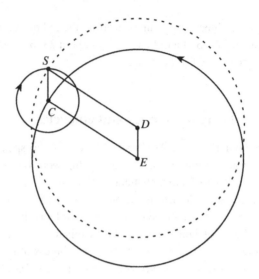

Fig. 2.11. Eccentric circles and epicycles for the Sun.

In order to use these models it is necessary to determine the length and direction of the line ED so that the position of the Sun relative to the Earth matches the observations. This requires the ability to solve the triangle DES which in turn requires trigonometry. It was precisely this sort of need that led to the later development of that subject by the astronomers Hipparchus and Ptolemy.

Apollonius also made the crucial realization that the epicyclic model could be used to account for the retrograde motions of the planets if the epicycle rotates in the same sense as that with which its centre rotates around the deferent; this is illustrated schematically in Figure 2.12. If the radii and the speeds of rotation of the two circles are chosen suitably, the combined effect produces motion that is predominantly anticlockwise, but with short periods during which the planet appears from the Earth to move in the opposite direction. It was well known in Apollonius' time that the superior planets appear brighter during their periods of retrograde motion and the epicycle model explained this quite naturally, because these were precisely the places at which the planets were closest to the Earth. The epicycle–deferent construction was a significant improvement over Eudoxus' homocentric spheres, but in its basic form it was still incapable of reproducing the complex motions of the planets with any real accuracy.

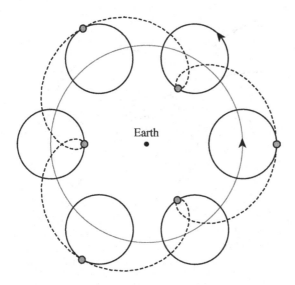

Fig. 2.12. Epicycles and retrograde motion.

Apollonius did more than just indicate that epicycles can be used to produce retrograde motion. He also proved an important theorem that provides a condition that the rotation rates and radii of the epicycle and deferent must satisfy for stationary points (i.e. points where the planet appears to change direction) to occur, and this is illustrated in Figure 2.13. The epicycle rotates anticlockwise about its centre C, with angular speed ω_2, while the point C rotates in the same sense with angular speed ω_1 around the deferent, the centre of which is the Earth, E. The resulting motion for the planet is the dashed curve, and there are two points at which the motion appears stationary as seen from the Earth, one of which is the point P.

The velocity of the planet at P can be decomposed into two parts, one due to the rotation of C around the deferent (the vector with magnitude u) and one due to the rotation of the epicycle (the vector with magnitude v). Clearly $u = \omega_1|EP|$ and $v = \omega_2|CP|$. For the point P to be stationary as seen from the Earth, the sum of these two velocities must lie along EP and, hence, we must have $u = v \cos \angle APD$. Now, since $\angle APC$ and $\angle BPD$ are both right angles it follows that $\angle APD = \angle CPB$ and, hence, that $\cos \angle APD = |BP|/2|CP|$. Putting all this together gives the condition for stationary points as

$$\omega_1|EP| = \tfrac{1}{2}\omega_2|BP|.$$

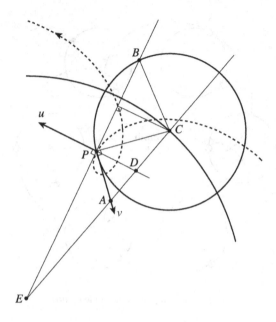

Fig. 2.13. Apollonius' theorem on stationary points.

Apollonius' proof was rather different (and rather more complicated) because he did not have modern trigonometrical methods at his disposal.[31] Now, $1/2|BP| < |CP|$ and $|EP| + |CP| > |EC|$, so a consequence of this theorem is that retrograde motion can occur only if the angular speeds and the radii of the epicycle and deferent satisfy the inequality[32]

$$\frac{\omega_1}{\omega_2} < \frac{|CP|}{|EC| - |CP|}.$$

In the hands of the ancient Greeks, the epicycle–deferent mechanism was employed skillfully to account for many different irregularities in the motions of the heavenly bodies. By altering the relative sizes of the two circles and their directions of rotation, different effects could be produced, and by placing epicycles upon epicycles, extremely complex motions could be generated. This basic mechanism was the cornerstone of all quantitative planetary theories until the time of Kepler. Even Copernicus, who succeeded in removing the need to

---

[31] More details can be found in Neugebauer (1955, 1959).
[32] It is not known whether Apollonius used his theorem to determine the numerical values of the parameters necessary to reproduce the observed phenomena, but Ptolemy later used it in Book XII of the *Almagest* to construct a table of stationary points even though it is not strictly valid for Ptolemy's more complicated scheme.

model retrograde motion using epicycles, still made extensive use of them to account for other irregularities.[33]

---

[33] The power and versatility of the epicycle is described in Hanson (1960). Hanson shows that an orbit of any given shape can be reproduced to any given accuracy by a model built up from a finite number of epicycles placed one upon the other. Mathematically this is equivalent to approximating an arbitrary continuous function by a finite truncation of a Fourier series.

3

The Ptolemaic universe

Hipparchus

Hipparchus, who lived in the second century BC,[1] built an observatory and performed most of his work on the island of Rhodes and was perhaps the greatest astronomer of antiquity. He used observations to produce geometrical models with real quantitative predictive power. His theory of the motion of the Sun was extremely accurate and he produced a model for the Moon that worked well at new and full moons, thus enabling him to produce a theory of eclipses which, in the case of lunar eclipses, was very successful.

All of Hipparchus' works are lost except for his relatively unimportant *Commentary on the Phaenomena of Eudoxus and Aratus*,[2] though Ptolemy quotes his work often, sometimes verbatim. We also know of Hipparchus' work on the Sun through an introduction to astronomy written in AD first century by Geminus of Rhodes. Part of the reason for the lack of extant work by Hipparchus may well be the fact that Ptolemy's subsequent writings superseded those of his predecessor so totally, just as the existence of Euclid's *Elements* rendered obsolete all previous works on geometry.

Hipparchus attempted to use the eccentric circles and epicycles of Apollonius to develop models for the motion of the heavenly bodies that, in contrast to Babylonian theories, would enable future positions to be calculated for all times. For the Sun and the Moon he found that he could use just one such device, but for the planets he needed to combine the two. He was a great

---

[1] Ptolemy refers to observations made by Hipparchus between 161 and 126 BC.
[2] In about 275 BC, Aratus of Soli wrote a very popular poem (the *Phaenomena*, inspired by a more technical, but now lost, work of Eudoxus) describing the risings and settings of stars and weather signs in both heavenly and natural phenomena. It was later translated into Latin and remained widely read for over 1000 years. Kidd (1997) contains a translation and commentary.

observational astronomer who improved the design of the instruments used for observing the skies and used these instruments to compile a catalogue of about 850 stars.

As we have seen, quantitative calculations that arose from astronomical problems often involved the solution of triangles, and it was for this reason that the subject of trigonometry developed. In fact, trigonometry did not become a branch of mathematics separate from astronomy until the fifteenth century. Hipparchus, who is considered to be the founder of trigonometry, constructed a table of chords (equivalent to a table of sines) though we do not know how he did this.[3] He subdivided the circle into 360°, an idea introduced into Greek astronomy from Babylonia through the work of his contemporary, Hypsicles of Alexandria.

One of Hipparchus' greatest achievements was his discovery of the precession of the equinoxes. He discovered that the points at which the ecliptic crosses the celestial equator move slowly with respect to the stars. He was able to do this because Babylonian astronomical data became available in the Greek world and he examined systematically old observations and compared them with his new ones in order to discover changes that were too slow to be detected by astronomers using only data gathered during their own lifetimes. The arrival of large quantities of data covering observations made over many centuries helped transform Greek astronomy. It was now possible to take the geometrical models that had been developed and use them to produce procedures for accurate quantitative prediction. The work of Hipparchus, based as it was on a merging of Greek and Babylonian approaches, marks the transition between qualitative and quantitative mathematical astronomy.[4]

Hipparchus' contributions to astronomy were enormous. In the words of the French astronomer J.-B. J. Delambre:

> When we consider all that Hipparchus invented or perfected, and reflect upon the number of his works and the mass of calculations which they imply, we must regard him as one of the most astonishing men of antiquity, and as the greatest of all in the sciences which are not purely speculative, and which require a combination of geometrical knowledge with a knowledge of phenomena, to be observed only by diligent attention and refined instruments.[5]

---

[3] Toomer (1974) contains a plausible reconstruction. Ptolemy also constructed a table of chords and his method is preserved. This will be described later.

[4] Babylonian arithmetical methods continued to be used in the Greek world right up to Ptolemy's time. Indeed, Hipparchus used them to compute both solar and lunar longitudes while developing his geometrical schemes (see Jones (1991a)).

[5] DELAMBRE *L'histoire de l'astronomie ancienne*, I (1817). Translation from Berry (1961).

This is probably a bit over the top. Delambre tends to credit Hipparchus with things that most modern scholars attribute to his successor, Ptolemy.

The distances of the Sun and Moon

In *On Sizes and Distances*, a work described by Pappus of Alexandria in his commentary on Ptolemy's *Almagest*, Hipparchus calculated the distances of the Sun and Moon from the Earth, measured in Earth radii. His method was based around the concept of diurnal solar parallax.

Just like the stellar parallax in a heliocentric universe (see Figure 2.6, p. 40), the diurnal parallax is the change in the apparent position of an object (in this case the Sun) as a result of the position of the observer. With stellar parallax it is the position of the Earth in its orbit that causes the difference, whereas for diurnal parallax it is the position of the observer on the Earth that is the cause. Thus, the longitude of the object O in Figure 3.1 (which can be thought of as a view from the celestial north pole) differs depending upon whether it is viewed from P_1 or P_2, two points at different positions on the Earth. Two observers are not required to determine this difference, however, since the object O takes part in the daily rotation of the heavens and its longitude will thus be affected by the time of day at which measurements are taken. Thus, if one makes observations of the sun 6 h apart, for example, the longitudes will not differ by the amount due to the motion of the Sun alone (which can, in principle, be calculated), but there will be a small error (the angle marked in the diagram) corresponding to the diurnal parallax. The solar parallax is a direct measurement of the distance of the Sun from the Earth in terms of the size of the Earth – the greater the distance of the Sun the smaller the parallax – but unfortunately Hipparchus was unable to measure it due to its very small magnitude.

While it is true that Hipparchus could not measure any solar parallax, he knew this did not mean it did not exist. Accordingly, he assumed a solar parallax of 7′, on the basis that if it were bigger he would be able to measure it. From Figure 3.1,

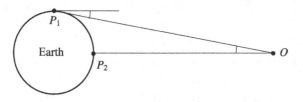

Fig. 3.1. Diurnal parallax.

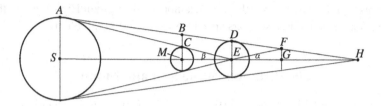

Fig. 3.2. Hipparchus' method for determining the relative distances of the Sun and Moon.

we see that if the marked angle is $7'$, the distance to the Sun is $1/\tan 7' \approx 490$ Earth radii, and this is the value Hipparchus used. He then obtained, using data from lunar eclipses, a lunar distance of $67\frac{1}{3}$ Earth radii (which is just over 10 per cent too big) and hence, a ratio of the Sun's distance to the Moon's distance of about $7\frac{1}{4}$ (which is worse than Aristarchus' estimate). He also obtained the lunar and solar radii as $\frac{1}{3}$ and $2\frac{1}{3}$ Earth radii, respectively, the former being quite close to the modern value of 0.27, although the latter is about 50 times too small.

The method Hipparchus used is described by Ptolemy and is illustrated in Figure 3.2.[6] The centres of the Sun, Moon and Earth are S, M and E, respectively, and the point G is on SH such that $|EG| = |EM|$. The lines SA, MC and ED are all perpendicular to SH and intersect the circles representing the Sun, Moon and Earth in A, C and D, respectively. The lines ADH and ACE are then good approximations to the lines tangent to the Sun and Earth, and Sun and Moon. The angle β (the apparent radius of the Moon) was determined through observation, and then Hipparchus determined the length of the Earth's shadow from the length of lunar eclipses. He took $\alpha/\beta = 2\frac{1}{2}$, where previously Aristarchus had used 2.

It is convenient to work in units of Earth radii so that $|ED| = 1$. First, we note that in reality α and β are very small, so that to a good approximation $\alpha/\beta = \tan\alpha/\tan\beta = |FG|/|MC|$. Now, MB, ED and FG are parallel so that $|MB| + |FG| = 2|ED| = 2$, and, hence,

$$|BC| = |MB| - |MC| = 2 - |FG| - |MC| = 2 - \left(\frac{\alpha}{\beta} + 1\right)|MC|.$$

$$(3.1)$$

Finally, we use similar triangles to obtain the result

$$\frac{1}{|BC|} = \frac{|EA|}{|CA|} = \frac{|ES|}{|MS|} = \frac{|ES|}{|ES| - |EM|}$$

[6] PTOLEMY *Almagest*, Book V, 11. Further details can be found in Swerdlow (1969).

from which, using Eqn (3.1) and $|MC| = |EM| \tan \beta \approx |EM| \sin \beta$, we obtain

$$|ES| = \frac{|EM|}{\left(\frac{\alpha}{\beta} + 1\right)|EM|\sin\beta - 1}.$$

This expression is rearranged easily so as to give $|EM|$ in terms of $|ES|$. Hipparchus substituted into this expression the vales $|ES| = 490$, $\beta = 0; 16, 37°$ and $\alpha/\beta = 2\frac{1}{2}$, from which he obtained the result $|EM| = 67\frac{1}{3}$ Earth radii.

The motion of the Sun and precession

Perhaps Hipparchus' greatest contributions concerned the motion of the Sun. In *On the Length of the Year* he claimed that the length of the tropical year (the time between identical equinoxes or solstices) was constant, and he measured it at $365 + \frac{1}{4} - \frac{1}{300}$ days (365 days 5 h 55 min 12 s) which exceeds the modern value by about $6\frac{1}{2}$ minutes but, nevertheless, represents a significant improvement over the previous value of $365\frac{1}{4}$ days. This probably was done by taking the Babylonian value for the synodic month (i.e. $29; 31, 50, 8, 20$ days) and using the approximate equality given by the Metonic cycle (i.e. 19 years = 235 months) and then checking the result with observations.[7]

Hipparchus was the first to attempt to calculate the parameters needed for the eccentric circle theory of Apollonius to agree with observations of the Sun's position, and his model of the solar motion, and the basic principles by which the parameters for the model were deduced, remained standard until the seventeenth century. The success of Hipparchus' solar model was due to being mathematically simple and yet very accurate. Provided the parameters in the model are calculated accurately, the errors in the predicted solar longitudes will not be detectable from naked-eye observations. Hipparchus' method for determining the parameters is illustrated in Figure 3.3, in which E is the Earth and the Sun S rotates around an eccentric circle centre O with an angular speed of $\omega = 1$ revolution per year $\approx 59'\ 8''$ per day. The points P_1, P_2, P_3, and P_4 represent the position of the Sun at the vernal equinox, the summer solstice, the autumnal equinox and the winter solstice, respectively. In order to be able to use the model to compute the position of the Sun, we need to calculate the longitude λ of the apogee A of the orbit of the Sun (the apogee is the point on the orbit furthest from the Earth), which we choose to measure from the vernal equinox, and the ratio of the eccentricity $|EO|$ to the radius $|OS|$ of the orbit of the Sun.

---

[7] See Swerdlow (1979).

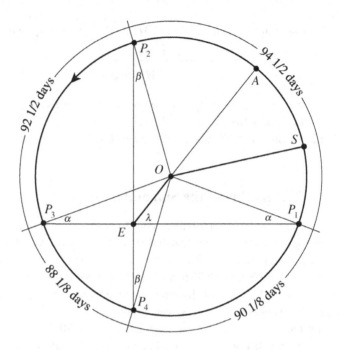

Fig. 3.3. Hipparchus' solar theory.

The technique Hipparchus used to do this can be described using modern trigonometry as follows. First, consider the triangles EOP_1 and EOP_2. The sine rule gives

$$\frac{|OP_1|}{\sin \lambda} = \frac{|OE|}{\sin \alpha} \qquad \text{and} \qquad \frac{|OP_2|}{\cos \lambda} = \frac{|OE|}{\sin \beta}.$$

From these equations it follows, since $|OP_1| = |OP_2| = |OS|$, that

$$\tan \lambda = \frac{\sin \alpha}{\sin \beta} \qquad \text{and} \qquad \frac{|EO|}{|OS|} = \frac{\sin \alpha}{\sin \lambda}. \qquad (3.2)$$

To obtain λ and $|EO|/|OS|$ from these equations, Hipparchus needed to know the time between the vernal equinox and the summer solstice and that between the summer solstice and the autumnal equinox, and from his observations he was able to improve the known values for the lengths of the seasons. He measured the time taken for the Sun to travel from P_1 to P_2 as $94\frac{1}{2}$ days and from P_2 to P_3 as $92\frac{1}{2}$ days (the times from P_3 to P_4 and from P_4 to P_1 being $88\frac{1}{8}$ and $90\frac{1}{8}$ days, respectively). Hence, since $\angle P_1 O P_2 = \alpha + \beta + 90°$ and $\angle P_2 O P_3 = \alpha - \beta + 90°$,

$$\alpha + \beta + 90° = \frac{189\omega}{2} \qquad \text{and} \qquad 2\alpha + 180° = 187\omega,$$

simultaneous equations that can be solved for α and β. Finally Eqn (3.2) can be used to obtain the needed parameters. Hipparchus obtained the values

$$\lambda = 65° \, 30' \qquad \text{and} \qquad \frac{|EO|}{|OS|} = \frac{1}{24}$$

which, given the rudimentary trigonometry at his disposal, are fairly impressive ($65° \, 25' \, 39''$ and $1/24.17$ are more accurate solutions to the equations). Modern computations show that in Hipparchus' time the true value of the longitude of the Sun's apogee was nearer to $66°$.[8]

Hipparchus made another hugely significant discovery concerning the Sun's motion. By comparing his own observations of the longitudes of certain stars with those made by Timocharis some 150 years previously, he discovered the phenomenon now known as the 'precession of the equinoxes'.[9] He noticed that the sidereal year (the time for the Sun to return to a particular fixed star) was slightly greater than the tropical year, and attributed this to a slow rotation of the stars, from west to east, about the poles of the ecliptic. A consequence of this is that the positions of the equinoxes on the celestial sphere gradually shift with time. According to Ptolemy,[10] Hipparchus gave the value of this motion as $1°$ per century (and Ptolemy used this value), although the actual value is about $1°$ every 72 years. This numerical inaccuracy was to have a significant influence over subsequent theories since, when precession was measured more accurately about 1000 years later, the differing values led many astronomers to believe that the rate of precession was a variable quantity.

The motion of the Moon

The motion of the Moon is much more complex than that of the Sun and it cannot be described by a simple eccentric circle mechanism. The reasons for this were known to Hipparchus and are stated by Ptolemy:[11] they are that first the Moon moves with varying speed in such a way that over the course of time it achieves its maximal speed for every value of its longitude λ, and, second, that

---

[8] More details of how Hipparchus arrived at his solar model can be found in Jones (1991b), and Maeyama (1998) has analysed the effect of observational errors on the underlying parameters and hence on the predictions of the model. According to Jacobsen (1999) the maximum error in Hipparchus' theoretical values for the Sun's longitude was about $22'$.

[9] A theory developed in the 1920s that credits the Babylonians with the discovery of precession was shown to be false by Neugebauer (1950).

[10] PTOLEMY *Almagest*, Book VII, 2. Ptolemy stated that Hipparchus wrote a work (now lost) entitled *On the Displacement of the Solstitial and Equinoctial Points*. Surprisingly perhaps, Hipparchus' discovery was mentioned by only a few Greek writers (see Dreyer (1953), p. 203); it took on a much greater significance in later centuries.

[11] PTOLEMY *Almagest*, Book IV, 2.

the Moon does not move on the ecliptic, but instead has latitudes which vary between ± 5° of the ecliptic in such a way that the Moon achieves its maximal latitude for every value of λ. It is a simple matter to see that a model like that used by Hipparchus for the Sun will lead always to the maximal speed occurring for the same value of the longitude (at the perigee – the point on the orbit closest to the Earth) and so cannot account for the first of these observations. The second observation suggests that the orbit of the Moon is inclined at an angle of about 5° to the ecliptic. However, if the line joining the intersections of the Moon's orbit and the ecliptic is fixed, the maximum latitude always will occur for the same value of the longitude which is observed not to be the case. In order for an eclipse to occur, the Moon must be near one of its nodes and thus observations of eclipses can be used to give fairly accurate information on the nodes of the Moon's orbit. Lunar periods had been computed accurately by the Babylonians, and Hipparchus, by comparing his own observations with earlier ones, confirmed the Babylonian lunar periods as:[12]

1 synodic month = 29; 31, 50, 8, 20 days,
251 synodic months = 269 anomalistic months,
5458 synodic months = 5923 draconitic months.

Hipparchus' lunar model is illustrated in Figure 3.4. It consists of an epicycle carrying the Moon, M, the centre C of which rotates around a deferent circle, the deferent–epicycle system being inclined at an angle of 5° to the ecliptic and intersecting that circle in the two nodes A and B. The nodal line AOB was made to rotate in order to account for the fact that the longitude of the position of maximum latitude of the Moon changes gradually. The motion on the epicycle ensures that the Moon's speed is variable and by making the period of revolution of M around the epicycle different from the period of C around the deferent, we ensure that the longitude of the position of maximal speed (which occurs when M is at its closest to O) varies over time.

Hipparchus made the simplifying assumption that the motions in latitude caused by the 5° angle of the orbit could be treated separately from the motion in longitude due to the epicyclic system. This introduces only very minor errors. The rotation of the nodal line thus becomes irrelevant for the longitude theory, which is reduced to a simple two-dimensional deferent–epicycle system. In order to use the theory, various parameters have to be computed. Thus, for the longitude calculations, we require the rates of rotation of C around the deferent and M around the epicycle as well as the ratio of the radii of the epicycle and

---

[12] See Swerdlow and Neugebauer (1984), I, p. 198.

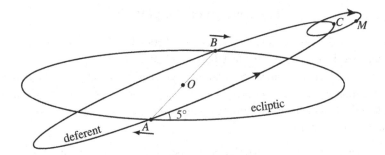

Fig. 3.4. Hipparchus' lunar theory.

deferent, whereas for the latitude calculations we require simply the rate of rotation of the nodal line. This latter parameter can be determined from direct observations of eclipses. It turns out that in order to reproduce the observed behaviour of the Moon, the nodal line must rotate approximately once every 18 years (the so-called 'Saros period').

The rates of rotation of the deferent and the epicycle also are easily obtained. The former is simply the rate required for C to complete 1 revolution in 1 sidereal month which turns out to be about $13°\,10'\,35''$ per day, and the latter is chosen to ensure that the epicycle rotates once in each anomalistic month, which implies a rate of $13°\,3'\,54''$ per day. The final parameter, the ratio of the radii of the epicycle and deferent, presents a far harder problem. Hipparchus developed a geometrical method that enabled this ratio to be obtained from three observations of the Moon[13] and he used it on two different sets of three observations, obtaining different answers. There has been some debate as to what value Hipparchus actually used[14] although, when the lunar model was taken over later by Ptolemy, he computed a ratio of $5\frac{1}{4} : 60$.

Hipparchus' parameters were based on Babylonian observations of lunar eclipses. As a result, the model worked well at full moons but, as Ptolemy demonstrated 300 years later with observations of the Moon at other points in its orbit, it did not work well away from the syzygies. With his solar and lunar theories, Hipparchus created the first coherent theory of eclipses. Durations could be determined about as accurately as they could be measured, but the time at which the eclipse would occur was predicted less well. This situation did not improve substantially until the work of Tycho Brahe at the end of the sixteenth century.

[13] See, for example, Pedersen (1974), p. 172.
[14] See Neugebauer (1959), Toomer (1967), Pedersen (1974).

The motion of the planets

Hipparchus did not achieve a satisfactory theory for the motion of the five planets. Partly this is due to the fact that he, unlike many of his predecessors, had examined sufficiently many observations over a period of many years to realize that the motions were exceedingly complex, with, for example, retrograde arcs, the lengths of which vary according to the position of the planet in its orbit. As Ptolemy later put it:

> ... Hence it was, I think, that Hipparchus, being a great lover of truth, for all the above reasons, and especially because he did not yet have in his possession such a groundwork of resources in the form of accurate observations from earlier times as he himself has provided to us, although he investigated the theories of the sun and moon, and, to the best of his ability, demonstrated with every means at his command that they are represented by uniform circular motions, did not even make a beginning in establishing theories for the five planets, not at least in his writings which have come down to us. All that he did was to make a compilation of the planetary observations arranged in a more useful way, and to show by means of these that the phenomena were not in agreement with the hypotheses of the astronomers of that time.[15]

Ptolemy and the *Almagest*

Ptolemy (Latinized as Claudius Ptolemaeus) was one of the great scholars of antiquity, and mathematical astronomy was dominated by his ideas for nearly 1500 years following his death. Little is known of his life, but he taught in Alexandria and quoted the results of his observations made between AD 127 and 141. He was responsible for a number of great works, each of which place him among the most important ancient authors. The earliest of these is his masterpiece of mathematical astronomy, the *Almagest*,[16] and others include the *Tetrabiblos* (on astrology) and the *Geography* (on mathematical geography). These works exercised a colossal influence over mankind for the next 2000 years. Ptolemy's very wide range of interests is indicated by his works on other subjects, e.g. music, optics, and logic. His *Harmony*, in which he described musical consonances and their relationship to an underlying universal harmony, was later an inspiration to Kepler.[17]

The name *Almagest* – which is the name by which his astronomical treatise is usually known – is a corruption by medieval Latin translators of the Arabic

---

[15] PTOLEMY *Almagest*, Book IX, 2. Translation from Toomer (1984).
[16] All quotations from the *Almagest* are taken from the translation by Toomer (1984).
[17] See Martens (2000), Chapter 6.

word for 'the greatest', and was given to the book long after it was written. The original Greek title translates as *Mathematical Synthesis* or *Mathematical Collection* and the work, which survives in the original Greek, is sometimes referred to as the *Syntaxis*. Most of our knowledge of Greek astronomy is derived from this work, which was originally written around AD150, translated into Latin in the twelfth century, and first printed in the sixteenth century. The huge success of the *Almagest* resulted in the loss of most of the work of Ptolemy's predecessors, notably that of Hipparchus.

Ptolemy was, as far as we know, the first person to show how to convert observational data into numerical values for the parameters so as to make existing planetary models fit the observations. Unlike mathematical astronomers before him, who used unspecified observational data, Ptolemy used specific dated observations – indeed, many of the observations Ptolemy used, which span over 800 years, are only preserved through his work. The rigour with which Ptolemy developed his theories in the *Almagest* set a very high standard for future astronomical works.

The model of the Solar System as set out in the *Almagest* will be described in detail below. The various geometrical models he used are quite complicated and so, before describing these, we will begin with a brief description of the overall structure. In Ptolemy's universe, the spherical Earth is situated at the centre of the heaven, which is itself spherical. The Sun, Moon and planets all orbit the Earth and as to their order, Ptolemy wrote:

> ... we see that almost all the foremost astronomers agree that all the spheres are closer to the earth than that of the fixed stars, and farther from the earth than that of the moon, and that those of the three [outer planets] are farther from the earth than those of the other [two] and the sun, Saturn's being greatest, Jupiter's the next in order towards the earth, and Mars' below that. But concerning the spheres of Venus and Mercury, we see that they are placed below the sun's by the most ancient astronomers, but by some of their successors these too are placed above [the sun's], for the reason that the sun has never been obscured by them [Venus and Mercury] either. To us, however, such a criterion seems to have an element of uncertainty, since it is possible that some planets might indeed be below the sun, but nevertheless not always be in one of the planes through the sun and our viewpoint, but in another [plane], and hence might not be seen passing in front of it, just as in the case of the moon, when it passes below [the sun] at conjunction, no obscuration results in most cases.[18]

Thus, Ptolemy decided to side with those astronomers who put Mercury and Venus nearer than the Sun as this naturally separates those planets which can be seen at any longitude with respect to the Sun and those which always remain

---

[18] PTOLEMY *Almagest*, Book IX, 1.

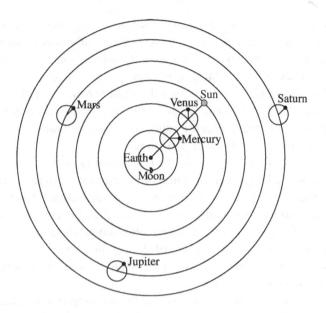

Fig. 3.5. A simplified view of Ptolemy's world system.

close to the Sun. Ptolemy's authority was such that the order of the planets that he proposed (Moon, Mercury, Venus, Sun, Mars, Jupiter, Saturn) was accepted by virtually all subsequent astronomers until the sixteenth century.

In the *Almagest*, the motion of each celestial body is considered in turn and, if all the models are put together, we get the world system sketched in Figure 3.5. The diagram illustrates the epicyclic nature of the planetary models, but not the many other elements used by Ptolemy to reflect more accurately the observational evidence. An important thing to note is the fact that, for Mercury and Venus, the line joining the Earth to the centre of their epicycles is the same as that joining the Earth to the Sun, whereas for the outer planets, which may appear anywhere with respect to the Sun, the radius connecting the planet to the centre of its epicycle is parallel to the Earth–Sun line. Thus, it is evident that the Sun does not simply orbit the Earth like the other planets, but has a much more significant role. The motion of the Sun also plays a role in the lunar theory.

As the Sun has an important function in the theories of all the heavenly bodies, it is logical to begin with a solar theory, and this is what Ptolemy does. He goes on then to consider the Moon and, finally, the planets. For each celestial body, Ptolemy describes the type of phenomena that must be accounted for, goes on to propose a geometrical model suitable for the purpose, shows how to use observations to derive the numerical values of the various geometrical parameters and, finally, produces tables to enable others to determine the position of the body on a given date. Ptolemy obtained data easily for the Sun, Moon, and all the planets from Venus to Saturn, but his over-reliance on the poor available data for Mercury – much the most difficult of the then-known planets to observe – led him to introduce a complicated geometrical device in order accurately to reproduce erroneous data.

The *Almagest* is a complete exposition of Greek mathematical astronomy divided into thirteen books. As Ptolemy himself says:

> We shall try to note down everything which we think we have discovered up to the present time; we shall do this as concisely as possible and in a manner which can be followed by those who have already made some progress in the field. For the sake of completeness in our treatment we shall set out everything useful for the theory of the heavens in the proper order, but to avoid undue length we shall merely recount what has been adequately established by the ancients. However, those topics which have not been dealt with [by our predecessors] at all, or not as usefully as they might have been, will be discussed at length, to the best of our ability.[19]

The first two books deal with the assumptions upon which the work is based and with mathematical methods. Books III and IV deal with the motion of the Sun and Moon, respectively. Book V contains a more advanced lunar theory (and a discussion, among other things, of the construction of an astrolabe) and in Book VI, Ptolemy presents his theory for eclipses. Books VII and VIII are taken up largely with his star catalogue[20] and, finally, Books IX–XIII deal with planetary motions. The organization is strictly logical, each book dependent only on those preceding it.

By late antiquity, the *Almagest* had become the standard textbook on astronomy, and remained as such for more than 1000 years. Perhaps the only scientific work to achieve greater dominance in history is Euclid's *Elements*. Ptolemy does not mention any physical interpretation of his system: his aim

---

[19] PTOLEMY *Almagest*, Book I, 1.

[20] In all, Ptolemy tabulated 1022 stars in 48 constellations, giving the longitude, latitude and magnitude of each. He was very interested in the question of whether the stars move in relation to each other and so he gave an extensive list of stars that lay in straight lines so that future observations easily could reveal any relative motion that exists. There is evidence to suggest that Ptolemy took the star catalogue of Hipparchus and added the correction in longitude given by Hipparchus' value for the precession. For the $2\frac{2}{3}$ intervening centuries this would amount to 2° 40′ which gives a good explanation of why Ptolemy's longitude values are consistently about 1° too small, since, had he used the correct value of the precession, he would have had to add 3° 40′. However, there is no direct evidence that Hipparchus ever produced a systematic catalogue of about 1000 stars as Ptolemy did. More details on this question can be found in Evans (1987) and Shevchenko (1990) and an excellent historical survey of the debate on this question, which has gone on for over 100 years, can be found in Evans (1998); (see also note 25 on p. 70).

solely is to represent the heavenly phenomena by purely kinematic hypotheses. He returned to this question in a later work, the *Planetary Hypotheses*. However, Ptolemy did use physical arguments to justify his choice of a fixed Earth at the centre of the Universe. He realized that the diurnal rotation of the heavens could equally well be accounted for by a rotating Earth, as espoused by Heraclides, but this was ruled out on the grounds that it would contradict Aristotelian physics.

Mathematics in the *Almagest*

It is unclear who first introduced the Babylonian system of numeration into Greek work, but certainly Hipparchus was familiar with the sexagesimal positional system. Sexagesimal numeration is used throughout the *Almagest*, with minor modifications to the basic Babylonian system, though Ptolemy used the traditional Greek form for fractions where precision was unnecessary. There is very little mathematical formalism in the *Almagest*; instead, Ptolemy gives detailed accounts of the procedures used in actual calculations. The formulas given below are therefore a modern shorthand way of representing what Ptolemy wrote out in words. The main mathematical interest in the *Almagest* comes from Ptolemy's use of trigonometry, a subject developed by the Greeks specifically for the solution of problems arising in astronomy.

The only trigonometric function used in the *Almagest* is the chord, and Ptolemy constructs a table of chords in Book I. It is almost certain that such tables existed long before Ptolemy (Theon of Alexandria tells us that both Hipparchus and Menelaus had written works on chords) but Ptolemy's table is the first surviving specimen and his account of its construction is the first treatise on trigonometry known to us. The Greek chord function is illustrated in Figure 3.6. Ptolemy's aim is to calculate the length of the chord AC for a given angle α, denoted in Eqn (3.3) by $\mathrm{ch}\,\alpha$ and, in order to facilitate the use of the sexagesimal system, he used a circle of radius 60. Since one-half of the chord divided by the radius is just the sine of half the angle α, this chord function is related to the modern sine function through

$$\mathrm{ch}\,\alpha = 120 \sin \tfrac{1}{2}\alpha, \tag{3.3}$$

though it should be remembered that the chord is a length, whereas the sine is a ratio.

Ptolemy's first step was to use well-known properties of regular polygons to evaluate some special values of the chord function. In this way he obtained

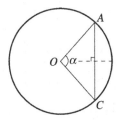

Fig. 3.6. The Greek chord.

$$\text{ch}\,36° = 37; 4, 55, \qquad \text{ch}\,60° = 1, 0; 0,$$
$$\text{ch}\,72° = 1, 10; 32, 3, \qquad \text{ch}\,90° = 1, 24; 51, 10,$$

from the properties of the decagon, hexagon, pentagon and square, respectively. Pythagoras' theorem and the fact that the angle inside a semicircle is a right angle[21] show that $\text{ch}^2(180° - \alpha) = 120^2 - \text{ch}^2\,\alpha$, and from this Ptolemy determined

$$\text{ch}\,120° = 1, 43; 55, 23, \qquad \text{ch}\,144° = 1, 54; 7, 37.$$

Next, Ptolemy proved what is now known as 'Ptolemy's theorem' but which, since it is so elementary, probably dates from an earlier period. Using arguments based on similar triangles Ptolemy showed that in a cyclic quadrilateral (a quadrilateral inscribed in a circle) the product of the diagonals is equal to the sum of the products of the opposite sides, or (see Figure 3.7(a))

$$|DB| \cdot |AC| = |AB| \cdot |DC| + |AD| \cdot |BC|.$$

Next, consider a cyclic quadrilateral with one side as diameter, as shown in Figure 3.7(b). An application of Ptolemy's theorem yields

$$\text{ch}\,\beta\,\text{ch}(180° - \alpha) = 120\,\text{ch}(\beta - \alpha) + \text{ch}(180° - \beta)\,\text{ch}\,\alpha.$$

Using Eqn (3.3), we see that this is equivalent to the modern trigonometric formula

$$\sin(x - y) = \sin x \cos y - \cos x \sin y,$$

where $x = \beta/2$, $y = \alpha/2$. In a similar way, Ptolemy derived the equivalent of the formulas

---

[21] EUCLID *Elements*, Book I, 47 and Book III, 31, respectively.

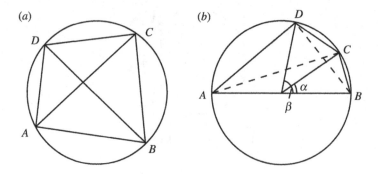

Fig. 3.7. Ptolemy's theorem.

$$\sin(x + y) = \sin x \cos y + \cos x \sin y$$
$$2 \sin^2 x = 1 - \cos 2x.$$

With these formulas and the chords already computed, it is possible to construct a table of chords in steps of $3°$ (or $3/2°$ or $3/4°$, etc.). However, Ptolemy's aim was to construct a table of chords in intervals of $1/2°$ and he achieved this with an ingenious argument: that for acute angles α and β with $\operatorname{ch}\alpha > \operatorname{ch}\beta$,

$$\frac{\operatorname{ch}\alpha}{\alpha} < \frac{\operatorname{ch}\beta}{\beta}, \tag{3.4}$$

a result known to Aristarchus and, since he had previously calculated

$$\operatorname{ch}\tfrac{3}{2}^\circ = 1; 34, 15, \qquad \operatorname{ch}\tfrac{3}{4}^\circ = 0; 47, 8,$$

Eqn (3.4) implies that

$$\tfrac{2}{3}\operatorname{ch}\tfrac{3}{2}^\circ < \operatorname{ch}1^\circ < \tfrac{4}{3}\operatorname{ch}\tfrac{3}{4}^\circ.$$

But to two sexagesimal places both $\tfrac{4}{3}\operatorname{ch}\tfrac{3}{4}^\circ$ and $\tfrac{2}{3}\operatorname{ch}\tfrac{3}{2}^\circ$ are equal to $1; 2, 50$. Hence, Ptolemy knew $\operatorname{ch}1°$ and could compute

$$\operatorname{ch}\tfrac{1}{2}^\circ = 0; 31, 25.$$

Ptolemy was now able to construct a table of chords from $1/2°$ to $180°$ in steps of $1/2°$ (equivalent to a table of sines from $1/4°$ to $90°$ in steps of $1/4°$) accurate to two sexagesimal places.

With his table of chords, Ptolemy could use algorithms equivalent to the modern formulas (following the standard convention that the side opposite angle A is labelled a, etc.):

$$\sin A = \frac{a}{c}, \qquad \cos A \equiv \sin(90° - A) = \frac{b}{c}, \qquad \tan A \equiv \frac{\sin A}{\cos A} = \frac{b}{a}.$$

Only right-angled triangles are solved in the *Almagest*, with oblique triangles being decomposed into right-angled ones, and in this way Ptolemy can solve any triangle (though somewhat clumsily by modern standards). Ptolemy's style throughout the *Almagest* suggests that this type of computation was fairly standard.

Of course, many of the astronomical calculations Ptolemy needed to perform concerned the angular distances between celestial bodies or, in other words, the positions of bodies on a spherical surface, for which spherical trigonometry is appropriate. Here, too, Ptolemy could use his table of chords. The geometry of the sphere, particularly with reference to astronomy, was one of the subjects taught within the Pythagorean quadrivium – and the subject was well advanced long before Ptolemy's time – but spherical trigonometry really came to prominence with the work of Menelaus of Alexandria in about AD100. The treatment of spherical trigonometry in the *Almagest* is based largely on the *Sphaerica* of Menelaus. In the first of three books, Menelaus introduced the concept of a spherical triangle – a figure formed by three arcs of great circles on a sphere, each arc being less than a semicircle – and proved some theorems about such triangles analogous to those Euclid had proved for plane triangles. The second book is concerned chiefly with astronomy and only indirectly with spherical geometry, while spherical trigonometry was the subject of the third book.

Ptolemy's use of spherical trigonometry is based on two results (see Figure 3.8(a)). Suppose we have a spherical triangle ABE and another great circle intersecting the sides of this triangle (produced where necessary) at D, F, and C as in the figure, then

$$\sin CE \cdot \sin DF \cdot \sin BA = \sin AE \cdot \sin CF \cdot \sin DB,$$
$$\sin CA \cdot \sin DF \cdot \sin BE = \sin AE \cdot \sin CD \cdot \sin BF.$$

Here, $\sin CE$ means the sine of the angle subtended at the centre of the sphere by the arc CE, etc., and therefore is related directly to the Greek chord function. There are two other similar relations that can be derived, but Ptolemy did not mention this. The proof of this theorem rests upon the corresponding theorem for plane triangles, which is still referred to as Menelaus' theorem, although it probably predates Menelaus. The theorem (which is proved by Ptolemy) states

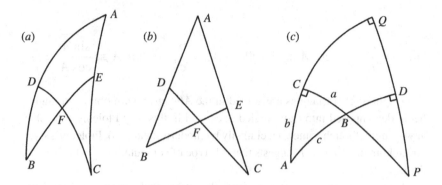

Fig. 3.8. Menelaus' theorem (*a*) for spherical triangles, (*b*) for plane triangles, and (*c*) Ptolemy's use for right-angled spherical triangles.

(see Figure 3.8(*b*)), that

$$|CE| \cdot |DF| \cdot |BA| = |AE| \cdot |CF| \cdot |DB|,$$
$$|CA| \cdot |DF| \cdot |BE| = |AE| \cdot |CD| \cdot |BF|.$$

As in the case of plane triangles, Ptolemy actually considered only right-angled spherical triangles, in which two of the great circles meet at 90°. By subdividing other triangles into two or more right-angled ones, he was able to solve them using the above theorem.[22]

For right-angled spherical triangles, Menelaus' theorem can be reduced to a relationship between three quantities. Thus, in Figure 3.8(*c*), *ABC* is a spherical triangle with right angle at *C*. The point *A* is the pole of the great circle *PQ*, and the arcs *a*, *b* and *c* are extended to meet this great circle in the points *P*, *Q* and *D*, respectively. From the second version of Menelaus' theorem for spherical triangles given above, using the facts that $QP = CP = AD = 90°$, we obtain

$$\sin BC = \sin QD \cdot \sin AB.$$

Since *A* is the pole of *QP*, it follows that $\sin A = \sin QD$,[23] and so this equation can be written

$$\sin a = \sin A \sin c.$$

Similarly, Ptolemy used algorithms equivalent to the formulas

$$\tan a = \sin b \tan A, \qquad \cos c = \cos a \cos b, \qquad \tan b = \tan c \cos A.$$

[22] Details of Ptolemy's procedures can be found in Pedersen (1974) and Katz (1998) as well as in the *Almagest* itself.
[23] See, for example, Smart (1960).

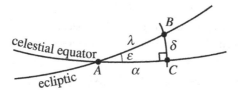

Fig. 3.9. The determination of the declination δ and right ascension α of the Sun; A is the vernal equinox and the Sun is at B.

As an example, consider Ptolemy's determination of the declination δ and right ascension α of the Sun for a given ecliptic longitude λ. In Figure 3.9, A is the vernal equinox, the Sun is at B, and the angle ε is the obliquity of the ecliptic, which Ptolemy took to be $23°\ 51'\ 20''$.[24] The first and last of the four formulas listed above then show that

$$\sin \delta = \sin \varepsilon \sin \lambda, \qquad \tan \alpha = \tan \lambda \cos \varepsilon.$$

Solar theory

After the mathematical preliminaries, Ptolemy discusses the motion of the Sun, and here he uses exactly the same model as that developed by Hipparchus. Ptolemy made some observations of the dates of the equinoxes so as to see whether Hipparchus' value for the length of the tropical year was still correct after a 300-year period, and he concluded that it was and that it has the value ascribed to it by Hipparchus, i.e. 365 days 5 h 55 min 12 s – about $6\frac{1}{2}$ minutes too long.

Ptolemy also used his observations of the equinoxes to recompute the length of the seasons and again found agreement with Hipparchus' values of $94\frac{1}{2}$ days for the length of the spring season and $92\frac{1}{2}$ days for the length of summer. He thus obtained the same numerical parameters for the eccentric model of the Sun, i.e. $65°\ 30'$ for the longitude of the solar apogee as measured from the vernal equinox, and 1/24 for the eccentricity of the orbit. The fact that Ptolemy arrived at the same result for the solar apogee as Hipparchus had done

---

[24] In the middle of Book I of the *Almagest*, between his construction of a table of chords and his discussion of spherical trigonometry, Ptolemy discussed the angle that the ecliptic makes with the celestial equator, the obliquity ε. He described how it can be calculated and claimed to have found that $23\frac{5}{6}° < \varepsilon < 23\frac{7}{8}°$. Since the value he attributes to Eratosthenes, 22/83 of a right angle ($23°\ 51'\ 20''$), lies in this range, this is the value he adopted (see Goldstein (1983)). The actual value in Ptolemy's time was about $10'$ less than this (Wilson (1980), p. 63) and Ptolemy's failure to find a more accurate result, and hence perhaps discover the slow decrease in the obliquity over time, can be explained by the crudity of his measuring devices (see Britton (1969)).

300 years previously led him and other ancient astronomers to believe that this was an astronomical constant, whereas in fact, because of the precession of the equinoxes, in Ptolemy's time the true value was approximately $70°$.[25]

Ptolemy used the geometrical solar theory to produce a table so that the position of the Sun at a given time can be calculated. From Figure 3.10 it is clear that, viewed from the Earth E, the Sun's longitude measured from its apogee A is α (the so-called true anomaly) and the mean anomaly $\bar{\alpha}$ is known, since the Sun S rotates uniformly around the centre of the eccentric circle, O, completing 1 revolution in 1 tropical year. It is also clear that $\alpha = \bar{\alpha} \pm \delta$ (the choice of sign depending on where the Sun is in its orbit) and so the position of S can be found once δ is known as a function of $\bar{\alpha}$. This is what Ptolemy tabulates. Ptolemy calls δ the *prosthaphaeresis*, which translates as the 'amount to be added and subtracted' but, following medieval usage, it is more commonly referred to as 'the equation of centre'.

In general, the term 'equation' is used to refer to any angle that must be added or subtracted from a mean motion in order to account for a particular geometrical feature. This is Ptolemy's style throughout – first the mean motions are described, and then the various small corrections, the equations, are calculated. In the case of the Sun, there was one such equation, but for the Moon and the planets there were more. Ptolemy's quantitative solar theory was used not only to determine the position of the Sun but was also an essential part of his theory for the other planets.

Because the Sun moves non-uniformly around the ecliptic, which is itself inclined to the celestial equator, the length of the solar day (the time between local noon on successive days) is not constant. Thus, when calculating the time difference in days between two events (such as a pair of eclipses) it is necessary to correct for this variation. The effect, however, is quite small and, until Ptolemy's accurate quantitative astronomy, the resulting discrepancies

---

[25] Some authors, notably Newton (1977), claim that Ptolemy must have fudged his data so as to reproduce Hipparchus' parameter values, but in fact Ptolemy's errors are not large when one takes into account that an error of 6 h in the length of the spring season can lead to an error of about $7°$ in the solar apogee (Peterson and Schmidt (1967), Maeyama (1998)). In fact, this demonstrates that the great accuracy achieved by Hipparchus was fortuitous. North (1994) goes so far as to say 'a modern tradition that Ptolemy was little more than a plagiarist of Hipparchus is hardly worth refuting' and Hamilton and Swerdlow's review of Newton's work in the *Journal for the History of Astronomy*, **12** (1981), 59–63 is deeply critical of the latter's approach. On the other hand, Britton (1992) has demonstrated convincingly that Ptolemy must have had access to rather more observations than are mentioned in the *Almagest* and that he sometimes used these selectively so as to reproduce agreement with predetermined values. Sheynin (1973b) suggests that Ptolemy might have selected those observations he believed to be the least susceptible to random or systematic error. Whatever the truth of the matter – and the debate continues (see Thurston (2002), Gingerich (2002)) – astronomers from the eighteenth century onwards certainly realized that Ptolemy's observations were not to be relied upon (see, for example, Wilson (1984), also note 20 on p. 63 of this text).

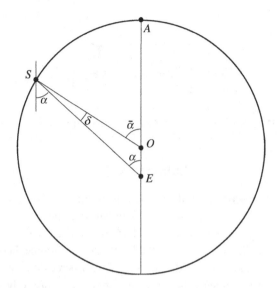

Fig. 3.10. The *prosthaphaeresis* angle δ.

(which never amount to more than about half an hour) were of little significance. Local time is determined by the position of the Sun with respect to
meridians (great circles perpendicular to the celestial equator) and so the simplest way to appreciate the cause of the variation in the length of the day
is to introduce the concept of the equatorial mean sun, a point that travels around the celestial equator at a uniform rate, once per tropical year.[26]
This then leads to the idea of the mean solar day, a concept Ptolemy
introduced.

   Thus, in Figure 3.11, when the actual Sun is at S, the equatorial mean sun
will be at $\bar{S}$ which, because S moves non-uniformly around the ecliptic and α
is nonlinearly related to λ (see Figure 3.9), will at different times of the year
be sometimes ahead of and sometimes behind A. The arc $A\bar{S}$ (the difference
between the right ascension of the equatorial mean sun and that of the Sun
itself) is known as the 'equation of time'. The discovery of the equation of time
by Greek astronomers is just one example of the high level of sophistication
they achieved, as it was deduced as a theoretical consequence of Hipparchus'
solar theory, rather than from observation.

---

[26] The equatorial mean sun should not be confused with the mean sun, which is another
important astronomical concept. The latter rotates uniformly around the ecliptic once each
tropical year. The two clearly are related since they move around their respective circles at
precisely the same rate. In fact, the right ascension of the equatorial mean sun is equal to the
ecliptic longitude of the mean sun.

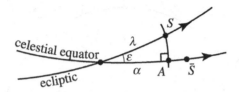

Fig. 3.11. The cause of the equation of time. The Sun is at S and $\bar{S}$ is the equatorial mean sun.

Lunar theory

By virtue of its proximity, small irregularities in lunar motion are discernible more easily than those of the planets, and Ptolemy was the first to discover that the Moon was subject to an anomaly that had not been detected by Hipparchus. As we have seen, Hipparchus treated the longitude and latitude theories for the Moon separately, and Ptolemy did the same but, whereas the latitude theory was not modified by Ptolemy, he found the longitude theory to be inadequate away from the syzygies.

> As far as concerns the [moon's] syzygies ... we find that the hypotheses set out above for the first, simple anomaly is sufficient, even if we employ it just as it is, without any change. But for particular positions [of the moon] at other sun-moon configurations one will find that it is no longer adequate, since ... we have discovered that there is a second lunar anomaly, related to its distance from the sun.[27]

Ptolemy showed that the discrepancy between the observed longitude of the Moon and that predicted by Hipparchus' theory depended on the position of the Moon relative to the Sun and was greatest at the quadratures, i.e. half moons. Thus, the Moon was subject to a second, independent irregularity in its motion which has become known as 'evection'.[28] In order to rectify the theory, Ptolemy introduced a mechanism by which the Moon's epicycle was brought closer to the Earth at the quadratures than at new and full moon. This device had the effect of incorporating the theory of the Sun into that for the Moon.

His solution is illustrated in Figure 3.12, and involves an epicycle on a deferent which is no longer centred on the earth E but is instead placed eccentrically. Moreover, the centre of the deferent O rotates around the Earth in a small circle with a constant angular speed with respect to the position of the mean sun, $\bar{S}$. This has the effect of moving the apogee of the deferent A from east to west

---

[27] PTOLEMY *Almagest*, Book V, 1.
[28] The terminology is due to Ismael Boulliau in 1645.

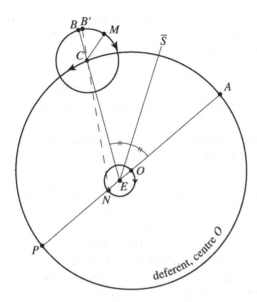

Fig. 3.12. Ptolemy's lunar theory.

relative to the mean sun. The centre of the Moon's epicycle C rotates around the deferent circle in such a way as to keep the angles $AE\overline{S}$ and $\overline{S}EC$ equal. Thus, relative to the mean sun, C rotates once around the deferent in an anti-clockwise direction (i.e. from west to east) in a synodic month, while O rotates around a circle, centred at E, at exactly the same rate in the opposite direction. Equivalently, if one considers the motion of the epicycle relative to the line EA, C rotates around the deferent twice each synodic month. At the syzygies, C and A coincide, whereas at the quadratures C and P coincide. It is clear that in the lunar model, the deferent rotates with uniform speed around the Earth, a point that is not its centre, and so Ptolemy was violating one of the fundamental principles on which Greek mathematical astronomy had been built, i.e. uniform circular motion. This radical shift, which Ptolemy did not even mention, was later a major source of criticism.

From observations[29] Ptolemy was able to determine the relative sizes of the circles in his lunar theory needed accurately to reproduce the phenomena. Letting $|EA|$ be 60 units, Ptolemy's skill with trigonometry enabled him to show that $|OE|$ should be 10; 19 and the radius of the epicycle was 5; 15.[30] In

---

[29] The claim that Ptolemy never made any lunar observations, in support of the general thesis of Newton (1977), is made by Goldstein (1982).
[30] Details can be found in Pedersen (1974).

the first version of his lunar theory, the Moon M was made to rotate around its epicycle at a uniform rate with respect to the line ECB (see Figure 3.12) once in each anomalistic month, exactly as in Hipparchus' theory. The angle BCM thus was a given linear function of time. In order to calculate the longitude of the Moon for a given time, Ptolemy had first to compute the longitude of the point C (the mean moon) from the mean lunar motion. Then he could calculate the angle $CE\bar{S}$, since the solar theory provided the longitude of the mean sun. Next, the distance $|CE|$ can be computed by solving the triangle EOC and, finally, $\angle CEM$ (the *prosthaphaeresis* angle for the Moon) can be found by solving the triangle CEM.[31]

The overall effect of Ptolemy's scheme was to leave the longitude at conjunction and opposition unchanged from that given by Hipparchus (since $\angle CEA = 0$ in these cases) and the theory accounted well for the observed position of the Moon at the quadratures. However, Ptolemy found that there were still noticeable errors at the octants – the points midway between the quadratures and the syzygies – and decided that a further modification was necessary. He did this by making the Moon rotate around its epicycle at a non-uniform rate with respect to ECB, but uniform with respect to NCB', where the point N is such that E is the midpoint of ON. For this new scheme, it is the angle $B'CM$ that is a given linear function of time, so in order to compute the *prosthaphaeresis* angle, Ptolemy first had to calculate $\angle BCB' = \angle NCE$. This he could do by solving the triangle NCE once the length $|EC|$ had been determined. This final peculiarity in Ptolemy's lunar theory, which has no effect on the position of the Moon at the syzygies or at quadrature and which was another source of criticism by later astronomers, is known as *prosneusis*. In actual fact, the effect of *prosneusis* was to make things better some of the time but worse at others. Ptolemy did not test his final theory against new observations away from the octant points.[32]

The lunar theory is not without its problems. One defect is that the ratio of the maximum to minimum lunar distances implied by the model is $64;15/34;7 \approx 1.9$ and so one should observe a doubling of the Moon's apparent diameter as it circles the Earth, whereas in actual fact a change of only 14 per cent is observed. Another objection that was raised by later astronomers was the fact that the epicyclic model seems to be incompatible with the phenomenon – mentioned

---

[31] The calculations involved in applying this procedure are very laborious and in order to make the application of his theory more straightforward Ptolemy produced (not just for the Moon but for all the celestial bodies) tables of values corresponding to the modern concept of functions of one or two variables. To make the computation of the tables less time-consuming he used various methods of interpolation (see Pedersen (1974), van Brummelen (1994)).

[32] The accuracy of Ptolemy's lunar theory is described in detail in Peterson (1969).

by Aristotle in *On the Heavens* – that the same side of the Moon is always visible to us. One plausible reason why neither of these 'defects' concerned Ptolemy is that he regarded his theory simply as a device for computing the latitude and longitude of the Moon rather than as a physical model of reality. However, this seems to be contradicted by the way Ptolemy later treated his models in the *Planetary Hypotheses* (see p. 81).

Two aspects of the lunar model are of particular interest. The coupling between Sun and Moon is a major shift from previous theories, and so is Ptolemy's abandonment of uniform circular motion. Prior to Ptolemy, uniform circular motion always had implied angular rotation around a circular path uniform with respect to the circle's centre. But in Ptolemy's final lunar scheme, the Moon revolves with uniform speed with respect to a different point. As we shall see, many later astronomers objected to this, the most notable being Copernicus.

Having now described his theories for the motion of the Sun and Moon, Ptolemy was in a position to give a detailed theory of lunar and solar eclipses, which he did in Book VI of the *Almagest*. He began with a discussion of lunar parallax, which is significant when making observations of the Moon, and from his measurements he arrived at the conclusion that the radius of the deferent in his lunar model is about 49 Earth radii. Once he had a value for the distance to the Moon in terms of the size of the Earth, he was in a position to calculate the distance to the Sun by reversing the procedure used by Hipparchus for calculating the lunar distance from an assumed solar distance.[33] His final result was that the mean distance to the Sun is 1210 Earth radii (corresponding to a parallax of about 3′), which is about 19 times too small.

Planetary theory

Books IX–XIII of the *Almagest* are devoted to the motion of the planets, with longitudes and latitudes being considered separately. For the longitude theories, there are two anomalies to model. The first is manifested by the varying speed of the planet as it travels round the ecliptic and is thus similar to the anomaly in the Sun's motion. This suggests an eccentric deferent as a suitable geometrical scheme to account for the irregularity. The second anomaly is the phenomenon of retrograde motion and this ultimately is linked to the motion of the Sun. The superior planets always reach the centre of their retrograde arcs when they are

---

[33] See p. 54. Ptolemy's method is described in detail in van Helden (1985). Unlike the procedure adopted by Hipparchus, Ptolemy's approach is extremely sensitive to errors in the measured parameters.

in opposition to the Sun, whereas this happens at conjunction for the inferior planets. As Apollonius had shown, retrograde motion can be modelled by an epicyclic theory, and so some combination of eccentric deferent and epicycle suggests itself.

However, it turns out that this is insufficient by itself accurately to predict the correct positions at which the retrograde motions begin and the correct angular widths of the retrograde arcs.[34] To solve the problem, Ptolemy introduced a further modification to Apollonius' scheme and separated the centre of the deferent circle from the centre of uniform rotation. Thus, he introduced a new point – the equant – about which the centre of the deferent rotated with uniform angular speed. Nowhere does Ptolemy state how he devised this construction, but it works astonishingly well – a Ptolemaic equant produces planetary longitudes differing from modern theory by less than 10' of arc, even for the comparatively large eccentricity of Mars. The discovery of the equant mechanism thus represents one of the major achievements of Greek mathematical astronomy.[35] Brilliant it may have been, but the incorporation of the equant introduced a major problem into the science of astronomy because it violated the principle of uniform circular motion. To Ptolemy, it clearly was more important to reproduce accurately the phenomena than to stick rigidly to accepted philosophical dogma. Others did not necessarily share Ptolemy's attitude and the status of the equant was one of the main concerns of future astronomers.

Although Ptolemy described his theories for the inferior planets first, it is perhaps more helpful to begin with his scheme for the superior planets – Mars, Jupiter and Saturn – since the Mercury theory is relatively complicated. The geometrical model that Ptolemy finally arrived at is shown in Figure 3.13. Each planet P is carried round a deferent circle, centre O, on an epicycle, centre C, which rotates in the same sense as the motion of C around the deferent. Just as in the theory of the Moon, the Earth E is not situated at O, and the distance between the Earth and the deferent centre $|EO|$ is known as the 'eccentricity' of the model. The motion of C around the deferent is uniform with respect to the new point Q – the equant – which is chosen so that O bisects EQ.[36] The mean longitude $\bar{\lambda}$ (measured from the vernal equinox V) increases at a uniform

[34] A nice explanation of why this is the case is given in Evans (1984).

[35] Did Ptolemy devise it? It is not possible to say with certainty and he certainly does not claim explicitly that it is his idea. On the other hand, Ptolemy is good at giving credit where credit is due and he does not mention anyone else as the discoverer of the equant idea. For a strident defence of the idea that this construction was not due to Ptolemy, see Rawlins (1987).

[36] It is unclear as to why Ptolemy chose to put the centre of the deferent exactly equidistant from the Earth and the equant. It seems most likely (see Wilson (1973), Pedersen (1974)) that the observations suggested that this was the best position for the Venus model and he extended it to the superior planets by analogy.

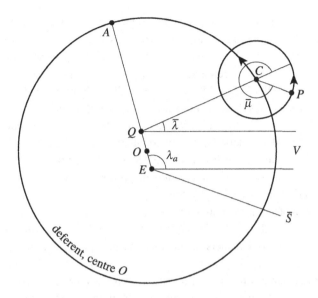

Fig. 3.13. Ptolemy's theory for the superior planets.

rate, with the epicycle rotating once around the deferent in the time it takes the planet to make one circuit around the ecliptic (its zodiacal period), i.e. 687 days for Mars, 11.86 years for Jupiter and 29.46 years for Saturn. The motion of the planet around its epicycle is uniform relative to the line QC, so that the angle $\bar{\mu}$ in Figure 3.13 (the mean epicyclic anomaly) increases uniformly with time t.

For each of the superior planets, observations show that the zodiacal and synodic periods, T_c and T_p respectively, satisfy (see Table 1.2, p. 9)

$$\frac{1}{T_c} + \frac{1}{T_p} = \frac{1}{T},$$

where T is the tropical year. In order to incorporate this feature into his model, Ptolemy made the planet rotate once around its epicycle each synodic period. We have then (ignoring irrelevant constants) $\bar{\lambda} = t/T_c$ and $\bar{\mu} = t/T_p$ and, hence, the longitude of the mean sun $\bar{S}$ is given by

$$t/T = \bar{\lambda} + \bar{\mu}.$$

Some elementary geometry reveals that this ensures that the line CP is always parallel to the line connecting the Earth E to the mean sun. The point A at which the line EOQ extended intersects the celestial sphere is the apogee of the deferent, and in Ptolemy's model it is assumed fixed with respect to the stars. In other words, its longitude λ_a increases slowly due to the effect of precession.

Finally, Ptolemy needed to determine the relative dimensions of his model, and he found that, with a deferent radius of 60, the required double eccentricities were given by $|EQ| = 12$ for Mars, $|EQ| = 5; 30$ for Jupiter and $|EQ| = 6; 50$ for Saturn. The values Ptolemy used for the radii of the epicycles were $39; 30$ for Mars, $11; 30$ for Jupiter, and $6; 32$ for Saturn. By using a complicated series of trigonometrical calculations, Ptolemy could then compute the true longitude of the planet (the angle VEP in Figure 3.13) at any given time.

The above geometrical arrangement is inappropriate for the inferior planets (Mercury and Venus) since they follow the Sun as it travels around the heavens, never deviating from it by more than about 29 and $47°$, respectively. The mean positions of these planets are the same as that of the Sun, and so in Ptolemy's models for the inferior planets the centre C of the epicycle rotates uniformly around an equant Q, such that QC is parallel to the line connecting the Earth E to the mean sun $\overline{S}$. Ptolemy's model for Venus is shown in Figure 3.14, and it can be seen that, apart from the different treatment of the coupling to the motion of the Sun, the scheme has the same essential features as his theories for the superior planets. The epicycle is carried around a deferent, the centre O of which is not the Earth, and the motion is uniform with respect to an equant Q which again is chosen so that O is the midpoint of EQ. It is interesting to note that in his model for Venus, Ptolemy in effect introduced an equant for the motion of the Sun, something he had not found to be necessary in the solar theory itself. He was, of course, unaware that the solar deferent and the deferent of Venus represent one and the same motion – the rotation of the Earth around the Sun. The apogee of the deferent A is assumed fixed with respect to the stars, exactly as before.

With a deferent radius of $R = 60$, the epicycle's radius was taken to be $r = 43; 10$ and the double eccentricity as $|EQ| = 2; 30$. The required size of the epicycle can, at least approximately, be determined from the maximum elongation (i.e. angular distance from the Sun) of the planet. Assuming zero eccentricity so that E and O coincide, the maximum elongation θ occurs when $\angle OPC$ is a right angle, i.e. when $r/R = \sin\theta$. Ptolemy's values give

$$\theta = \sin^{-1}\frac{43; 10}{60} \approx 46°,$$

in agreement with the value given by the Roman writer Pliny in AD first century. The effect of a non-zero value for the eccentricity on this simple argument is, however, quite complex.

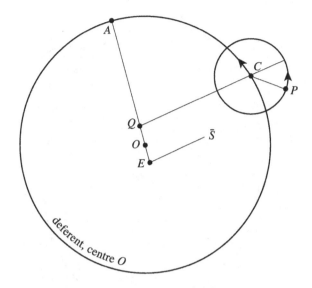

Fig. 3.14. Ptolemy's theory for Venus.

When it came to Mercury, however, Ptolemy found that he could not use the same geometrical scheme. Of the planets known to Ptolemy, Mercury is by far the hardest to observe, and its orbit deviates significantly from a circle. The observations Ptolemy used for his theory of Mercury – some of which were very inaccurate – led him to believe that Mercury's orbit had two perigees. Thus, he came up with a geometrical device that modelled this phenomenon, and this is shown in Figure 3.15. The line EQO points to the apogee of Mercury's orbit, the longitude of which is fixed with respect to the stars, and the equant Q is the midpoint of EO. The centre of the epicycle C rotates uniformly around the equant, following the mean sun, in such a way that it lies always on a deferent circle, the centre D of which is rotating in the opposite sense around a small circle centred on O. The rates of rotation are chosen so that the angles DOA and AQC always are equal. This construction causes C to move in an oval orbit that has two points of closest approach to the Earth, P_1 and P_2. For a deferent radius of 60, Ptolemy calculated the radius of Mercury's epicycle as 22; 30 and $|EQ|$ as 6; 0. Based on the simple argument described above for Venus, an epicycle radius of 22; 30 corresponds to a maximum elongation of 22°, which is again the value quoted by Pliny.[37] In his later work, the *Planetary Hypotheses*,

---

[37] Detailed discussions of the empirical basis for Ptolemy's theories of the inferior planets can be found in Wilson (1973), Swerdlow (1989), and the accuracy of the Mercury model is investigated in Nevalainen (1996).

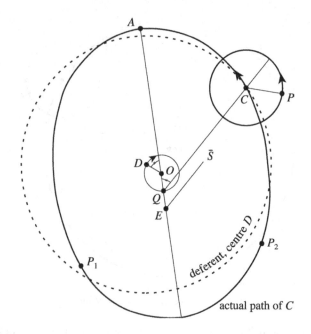

Fig. 3.15. Ptolemy's theory for Mercury.

Ptolemy justified the increased complexity of the motion of Mercury, as well
as that of the Moon, by pointing out that these two celestial bodies are closest
to the Earth and, hence, closest to the changeable air.

The planetary theories described above are all dedicated to the determination
of longitudes; latitudes are treated separately in Book XIII of the *Almagest*.
Unlike the simple latitude theory for the Moon, Ptolemy's theory for planetary
latitudes is cumbersome and complicated. This is hardly surprising since the
plane of the orbit of the Moon passes (very nearly) through the centre of the
Earth, whereas the planes of the planetary orbits pass through the centre of
the Sun, and Ptolemy was attempting to construct a geocentric latitude theory.
For the superior planets, Ptolemy accomplished this by tilting the deferent with
respect to the ecliptic and also inclining the plane of the epicycle with respect
to the deferent. Now, for these planets we know that the motion of the planet
around the epicycle actually is modelling the motion of the Earth around the
Sun, so the epicycle should be in a plane parallel to the ecliptic. Ptolemy had no
such knowledge, of course, and so had to determine two inclinations for each
planet. The system Ptolemy used for the inferior planets was similar, except

in this case the inclination of the deferent with respect to the ecliptic was an oscillatory function of time.[38]

The *Almagest* represents the culmination of 500 years of Greek mathematical astronomy, and the result is ingenious, mathematically sophisticated and tolerably accurate. It formed the foundation for subsequent developments and for the teaching of astronomy, and was the dominant influence in theoretical astronomy until the sixteenth century. As a scientific theory it is still *ad hoc* in that it explains only those phenomena built into the model, and new observations can be accommodated simply by tweaking the parameters.

The *Handy Tables* and the *Planetary Hypotheses*

After writing the *Almagest*, Ptolemy wrote two further astronomical works, the *Handy Tables* and the *Planetary Hypotheses*. The former contains the procedures that need to be followed to compute the positions of heavenly bodies from Ptolemy's theories, but with no discussion of the theoretical models on which they are based, while the latter contains his physical interpretations of the mathematical models developed in the *Almagest*. In the *Planetary Hypotheses*, parts of which have only become well known in the West fairly recently,[39] Ptolemy made two assumptions. The first was that the geometrical devices he had devised accurately to predict the observed phenomena actually exist in the heavens. The second was that he assumed there were no empty spaces between the mechanisms for each of the heavenly bodies. As some of the geometrical schemes Ptolemy had used in the *Almagest* contradicted Aristotelian natural philosophy – notably the equant construction – the first of his assumptions came under repeated attack during the Middle Ages. The second was not so controversial, but Ptolemy himself actually violated it by leaving an empty space between Venus and the Sun. Many later astronomers repeated Ptolemy's calculations to determine the dimensions of the Universe and managed, one way or another, to remove this void from the theory. Ptolemy's world view was as influential in the field of cosmology as the *Almagest* was in theoretical astronomy.

In the *Planetary Hypotheses*, Ptolemy deliberately simplified some of his geometrical constructions so as to make them easier to understand; some of the simplifications just take the form of approximating parameters by more convenient values, but there are also some changes of substance, notably in the latitude theories for the planets, which actually were improved in the process.

---

[38] Details of Ptolemy's latitude theories for the planets can be found in, for example, Pedersen (1974), Riddell (1978), Jacobsen (1999).
[39] See Goldstein (1967).

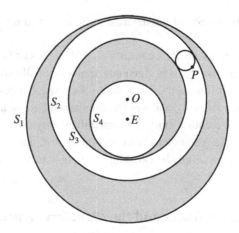

Fig. 3.16. Ptolemy's physical interpretation for a superior planet.

A schematic illustration of Ptolemy's physical interpretation of the motion of a superior planet is shown in Figure 3.16. The figure depicts four spherical surfaces: S_1 and S_4 centred on the Earth E (the centre of the Universe) and S_2 and S_3 centred on O (the centre of the deferent (see Figure 3.13)). The planet P is situated on a sphere (the epicycle) which is confined to the spherical annulus between S_2 and S_3. The shaded regions between S_1 and S_2 and between S_3 and S_4 are filled with the ether, which serves to transmit the required motion between the various spheres.

If we consider the diagram as representing the motion of Saturn, then Ptolemy now could fit his whole planetary scheme together without spheres intersecting simply by placing Jupiter inside S_4, and so on. On the assumption that the spheres fit as closely together as possible, we thus have a method for calculating the size of the Universe and this was done by Ptolemy (see Table 3.1).[40] In the *Almagest*, Ptolemy had already calculated the distances of the Sun and the Moon, and so he began with the Moon and worked outwards arriving at a maximum distance for Venus of 1079 Earth radii. But the minimum distance to the Sun was 1160 Earth radii, and so Ptolemy was forced to leave a gap between the spheres of Venus and the Sun.[41] Working outwards from the Sun, he arrived at a distance to the outer sphere of Saturn of 19 865 Earth radii (about 120 000 000 km in modern units, less than the true value for the radius of the Earth's orbit) and he rounded this to 20 000 Earth radii as the distance

---

[40] The figures are quoted from Goldstein and Swerdlow (1970).

[41] Ptolemy was aware that minor adjustments to some of the measured parameters could be used to close the unwanted gap, but he did not bother to make them (see van Helden (1985), p. 23).

Table 3.1. *Greatest and least distances,*
measured in Earth radii, according to Ptolemy.

	d_{min}	d_{max}
Moon	33	64
Mercury	64	166
Venus	166	1 079
Sun	1 160	1 260
Mars	1 260	8 820
Jupiter	8 820	14 187
Saturn	14 187	19 865
Fixed stars	20 000	

to the fixed stars. From a modern perspective, Ptolemy's value for the size of the Solar System is hopelessly wrong but, seen in the context of his own time, he was actually the first person to suggest that the dimensions of the Universe were unimaginably large. The number of spheres necessary to construct the whole system was thirty-four, and Ptolemy thus claimed that he had produced a simpler system than any of his predecessors. Despite its many deficiencies, Ptolemy's cosmology reigned supreme – with modifications as to its detail – until the demise of geocentric astronomy in the seventeenth century.

No Greek astronomer after Ptolemy made any significant advance on Ptolemy's work. People were becoming more sceptical about the value of this type of endeavour, and working conditions for those involved in rational scientific enquiry gradually deteriorated. Commentaries on the *Almagest* were written in AD fourth century by Theon of Alexandria and by Pappus, but they added little. Progress in mathematical astronomy had to wait until the revival of scholarly activity in the Islamic civilization that grew up following the death of Mohammed in AD 632.

4

Developments in geocentric astronomy

Astronomy in India

Indian astronomy has a rich history, right up to modern times.[1] The Vedic religion, from which modern Hinduism has developed, is one of the earliest religions recorded in written form (the language being Sanskrit) and the Vedic literature contains many references to the heavens and their divine qualities. The earliest astronomical text – the *Vedāṅga Jyotiṣa* – dates back to around 1200 BC. It is clear from the use of numerical periods determined by the Babylonians that the two civilizations were in contact and, from the use of epicyclic mechanisms, that at some time between Hipparchus and Ptolemy aspects of Greek astronomy were transmitted to India, though the precise mechanism is uncertain. That this happened before Ptolemy can be deduced from the fact that the lunar theories of the Hindus show no evidence of Ptolemy's modifications to Hipparchus' theory.

Two different approaches are evident in early Hindu astronomy. First, there were arithmetical methods similar to those developed by the Babylonians and, second, there were the trigonometric methods based on Greek geometrical constructions. Examples of the former type can be found in the *Pañca Siddhāntikā*, written in about AD 550 by the Indian philosopher, astronomer and mathematician Varāha Mihira. Just as in Babylonian astronomy, these Hindu models could be used to compute longitudes at discrete times and underlying the techniques were zigzag functions, but the actual formulation was quite distinctive. In order to illustrate these Indian arithmetic models we will consider the motion of the Moon.[2]

---

[1] A detailed account of the history of mathematical astronomy in India from its beginnings through to the sixteenth century is given in Pingree (1975) and Balachandra Rao (2000) provides a comprehensive summary of the algorithms that were actually used to determine the positions of the Sun, Moon, and planets.

[2] For further details, including theories for Jupiter and Saturn, see Abraham (1982).

The anomalistic month (the average time it takes for the Moon to return to the same speed) was taken to be 248 *pada*s (a *pada* being one-ninth of a day). This value is extremely accurate (see Table 1.1, p. 7) and was also used by Babylonian astronomers. This month was divided into two equal parts, with the Moon's speed increasing during the first half and decreasing during the second. For the first half, the longitude of the Moon after each *pada* was given (in degrees) by the quadratic formula

$$\lambda(p) = p\left(1 + \frac{1094 + 5(p-1)}{3780}\right), \qquad p = 0, 1, \ldots 124,$$

and then for the second half of the anomalistic month by

$$\lambda(124 + q) = \lambda(124) + q\left(1 + \frac{2414 - 5(q-1)}{3780}\right), \quad q = 0, 1, \ldots 124.$$

Since there are 9 *pada*s in 1 day, the daily motion of the Moon in the first half of the month is given by

$$D(p) = \lambda(p+9) - \lambda(p) = \frac{117}{10} + \frac{p}{42},$$

an increasing linear function of p, and so the difference between successive daily motions is constant ($D(p+9) - D(p) = 3/14$). A similar calculation for the second half of the month shows that the daily motion is then a decreasing linear function of q (though there are minor complications around the end and the middle of the month) and so the effect of the model is to produce a zigzag function for the variation in the daily motion of the Moon. The maximum and minimum daily motions according to this theory are $14°\,39'\,9''$ and $11°\,42'$, respectively.

Early Indian astronomy based on Greek geometrical models is best known through the *Sūrya Siddhānta*, one of several similar works of unknown authorship written around the fourth century. The basic mechanism of Hindu planetary astronomy in works such as this was an epicycle on an eccentric deferent, . which is consistent with the idea that the knowledge came from Greece between the time of Hipparchus – who had no geometrical planetary theory – and of Ptolemy – who used epicycles on eccentric deferents together with the equant construction.[3] However, we also find an imaginative modification that seems to have been an indigenous invention. In Āryabhaṭa's *Āryabhaṭīya* (written in about 500), the author makes the eccentricity of the deferent and the radius of

---

[3] A thorough discussion of the transmission of Greek planetary models to India can be found in Pingree (1971). The parameters that actually were used in these geometrical models were probably of Babylonian origin (see, for example, Abhyankar (2000)).

the epicycle vary periodically between two (not greatly different) values. Hindu latitude theories were very basic, indicating that the complex latitude theories of the *Almagest* did not develop over a long period prior to Ptolemy.

The biggest influence of the Indian civilization on the evolution of Western mathematical astronomy, however, came not from the arithmetical or geometrical models that were employed, but from the development of new mathematical tools. When using the Greek chord function in astronomical calculations, astronomers often had to deal with half-chords of double angles. Indian mathematicians realized that it would be simpler to tabulate the half-chords themselves; these are our modern sines (though, of course, they still referred to lengths rather than ratios). The *Āryabhaṭīya* contains a description of the construction of a sine table for the angles 0 to 90° in steps of 3° 45', and Brahmagupta (seventh century) devised fairly accurate methods of interpolation based on what we would now call 'second-order differences'.[4] Of perhaps even greater long-term significance was the emergence of the decimal place-value number system for integers, including the use of a symbol for zero. The origins of this system are unclear, but it was in use in India in the eighth century.

Islamic astronomy

Ptolemy was one of the last great scientists of antiquity, and in the centuries following his death Greek scientific activity declined in both quality and quantity. When the Muslims conquered the lands surrounding the Mediterranean in the seventh century, they would have found the records of 1000 years of Greek intellectual thought, but they would not have encountered much ongoing scientific inquiry.

After a couple of centuries of rapid expansion, the Islamic civilization settled down and scholarship began to flourish. From the eighth to the fourteenth centuries, most of the advances in astronomy were achieved by scholars in the Middle East, North Africa, and Moorish Spain. This work crossed religious and ethnic boundaries, with contributions from, among others, Arabs, Iranians, and Turks, and from Muslims, Jews, and Christians. Islamic scholarship explored all branches of knowledge and built on not only the traditions of Greek science and philosophy, but also those of Persia, India, central Asia, and to some extent, China. The unifying feature of this endeavour was the Arabic language, which was very flexible so that it was possible for translators to create the

---

[4] Described in Katz (1998), pp. 212–15, for example.

necessary technical vocabulary. Greek works were translated into Arabic from the late eighth century onwards, including those of Galen, Aristotle, Euclid, Archimedes, and Apollonius. The *Almagest* was translated several times, an excellent translation being made by Isḥaq ibn Ḥunain – the son of one of the first leaders of the Academy set up in Baghdad called the 'House of Wisdom' – and this was edited later by Thābit ibn Qurra in the latter half of the ninth century. Ptolemy's *Planetary Hypotheses*, *Tetrabiblos*, and substantial parts of the *Handy Tables* were also translated.[5]

The Islamic civilization contributed a great amount of theory, computation and instrumentation to astronomy, but it did not provide many observations for later use because, by and large, Islamic astronomers followed Ptolemy's procedures for obtaining planetary parameters from a limited number of selected observations. Many astronomers produced works, each known as a *zīj*, a word used originally to mean 'a set of astronomical tables' (of which the *Handy Tables* was the prototypical example) but which was later used to refer to any astronomical treatise. Many of these works survive to this day.[6]

Perhaps the first *zīj* to appear in Arabic was based on the *Siddhānta* of Brahmagupta, which was probably translated in the 770s. It is, however, doubtful that the Arabs would have had the necessary expertise to make much use of the tables until they had become familiar with Greek geometrical methods. Other Sanskrit works such as the *Āryabhaṭīya* of Āryabhaṭa were translated around this time and would have been the authority on astronomy until the Greek works became available. From these texts the Arabs became aware of Hindu mathematical advances in areas such as trigonometry and numeration.

Arabic mathematics came of age with the work of al-Khwārizmī, an early member of the House of Wisdom. His great influence on the development of mathematics comes from his *Ḥisāb al-jabar wa-l-muqābala*, written in about 825, which contains the beginnings of the subject now known as algebra, *al-jabar* referring to the operation of taking a subtracted quantity from one side of an equation to the other and *al-muqābala* referring to the operation of subtracting the same quantity from both sides of an equation. Al-Khwārizmī is also responsible for introducing the Hindu decimal numeration system, including a symbol for zero, to the Arabic speaking world. He described algorithms (the word being derived from his name) for using these numbers to perform the basic operations of arithmetic. Al-Khwārizmī compiled a *zīj* containing tables constructed largely from Ptolemy's theories, though also incorporating aspects

[5] The period of translation is described in detail in O'Leary (1948), Chapter XII (see also Kunitzsch (1974), Toomer (1984), p. 2, for more information on translations of the *Almagest*).
[6] They have been catalogued by Kennedy (1956b).

of Indian and Persian astronomy. This work set the style for future Islamic astronomy and was influential in Europe in the Middle Ages after it was translated into Latin in the twelfth century by Adelard of Bath.

The *Almagest* was very difficult to get to grips with, and a number of new manuals of theoretical astronomy appeared, one of the most influential of which was by al-Farghānī (known in Latin as Alfraganus), who was employed by the caliph al-Ma'mūn, the ruler in Baghdad between 813 and 833. Al-Farghānī's work usually is known by the title *Elements of Astronomy* and was a comprehensive summary of Ptolemaic astronomy that was entirely descriptive and non-mathematical. It became very popular as a textbook and was the primary medium by which knowledge of Ptolemy's work spread until the sixteenth century. Al-Farghānī recomputed the distances of the planets from the Earth based on the parameters in the *Almagest*, exactly as Ptolemy had done in the *Planetary Hypotheses*. He arrived at values similar to Ptolemy's but, conveniently, the philosophically unacceptable gap between Venus and the Sun had disappeared. It was Gerard of Cremona's translation into Latin of al-Farghānī's work in the twelfth century that was the main source of knowledge about Ptolemaic astronomy Dante used when writing his *Divine Comedy*, and John of Seville's translation a few decades earlier formed the basis of Johannes de Sacrobosco's *On the Sphere*, which went through more than 200 editions and was used at universities throughout Europe until the early seventeenth century.[7]

Technical modifications to Ptolemaic astronomy

In Ptolemy's theory, the motion of the Sun plays a fundamental role in determining the motion of all the other heavenly bodies, and so all of the calculations in the *Almagest* are based on the parameter values Ptolemy used for his solar theory. Ptolemy failed to improve on the values of Hipparchus and so used $23° 51' 20''$ for the obliquity of the ecliptic, $365; 14, 48$ days for the length of the tropical year, $65° 30'$ for the longitude of the solar apogee, and $1°$ per century for the value of precession. Islamic astronomers had one great advantage over their Greek predecessors: they could compare their observations with those of others who had lived over 1000 years previously. Thus, they had a much better chance of discovering small irregularities only discernible from observations collected over a long period of time. Two of these irregularities discovered in the ninth century were the variation in the obliquity of the ecliptic and the change

---

[7] The size of the cosmos, as described by al-Farghānī, can be found in a number of medieval works of popular literature (see van Helden (1985), pp. 37–9).

in the ecliptic longitude of the solar apogee, and the astronomers responsible for the discoveries were al-Battānī (known in Latin as Albategnius), and Thābit ibn Qurra, two of the most influential of the early Islamic astronomers. Both were quoted by a number of later Latin writers.

The interest of Islamic astronomers in the variability of the obliquity goes back to the mid eighth century – which is quite surprising, since the change (approximately $1/2''$ per year) was of no practical use. Al-Battānī measured the obliquity in his day as $23° 35'$, which is in line with modern theory, and Thābit used $23° 33'$. If the value used by Ptolemy was accurate, and most astronomers believed that it was, the obliquity clearly had decreased since the second century, but the implied rate at which it had changed was exaggerated by the fact that Ptolemy's value was about $10'$ too large.

In 831, Thābit ibn Qurra found the longitude of the solar apogee to be $82° 45'$ and recognized that the change in apogee could be attributed to precession. Since the increase in longitude of the stars since Ptolemy's time as a result of precession was similar to the change in longitude of the solar apogee, he believed that the solar apogee, like that of all the planets, remains fixed with respect to the stars, and Thābit thus used the sidereal year rather than the tropical year as the basic period of his solar theory.[8] Similar conclusions were reached by al-Battānī, who found $82° 17'$ for the solar apogee. Both astronomers also found new values for the eccentricity of the deferent of the Sun and constructed new solar tables.

Due to the fact that different astronomers had computed different values for precession, and in particular due to the inaccurate value used by Ptolemy, Thābit believed that the precession of the equinoxes was not a linear function of time as Ptolemy had supposed, but that it varied periodically. This led him to propose his influential 'trepidation' theory, a consequence of which is that, not only is the rate of precession variable, but the obliquity of the ecliptic is also a periodic function of time.[9]

Thābit's theory is illustrated schematically in Figure 4.1. The celestial equator is ARB, A and B being diametrically opposite points, and the fixed mean ecliptic is AQB, Q being midway between A and B. The mean ecliptic makes an angle $\varepsilon = 23° 33'$ with the equator (it is exaggerated greatly in the figure).

---

[8] The solar apogee actually is not fixed with respect to the stars but possesses a very slow steady motion. This was discovered in the eleventh century (see p. 97).

[9] His theory is described in his treatise entitled *On the Motion of the Eighth Sphere*, that has been translated into English with a commentary by Neugebauer (1962) and is discussed in Goldstein (1964). Thābit was not the first to suggest that the rate of precession was variable, theories incorporating non-constant precession were known to, for example, Theon of Alexandria (fourth century) and Proclus (fifth century) and were incorporated into early Indian astronomy (Pingree 1972).

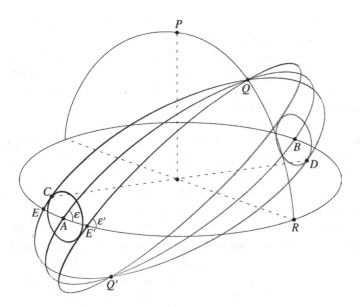

Fig. 4.1. Thābit ibn Qurra's theory of trepidation.

Two small circles, each with a radius of $4°\,18'\,43''$, are centred at the points A and B and a point C is chosen on the circumference of the circle centred at A, with the diametrically opposed point D on the other small circle. The great circle CQD is the movable ecliptic and, as C rotates around A, the ecliptic plane oscillates, Q and Q' being fixed points. The points at which the movable ecliptic crosses the equator are the equinoxes, and the position of the equinox, the mean position of which is A, varies between E and E'. Similarly, the obliquity has its mean value when the equinox is at A and has its maximum value ε' when the equinox is at E'. The stars have fixed longitudes with respect to the movable ecliptic.[10]

The theory of trepidation was used extensively by later Islamic and medieval astronomers, but it was not to everybody's liking. Al-Battānī devoted a special chapter of his *zīj* to refuting it, instead arguing for the traditional linear theory of precession at 1° in 66 years, or $54.5''$ per year. Al-Battānī's *zīj* is regarded by some as one of the most important works on astronomy between Ptolemy and Copernicus, and was translated into Latin by Robert of Chester in the twelfth century. The first part is modelled closely on the *Almagest*, the second part on

---

[10] Implementing such a scheme in astronomical tables was no easy task and it appears that most astronomers, Thābit included, used procedures which only approximated the true geometrical picture (see North (1967)).

the *Handy Tables*. As was fairly typical in Islamic astronomy, the solar theory of Ptolemy was refined considerably, but the lunar and planetary theories were left untouched. Al-Battānī improved on Ptolemy's value for the obliquity of the ecliptic, the solar mean motion (and, hence, the precession of the equinoxes), the eccentricity of the Sun, and the longitude of its apogee. He also introduced developments in trigonometrical techniques by replacing Greek chords with the sines (imported from India) and introducing cosines.

The method used by Ptolemy for finding the distance of the Sun in terms of the radius of the Earth was the basis of all future attempts for well over 1000 years. The earliest surviving redetermination of this distance is that of al-Battānī, and his result for the mean Earth–Sun distance is 1108 Earth radii, which compares with Ptolemy's value of 1210.[11]

The influence of al-Battānī's work, which has probably been studied more carefully than that of any other Islamic astronomer, can be seen from the fact that Peurbach in the fifteenth century and Copernicus in the sixteenth quote him often, particularly on matters of solar motion and precession. References to al-Battānī can be found also in the works of Tycho Brahe, Kepler and Galileo.[12] The close relation that still existed between astronomy and astrology is evident from the fact that al-Battānī also wrote a commentary on Ptolemy's *Tetrabiblos*.

The work of Thābit ibn Qurra and al-Battānī raised the level of awareness among Islamic astronomers. In particular, they began to appreciate the inadequacies of the parameters used in the *Almagest*. This led to numerous attempts to improve on Ptolemy's values so as to produce more accurate tables, and also to a much greater interest in the theoretical aspects of Ptolemy's geometrical schemes.

Developments in trigonometry

The last great representative of the House of Wisdom in Baghdad was the tenth-century astronomer and mathematician Abū al-Wafā.[13] He made significant

---

[11] See van Helden (1985), p. 32. Copernicus later based his own determination on parameters adapted from al-Battānī (see Swerdlow (1973)).

[12] In his *L' histoire de l'astronomie du moyen âge* (1819), Delambre devotes fifty-three pages to a thorough analysis of al-Battānī's *zīj*, but the best modern work is held widely to be the Latin translation and commentary written by C. A. Nallino between 1899 and 1907.

[13] Abū al-Wafā wrote a major astronomical work that was modelled on the *Almagest* but which did not introduce anything essentially new to theoretical astronomy. He was, however, credited by the nineteenth-century French scholar L. Sédillot, with discovering the so-called variation of the Moon, though this view has subsequently been shown to be false and Tycho Brahe has been reinstated as the true discoverer of this phenomenon.

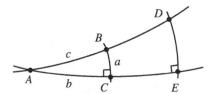

Fig. 4.2. The rule of four quantities.

achievements in the development of spherical trigonometry and the construction of trigonometric tables that were more accurate than those of Ptolemy. One result that facilitated calculations with right spherical triangles was what has become known as the 'rule of four quantities', illustrated in Figure 4.2. If ACB and AED are right-angled spherical triangles with a common angle at A and right angles at C and E, then

$$\frac{\sin BC}{\sin AC} = \frac{\sin DE}{\sin AE}.$$

A result of perhaps greater significance was the sine theorem for spherical triangles, which states that in any spherical triangle ABC,

$$\frac{\sin a}{\sin A} = \frac{\sin b}{\sin B} = \frac{\sin c}{\sin C}.$$

This result (which interestingly predates the sine theorem for plane triangles) simplified greatly the solution of many problems involving oblique spherical triangles, because it removed the need for decomposing them into a number of smaller right-angled ones.

The sine formula for spherical triangles was used to good effect by the famous Islamic scholar al-Bīrūnī with his solution to the *qibla* problem, this being to determine the direction in which Mecca was closest from a given location on the Earth, i.e. along a great circle. Thus in Figure 4.3, in which P represents one's own position, M is Mecca, and P_N the north pole of the Earth, the required angle is θ.

One of al-Bīrūnī's solutions to this problem was as follows.[14] The latitude and longitude of one's own position (α, γ) and of Mecca (β, δ) are assumed known. We then have $P_N P = 90° - \alpha$, $P_N M = 90° - \beta$ and $\angle P P_N M = \delta - \gamma$. The sine formula is not immediately applicable to the spherical triangle $P P_N M$, but al-Bīrūnī devised a solution procedure that involved a sequence of triangles for which the sine formula could be used. The technique is illustrated in Figure 4.4, which is a view looking down on the Earth from directly above the position

---

[14] See Katz (1998), p. 279 (see also Kennedy (1984) and references cited therein).

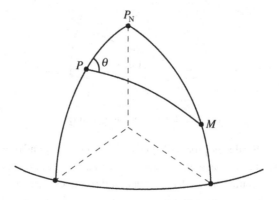

Fig. 4.3. The *qibla*. The observer is at P, M is Mecca, P_N is the north pole.

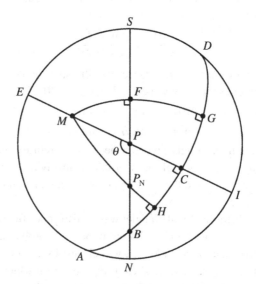

Fig. 4.4. Al-Bīrūnī's solution to the *qibla* problem.

P. Thus, $ANIDSE$ is the horizon circle for which P is the pole. The horizon circle for which M is the pole is $ABHCGD$. Two further great circles are shown both passing through M. First, MFG is the horizon circle for which B is the pole and, second, MP_NH is the great circle passing through Mecca and the north pole. The solution procedure involves three applications of the sine formula and one application of the rule of four quantities.

We begin by applying the sine formula to the triangle P_NFM, noting that $\angle MP_NF = \delta - \gamma$, $P_NM = 90° - \beta$ and $\angle MFP_N = 90°$. Thus,

$$\sin MF = \sin(\delta - \gamma)\cos\beta.$$

Hence, arc MF is known, and then so is $FG = 90° - MF$. Since FG is part of the horizon circle the pole of which is B, we know that $\angle P_N BH = \angle FBG = FG$. This then permits the application of the sine formula to the right-angled triangle $P_N BH$ (in which $P_N H = \beta$):

$$\sin P_N B = \frac{\sin \beta}{\sin FG}.$$

Thus, $P_N B$ is now determined, and since $P_N N = \alpha$, immediately we get $BN = \alpha - P_N B$ and $BP = 90° - \text{BN}$. Next, we use the rule of four quantities on the triangles BPC and BFG, noting that $BF = 90°$,

$$\sin PC = \sin PB \cdot \sin FG.$$

The quantities on the right-hand side are all known, so PC and then $CI = 90° - PC$ can be determined. Finally, we use the sine formula on triangle BAN. As A is the pole of $EMPCI$, we have $\angle BAN = \angle CAI = CI$, and so

$$\frac{\sin AN}{\sin \angle ABN} = \frac{\sin BN}{\sin CI}.$$

Now $\angle ABN = \angle P_N BH$ and the right-hand side is known, so AN can be determined. The *qibla* can then be calculated since $\theta = \text{NE} = 90° + AN$.

Al-Bīrūnī's interests ranged over virtually all the branches of science known in his time. He studied Aristotle closely at an early age and engaged in a fairly acrimonious exchange of letters with the philosopher Ibn Sīnā (known in Latin as Avicenna), arguably the most famous of all Islamic scientists. Al-Bīrūnī sent Ibn Sīnā a series of questions attacking Aristotelian natural philosophy, pointing out that many of its tenets had scant justification. As just one example, al-Bīrūnī wrote: 'There is nothing wrong in imagining the forms of Heavens as elliptic. Aristotle's reason for making them spherical is hardly convincing.'[15] How true!

Al-Bīrūnī wrote eight major astronomical works, the most comprehensive being his *Canon*, which was one of the most important astronomical encyclopedias, dealing with such subjects as cartography as well as theoretical astronomy. His models for the heavenly bodies essentially were Ptolemaic, though many of his parameters were derived independently. The *Canon* was not translated into Latin, and al-Bīrūnī's work remained unknown in the medieval West.

Another great Muslim astronomer from this period, whose works remained largely unknown in Europe during the Renaissance, was Ibn Yūnus. He lived and worked in Cairo and is reported to have been an eccentric figure who devoted

---

[15] Said and Khan (1981).

his time to astronomy, astrology and poetry. He wrote a major astronomical handbook (the *Hakemite Tables*) that, unlike those of many of his contemporaries, contained a large number of observations of eclipses and conjunctions. This work was used widely in the Islamic world but only became known in the West in the eighteenth century.[16] Ibn Yūnus also made significant contributions to trigonometry and introduced the idea of *prosthaphaeresis*, which was used by astronomers to shorten calculations before the invention of logarithms in the seventeenth century. He realized that the recently discovered identity

$$2\cos x \cos y = \cos(x + y) + \cos(x - y)$$

could be used in conjunction with trigonometric tables to simplify the process of multiplication by converting it into one of addition and dividing by 2. This method was one of the reasons why tables were often computed to over twelve significant figures.

Islamic religious observances presented quite a few problems in mathematical astronomy, the *qibla* being just one example, and this was one of the factors that encouraged the study of such problems. This influence is very evident in Ibn Yūnus's work. He produced a set of tables for time-keeping by the Sun and regulating the astronomically defined times of Muslim prayer and, up until the nineteenth century, virtually all Egyptian Muslim prayer tables were based on his work.[17] He also investigated the problem of determining just when the lunar crescent became visible in the evening sky following a conjunction of the Sun and Moon. This latter problem was of great importance because the beginning of each month was determined by the sighting of the crescent, and it is a complex problem involving, among other things, the relative positions of the Sun, Moon and horizon.[18]

Criticisms of Ptolemy

In terms of the physical structure of the Universe, it was the views of Aristotle that dominated, though an attempt to unify the ideas of Aristotle with those of Ptolemy, as espoused in his *Planetary Hypotheses*, was made by Ibn al-Haitham (known in Latin as Alhazen and famous primarily for his work on optics) in *On the Configuration of the World*, a work that had a considerable influence in Europe during the Renaissance. The fundamental principles that he believed had to underlie any such unification were that there was no empty

---

[16] Further details can be found in King (1974, 1975b).   [17] See King (1973).
[18] See, for example, Hogendjik (1988), King (1988).

space in the Universe, and celestial bodies move with uniform circular motion. He thought there had to be a unique spherical body corresponding to each motion that Ptolemy had introduced in the *Almagest*. In a later work, *Doubts Concerning Ptolemy*, Ibn al-Haitham noted that Ptolemy had set himself the task of accounting for the phenomena using uniform circular motions and that, since he had introduced the equant mechanism, he could not be considered to have succeeded. He objected to Ptolemy's lunar theory because it involved an imaginary point opposite the centre of the deferent controlling the motion of the lunar apogee, and this seemed physically impossible. Above all, Ibn al-Haitham argued, astronomy should deal with real bodies and not imaginary ones.

Ibn al-Haitham's work identified clearly those aspects of Ptolemaic astronomy that were considered unsatisfactory and indicated what would be required in any new formulation of theoretical astronomy. The first people who tried to reform astronomy along these lines emerged from a group of Islamic scholars from Andalusia, southern Spain, who in the twelfth century carefully studied and developed Aristotle's philosophy. The founder of the Spanish Aristotelian School was Ibn Bājja (known in Latin as Avempace), but his most famous advocate was Ibn Rushd (Averroës), who wrote numerous commentaries on Aristotle and became known simply as 'The Commentator'. Ibn Rushd's criticisms of Ptolemy were more extreme than those of his predecessors:

> To assert the existence of an eccentric sphere or an epicyclic sphere is contrary to nature ... The astronomy of our time offers no truth, but only agrees with the calculations and not what exists.[19]

Ibn Rushd and his followers produced a pantheistic philosophy which spread widely and was condemned as heresy by the Church.

Two other Spanish astronomers who were critical of Ptolemy, though not because of any adherence to Aristotelian principles, were Al-Zarqālī (Latinized as Azarquiel) and Jabīr ibn Aflah (Geber[20]). Al-Zarqālī, one of a small but important group of astronomers that grew up in Toledo in the second half of the eleventh century, is credited with being the first astronomer to recognize the slow steady motion of the Sun's apogee with respect to the fixed stars. He estimated this at $1°$ in 279 years, or $12.9''$ per year, which is close to the modern value of $11.6''$.[21] The very fact that there were observable phenomena that were not modelled by Ptolemy cast doubt on the whole of Ptolemaic theory. Al-Zarqālī is also often cited as the author of the collection of astronomical tables drawn

---

[19] Quoted from Gingerich (1992), p. 54.
[20] Not to be confused with the alchemist Jabīr ibn Hayyān (eighth century) who was also known as Geber in the West.
[21] Toomer (1969). Given the data he had to go on, the accuracy of the result was rather fortuitous.

up in Toledo, though it is more likely that he wrote some but by no means all of them. The tables were based on the treatises of many previous astronomers, including al-Khwārizmī and al-Battānī, and they utilized Thābit ibn Qurra's theory of trepidation. The original Arabic version of the tables is lost, but two Latin translations survive, and they were very successful in the Latin world until superseded by the *Alfonsine Tables* in the late thirteenth century.[22]

Jabīr ibn Aflaḥ's influence on future mathematical astronomy comes from his work *Corrections of the Almagest*, which was translated into Latin in 1175 by Gerard of Cremona, and from his work on trigonometry. As far as the latter is concerned, he simplified Ptolemy's treatment of spherical trigonometry by replacing Menelaus' theorem with simpler theorems on right spherical triangles. Jabīr was not the first to do this – Abū al-Wafā had made similar (and better) simplifications two centuries previously – but Abū al-Wafā's work was not translated subsequently into Latin. Jabīr criticized Ptolemy on a number of counts. He was concerned with the lack of mathematical rigour in some of Ptolemy's arguments; in particular, he felt that the assumption that the centre of the deferent in the models of the superior planets was exactly half-way between the centre of the Earth and the equant point, required proof. He also criticized Ptolemy's ordering of the planets. Concerning Mercury and Venus he wrote:

> Since, therefore, no parallax worth bothering about (according to Ptolemy) is to be found in either of them, and the sun does have a sensible parallax worth bothering about, how can they be below the sun?[23]

The view that Mercury and Venus were further away than the Sun had many adherents during the Middle Ages.

The most extensive attempt to reform theoretical astronomy by a member of the Spanish Aristotelian School was that of al-Biṭrūjī (Alpetragius) who flourished in the twelfth century and constructed a new planetary system in line with Aristotelian principles, that he hoped would displace the philosophically unsatisfactory Ptolemaic system. He was aware of the accuracy of Ptolemy's mathematical constructions and recognized that a philosophically acceptable system must also be able accurately to model the phenomena. Underlying the Aristotelian cosmos was the idea of concentric spheres centred on the Earth, and al-Biṭrūjī's reform involved replacing Ptolemy's planar geometric models with models on the surface of a sphere. By this process, eccentric circles and epicycles were removed from the theory, both of these mechanisms violating the Aristotelian principle of a single centre of rotation – the centre of the Earth.

---

[22] Details of the Toledan tables can be found in Toomer (1968).
[23] Quoted from Lorch (1975).

The fundamental idea in the new system was that each planet was supposed to be attached to a sphere, the motion of which was governed by the motion of its pole (a point with an angular distance from the planet of 90°) and as a consequence the planets in the new scheme remained equidistant from the Earth.[24]

A ninth sphere was added to the standard Aristotelian model in order to model the precession of the equinoxes. Al-Biṭrūjī believed that the rate of precession (fixed by Ptolemy at 1° per century) was variable, but unlike Thābit ibn Qurra's theory of trepidation, precession in al-Biṭrūjī's system was always in the same direction it was just its rate that oscillated. The ninth sphere carried the diurnal motion of the heavens, and the eighth sphere – the sphere of the fixed stars – was attached to the ninth, but its poles (i.e. the poles of the ecliptic) described two small circles around the poles of the ninth. As a consequence, the paths of the fixed stars were not circular but curved, these curves being interpreted traditionally as spirals; hence, al-Biṭrūjī's theory is often referred to as the 'spiral-motion theory'. In fact, the theory was never worked out in sufficient detail for anybody to be sure of its mechanisms, though it is clear that it was a system based on concentric spheres centred on the Earth, al-Biṭrūjī's order for the planets being Moon, Mercury, Sun, Venus, Mars, Jupiter and, finally, Saturn. This theory certainly was never able to predict quantitatively the positions of heavenly bodies and, hence, it could not be used to construct astronomical tables. As a result, it never become a serious rival to Ptolemaic theory, but from al-Biṭrūjī's time onwards, no serious astronomer could ignore the philosophical objections to Ptolemaic astronomy. The fact that al-Biṭrūjī's theory was not very successful did not stop it attracting the attention of philosophers and astronomers; it was translated into Latin in 1217 and spread through Europe during the thirteenth and fourteenth centuries. It continued to be cited throughout the fifteenth and sixteenth centuries, and notably by Copernicus.

Further opposition to Ptolemaic astronomy can be found in the writings of the Jewish astronomer and philosopher Maimonides who lived in Egypt in the twelfth century. Like al-Biṭrūjī, he objected to epicycles on philosophical grounds, since the planet carried on the epicycle does not move toward, away from, or around a centre. He also objected to eccentric orbits for the planets, and pointed out that in the *Almagest* some of the planet's deferents are centred outside the orbit of the Moon. Indeed, the centre of Jupiter's deferent is 525 Earth radii from the Earth and, hence, lies between the spheres of Mercury and Venus, something Maimonides considered quite improbable. Maimonides was also

---

[24] Al-Biṭrūjī's models were described in his treatise *On the Principles of Astronomy*, a translation of which is given in Goldstein (1971).

unhappy with Aristotelian models for the region above the Moon, since they failed to predict the observed motions but, rather than attempt to resolve these difficulties, he took the view that an understanding of the heavens was not within man's grasp.

> These difficulties do not concern the astronomer; for he does not profess to tell us the existing properties of spheres, but to suggest, whether correctly or not, a theory in which the motion of the planets is circular and uniform and yet in agreement with observation . . . Man's faculties are too deficient to comprehend even the general proof the heavens contain for the existence of Him who sets them in motion. It is in fact ignorance or a kind of madness to weary our minds with finding out things which are beyond our reach, without the means of approaching them.[25]

Although Maimonides was critical of Ptolemaic astronomy as a physical model for the Universe, he was quite happy to use it as the basis for astronomical calculations.[26]

The great conflict between Ptolemaic astronomy and Aristotelian cosmology – which continued right up until the sixteenth century – did not exist when originally Ptolemy wrote the *Almagest*. Ptolemy wrote his masterpiece 500 years after Aristotle, and the whole issue of the physical interpretation of his theory clearly was of secondary importance to him, and presumably to his contemporaries. However, to those rediscovering these works, the time between Aristotle and Ptolemy was not significant – they were both part of the 'ancient learning' that was being resurrected – and taken together they appeared full of contradictions. Eventually, these contradictions cast doubt on the whole edifice of ancient astronomy.

The Marāgha School

Theoretical developments in astronomy were not restricted to the astronomers of southern Spain. In the thirteenth century, another centre of theoretical activity grew up in the eastern part of the Islamic world, the first major figure being Naṣīr al-Dīn al-Ṭūsī (better known in the West as Nasir Eddin). Naṣīr al-Dīn was an advisor to the Mogul conqueror Hülegü, a grandson of Genghis Khan and (despite his other excesses) a patron of the sciences, and was a man of enormous influence, both during his own lifetime and subsequently. He was committed

[25] Quoted from Goldstein (1980).
[26] Details of Maimonides' mathematical astronomy, which is closely related to that of al-Battānī, can be found in Neugebauer (1949) and for a discussion of his attitude toward astronomy (see Kellner (1991)). For more on the history of Jewish mathematical astronomy, see Beller (1988).

to the Greek ideal of the pursuit of knowledge and became the first to head the Marāgha observatory built, beginning in 1259 in what is now northwestern Iran, at Hülegü's instigation. One of the primary functions of the observatory was to determine new parameters for solar, lunar and planetary models so as to improve the accuracy of astrological predictions. The observatory, which existed in one form or another until at least 1316, became a large research institution with a well-stocked library, a staff of at least ten, and a wide selection of expensive observing instruments.[27]

Naṣīr al-Dīn made contributions in virtually all fields of Islamic scholarship and wrote a great many manuscripts. In astronomy, his major contribution was his *Memoir on Astronomy*, the final revisions of which were made shortly before he died in 1274.[28] The *Memoir on Astronomy* was a narrative commentary on the *Almagest* which attempted to give Ptolemy's mathematical models a physical meaning. It was hugely influential, both in the Islamic world (where it became part of the school curriculum and was itself the subject of numerous commentaries) and farther afield. In general, the work does not contain geometrical proofs (the reader is directed to the *Almagest* for these), the notable exception being al-Ṭūsī's development of a geometrical device (now named after him) designed to remove some of the philosophical objections to Ptolemaic theory.

Naṣīr al-Dīn discovered that motion in a straight line could be produced by a combination of two circular motions, and he invented what has become known as the 'Ṭūsī couple' to achieve this.

> Let us set forth . . . a lemma, which is as follows: if two coplanar circles, the diameter of one of which is equal to half the diameter of the other, are taken to be internally tangent at a point, and if a point is taken on the smaller circle – and let it be the point of tangency – and if the two circles move with simple motions in opposite directions in such a way that the motion of the smaller [circle] is twice that of the larger so the smaller completes two rotations for each rotation of the larger, then that point will be seen to move on the diameter of the large circle that initially passes through the point of tangency, oscillating between its endpoints.[29]

The circles across the top of Figure 4.5 illustrate one cycle of the motion with the point C on the circumference of the smaller circle oscillating along the diameter through the initial point of tangency. A proof that the motion resulting

[27] The term 'Marāgha School' is used commonly to refer to all the astronomers who contributed to the research programme begun at the observatory, regardless of whether the individual actually worked there or not. For a general discussion of the astronomical achievements of this School (see, for example, Saliba (1987a, 1991).

[28] This work has been translated, with an extensive commentary, in Ragep (1993).

[29] From Naṣīr al-Dīn al-Ṭūsī's *Memoir on Astronomy*. Translation from Ragep (1993). Naṣīr al-Dīn also developed a version of the couple which was designed to produce oscillations along circular arcs (see Saliba and Kennedy (1991)).

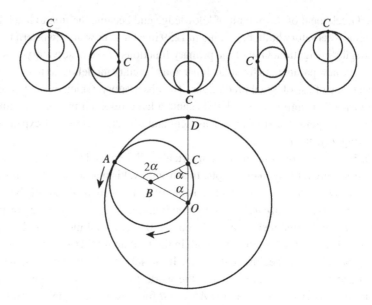

Fig. 4.5. The Ṭūsī couple.

from the two oscillations described above is, indeed, rectilinear can be given
with reference to the lower part of the figure. The radius *OBA* is considered to be
fixed to the larger circle, which rotates anticlockwise carrying the smaller circle
around with it. At the same time, the smaller circle, carrying the point *C*, rotates
clockwise at twice the rate of rotation of the larger circle. The motion starts when
A and *C* coincide (at *D*) and $\angle DOA = \alpha$ is zero. When the larger circle has
turned through α, the smaller one has turned through 2α and so $\angle ABC = 2\alpha$. It
follows that $\angle OBC = 180° - 2\alpha$ and, hence, that $\angle BCO = \alpha$. This shows that
C always lies on the diameter through *D*. The speed of the point on the smaller
circle which is instantaneously at *A* is proportional to $|OA| - 2|OB|$, which
is zero by construction. Hence, another way of visualizing the mechanism –
though not one that would have been acceptable philosophically to Naṣīr al-
Dīn – is to think of the smaller circle as rotating around the inside of the larger
one.

The Ṭūsī couple is very simple geometrically, but in the context of its day
it had profound consequences, since it undermined the Aristotelian distinction between circular celestial motions and rectilinear terrestrial motion. Using
this device, the astronomers at the Marāgha Observatory were able to modify
Ptolemy's geometrical models so as to remove the need for either an eccentric deferent or an equant. The most sophisticated versions of these new models
were produced by Ibn al-Shāṭir, who was employed as a *muwaqqit* (time keeper)

at the Umayyad mosque in Damascus. The office of *muwaqqit* came into being
around the eleventh century, at a time when theoretical astronomy began to
be accepted widely in religious society, and provided a secure and respected
position for many Islamic astronomers.

Ibn al-Shāṭir was the most distinguished Muslim astronomer of the fourteenth
century, but his work remained largely unknown in medieval Europe, only being
rediscovered in the 1950s.[30] He constructed new models for the Sun, Moon, and
planets which incorporated a number of ingenious modifications to Ptolemaic
theory. The reasons behind the changes were twofold. First, Ibn al-Shāṭir wanted
to remove some of those aspects of Ptolemy's astronomy that were unacceptable
philosophically, (e.g., the equant mechanism) though as we shall see, he had no
objection to using epicycles. Second (in the case of the Sun and Moon), there
were observed phenomena that Ptolemy had made no attempt to reproduce.

Ibn al-Shāṭir's modified solar theory is described in *The Final Quest Concerning the Rectification of Principles*. The need to modify Ptolemy's theory came from two observed phenomena not reproduced in Ptolemy's original
scheme. The first was the steady motion of the solar apogee, discovered by al-
Zarqālī, which Ibn al-Shāṭir modelled by simply adding an additional sphere:

> The repeated accurate observations have confirmed that the motion of the apogee is
> faster than the motion of the eighth [sphere]. It was therefore necessary to add
> another orb which would move the apogee with the observed motion.[31]

The other phenomenon that concerned Ibn al-Shāṭir was the change in the
apparent diameter of the Sun. According to Ptolemy, the diameter could be
regarded as a constant, and he gave the value 0; 31, 20° though, in his solar
theory, the ratio of maximum to minimum distance is about 0.92. From the
observations available to him, Ibn al-Shāṭir came to the erroneous conclusion
that the diameter at apogee was 0; 29, 5° and at perigee it was 0; 36, 55°, with
a mean value of 0; 32, 32°. This implied that the ratio of maximum to minimum
apparent diameter and, hence, the ratio of the solar distance at apogee to that at
perigee, was about 0.788, significantly different from the value used by Ptolemy.

Ibn al-Shāṭir's solution is illustrated in Figure 4.6. The Earth is E and the
point D moves around a circle centred on E such that the line ED always points
toward the mean sun $\bar{S}$. This circle is called the 'parecliptic' and is taken to
have a radius of 60 units. The parecliptic carries a smaller circle – the deferent –
centred at D, that rotates in the opposite direction at the same rate, so that the
line DB connecting the centre of the deferent to a point on its circumference

[30] See Roberts (1957), Kennedy and Roberts (1959), Abbud (1962), Roberts (1966).
[31] Quoted from Saliba (1987b).

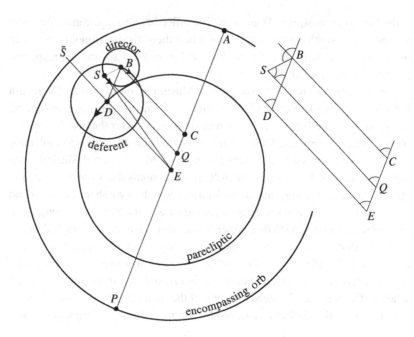

Fig. 4.6. Ibn al-Shāṭir's solar theory.

always is parallel to the line EA which points toward the solar apogee. If BC is drawn parallel to DE, then $EDBC$ is a parallelogram. If B were taken to represent the Sun, then the epicycle model just described would be equivalent to an eccentric circular motion centred at C (see Figure 2.11, p. 46) and, hence, equivalent to Ptolemy's solar theory.

However, Ibn al-Shāṭir introduced a third circle, centred at B, called the 'director', that carries the Sun S. This rotates at twice the angular speed of the deferent and parecliptic and in the opposite sense to that with which B moves around the deferent. This geometrical arrangement ensures that if SQ is drawn parallel to DE, the point Q remains fixed, with $|EC|$ equal to the radius of the deferent and $|QC|$ the radius of the director, and the Sun rotates uniformly around Q. This is demonstrated in the diagram accompanying the main figure, which shows an enlarged version of the parallelogram $EDBC$ in which all the angles marked are equal. Ibn al-Shāṭir thus managed, in effect, to incorporate an equant into the solar theory using only uniform circular motions. The parameters he found to be appropriate for his model were $|EC| = 4; 37$, $|QC| = 2; 30$ and so $|EQ| = 2; 7$ which is close to Ptolemy's value of the eccentricity of $60/24 = 2; 30$, so that Ibn al-Shāṭir's model predicts solar longitudes that are

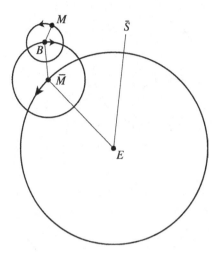

Fig. 4.7. Ibn al-Shāṭir's lunar theory.

close to those predicted by Ptolemy. The variation in apparent diameter is now

$$\frac{60 - 4;\ 37 - 2;\ 30}{60 + 4;\ 37 + 2;\ 30} \approx 0.788,$$

in agreement with Ibn al-Shāṭir's, albeit erroneous, value. The whole mechanism then is attached to a larger geocentric sphere, the encompassing orb, which has the observed steady motion of the solar apogee. Ibn al-Shāṭir stated explicitly that he believed the theory of trepidation to be unsound because it was not supported by observational evidence.

For the Moon, Ibn al-Shāṭir again noted that one reason for modifying Ptolemy's scheme was better to predict the changes in its apparent diameter:

> [the model of Ptolemy also] requires that the diameter of the moon should be twice as large at quadrature than at the beginning, which is impossible, because it was not seen as such.[32]

The modified lunar theory is illustrated in Figure 4.7 and can be seen to be similar in structure to the solar theory, with an epicycle on an epicycle. The mean moon is $\bar{M}$ and this rotates around the Earth once per synodic month (relative to $E\bar{S}$), the radius of this deferent circle being taken as 60. The point B, which is the centre of the epicycle that carries the Moon M rotates in the opposite sense around $\bar{M}$ once in each anomalistic month. Finally, the Moon rotates around its epicycle in such a way that the angle $\bar{M}BM$ is twice the angle $\bar{M}E\bar{S}$. Ibn

---

[32] Quoted from Saliba (1987b).

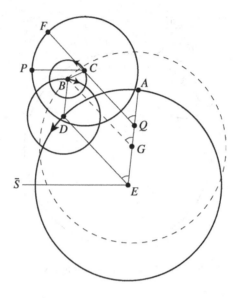

Fig. 4.8. Ibn al-Shāṭir's theory for the superior planets.

al-Shāṭir took the radii of the two epicycles to be $r_1 = 6$; 35 and $r_2 = 1$; 25, and then this mechanism implies that at the syzygies the distance of the Moon from the Earth varies between $r \pm (r_1 - r_2)$ (i.e. between 54; 50 and 65; 10) and at the quadratures between $r \pm (r_1 + r_2)$, i.e. between 52; 0 and 68; 0.

Ibn al-Shāṭir also devised a method for incorporating similar mechanisms into the models for planetary longitudes. Here, the motivation would appear to have been entirely aesthetic, since, as far as accuracy was concerned, his models did not improve on those of Ptolemy, but simply removed the unpleasant equants. Indeed, Ibn al-Shāṭir chose his parameters so as to make his models equivalent mathematically to those of Ptolemy. However, he was very pleased with the result, as in the introduction to his *zīj* (*The New Astronomical Handbook*), he wrote:

> I therefore asked Almighty God to give me inspiration and help me to invent models that would achieve what was required, and God—may He be praised and exalted, all praise and gratitude to Him—did enable me to devise universal models for the planetary motions in longitude and latitude and all other observable features of their motions, models that were free—thank God—from the doubts surrounding previous models.[33]

The mechanism that Ibn al-Shāṭir used to model the superior planets is shown in Figure 4.8. As in the solar and lunar schemes, the Earth is at the centre, but

---

[33] Quoted from King (1975a).

now there are three, rather than two, epicycles. The point D rotates around a circle, radius 60, and B rotates around D in the opposite sense and at the same rate, so that DB is always parallel to EA, A being the apogee of the orbit. This is equivalent to B moving anticlockwise around an eccentric circle (the dashed circle in the figure, centre G) with eccentricity equal to $|EG|$, which is the same as $|BD|$. The centre of the third epicycle, C, rotates around B so that $\angle DBC$ is twice $\angle DEA$ and, just as in the solar theory, this ensures that C rotates uniformly about a fixed point Q. Finally, the planet, P, rotates around C on this third epicycle. The uniform rotation of P is measured with respect to the line QCF and is such that CP always is parallel to $E\bar{S}$. In order to make this geometrical arrangement match closely Ptolemy's system, it is necessary to choose the radii of the first and second epicycles, $|BD| = r_1$ and $|BC| = r_2$, so that they satisfy

$$r_1 - r_2 = e \qquad \text{and} \qquad r_1 + r_2 = 2e,$$

where e is the eccentricity in Ptolemy's model (i.e. $e = |EO|$ in Figure 3.13, p. 77) and this is precisely what Ibn al-Shāṭir did. Ibn al-Shāṭir's models for the inferior planets were similar though, not surprisingly, that for the motion of Mercury was more complicated, involving the use of yet another epicycle.

The work of Ibn al-Shāṭir represents the culmination of a programme of research that was driven by the desire to correct certain deficiencies in Ptolemaic astronomy. In the main, the perceived defects were at a philosophical level – astronomers objected to some of Ptolemy's geometrical models because they could not physically be realized without violating some well-established principle. Most Islamic astronomers were quite happy to take Ptolemy's observations as they were and, apart from correcting his values for things like precession and obliquity, wanted to produce theories that reproduced exactly the same phenomena as Ptolemy's did. Thus, for example, Mu'ayyad al-Dīn al-'Urḍī, an astronomer from Damascus who developed alternative lunar and planetary theories, explicitly stated that he had made no observations of his own and stressed that his criticisms of Ptolemy were at a higher conceptual level.[34]

The new models that were devised by the Marāgha School of astronomers were a great success in that they produced longitude predictions as accurate as those from the *Almagest*, but without the use of the physically impossible devices introduced by Ptolemy. The majority of practising astronomers, on the

---

[34] Discussions of various aspects of al-'Urḍī's work can be found in a number of articles that have been reprinted in Saliba (1994) and his ideas on the size of the Cosmos, which differed greatly from those of Ptolemy, are described in van Helden (1985), pp. 32–3.

other hand, stuck with the tried and tested procedures that had been in existence for centuries.

The Marāgha Observatory was not the only significant establishment of its type in the history of Islamic astronomy. There was another observatory, built by Ulugh Beg in Samarkand. Ulugh Beg (this is not his original name, but means 'great prince') was, from the age of 15, a provincial governor in Samarkand but, unlike his grandfather, the Mongol conqueror Tamerlane, he was not interested in conquest but in science. He succeeded his father as ruler of the Mongol empire in 1447, but was murdered by his son 2 years later. During the quarter century leading up to Ulugh Beg's assassination in 1449, Samarkand was the most important scientific centre in the East and in 1420 he founded an institute of higher learning there (in which astronomy was the most important subject). In 1424, he built an observatory, which was destroyed in the sixteenth century, its location being rediscovered in 1908 by the archaeologist V. L. Vyatkin. Ulugh Beg's observatory was responsible for many accurate determinations of solar and planetary parameters, as well as the construction of extremely accurate trigonometric tables. However, it is most famous for the production of the first star catalogue not based on that of Hipparchus and Ptolemy. The influence of this catalogue on later developments in the West was minimal, though, since it only became well known in Europe in the mid seventeenth century, after the publication of the more accurate observations of Tycho Brahe.

One of the chief scientists at Ulugh Beg's observatory was al-Kāshī, who wrote a number of works, the most famous of which was an encyclopedia of elementary mathematics that included large sections on methods of calculation for astronomers.[35] Al-Kāshī also achieved a significant computational feat when he computed 2π to sixteen decimal places, greatly exceeding all previous results. His stated motivation for doing this was so that he could calculate the circumference of the Universe to within the thickness of a horse's hair! Rather more useful was his computation of sin 1° to ten sexagesimal places.

The revival of learning in Western Europe

While Islamic scholarship flourished, learning in Western Europe stagnated. The disintegration of the Roman Empire resulted in the almost total disappearance of the Greek language and, thus, most of the great Greek scientific works were inaccessible to European scholars. A few secondary works had been translated into Latin, including the first part of Plato's *Timaeus* – which

---

[35] Described in detail in Kennedy (1990).

represented the main source of knowledge concerning Greek cosmology – and a few Latin works (e.g. Pliny the Elder's *Natural History* (first century) and Martianus Capella's popular mythological allegory *The Marriage of Philology and Mercury* (fifth century)) described some basic astronomy. Euclidean geometry was available only through an incomplete sixth-century translation by Boethius, and the detailed Ptolemaic theory of the heavens appears to have been completely unknown. Time-keeping, both on a daily basis and for fixing religious observances within the calendar was, of course, very important and the ability to do this had to be maintained. In the eighth century the Venerable Bede, who lived in northern England, wrote works on this subject that were used for over 1000 years.[36]

Trade routes between Western Christendom and the Arab world had been established by the ninth century, but the transmission of scientific knowledge to the West did not really begin until contact with Islamic scholarship was made when European scholars started visiting Spanish monasteries in the eleventh century. Latin translations of previously unknown works began to be made from about 1000, the rate at which they appeared reaching a peak in the twelfth century – often referred to as the 'century of translation'. The Spanish city of Toledo was conquered by Alfonso VI in 1085 and, in the twelfth century, became the main centre for translation from Arabic into Latin. Many of the works that were translated were highly technical, and specialized words which were not understood were often simply transliterated. Some of these words now form part of our technical vocabulary, such as 'zenith' and 'nadir' (the nadir is the point on the celestial sphere diametrically opposed to the zenith). Aristotelian natural philosophy was rendered into Latin through translations of his *Meteorology*, *Physics* and *On the Heavens*; Europe became aware of the enormous achievements of the Greeks in mathematics through translations of Euclid, Archimedes and Apollonius, and as far as astronomy was concerned, scholars at last were able to appreciate the sophistication of Ptolemy's *Almagest* – translated in 1175 by Gerard of Cremona – as well as the results of the previous 400 years of Islamic endeavour.[37] From the end of the twelfth century, more and more translations were made direct from the original Greek texts, rather than through the intermediate Arabic translations.

Also significant was the introduction into Europe of the Hindu-Arabic numerals by Leonardo of Pisa (better known as Fibonacci). Leonardo spent a great deal of his early life travelling around the Mediterranean and came into

---

[36] For details of early medieval astronomy, see, for example, Eastwood (1997), McCluskey (1998).
[37] A list of translations made during this period can be found in Crombie (1959).

contact with the mathematics of the Islamic world. It was problems that arose in the world of commerce that provided the motivation for his own work but its influence was much broader. In his widely read *Book of Calculation* (1202), he described algorithms by which all the basic operations of arithmetic could be performed on numbers written using the decimal positional system. (This was for integers only; it would be another 400 years or so before decimal fractions were introduced.) Fibonacci also introduced Europeans to the algebraic techniques of men such as al-Khwārizmī.

Works such as the *Almagest* intrinsically were difficult, and the problems often were exacerbated in the Latin translations. As a result, a number of greatly simplified versions began to appear, e.g. the many manuscripts going by the name *Theory of the Planets* but only containing simple descriptions of epicyclic motion, and the hugely popular *On the Sphere* by Johannes de Sacrobosco.[38] Sacrobosco's book, which was used in schools until the early seventeenth century, was very short and contained the basic results of geocentric astronomy. The work has four sections, but only the final part contained information on the Ptolemaic theories for the Sun, Moon and planets, and the treatment was extremely rudimentary. Many commentaries on *On the Sphere* were written, some of which were attempts to supplement the brief details of the original work, but the first complete account of Ptolemy's astronomical system that was actually written in the West, which included numerical details from the *Almagest* as well as the physical interpretation from the *Planetary Hypotheses*, was written by Campanus of Novarra in the mid thirteenth century.[39]

During the Middle Ages, all technical astronomy was based on that of Ptolemy, whether from the *Almagest* or from the large body of literature that grew up around it. The battle between Ptolemaic astronomy and Aristotelian natural philosophy continued but was overshadowed largely by the conflict between Aristotle and Christianity. One of the few attacks on the technical aspects of Ptolemy's theory was the broadside made in 1364 by Henry of Hesse (Henry of Langenstein), who went to great lengths to try and pick holes in the details of Ptolemaic astronomy. He was wide of the mark, though, since Henry spent most of his time attacking theories that had been attributed to Ptolemy by later commentators and interpreters, rather than the actual contents of the *Almagest*.

---

[38] Although Johannes de Sacrobosco wrote what became a standard text on astronomy, virtually nothing is known about him. He was probably English and he is often referred to by the name John of Holywood, this surname being the modern English for which Sacrobosco is the Latinized form. As to when he lived, all that can be said with any certainty is that he was active in Paris between about 1220 and 1240. For more details, see Pedersen (1985). A translation of the *De sphaera* can be found in Thorndike (1949).

[39] Benjamin and Toomer (1971). Details of textbooks used in the teaching of astronomy at medieval universities can be found in Pedersen (1981).

The fact that Henry, an advocate of homocentric astronomy, had a great reputation for his astronomical knowledge, just goes to show that such expertise was in fairly short supply in the fourteenth century.[40]

As far as the history of astronomy is concerned, one of the most significant events of the Middle Ages took place in thirteenth-century Spain. The Christian King Alfonso X (the Wise) ruled the kingdoms of Leon and Castile from 1252 to 1284, and established and presided over a group of predominantly Christian and Jewish astronomers charged with translating a number of astronomical texts into the Castilian language.[41] The most important output from this enterprise was the set of astronomical tables that were produced. The original text of the *Alfonsine Tables* is lost, but it is presumed widely that they were written in Toledo in about 1270 to replace the previous tables produced there in the eleventh century (see p. 97–8).[42] The tables circulated in many forms (in Latin translation), the most popular of which was the version composed by John of Saxony in 1327, and they formed the basis of practically all astronomical calculations until the mid sixteenth century. The tables utilize a theory of trepidation which is a modification of Thābit's in which precession has both a uniform and a periodic component.

The work of the Spanish Islamic astronomers was translated into many languages, not only Latin and Castilian. By the mid thirteenth century, most of the major astronomical works also had been translated into Hebrew and, in the fourteenth century, Hebrew astronomy flourished in southern France. The major figure from this region was Levi ben Gerson (known also as Gersonides and Leo de Balneolis). Levi was a mathematician, making contributions in algebra, trigonometry and combinatorics, a philosopher and a biblical commentator, as well as an astronomer. He is perhaps most famous in the West for inventing the cross staff (Jacob staff), an instrument used for centuries to measure the angular separation between celestial bodies. Levi's *Astronomy* forms Part 1 of Book V of his great work on religious philosophy, *Wars of the Lord*.[43]

---

[40] See Kren (1968, 1969).

[41] A legend has grown up around King Alfonso in which he is said to have had his doubts over Ptolemaic astronomy. For example, Dreyer (1953) mentions Alfonso's 'well-known saying' that if God had consulted him when creating the world, he would have given Him good advice. However, as Franseen (1993) has shown, this legend only began in the late seventeenth century. Alfonso was in the habit of blaspheming against God and saying that he would have made the world differently (i.e. better), but the linking of such remarks with the mathematical astronomy of Ptolemy would appear to be unjustified.

[42] It has been suggested that the *Alfonsine Tables* were written in Paris in the 1320s and were named in honour of the Castilian king (Poulle 1988).

[43] Because of its great length, the *Astronomy* was not included in manuscript or printed versions of his philosophical treatise. The first twenty chapters (there are 136!) are translated in Goldstein (1985).

There are a number of ways in which the astronomy of Levi ben Gerson differed from that of other medieval astronomers. In the first place, he made his own observations and was enlightened sufficiently to believe that these should form the ultimate test of his theories. Like many before him, he was critical of Ptolemy, but his overriding concern seems to have been that the models of the *Almagest* simply were not good enough at representing the celestial motions. He was extremely critical of al-Biṭrūjī's attempted reform of Ptolemaic astronomy because it failed to perform the fundamental function of any astronomical theory, i.e. to reproduce the observed phenomena. In contrast to Ptolemy and independently from Ibn al-Shāṭir, Levi decided that the apparent sizes of celestial bodies was part of the observational data that a theory should reproduce. Rather than accept that the obliquity of the ecliptic and the rate of precession were variable – views which had been almost universally held since the eleventh century – he argued that it was much more likely that Ptolemy's values for these quantities were erroneous and that they had remained constant since ancient times.[44] His independence of mind is also apparent from the fact that, unlike most astronomical writing of his day, his *Astronomy* does not follow the arrangement of the *Almagest*. Original as they were, Levi's ideas do not appear to have had much influence other than on some later Hebrew writers.

Levi ben Gerson's greatest achievement in theoretical astronomy was his lunar theory, which, he claimed, better fitted with observations than previous models. He was critical of Ptolemy's theory for the obvious error in the predicted variation in apparent diameter, and he also argued against the use of an epicycle because the same side of the Moon was always seen from the Earth, which would not be the case if it were attached rigidly to an epicycle. As for the accuracy of lunar longitudes, Levi concluded from his own observations that Ptolemy's model was fine at syzygy and quadrature, but not at the octants. The theory that Levi proposed managed to avoid the use of epicycles and yet reproduce his own lunar observations better than Ptolemy's model.[45] For example, Levi's observations revealed very little change in the apparent diameter of the Moon between quadrature and opposition, and his model predicted that the ratio of the distances of the Moon at these points should be $0; 58 \approx 0.97$. Unfortunately, not all the observations against which Levi tested his model were accurate, and not all the changes he introduced were improvements. For example, he took the angle of inclination of the Moon's orbit to be $4\frac{1}{2}°$ instead of Ptolemy's more accurate value of $5°$.

---

[44] For details of Levi ben Gerson's discussion of precession, which forms Chapter 61 of his *Astronomy*, see Goldstein (1975).
[45] Details can be found in Goldstein (1972, 1974).

Scholasticism

Throughout the Middle Ages in Western Europe, intellectual thought was dominated by the Catholic Church. Prior to about the tenth century the Church was, by and large, opposed to scientific endeavour, not unnaturally since the early Christians had had to fight for the survival of their religion by emphasizing the importance of its theology at the expense of pagan learning. One of the more liberal early Christian thinkers, St Augustine, wrote in his handbook for Christians:

> When, then, the question is asked what we are asked to believe in regard to religion, it is not necessary to probe into the nature of things, as was done by those whom the Greeks called *physici*; nor need we be in alarm lest the Christian should be ignorant of the force and number of the elements,– the motion, and order, and eclipses of the heavenly bodies; the form of the heavens; ... It is enough for the Christian to believe that the only cause of all created things, whether heavenly or earthly, whether visible or invisible, is the goodness of the Creator, the one true God; and that nothing exists but Himself that does not derive its existence from Him.[46]

By the time Christian Europe came back into contact with ancient learning in the tenth century, things had changed. The authority of the Church was now complete, and provided that it maintained control over it, pagan learning was no longer a threat. Indeed, many churchmen devoted considerable time toward the rediscovery of ancient knowledge. Schools of higher learning attached to cathedrals and monasteries began to appear and eventually these developed into universities – Bologna in 1088 being the first – the universities of Paris, Oxford, and Cambridge being founded a little later, around 1200. The universities produced an élite with an education in such subjects as law, medicine and theology, and the study of mathematics became codified into the standard format of the *quadrivium*: arithmetic, geometry, music, and astronomy.

By the twelfth century, the study of cosmology and natural philosophy once again became acceptable and, by the thirteenth century, educated Christians were familiar with the basic principles of the Aristotelian cosmos. The conflicts between Aristotle and the Scriptures still existed, of course, and the study of Aristotelian physics and metaphysics was not always welcomed, but over a period of time Christian theology and ancient Greek ideas about the Universe gradually were melded together into a unified whole, known as 'scholasticism'.

Perhaps the person most influential in determining the ultimate nature of this Christian cosmology was St Thomas Aquinas, who believed that a complete

---

[46] St Augustine *Works*. Quoted from Kuhn (1957), p. 107.

understanding of the world could be obtained only through both revelation and reason. Many of Aristotle's ideas were taken over unchanged, including the perfect circular motion of the heavens, but others (e.g. the continual existence of the Universe) were opposed so fundamentally to Christian thinking that they had to be discarded. In other cases, the inherent contradictions between Scripture and Aristotelian philosophy were removed by the device of claiming that the actual words used in the Bible had been simplified deliberately so that they would be understood by ordinary people. Through his compendium of Christian knowledge, the *Summa theologica*, Aquinas enabled Aristotle's world view to become a constituent part of Christian thought.

This medieval view of the Universe was enshrined in poetry by the Italian Dante Aligheri, whose writings contain many references to astronomy. Dante revered Aristotle, whom he described as the supreme and highest authority, but he did not feel the need to follow him when it came to matters of astronomical detail. Here, he would turn to al-Farghānī's description of Ptolemaic astronomy. Dante's *Divine Comedy* follows the author on his journey as he passes through the centre of the Universe and then from planet to planet until he reaches the outermost sphere of the stars. He begins by descending into Hell, which is an inverted cone situated directly beneath the centre of the inhabited world – Jerusalem – and whose apex is the centre of the Earth (and, hence, of the Universe). He emerges in Purgatory, an island diametrically opposite Jerusalem on the Earth's surface, and on this island is a mountain that extends to the upper reaches of the atmosphere. From here, Dante enters Paradise, which is made up of the planetary spheres of medieval astronomy. The journey has a very precise chronology, beginning on the vernal equinox and lasting 8 days. At each stage, Dante marks his progress by the positions of the heavenly bodies as they would have appeared from his current position, and it is through these descriptions that the author displays his knowledge of technical astronomy.[47]

Throughout the Middle Ages the belief that the heavenly bodies influence what happens on Earth was almost universal; St Augustine had described astrology as impious superstition and, in the early Middle Ages, the study of astrology was frowned upon by the Church, as it had been by Islamic religious leaders. But by the end of the fourteenth century, astrology was practised so widely that it could no longer be resisted. 'Practical astronomy' became very important; there were chairs of astrology at several major universities, and astrologers were appointed to high offices in the courts of kings and princes. Many arguments against the validity of astrology were put forward but largely they were ignored, and astrology became a major stimulus for astronomy, particularly in encouraging people to construct tables of planetary positions.

---

[47] The details of Dante's astronomy are described in Orr (1956).

One of those who spoke out against astrology was Nicole Oresme, an advisor to Charles V of France and then tutor to Charles VI. Together with Jean Buridan and Albert of Saxony at the University of Paris, Oresme mounted a powerful attack on Aristotelian physics, particularly his theory of motion, but their influence waned when France was ravaged by the Hundred Years War with England.[48]

Following ideas of William of Ockham, Buridan developed an 'impetus theory' of motion in which, contrary to Aristotle's teaching, it was not necessary for a body to be acted upon continually by a force for it to remain in motion.[49] This was a major step on the long road toward a modern principle of inertia, but equally significant for astronomy was Buridan's suggestion that God had given impetus to the heavenly bodies which, since they encountered no resistance, accounted for their motion. Here, we see one of the first attempts to apply the same laws of physics to celestial and terrestrial phenomena.

Oresme produced a French translation and commentary of *On the Heavens* in which, while he pretty much agreed with all of Aristotle's conclusions except those that contradicted the Creation, he criticized strongly many of Aristotle's arguments. He pointed out that the proof that the Universe was geocentric presupposed Aristotle's theory of motion in which the element earth naturally moved toward the centre of the Universe and, hence, the argument was circular. Maybe the natural motion of the element earth was toward the centre of the Earth, in which case that could be anywhere in the Universe. In the fourth century BC, Heraclides had suggested that the perceived daily rotation of the stars was due to a rotation of the Earth about its own axis, and Aristotle had argued that this was impossible. Oresme agreed with Aristotle that the Earth was stationary, but he criticized Aristotle's refutation of Heraclides. Aristotle had argued that if the Earth rotated, there would be a continuous wind from the east which was not observed. Oresme pointed out that one only had to assume that the atmosphere also took part in the daily rotation to counter this argument. In the end, Oresme believed, the choice between believing in a stationary or a moving Earth had to be a matter of faith.

Oresme was also one of a number of men who questioned Aristotle's insistence on a finite Universe. Perhaps the sphere of the fixed stars was surrounded

---

[48] Oresme wrote on numerous other topics relevant to astronomy. He was fascinated by the subject of celestial commensurability, i.e. whether or not ratios between celestial motions could be expressed as ratios of integers (see Grant (1971)).

[49] Buridan was by no means the first to criticize Aristotle's theory of motion. In the sixth century, the Greek Christian philosopher John Philoponus suggested that projectiles move due to some 'incorporeal power' rather than as a result of forces imparted due to the disturbance of the surrounding air. Here, we have an early forerunner of our concept of momentum (see, for example, Toulmin and Goodfield (1965), p. 118, and Cushing (1998), pp. 74–5).

by an infinite expanse of empty space.[50] This separation of space and matter was opposed diametrically to Aristotelian philosophy and was too radical to attract many adherents. Not until Newton did the independent existence of material bodies and the space they occupy become accepted.

Peurbach and Regiomontanus

The history of astronomy in the fifteenth century is dominated by two men, Georg Peurbach and Regiomontanus. The significance of their contributions comes, not so much from the technical content (their astronomy continued the medieval tradition), but from the fact that with the introduction of printing into Europe, their books became the first astronomy textbooks to achieve what might be described as a mass circulation.[51]

After receiving his master's degree from the University of Vienna in 1453, Georg Peurbach accepted the position of court astrologer to King Ladislas V of Hungary and later became the imperial astrologer to the Holy Roman Emperor Frederick III. He is most famous for his *New Theories of the Planets*, an astronomical textbook he wrote following a series of lectures he gave in Vienna in 1454.[52] The first printed edition was published by Peurbach's student Johannes Müller (better known as Regiomontanus, Latin for his place of birth, Königsberg) in 1472, and the book became very popular, going through nearly sixty editions in the fifteenth and sixteenth centuries. Peurbach's work is based on Ptolemy's *Planetary Hypotheses* and is also influenced heavily by Ibn al-Haitham's *On the Configuration of the World*. He described theories for the Sun, the Moon and the planets, in which each component of the geometrical mechanism was produced by the motion of a separate celestial sphere, and then provided a theory of precession based on the work of al-Battānī and al-Farghānī. A section on Thābit ibn Qurra's theory of trepidation was added in about 1460. The fact that Peurbach was aware of the significant role of the Sun in the geocentric theory is evident from the following passage:

> ... it is evident that the six planets share something with the sun in their motions and that the motion of the sun is like some common mirror and rule of measurement to their motions.

---

[50] Different medieval explanations of what lay beyond the fixed stars are described in Grant (1994), Chapter 9, an expansive study of medieval cosmology.

[51] The first printed edition of the *Almagest*, a rather unsatisfactory medieval Latin version, appeared in 1515. A new Latin text was printed in 1528 (Boas Hall (1994)).

[52] The Latin title was *Theoricae novae planetarum*; it is translated in Aiton (1987).

In 1460, at the request of Cardinal Bessarion, the papal legate to the Holy Roman Empire and himself a scholar of distinction, Peurbach began work on an abridgment of Ptolemy's *Almagest* which, according to Regiomontanus, he knew almost by heart, and he completed the first six books of his work before he fell ill and died. Before his death he persuaded Regiomontanus (who took over Peurbach's professorship in Vienna) to complete the work, which he did over the next couple of years. This work, the *Epitome of Ptolemy's Almagest*, which provided a relatively simple summary of Ptolemy's treatise and served subsequently as Copernicus' guide to Ptolemaic astronomy, was first published 20 years after Regiomontanus' death, in 1496. Although the *Almagest* had been available in Latin translation for over 300 years, it is clear that it was not read widely and understood until the printing and wide circulation of Regiomontanus' *Epitome*, which also included material from later Arabic sources such as Jābir.

Before his death, Peurbach had been planning to travel to Italy so that he could study original Greek manuscripts, and this is precisely what Regiomontanus did for a number of years starting in 1462. On his return, he settled in Nuremberg and began an ambitious project of translating and publishing the great scientific works of antiquity. He did translate some Greek works, including Apollonius' *Conics*, but his untimely death put an end to his laudable endeavour. Regiomontanus was also the first in the Latin West to produce a systematic treatment of trigonometry, and he was well aware of how important this subject was for the study of astronomy. In his work *On Triangles of Every Kind*,[53] which was written in about 1463 but not published until 1533, he wrote:

> You, who wish to study great and wonderful things, who wonder about the movement of the stars, must read these theorems about triangles. Knowing these ideas will open the door to all of astronomy ...

Following Islamic scholars, and also Peurbach's work, Regiomontanus based his trigonometry on the sine rather than the Greek chord, and considered both plane and spherical triangles. He also began the move toward the use of the decimal system by basing his tables on a circle, the radius of which was a power of 10. There is little conceptually new in *On Triangles*, much of the work being taken directly from Arabic sources,[54] but the quality of the exposition and the inclusion of clear numerical examples made the work extremely influential.

With the work of Peurbach and Regiomontanus, mathematical astronomy in Western Europe reached the level of sophistication that the Greeks had achieved some 1200 years previously. But now there was a stimulus to press on. First, it

---

[53] In Latin, *De triangulis omnimodis*. A translation can be found in Hughes (1967).
[54] See Hairetdinova (1970).

was recognized that Ptolemy was far from being the final word on the subject; even with the refinements of the Islamic astronomers, the models of the *Almagest* simply were not accurate enough to match naked-eye observations. Second, with the advent of the printing press, interest in astronomy was becoming more widespread. The growing need for sophisticated tools to aid navigation, and an increasing interest in astrology, created a thirst for knowledge.

Not all astronomers from this period were forward-looking, however. Taking their lead from Ibn Rushd (Averroës), who had declared the Ptolemaic universe incompatible with Aristotelian physics, two Italians in the early sixteenth century, Girolamo Fracastoro[55] and Gianbattista Amico, attempted to resurrect the homocentric model of Eudoxus–Callippus–Aristotle and make it predict more accurately the known phenomena. It is quite likely that Fracastoro knew Copernicus as they both studied at Padua at the same time and they may well have discussed the problems of Ptolemaic astronomy. Fracastoro's proposed alternative, described in his *Homocentrica* (1538), was an attempt to revive the dead. He ended up with a complex system of seventy-nine spheres which, of course, retained the same fundamental flaws present in Eudoxus' original model. Amico's scheme, contained in *On the Motions of the Heavenly Bodies according to Peripatetic Principles without Eccentrics or Epicycles* (1536), which utilized a similar number of spheres but differed in technical detail from that of Fracastoro, appears to have been arrived at independently.[56] Interestingly, both Fracastoro and Amico made use of geometrical devices equivalent to the Ṭūsī couple.[57]

The desire to replace the established system with something new was widespread in the early sixteenth century, but few had the patience or skill to turn their ideas into a computational scheme that could match Ptolemy for accuracy. One man, however, spent 40 years doing just that – Copernicus.

---

[55] Fracastoro is better known for his work on the spread of contagious diseases and for giving the disease syphilis its name.
[56] Some details of both models can be found in Dreyer (1953), Chapter XII.
[57] See Swerdlow (1972), di Bono (1995).

5

The heliocentric universe

Copernicus

The year 1543 marks the beginning of the end for geocentric astronomy. In this year, shortly before his death, Nicholas Copernicus published *On the Revolutions of the Heavenly Spheres*, in which the Sun, and not the Earth, lies at the centre of the Universe.[1]

Nicholas Copernicus, the youngest of four children, was born in the city of Toruń on the Vistula in Poland in 1473.[2] He entered the prestigious University of Cracow in 1491 where he stayed for 4 years studying the liberal arts, including a broad range of instruction in astronomy and mathematics. He purchased a number of fundamental works during this period, including the two books containing the most up-to-date tables necessary for astronomical calculation: the 1492 edition of the *Alfonsine Tables*, and the 1490 edition of Regiomontanus' *Tabulae directionum* (the latter a treatise on spherical astronomy). He bought, also, an edition of Euclid's *Elements*.

From 1496 to 1503, he studied at various universities in Italy, including Bologna, Ferrara, and Padua. Among other things, he studied medicine and

---

[1] The Latin title is *De revolutionibus orbium Caelestium*. At least three English translations have been published and their relative merits are discussed in Swerdlow and Neugebauer (1984), which is also the source of much of the technical detail concerning Copernican astronomy that appears here. All quotations from *On the Revolutions* are taken from Rosen's translation which first appeared in 1978 (Copernicus (1992)). Rosen rewrites *On the Revolutions* in a readable modern English, rather than sticking religiously to Copernicus' style, and as such has come in for some criticism (see Swerdlow (1981b) and the resulting letters (*Isis*, **72**, (1981) pp. 629–30); see also Swerdlow (1981a) and Toomer's review of Rosen's translation in *Journal for the History of Astronomy*. **12** (1981), 198–204).

[2] The brief biographical details following are based largely on the accounts given in Dobrzycki (1973a) and Swerdlow and Neugebauer ((1984)). Kesten (1945) has written a popular biography and Copernicus' life and work is the subject of a novel, *Doctor Copernicus*, by John Banville (Banville (1976)).

law, learned Greek, and became friendly with Domenico Maria Novara, who occupied the chair of astronomy at Bologna. He returned to Poland in 1503, initially assisting the Bishop in Heilsberg in his administrative duties, and then in 1510 he became a canon at the Frombork (Frauenburg) Cathedral in Warmia (a small territory also known as Ermland), where he spent the rest of his life.

Early in the sixteenth century, Copernicus was asked, like Regiomontanus before him, if he would advise the papacy about calendar reform, demands for which had been growing over the preceding centuries as the errors in the Julian calendar were recognized. He declined, believing that what was needed was a reform in astronomy based on new observations so that more accurate calculations could be made; only then would reforming the calendar be worth while.[3]

It was not Copernicus' intention completely to revolutionize astronomy – though his radical proposal eventually led to just such a revolution – but he did realize that a major technical change was necessary, since all the minor changes and alterations to Ptolemy's theory that had been introduced over the preceding 1000 or so years had still not produced a system accurate enough to conform with good naked-eye observations. That Copernicus was not trying to start a great debate about man's place in the Universe is evident from the fact that *On the Revolutions* is not aimed at the general public or, indeed, at educated lay people, but is written only for those few who were conversant fully with the technical details of Ptolemaic astronomy. The idea of placing the Sun rather than the Earth at the centre of the Universe had profound consequences, but the geometrical models that Copernicus devised so as to transport Ptolemy's mathematical astronomy into a heliocentric context did not.

Copernicus was, of course, not the first to suggest a moving Earth (the ideas of Aristarchus, Heraclides, and Oresme have been discussed earlier, for example) and, indeed, he did not claim originality in the idea. Many astronomers before Copernicus also had realized that the role of the Sun was not just like that of

---

[3] The Julian calendar, with a leap year every 4 years, implies a tropical year of 365.25 days, whereas in fact the tropical year is, to three decimal places, 365.242 days. Hence, in 1500 years, the time that had elapsed between Caesar and Copernicus, the calendar had advanced $1500 \times 0.008 = 12$ days with respect to the seasons. More seriously, as far as the Church was concerned, was the fact that computations of the date of Easter, which rely on cycles linking the motions of the Sun and Moon, were now also in error. The Gregorian calendar, in which 3 leap years are removed from each 400-year period (corresponding to a tropical year of 365.2425 days) came into use in Catholic countries in October 1582, following the decree of Pope Gregory XIII that Thursday, 4 October of that year would be followed by Friday, 15 October. Protestant countries resisted the change and the adoption of the Gregorian calendar took place at different times in different places (a list can be found in, for example, Richards (1998)). Britain and its colonies did not adopt the improved calendar until 1752 (Wednesday, 2 September was followed by Thursday, 14 September) and dates after 1582 from calendars that had not undergone the reform are usually referred to as 'Old Style'.

the other planets. Copernicus' achievement was to take the heliocentric idea of Aristarchus and incorporate it into a system of Ptolemaic geometry so that it predicted accurately the phenomena. To Copernicus, the motion of the Earth was a necessary consequence of his accurate solution to the problem of the planets: 'In so many and such important ways, then, do the planets bear witness to the earth's mobility.'[4] He was still under the spell of circular orbits with uniform speed, and throughout his work he retains the ancient idea of planets being carried on celestial spheres, though he avoids the question of whether these spheres are real or imaginary.[5] Copernicus might well be described as the last of the ancients, a spiritual companion of Aristarchus, Hipparchus, and Ptolemy. His theory was conceived as a technical modification (albeit a major one) of classical planetary astronomy, but over the two centuries following its publication it became the focus of great debates in religion and philosophy, as well as in science. By changing the role of the Earth in the overall scheme of things, Copernicus forced people to re-examine their ideas about the relationship between man and God.[6]

Was Copernicus aware of the heliocentric theory of Aristarchus? The first printed edition of Archimedes' *Sand Reckoner* did not appear until a year after Copernicus died. However, the following passage appears in a manuscript of *On the Revolutions* that survives to this day, though it was deleted prior to publication:

> The motion of the sun and the moon can be demonstrated, I admit, also with an earth that is stationary. This is, however, less suitable for the remaining planets. Philolaus believed in the earth's motion for these and similar reasons. This is plausible because Aristarchus of Samos too held the same view according to some people, who were not motivated by the argumentation put forward by Aristotle and rejected by him [*Heavens*, II, 13–14]. But only a keen mind and persevering study could understand these subjects. They were therefore unfamiliar to most philosophers at that time, and Plato does not conceal the fact that there were then only a few who mastered the theory of the heavenly motions.[7]

Thus, Copernicus certainly did know of the Greek heritage for the heliocentric hypothesis. It is, however, not clear whether he knew of Oresme's work on the rotation of the Earth.

---

[4] COPERNICUS *On the Revolutions*, Book I, Chapter 11.

[5] The question as to whether Copernicus believed in the physical reality of the celestial spheres has been a source of some controversy. For more details, see Westman (1980), Aiton (1981), Jardine (1982), Grant (1987).

[6] The argument of Kuhn (1957) that Copernicus was influenced by the Neo-Platonic tradition of Sun worship is refuted by Rosen (1983).

[7] COPERNICUS *On the Revolutions*, I, 11. Further details are given in Rosen's notes in Copernicus (1992) (see also Africa (1961)).

Copernicus made a number of his own observations, but they are not notable for their accuracy. In *On the Revolutions*, he described the instruments that he used for his observations, all of which were known in ancient times; indeed, Copernicus' descriptions of the instruments are based on those in Ptolemy's *Almagest*. He relied heavily also on the observations of others, particularly those of Ptolemy (in whom Copernicus put rather too much faith) and also Islamic astronomers such as al-Battānī and al-Zarqālī. In Copernicus' time, a theory was considered accurate if it produced results in agreement with observations within the limit of observational accuracy, which was about $10'$ of arc, and this was the sort of accuracy that Copernicus strove for. He did not succeed, but the tables constructed from Copernicus' mathematical astronomy by Erasmus Reinhold – the *Prutenic* or *Prussian Tables* (1551) – were an improvement over the preceding *Alfonsine Tables* from the thirteenth century, and remained a standard source of information until superseded by Kepler's *Rudolphine Tables* in 1627. This improvement was due, not to his geometrical models being intrinsically more accurate than Ptolemy's (they are not)[8] but simply because he recomputed many of the parameters that must be entered into the models in order to construct the tables and, since many of these had changed since the time of Ptolemy, the final results were more accurate.

The *Commentariolus*

Copernicus developed his heliocentric theory while in Italy. The first exposition of his theory was a short sketch (*commentariolus*) which was written in about 1507, circulated in handwritten form to a few scholars, but never printed during Copernicus' life. The theory presented in this brief document differs in several essential features from the one that appeared eventually in *On the Revolutions*, and thus represents a preliminary stage in Copernicus' development of a heliocentric theory. No mention of the *Commentariolus* is made in *On the Revolutions*, and no manuscripts have been found among his own books and papers, so it is possible that, by the time he wrote *On the Revolutions*, he did not wish to be associated with his earlier work. All the existing manuscripts of the *Commentariolus* are thought to be descended from a copy received in 1575 by Tycho Brahe.[9] Copernicus clearly laid out the assumptions on which he had based his work:

---

[8] See, for example, Babb (1977).

[9] The passage of this work from Copernicus to Tycho, via a number of other astronomers, has been traced by Dobrzycki and Szczucki (1989). The title appearing on existing manuscripts translates as *Nicholas Copernicus, Sketch of his Hypotheses for the Heavenly Motions*. The authenticity of this title has been doubted widely on the basis that Copernicus would not have

1. There is no one center of all the celestial circles or spheres.

2. The center of the earth is not the center of the universe, but only of gravity and of the lunar sphere.

3. All the spheres revolve about the sun as their mid-point, and therefore the sun is the center of the universe.

4. The ratio of the earth's distance from the sun to the height of the firmament is so much smaller than the ratio of the earth's radius to its distance from the sun that the distance from the earth to the sun is imperceptible in comparison with the height of the firmament.

5. Whatever motion appears in the firmament arises not from any motion of the firmament, but from the earth's motion. The earth together with its circumjacent elements performs a complete rotation on its fixed poles in a daily motion, while the firmament and highest heaven abide unchanged.

6. What appear to us as motions of the sun arise not from its motion but from the motion of the earth and our sphere, with which we revolve about the sun like any other planet. The earth has, then, more than one motion.

7. The apparent retrograde and direct motion of the planets arises not from their motion but from the earth's. The motion of the earth alone, therefore, suffices to explain so many apparent inequalities in the heavens.

His programme was to represent the motions of the heavenly bodies using uniform circular motions. This he considered to be the true goal of astronomy, and he noted that all previous attempts had been unsatisfactory in some way. The concentric spheres of Eudoxus failed accurately to predict the phenomena and, while Ptolemy had produced an accurate theory, it violated the principle of uniform circular motion through the introduction of the equant. The system of eccentrics and epicycles that Copernicus created was designed to rectify these problems. Of course, as Kepler later came to realize, it was precisely because the equant mechanism violated the principle of uniform circular motion that it was a step in the right direction.

The first three assumptions, in which the heliocentric nature of the system is stated clearly, put Copernicus in immediate conflict with the universally held Aristotelian view that the Earth was the centre of rotation for all the heavenly bodies, and assumption 4 accounts for the fact that no annual parallax had been observed for the fixed stars. The 'circumjacent elements' referred to in assumption 5 are the atmosphere and the waters that lie on the surface of the Earth, and the final two assumptions describe how the complex motion of the

presented his system as a mere hypothesis. However, a thorough investigation of Copernicus' use of the term 'hypothesis' throughout his astronomical writings led Rosen (1959) strongly to criticize this view. The work was first published in Warsaw in 1854 (Koyré (1973)) but it took some time before a proper understanding of its place in the development of Copernicus' ideas was reached. For example, in Dreyer's *History of Astronomy* (Dreyer 1953) written in 1906, the *Commentariolus* is treated as a summary of the heliocentric system written after *On the Revolutions*. Quotations from the *Commentariolus* are taken from the translation in Rosen (1959).

Sun and planets are just manifestations of the motion of the Earth. No detailed explanations were given in the *Commentariolus*, but it is clear from the text that when Copernicus wrote it he had already planned *On the Revolutions*. The *Commentariolus* is in many ways a rough outline of Copernicus' mathematical theory, and when it was written he may well have believed that it would be a simple matter to add the mathematical demonstrations to his sketched-out theory. As it turned out, however, *On the Revolutions* became a far more extensive undertaking than envisaged originally.

The *Commentariolus* had two major shortcomings. First, Copernicus did not attempt to provide proof for the fundamental assumptions on which the work was based, preferring simply to state them as axioms. He must soon have realized that this would not be sufficient for his audience and that he would have to provide supporting evidence. In the end, the evidence that he did come up with in *On the Revolutions* was hardly convincing, something of which he was all too painfully aware. Second, the numerical parameters on which the models in the *Commentariolus* were based were largely those of the *Alfonsine Tables*, and these were simply not accurate enough for Copernicus' purposes. Thus, he realized that new parameters would have to be derived from observations.

The actual models that Copernicus used in the *Commentariolus*, which were similar but not identical to those appearing later in *On the Revolutions*, will not be discussed here, but one aspect is worthy of note. The theories of the Moon and planets are pretty much identical to those of Ibn al-Shāṭir, though of course Ibn al-Shāṭir's planetary theories were set in the context of a geocentric universe and the numerical parameters are different. Is this a coincidence? From the number of similarities, it seems clear that Copernicus must have come across the work of the Marāgha observatory somehow, and there is evidence that this Islamic astronomy was known in Italy in the late fifteenth and early sixteenth centuries. Also, as we shall see below, Copernicus made extensive use of geometrical devices equivalent to the Ṭūsī couple. Thus, it appears likely that Copernicus came across these geometrical theories while on his travels.[10]

---

[10] According to Swerdlow and Neugebauer ((1984)) 'the question therefore is not whether, but when, where, and in what form he learned of the Marāgha theory' (see also Abbud (1962)). The problem of how Copernicus acquired the techniques he used is discussed by di Bono (1995), who suggests some alternative sources. For example, Veselovsky (1973) thought it rather more likely that Copernicus generated his equivalent of the Ṭūsī couple from remarks made by Proclus in his *Commentary on the First Book of Euclid*. Rosińska (1974) discusses how Islamic geometrical models could have found their way to Cracow in the fifteenth century, and Kren (1971) suggests that the Ṭūsī couple was described by Oresme (albeit badly) in his *Questiones de spera*, a series of questions concerning Sacrobosco's *On the Sphere* written some time before 1362.

After describing the motions of the Earth, Moon, and planets, Copernicus concluded proudly:

> Altogether, therefore, thirty-four circles suffice to explain the entire structure of the universe and the entire ballet of the planets.

Subsequently, however, the examination of more observational data led Copernicus to realize that the theory presented in the *Commentariolus* was insufficient to predict the positions of the heavenly bodies accurately over long periods, and he had to discard many of the geometrical constructions and replace them with more complicated devices.[11]

On the Revolutions

Copernicus' move to Frombork in 1510 led to a substantial increase in his administrative duties as a canon of the Chapter of Warmia, and he was also a respected physician, medicine being a subject he had studied in Italy. It was against this background that he began to write his extensive exposition of heliocentric astronomy that was to become *On the Revolutions*. He needed considerably more observational data, which he proceeded to collect over the years between 1512 and 1529. For example, a series of observations of the Sun performed by Copernicus in 1515 and 1516 led him to the conclusion that the eccentricity of the Sun's orbit was not constant as he had previously assumed.

Much of *On the Revolutions* essentially was complete by 1530, but Copernicus delayed publication. Part of the reason was his inability to find any direct evidence for the Earth's motion, and he realized that without this he was going to find it very difficult to convince the sceptics. In the *Letter of Dedication* to Pope Paul III he wrote:

> ... the scorn which I had reason to fear on account of the novelty and unconventionality of my opinion, almost induced me to abandon completely the work which I had undertaken.

Despite Copernicus' reticence, news of his work spread and aroused a great deal of interest. In 1539, Georg Rheticus, a young German professor, travelled to Frombork to find out more. He brought with him a number of recently published books including the 1538 Greek edition of the *Almagest* which was much more

---

[11] It is not easy to state precisely how many circular motions were required in Ptolemy's system or Copernicus' final theory. A summary of different opinions can be found in Cohen (1985), p. 119. Contrary to what is often alleged, the system that was described in *On the Revolutions* actually is marginally more complex than that in the *Almagest*.

accurate than the Latin version Copernicus had been using, and Regiomontanus' *On Triangles*.

Rheticus became an enthusiastic supporter of Copernicus' theory and he wrote an extensive summary, published in 1540 in the form of a letter to Johannes Schöner – an astronomer and publisher from Nuremberg – and which now is known usually by the abbreviated title *Narratio prima* or *First Report*.[12] The publication of this concise well-written summary of Copernicus' theory led to a surge in interest, and may well have been at least partly responsible for inducing Copernicus to publish his work. The *First Report* was a very useful introduction to Copernican astronomy. In it, Rheticus stressed that Copernicus was attempting to reproduce celestial motions in accordance with the Pythagorean principle of uniform circular motions, and that he had 'liberated' astronomy from the equant. The work, which also included biographical information about Copernicus, was nearly always included in published editions of *On the Revolutions*, and was later published with Kepler's *Secret of the Universe* to help the reader understand Copernican astronomy.

Rheticus left Frombork in 1541 with a manuscript of *On the Revolutions* and set about arranging for its publication. The fact that Copernicus – who was worried about being misunderstood and ridiculed for his views – handed over his manuscript to Rheticus shows that a very high degree of trust had developed between them. In October 1542, when he took up his professorship at Leipzig, Rheticus entrusted the responsibility for the printing of Copernicus' treatise to Andreas Osiander, a Lutheran theologian closely associated with the distinguished Nuremberg printer Johann Petreius. Osiander had previously in 1541 written to Copernicus suggesting that in order to 'mollify the peripatetics and theologians whose opposition you fear'[13] some words be added to the effect that the heliocentric theory was merely a hypothesis which, even if wrong, accurately reproduced the phenomena. Copernicus, of course, did not agree, since it had always been his aim to demonstrate the true structure of the planetary system, and as a computational scheme Ptolemy's astronomy, with parameters updated to take account of more recent observations, was perfectly adequate. However, Osiander took it upon himself to add an unsigned preface entitled *To the Reader Concerning the Hypothesis of this Work* in which he remarked:

> For these hypotheses need not be true nor even probable. On the contrary, if they provide a calculus consistent with the observations, that alone is enough.... For

---

[12] Translated in Rosen (1959).
[13] Quoted from Rosen's commentary on the front matter of *On the Revolutions* in Copernicus (1992).

this art, it is quite clear, is completely and absolutely ignorant of the causes of the apparent nonuniform motions. And if any causes are devised by the imagination, as indeed very many are, they are not put forward to convince anyone that they are true, but merely to provide a reliable basis for computation. . . . So far as hypotheses are concerned, let no one expect anything certain from astronomy, which cannot furnish it, lest he accept as the truth ideas conceived for another purpose, and depart from this study a greater fool than when he entered it.

It is not known whether Copernicus ever saw this unwanted addition to the text, but Rheticus was furious and tried to force the publication of a corrected edition. This action failed, however.[14] It was probably Osiander who was also behind the insertion on the title page of the warning: 'Let no one untrained in geometry enter here', a phrase reputed to have been inscribed above the entrance to Plato's Academy. The printing of *On the Revolutions* was completed in 1543 shortly before the author's death in May of that year.

Probably about 4–500 copies were printed originally, of which the whereabouts of as many as 250 are known today! Rheticus' copy is one of the known ones and contains a poem (in Greek) on the flyleaf, hand written by its author Joachim Camerarius from Leipzig. The poem describes the following conversation:[15]

STRANGER: What is this book?
PHILOSOPHER: A new one, with all kinds of good things in it.
STRANGER: O Zeus! How great a wonder do I see!
The earth whirls everywhere in aethereal space.
PHILOSOPHER: But, do not merely wonder, nor condemn a good thing
as the ignorant do before they understand,
but examine and ponder all these things.

Throughout *On the Revolutions*, Copernicus uses Ptolemy's, often difficult, procedures to turn observational data into the numerical parameters required in his geometrical models, and in this he was helped greatly by Regiomontanus' *Epitome of Ptolemy's Almagest*. Copernicus was a competent mathematician, but not a great one, and what Copernicus did not do, and what sets Ptolemy's achievements apart from those of Copernicus, was to deduce models from the observations. The theoretical models Copernicus used were, in

---

[14] The fact that Osiander, rather than Copernicus, was the author of the preface generally was not appreciated at the time. Kepler found out from a friend in Nuremberg and printed the information in 1609, though the authorship has been attributed wrongly to Copernicus on numerous occasions since then. Osiander's argument was not devised in response to Copernicus' revolutionary idea. The same beliefs concerning the nature of astronomical hypotheses had been held widely by ancient Greek philosophers and preserved through the works of, for example, Maimonides and St Thomas Aquinas (see Duhem (1969)).
[15] See Gingerich (1992), Chapter 10.

essence, the same as Ptolemy's and, instead of devising new ones, he concentrated on producing a new physical interpretation with the Sun at the centre of the Universe, and on correcting the underlying geometrical parameters.[16] Had Ptolemy been rather more forthcoming in explaining how he came up with his models, Copernicus might have been persuaded to spend time looking for new ways of fitting mathematical schemes to the known phenomena, but as it was, it was left to Kepler to discover, from a careful analysis of observational data, that a new mathematical model was required.

Copernicus did not make any original contributions to trigonometry, but the trigonometric theorems that he presented in Book I of *On the Revolutions* were well explained and quite wide-ranging, and Rheticus published them separately in 1542. Copernicus' trigonometry is based entirely on chords and sines, though he refers to the latter as half-chords subtending double arcs, and he notes that 'in demonstrations and calculations half-lines are used more frequently than whole lines'.[17] Ptolemy's table of chords in the *Almagest* was based on a circle of radius 60 and the use of sexagesimal fractions, but by the sixteenth century the Hindu-Arabic numerals had been introduced into Europe and Copernicus used these since 'this numerical notation certainly surpasses every other, whether Greek or Latin, in lending itself to computations with exceptional speed'.[18] However, decimal fractions did not come into common use until they were popularized by the Dutch mathematician and engineer Simon Stevin[19] and in *On the Revolutions* Copernicus constructed a table of sines in intervals of $10'$ (following closely the method used by Ptolemy) based on a circle of radius $100\,000$. By taking such a large radius, Copernicus could work entirely with integers in his table and still 'exclude any obvious error'.[20]

---

[16] The claim has been made that, apart from the philosophical section on the motion of the Earth, *On the Revolutions* is little more than a reshuffled version of the *Almagest* (see de Solla Price (1962)).

[17] COPERNICUS *On the Revolutions*, I, 12. [18] COPERNICUS *On the Revolutions*, I, 12.

[19] Stevin was not the only advocate of decimal fractions around this time, notable also was Viète. However, it was Stevin's little book *The Tenth*, published in both Dutch and French in 1585, that brought about the widespread use of the system. Stevin's notation was fairly clumsy; our modern notation is due to Napier in 1616 (see, for example, Boyer (1989), p. 354). Interestingly, Stevin was one of the few early advocates of Copernicus' heliocentric theory.

[20] COPERNICUS *On the Revolutions*, I, 12. The table of sines that appears in the short work published by Rheticus in 1542 is not the same as that which appears in *On the Revolutions*. It is constructed in intervals of $1'$ and based on a radius of 10^7. Rheticus' work on trigonometry did not stop here, though he died before he had completed his major work, a collection of tables for all six trigonometric functions built around a table of sines based on a circle of radius 10^{15} and constructed in intervals of $45''$. This work, the most important contribution to trigonometry written in the sixteenth century, was completed by Rheticus' pupil L. Valentine Otho and published posthumously in 1596 under the title *Opus Palatinum de triangulis* (the title acknowledged the financial help of the Count Palatine, Frederick IV).

The motion of the Earth

On the Revolutions begins with some well-established arguments for the sphericity of the Universe, the heavenly bodies, and the Earth, and then Copernicus reiterates one of the fundamental tenets of theoretical astronomy as created by the ancient Greeks, i.e. that 'the motion of heavenly bodies is uniform, eternal, and circular, or compounded of circular motions'.[21] His only justification for this view is the Aristotelian statement that the motion appropriate to a sphere is rotation in a circle. But, Copernicus went on to say, if the Earth is spherical should we not consider the possibility that it, too, possesses circular motion. After explaining the ancient arguments for a stationary Earth, he refuted them, largely on the ground that if the Earth does not move, the heavens, due to their immense size, would have to rotate at an implausibly fast rate.

> Why then do we still hesitate to grant [the earth] the motion appropriate by nature of its form rather than to attribute a movement to the entire universe, whose limit is unknown and unknowable.[22]

Copernicus also dismissed Ptolemy's objections to a rotating Earth, which are largely based on the effect such a rotation would have on atmospheric phenomena. His refutation involved a considerable reworking of Aristotelian physics: discarding those features which do not fit well with a rotating Earth, while retaining those he could use to support his argument. Copernicus concluded that it is more likely that the Earth moves than that it is at rest.

The motion of the Earth was a necessary consequence of Copernicus' mathematical astronomy, but his arguments justifying it are no more convincing than Aristotle's reasons for a stationary Earth. It is not surprising that the vast majority of people, who did not have the technical expertise to delve further into Copernicus' mathematical world, remained unconvinced.[23]

After explaining why it is sensible to consider a moving Earth, Copernicus then discussed the order of the heavenly spheres. There had been significant differences of opinion over the two preceding millennia as to the positions of the spheres of Venus and Mercury. For example, Plato put them both above the Sun, Ptolemy put them both below the Sun, and al-Biṭrūjī placed Venus above and

---

[21] COPERNICUS *On the Revolutions*, I, 4.
[22] COPERNICUS *On the Revolutions*, I, 8. To a believer in Aristotle's universe, which was not infinite but of finite size, the motion of the heavens did not cause a problem because that was their natural state, whereas the Earth, being heavy, would need a great force to set it in motion.
[23] It is very difficult to put oneself in the position of a geocentric astronomer faced with trying to decide between the Copernican idea and their own entrenched beliefs. Whatever the merits of a new approach, there is a cognitive barrier that needs to be negotiated before one can accept it. This and related ideas are explored in Margolis (2002).

Mercury below. Copernicus also mentioned Martianus Capella (see p. 109) who asserted that Mercury and Venus orbited the Sun rather than the Earth. These disagreements arose because the planets traditionally were ordered according to their zodiacal periods, and the average periods of revolution of Mercury, Venus, and the Sun around the zodiac are all the same, i.e. 1 year. Copernicus criticized Ptolemy's reasoning as follows:

> Ptolemy argues also that the sun must move in the middle between the planets which show every elongation from it and those which do not. This argument carries no conviction because its error is revealed by the fact that the moon too shows every elongation from the sun,[24]

and concluded that:

> Either the earth is not the centre to which the order of the planets and spheres is referred, or there really is no principle of arrangement nor any apparent reason why the highest place belongs to Saturn rather than Jupiter or any other planet.[25]

He was then in a position to present his schematic view of the Universe, with the planets ordered according to their periods of revolution about the Sun, and to proclaim that:

> In this arrangement, therefore, we discover a marvelous symmetry of the universe, and an established harmonious linkage between the motion of the spheres and their size, such as can be found in no other way. For this permits a not inattentive student to perceive why the forward and backward arcs appear greater in Jupiter than in Saturn and smaller than in Mars, and on the other hand greater in Venus than in Mercury. This reversal in direction appears more frequently in Saturn than in Jupiter, and also more rarely in Mars and Venus than in Mercury. Moreover, when Saturn, Jupiter and Mars rise at sunset, they are nearer to the earth than when they set in the evening or appear at a later hour. But Mars in particular, when it shines all night, seems to equal Jupiter in size, being distinguished only by its reddish colour. Yet in the other configurations it is found barely among the stars of the second magnitude, being recognized by those who track it with assiduous observations. All these phenomena proceed from the same cause, which is the earth's motion.[26]

By considering the Earth as a planet, rotating about its axis once a day while orbiting the Sun once a year, Copernicus could explain, at least qualitatively, many of the obvious irregularities in the motions of the other planets. He did not bother to elucidate the above remarks with further comment; this had to wait until the detailed quantitative explanation that appears later. The non-mathematical reader had to take Copernicus' conclusions on trust.

---

[24] COPERNICUS *On the Revolutions*, I, 10.   [25] COPERNICUS *On the Revolutions*, I, 10.
[26] COPERNICUS *On the Revolutions*, I, 10.

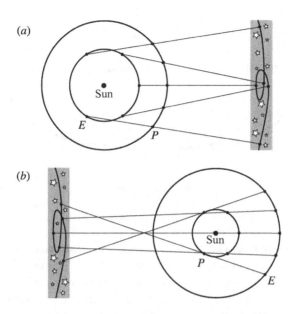

Fig. 5.1. Retrograde motion for (*a*) superior and (*b*) inferior planets in a heliocentric universe.

Figure 5.1(*a*) demonstrates schematically how the annual motion of the Earth leads to retrograde arcs in the path of a superior planet. As the Earth moves around the Sun (for simplicity we assume that its orbit is a circle) beginning from the point E, the planet moves around its orbit beginning from the point P. The observed position of the planet against the backdrop of the stars at a given time is determined by the direction of the line joining the Earth and the planet and, as the diagram illustrates this can lead to a retrograde loop typical of planetary motions. The diagram also shows why periods of retrograde motion always occur when the planet is in opposition to the Sun and why it is at these positions that the planets, being at their closest approach to the Earth, appear brightest. Figure 5.1(*b*) illustrates retrograde motion for inferior planets in a heliocentric universe, and shows why this occurs when the planet lies directly between the Earth and the Sun (at inferior conjunction).

The large irregularities in the motions of the planets are not observed in the fixed stars and so, Copernicus concluded, they must be a vast distance away. If the Earth is orbiting the Sun then, as Archimedes had made clear, one would expect to detect stellar parallax, but none had been observed in Archimedes' time and this had not changed by the mid sixteenth century when astronomers could make observations accurate to within $10-15'$ of arc. If we take the radius of the Earth's orbit to be 1200 Earth radii (Ptolemy's Earth–Sun distance) and

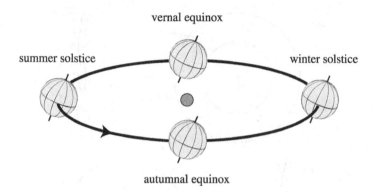

Fig. 5.2. The tilt of the orbit of the Earth.

insist that any stellar parallax is less than 10′ of arc, we are led to the conclusion that the fixed stars must be of the order of 1 000 000 Earth radii away. To Copernicus, this was a necessary consequence of his planetary theory, but to many others, brought up to believe that the stars were just outside the sphere of Saturn about 20 000 Earth radii away, this was yet another reason not to take Copernicus seriously. This situation only got worse in the years following the publication of *On the Revolutions*. By the end of the sixteenth century, Tycho Brahe was making observations accurate to within 4′ of arc and still did not observe a stellar parallax. In fact, the distance to the nearest star is over 6×10^9 Earth radii and the associated parallax is less than 1 arc-second.[27]

We now turn to the precise nature of the motion of the Earth which, according to Copernicus, is threefold. The first motion, the daily rotation of the Earth about its axis, is straightforward and accounts for the perceived rotation of the heavens once each day. Now the Sun, as viewed from the Earth, also possesses an annual motion around the ecliptic that is inclined at an angle ε to the celestial equator. Copernicus showed that the same effect results if the Earth orbits the Sun once per year with its equatorial plane making the angle ε with the plane of its orbit. This is illustrated schematically in Figure 5.2.

However, Copernicus viewed this annual motion as being made up from two rotations, due to his belief that the Earth was carried around the Sun on one of his celestial spheres. If the annual motion of the Earth around the Sun were simply a single rotation then, Copernicus argued, the axis of the Earth would necessarily take part in that rotation. But this is not what happens – the direction

[27] Stellar parallax was first measured, long after the invention of the telescope, in the 1830s (see note 7, p. 357). The first direct experimental confirmation of the motion of the Earth actually came from a quite different source – the discovery of the aberration of starlight in 1728 (see p. 307).

of the axis of the Earth remains constant with respect to the fixed stars – and so Copernicus introduced a third rotation about a line perpendicular to the orbital plane through the centre of the Earth that he called the 'motion of the inclination' and which was equal and opposite to the second, so that the axis of the Earth does not rotate. In fact, the rotation rates in Copernicus' scheme were only approximately equal, the rotation of the axis of the Earth being slightly greater than its annual rotation rate so as to account for the precession of the equinoxes. Since the axis of the Earth is, in effect, wobbling, Copernicus advocated the use of the fixed stars, rather than the equinoxes, as the correct frame of reference for the description of astronomical phenomena.

Nowadays, we are not constrained to think of Copernicus' second rotation as if the Earth were attached rigidly to a revolving sphere, and so it is more natural to think of the motion of the Earth, at least as far as we have discussed it so far, as being made up of the daily rotation about its axis, its annual motion around the Sun, and the slow rotation of its axis about the poles of the ecliptic once every 26 000 years or so.

Because of the discrepancy between Ptolemy's value for precession ($1°$ in 100 years) and those of others ($1°$ in 66 years according to al-Battānī and $1°$ in 71 years according to Copernicus), and the discrepancy between the values of the obliquity used by ancient and Islamic astronomers ($23° 51' 20''$ according to Ptolemy and $23° 35'$ according to al-Battānī), Copernicus believed that these quantities were variable:

> When [Hipparchus] was scrutinizing the length of the year more intensely, he found that as measured with reference to the fixed stars, it was longer than when measured with reference to the equinoxes or solstices. Hence he thought that the stars too had a motion in the order of the zodiacal signs, but a very slow motion which could not be perceived immediately. Now however, with the passage of time it has become absolutely clear. . . . Moreover the motion is found to be nonuniform. . . . Besides, another marvel of nature supervened; the obliquity of the ecliptic does not appear as great to us as it did before Ptolemy. . . . Now the measurement of this motion and the explanation of its variation were not known to earlier [astronomers]. The reason is that the period of its revolution is still undiscovered on account of its unforeseeable slowness. For in so many centuries since it was discovered by mortal man, it has completed barely 1/15 of a circle. Nevertheless, so far as I can, I shall clarify this matter by means of what I have learned about it from the history of the observations down to our time.[28]

Following the long-standing tradition begun by Thābit ibn Qurra, but with virtually no evidence to support the view, Copernicus believed that these two

---

[28] COPERNICUS *On the Revolutions*, III, 1.

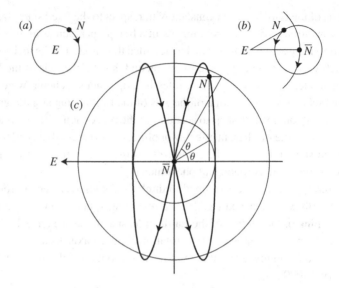

Fig. 5.3. Copernicus' theory of precession.

functions were oscillatory functions of time and, because the value for precession appeared to have a minimum around the time of al-Battānī but no such
minimum was observed in the obliquity, Copernicus chose to make the period
of the obliquity exactly twice the period of variation in precession, the latter
being 1717 years. Copernicus measured the obliquity as 23° 28′ 24″ and, even
though the data did not support his argument, he claimed that the maximum
and minimum values of ε over its 3434 year cycle were 23° 52′ and 23° 28′,
respectively. All the tables in *On the Revolutions* that are dependent on the value
of ε are computed for $\varepsilon = 23° 28′$ with a correction column provided to enable
the user to adjust for larger values of ε.

The actual mechanism that Copernicus used to model the variations in precession and obliquity is shown in Figure 5.3. A uniform rate of precession
corresponds to a steady revolution of the north pole N around the pole of the
ecliptic E as shown in Figure 5.3(a), and in this model the obliquity of the eclip-
tic, being determined by the distance $|EN|$, is constant. The various theories of
trepidation that were proposed by Islamic astronomers used, in effect, a polar
epicycle with the mean north pole $\bar{N}$ rotating around E and the actual north pole
rotating about a small circle centred at $\bar{N}$, as shown in Figure 5.3(b), and in this
case the periods of variation of precession (determined by the angle $NE\bar{N}$) and
of the obliquity (determined by the distance $|EN|$) are the same. But Copernicus
wanted a mechanism by which the period of the distance $|EN|$ was twice that
of $\angle NE\bar{N}$, and the one he came up with is shown in Figure 5.3(c). Instead of

moving around a circle centred on $\bar{N}$, the north pole follows a figure-of-eight-type path built up from two perpendicular oscillations: the motion parallel to $E\bar{N}$ is controlled by the angle θ and the inner circle, whereas the motion perpendicular to this line is governed by the angle 2θ and the outer circle.

In fact, things are rather more complicated than this because these figures should be drawn on a sphere rather than on a flat surface, but Copernicus argued that the motions are small enough for the approximation not to lead to noticeable errors. Copernicus used the equivalent of the Ṭūsī couple to produce the two rectilinear motions required, so that everything was built up from uniform circular motions. Interestingly, Naṣīr al-Dīn al-Ṭūsī had suggested earlier that his mechanism might be used for just such a purpose:

> The same method may also be used for trepidation and for the movement of the obliquity in latitude for the ecliptic orb if the fact of these two motions and their variability is ascertained.[29]

The period of the mean precession (i.e. the time it takes for the mean north pole to complete 1 revolution about the pole of the ecliptic) was calculated by Copernicus as 25 816 Egyptian years (an Egyptian year being 365 days) which is very close to the modern value. The values computed from Copernicus' elaborate theory give excellent agreement with the observational data that he was attempting to reproduce.

Finally, Copernicus provided his theory for the shape of the orbit of the Earth. This can be thought of as an extension of the solar theory of Hipparchus which, if transferred to a heliocentric viewpoint, consists simply of the Earth moving round the Sun on an eccentric circle. Copernicus believed that the orbit of the Earth was subject to two further inequalities. The first of these concerned the eccentricity of the orbit of the Earth, which from historical measurements he thought was variable, and he made the highly questionable assumption that it oscillated with the same period as the obliquity. Second, there was the motion of the aphelion of the Earth. As we have already seen (see p. 97) al-Zarqālī discovered that the apogee of the Sun had a slow steady motion with respect to the fixed stars, but Copernicus misunderstood al-Zarqālī's theory and, instead, thought that the apogee moved non-uniformly.[30] In order to model this non-uniform motion of the apsidal line and the variable eccentricity, Copernicus modified Hipparchus' scheme as shown in Figure 5.4.

---

[29] Naṣīr al-Dīn *Memoir on Astronomy*. Translation from Ragep (1993). It seems plausible that Copernicus got the idea of using the couple in this way from the astronomers of the Marāgha observatory (see note 10, p. 124).
[30] Copernicus knew of al-Zarqālī's work from the account in the *Epitome* of Peurbach and Regiomontanus. Unfortunately, this was rather muddled (Toomer (1969)).

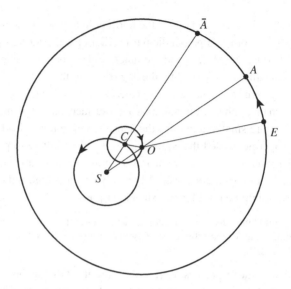

Fig. 5.4. Copernicus' theory for the orbit of the Earth.

The Earth E orbits uniformly about O, completing 1 revolution in 1 sidereal year, which Copernicus computes to be 365 days 6 h 9 min 40 s. The centre of the orbit of the Earth is no longer fixed but rotates around C with $|CO| = r_2$ say, completing 1 revolution in 3434 years, and C rotates around the Sun S with $|SC| = r_1$, completing 1 revolution in just over 53 000 years. The motion around the Sun is linked to the precession/obliquity mechanism shown in Figure 5.3 by assuming that SCO is a straight line when $\theta = 0$ and that this was the case in 65 BC. Copernicus took the radius of the Earth's orbit $|OE|$ to be 10 000 units and determined from the observed values of the eccentricity at various times that $r_1 = 369$, $r_2 = 48$. Thus, the eccentricity varies between $0.0369 + 0.0048 = 0.0417 \approx 1/24$ and $0.0369 - 0.0048 = 0.0321$. The aphelion A oscillates about the mean aphelion $\bar{A}$ with a period of 3434 years, while $\bar{A}$ takes 53 000 years to complete 1 revolution of the Sun.

In the *First Report*, Rheticus added his own astrological interpretation to the motion of the centre of the orbit of the Earth:

> I shall add a prediction . . . . Thus, when the eccentricity of the sun was at its
> maximum [i.e. when SCO was a straight line], the Roman government became a
> monarchy; as the eccentricity decreased, Rome too declined, as though aging, and
> then fell. When the eccentricity reached the boundary and quadrant of mean value
> [i.e. when SCO was a right angle], the Mohammedan faith was established;
> another great empire came into being and increased very rapidly, like the change in
> eccentricity. A hundred years hence, when the eccentricity will be at its minimum,
> this empire will complete its period. In our time it is at its pinnacle from which

equally swiftly, God willing, it will fall with a mighty crash. We look forward to the coming of our Lord Jesus Christ when the centre of the eccentricity reaches the other boundary of mean value, for it was in that position at the creation of the world. . . . Thus it appears that this small circle is in very truth the Wheel of Fortune, by whose turning the kingdoms of the world have their beginnings and vicissitudes.[31]

In describing the model illustrated in Figure 5.4 we have assumed tacitly that the Sun is at rest with the points C and O moving around small circles. However, it is possible to achieve precisely the same geometrical relationship between S and E by assuming that the centre of the orbit of the Earth O is motionless with S and C revolving around it on small circles. Copernicus realized that there was no way of saying which of O and S actually was at rest at the centre of the Universe, but it is quite clear that he believed this honour belonged to the Sun.

One of Copernicus' reasons for assuming a moving Earth is greatly to simplify the theories of the planets. But that would be of little use if it were not possible to devise a theory of the motion of the Earth that represented accurately the observed phenomena. This Copernicus achieved with a considerable degree of success, but the resulting scheme could hardly make any claim to simplicity. The theory is also full of holes. The values of many of the numerical parameters simply are assumed by Copernicus, particularly when the necessary mathematics would otherwise have been beyond him, and there are numerous arithmetical and historical inaccuracies.

The motion of the Moon

The lunar theory is much more straightforward than that for the Earth, since the observed motion of the Moon does not possess the same gross irregularities as the orbits of the planets. Copernicus agreed with the ancients that its motion takes place round the Earth; in fact, he went further. Since the Moon was now the only heavenly body orbiting the Earth, Copernicus attributed a certain similarity, or kinship, to the two bodies. This represents a sharp contrast with Aristotelian cosmology in which all celestial bodies were intrinsically different from the Earth. Copernicus objected to Ptolemy's theory of the Moon on precisely the same grounds as Ibn al-Shāṭir had done 200 years before him, i.e. that it violated the principle of uniform circular motion and it predicted a manifestly incorrect change in the apparent diameter of the Moon during the course of a month.

The model that Copernicus used for the longitude of the Moon is, in fact, identical to that of Ibn al-Shāṭir and the same as the one he had used previously

---

[31] RHETICUS *First Report*. Quoted from Rosen (1959).

in the *Commentariolus*, and the latitude theory is Ptolemy's. In his earlier
work, Copernicus simply had adapted numerical parameters from the *Alfonsine
Tables*, but in Book IV of *On the Revolutions* he set about deriving a new set
of parameters for the lunar theory. He also discussed the parallax of the Moon,
the distances and sizes of the Sun and Moon, and his theory of eclipses.

Whereas Ibn al-Shāṭir had used $r_1 = 6; 35$ and $r_2 = 1; 25$ for the radii of
the two lunar epicycles, based on a deferent radius of 60, Copernicus computes
$r_1 = 1097$ and $r_2 = 237$ based on a deferent radius of 10 000. These parameters
are almost identical, which is not surprising, since both men were attempting
to reproduce essentially the same phenomena and the lunar parameters are
not subject to the same long-term variations as those of the Sun. Copernicus
worked out the greatest and least distances of the Moon as $68\frac{21}{60}$ and $52\frac{17}{60}$ Earth
radii, respectively (Ibn al-Shāṭir had 68; 0 and 52; 0), from which the Moon's
apparent diameter should vary between $37' 34''$ and $28' 45''$. This, as Copernicus
pointed out, is a significant improvement over Ptolemy's theory.

Copernicus' planetary theory

Unlike solar and lunar theories, planetary theory in the sixteenth century had
changed little since the time of Ptolemy. The motion of the Sun and Moon was
of practical importance for the construction of calendars and the determination
of religious festivals, and these theories were fairly easy to test through eclipse
observations. On the other hand, the only use for planetary theory was in the
casting of horoscopes, and here great accuracy was not particularly relevant. It
was also a very difficult matter to carry out the observations required to verify
such a theory. The only significant contribution since AD 200 was Ibn al-Shāṭir's
geometrical construction, which removed the equant from Ptolemy's theory.

What Copernicus realized – indeed, it was the *raison d'etre* of his heliocentric theory – was that the so-called second anomaly in the motion of all
the planets (i.e. retrograde motion) that Ptolemy modelled using an epicycle
for each planet, could be reduced to a single cause. Thus, with all the planets
(including the Earth) orbiting the Sun, the occurrence of retrograde motion
was simply a consequence of viewing the planet from a moving observatory
(see Figure 5.1). This also explained the fact that retrograde motion happened
always at opposition for the superior planets and at conjunction for Mercury
and Venus, whereas Ptolemy had had to build these features into his model.
Copernicus' theory has a further advantage, as we shall see, since actually it
determines the distances of all the planets from the Sun in terms of the distance
of the Earth, and so it is no longer necessary to invoke any extra metaphysical

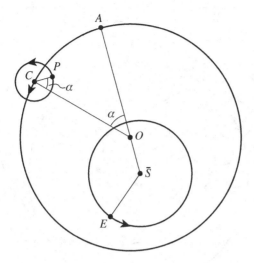

Fig. 5.5. Copernicus' theory for the superior planets.

principles (e.g. Ptolemy's set of touching spheres) to determine just where the planets are. However, Copernicus could not dispense with epicycles altogether, because his heliocentric structure did not explain the variation of the planets' speeds in their orbits – the so-called first anomaly.

In *On the Revolutions*, the planetary theories are modified versions of those used in the *Commentariolus*, based on the theories of Ibn al-Shāṭir. Copernicus began with the superior planets and the geometrical scheme he used is illustrated in Figure 5.5. The planet P rotates once per sidereal period around an epicycle centred at C. The centre of the epicycle rotates in the same sense and at the same rate around a deferent circle centred at O. This point is not the Sun, but is close to it. In fact, Copernicus did not use the actual Sun in the planetary theory, but rather the mean sun that is labelled $\bar{S}$ in Figure 5.5 and which corresponds to the point O in Figure 5.4. The Earth E moves in a circular orbit around $\bar{S}$. The motion is arranged so that $\angle PCO = \angle COA$, $\bar{S}OA$ being the apsidal line of the planet's orbit.

From a geometrical viewpoint, it is unimportant whether the Earth E or the Sun $\bar{S}$ is stationary, and Copernicus' model with the Earth stationary is shown in Figure 5.6 superimposed on his original scheme (shown by the dashed lines).[32] In Figure 5.5, the position of the planet with reference to the Earth is given by the vector $E\bar{S} + \bar{S}O + OC + CP$, whereas from the perspective of a stationary

---

[32] The comparison given here is based on that given in Neugebauer (1968).

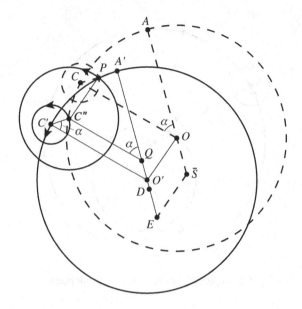

Fig. 5.6. Copernicus' theory for the superior planets from a geostationary point of view (solid lines) superimposed on elements of his heliocentric scheme (dashed lines).

Earth as shown in Figure 5.6, the planet is at $EO' + O'C' + C'C'' + C''P$, the two being identical, since $EO' = \bar{S}O$, $O'C' = OC$, $C'C'' = CP$ and $C''P = E\bar{S}$. So Copernicus' geometrical scheme is equivalent entirely to a geocentric model in which the planet moves on a double epicycle around an eccentric deferent or, since motion around an eccentric deferent can always be replaced by an epicycle, to a triple epicycle model like those of Ibn al-Shāṭir.

It is illuminating to compare Copernicus' theory with that of Ptolemy. The latter's success was based on his introduction of the equant and Copernicus wanted to achieve the effect of the equant without actually having one. If we construct the line $C''Q$, parallel to $C'O'$, intersecting the apsidal line at Q, elementary geometry implies that $O'Q$ is constant with $|O'Q| = |C'C''| = |CP|$, and $\angle C''QA' = \angle COA$ and, hence, increases uniformly with time. Thus the point Q plays the role of the equant in Copernicus' theory and if C'', which is the centre of the planet's epicycle, moved around O' in a circle, Copernicus' model would be identical geometrically to that of Ptolemy. It does not (though its path is almost circular given that $|C'C''|$ is small relative to $|O'C'|$) but if we were to try to match up with Ptolemy's model on the apsidal line, say, we would want to choose (remembering that Ptolemy has the centre of his deferent

Table 5.1. *The parameters in the models for the superior planets.*

	Ptolemy $|C''P|$	$|EQ|$	$|E\bar{S}|$	Copernicus $|\bar{S}O|$	$|CP|$	$|\bar{S}O|+|CP|$
Mars	6583	2000	6580	1460	500	1960
Jupiter	1917	917	1916	687	229	916
Saturn	1089	1139	1090	854	285	1139

D midway between E and Q):

$$R = |DC''| = \tfrac{1}{2}|EQ| - |O'Q| + R - |C'C''| = \tfrac{1}{2}|EQ| - 2|CP| + R,$$

where we have written R for the radius of both Ptolemy's and Copernicus' deferent, and:

$$|EQ| = |EO'| + |O'Q| = |\bar{S}O| + |CP|.$$

In order to satisfy both of these equations, we would choose $3|CP| = |\bar{S}O|$ and this is what Copernicus did, exactly in the case of Jupiter and Saturn and approximately in the case of Mars.[33]

It is evident from the parameters used by Copernicus and Ptolemy (given in Table 5.1 in terms of the length $|CO|$, which is fixed at 10 000) that, in a strictly geometrical sense, there is not much to choose between the two schemes. However, the Copernican scheme, as a scientific theory, has a great advantage over its Ptolemaic equivalent (though not one that was particularly influential in the sixteenth century). In the geostationary scheme shown in Figure 5.6, one must make the *ad hoc* assumption that $E\bar{S}$ and $C''P$ are parallel in order to link the planet's motion correctly to that of the Sun, and this must be done for all of the planets. On the other hand, in the heliocentric theory, all these assumptions are replaced by a single proposition – that the Earth orbits the Sun.

Another powerful way of making the same point is illustrated in Table 5.2. For each of the three superior planets, the table shows the zodiacal period T_c, the synodic period T_p, and the angular speeds to which they correspond, ω_c and ω_p, respectively. In each case, if we add ω_c and ω_p together, we get a rate of 1 revolution per year. Why? According to Ptolemy, $\omega_c + \omega_p$ is the rate at which the planet revolves around its epicycle relative to a fixed reference line (the planet rotates around the centre of its epicycle at the rate ω_p, but the centre of the epicycle is itself rotating at the rate ω_c). Since $\omega_c + \omega_p = 1$, Ptolemy makes

---

[33] The true orbit is an ellipse with eccentricity e, for example. A simple argument using complex numbers can be used to show that both Copernicus' and Ptolemy's theories give planetary positions accurate to first order in e (see, for example, Hoyle (1974)).

Table 5.2. *The periods of the superior planets.*

		Mars	Jupiter	Saturn
T_c	(yrs)	1.88	11.86	29.46
T_p	(yrs)	2.14	1.09	1.03
ω_c	(rev/yr)	0.53	0.08	0.03
ω_p	(rev/yr)	0.47	0.92	0.97
$\omega_c + \omega_p$	(rev/yr)	1	1	1

the line connecting the planet to the centre of its epicycle (CP in Figure 3.13, p. 77) remain parallel to the Earth–Sun line $E\bar{S}$ which by definition revolves at a rate of 1 revolution per year. However, had $\omega_c + \omega_p$ been different from unity, this would have caused no problem for Ptolemy – he could simply have made CP rotate relative to $E\bar{S}$ at the appropriate rate. Hence, for Ptolemy, the fact that $\omega_c + \omega_p$ is the same for all the superior planets is simply a coincidence – part of God's grand design. Not for Copernicus, though. In his theory, each superior planet rotates around the Sun at the rate ω_c and the Earth rotates around the Sun once each year. The angular speed associated with the synodic period of each planet (i.e. the rate at which it orbits the Earth) is then *necessarily* $1 - \omega_c$.

For the inferior planets, the simplification in terms of the number of *ad hoc* assumptions provided by the Copernican theory is more obvious, since it is the zodiacal period of those planets that is 1 year, and this is immediately implied by placing the Earth in orbit around the Sun between Venus and Mars.[34] In the heliocentric system, the orbital (sidereal) periods T_s of Venus and Mercury must be calculated from their observed synodic periods T_p. The synodic period of Venus is the time between successive inferior conjunctions, and so if the Sun, Venus and the Earth lie in a straight line at $t = 0$, they next do so (in the same order) at $t = T_p$. If we measure time in years, the Earth makes T_p orbits of the Sun in this time and so Venus must make $1 + T_p$ revolutions. It follows that the rate at which Venus orbits the Sun is $1/T_p + 1$ revolutions per year, and thus,

$$\frac{1}{T_s} = 1 + \frac{1}{T_p}.$$

For Venus, this gives a sidereal period of about 224.7 days, whereas for Mercury we get just under 88 days.

---

[34] Further discussion of the relative merits of Copernicus' and Ptolemy's theories, based on knowledge available in the sixteenth century, is given in Martin (1984).

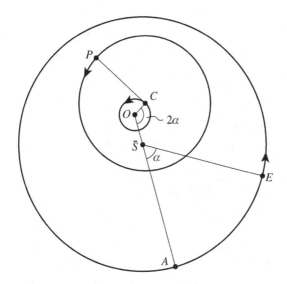

Fig. 5.7. Copernicus' theory for Venus.

Copernicus' model for the motion of Venus is shown in Figure 5.7 and is, in fact, very similar in structure to that for a superior planet. The planet P again moves around a circle centred at C once every sidereal period, the point C rotating around a circle centred at O at twice the rate that the Earth E orbits the sun $\bar{S}$. However, in this case the roles of the two circles are reversed, with the circle carrying the planet being the larger. Again, the Sun is displaced slightly from O. The point A in Figure 5.7 is the position of the Earth when it is at the greatest distance from the centre of the orbit of Venus. It is possible to perform a reduction to a geostationary model exactly as was done above for the superior planets, with the same conclusions. The circle centred at C and carrying the planet plays the role of Ptolemy's epicycle, and Copernicus obtained $|CP| = 7193$ (with $|E\bar{S}| = 10\,000$) where Ptolemy's model would imply a value of 7194. Ptolemy's value for the double eccentricity $|EQ|$ was 417, but Copernicus reduced this: 'Formerly it was all of 416 but now it is 350 as many observations show us.'[35]

Copernicus' theory for the motion of Mercury (Figure 5.8) follows closely the arrangement for Venus with two key changes. The first difference is that when $\alpha = \angle E\bar{S}A = 0$ in the model for Mercury, the point O lies between $\bar{S}$ and

---

[35] COPERNICUS *On the Revolutions*, V, 22. Had Copernicus distributed 416 in the same way as he distributed the eccentricity in the models for the superior planets he would have been led to $|OC| = 104$ and $|\bar{S}O| = 312$, but rather than reduce both these values in proportion so that their sum is 350, he simply reduced $|\bar{S}O|$ to 246.

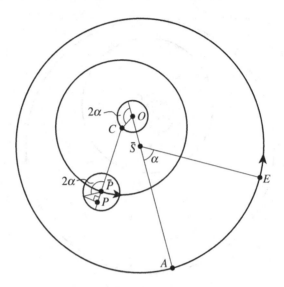

Fig. 5.8. Copernicus' theory for Mercury.

C, whereas in the case of Venus, C lies between $\bar{S}$ and O. This change reflects the fact that, in Ptolemy's model for Mercury, the equant Q lies between E and O rather than O lying between E and Q, as is the case for Venus. Second, in Copernicus' theory, the planet does not lie on the circle centred at C but rather oscillates backward and forward along the radius $C\bar{P}$, $\bar{P}$ rotating around C. This is Copernicus' method for reproducing Ptolemy's rotating deferent, and he pointed out that such a rectilinear motion can be produced from the sum of two uniform circular motions as he had done in his theory of precession, i.e. using the Ṭūsī couple. The dimensions of Copernicus' model are (with $|E\bar{S}| = 10\,000$) $|\bar{S}O| = 736$, $|OC| = 212$, $|C\bar{P}| = 3763$, and the radius of the circle used to produce the oscillation of P about $\bar{P}$ is 190. When we consider the technical detail involved here, it is hard to believe that Copernicus considered this as a model of physical reality.

The problems Copernicus faced in determining the appropriate numerical parameters for his models were immense. Thousands of tedious calculations were required – multiplications, root extractions, and interpolations from tables. It is not surprising that no one since Ptolemy had spent years of his or her life attempting to compute the elements of planetary orbits. For the superior planets, Copernicus could obtain fairly easily the observational data required to determine the necessary parameters, but for Mercury and Venus the situation was less satisfactory. In fact, for the inferior planets he relied pretty much on Ptolemy's observations.

The sixth and final book of *On the Revolutions* treats planetary latitudes. Unfortunately, as Kepler later noted, Copernicus' latitude theories were designed to reproduce the results of Ptolemy (which were not particularly accurate) rather than to represent nature. Part of the problem was that Copernicus' planetary theories were all linked to the mean Sun – the centre of the Earth's orbit – and not to the Sun itself. As a result, the variations in the inclinations of the orbital planes of the planets (which he termed obliquations) are linked in Copernicus' complicated theory to the longitudinal motion of the Earth. In Copernicus' defence it should be noted that he had little alternative but to base his latitude theory on Ptolemy's inaccurate data, since the relevant observations can be made only extremely rarely at certain configurations of each planet. Copernicus also was unfortunate in that he was not aware of the simplified (and improved) latitude theories that Ptolemy had introduced in the *Planetary Hypotheses*.

To find the solar distance, Copernicus used exactly the same method as had Ptolemy, and obtained virtually identical results. Prior to Copernicus, the main reason for computing the distance between the Earth and the Sun was to establish that the solar parallax was smaller than the accepted limits of observational error. In Copernicus' heliocentric universe, however, this distance is a fundamental unit in terms of which the mean distances of the planets from the Sun can be determined. In fact, in terms of predictive power, this is perhaps the most significant advantage of the Copernican system over its Ptolemaic predecessor. It is slightly surprising then that Copernicus did not choose to emphasize its significance, and he never used it to produce values for the distances of the other heavenly bodies.

The calculations can be done of course, based on the parameters derived by Copernicus for the planetary orbits, and the results of such a computation are shown in Table 5.3.[36] On the left are the values given by al-Battānī for the mean distances of the planets from the Earth in terms of the mean Earth–Sun distance, which are essentially the same as those given in Ptolemy's *Planetary Hypotheses* though the latter were unknown in the sixteenth century. In the second column we have the distances in Copernicus' theory, though these are now distances to the Sun rather than the Earth. What is immediately striking is that Copernicus' solar system is actually *smaller* than Ptolemy's universe. Of course, the distance to the stars is vastly different in the two systems. For al-Battānī the stars were just outside the sphere of Saturn at a distance of about

---

[36] The figures for al-Battānī and Copernicus are (rounded) decimal equivalents of values in Swerdlow and Neugebauer ((1984)). The orbits of Jupiter and Saturn are subject to large perturbations that result in significant variations in their orbital parameters over time. This accounts for the low accuracy of the modern values for these two planets.

Table 5.3. *Mean distances to the planets in terms of the Earth–Sun distance.*

	Al-Battānī (distance to Earth)	Copernicus (distance to Sun)	Modern
Mercury	0.1039	0.3764	0.3871
Venus	0.5578	0.7194	0.7233
Sun/Earth	1.0000	1.0000	1.0000
Mars	4.1372	1.5197	1.5237
Jupiter	9.45	5.22	5.20
Saturn	14.0	9.2	9.5–9.6

19 000 Earth radii, whereas, as mentioned previously, the lack of annual stellar parallax implied a distance of well over 1 million Earth radii to the stars in the Copernican universe. This enormous implied gap between Saturn and the stars containing no heavenly bodies at all with, as Tycho Brahe would later observe, no conceivable purpose, was yet another reason to be sceptical about the heliocentric theory. The modern values are shown in the third column, and it is clear that Copernicus' theory gave very good results for the relative dimensions of the planetary orbits.

The reception of Copernicus' theory

Since the work of Isaac Newton in the latter part of the seventeenth century, the theoretical calculation of the positions of heavenly bodies and the physical nature of the Universe have been linked inextricably through the theory of gravitation. But in the sixteenth century, most astronomers thought of these as entirely separate subjects, and the two parts of *On the Revolutions* – the first dealing with the physical reality of the Earth's motion and the second with geometrical methods for astronomical calculations – were received very differently.[37]

It is assumed often that there was an immediate conflict between Copernicanism and the Catholic Church. In fact, the educated Catholic clergy was not particularly interested in Copernicus' thesis, and it was accepted without comment by Pope Paul III, to whom *On the Revolutions* was dedicated. Since Osiander's unsigned preface asserted that the heliocentric system was just a convenient mathematical hypothesis, there was little cause for conflict.

---

[37] Detailed discussions of how the heliocentric idea fared around the world can be found in Dobrzycki (1973b).

Copernicus' great work was not entirely uncontroversial during the half century following its publication, though. Throughout the late Middle Ages, Christianity had coexisted fairly comfortably with scientific endeavour but, whereas the Catholic Church claimed the right to interpret the scriptures, the Reformation that created Protestantism was based on a literal reading of the Bible. The differences in modes of interpretation grew during the sixteenth century, and as a result the Lutheran Church did object to Copernicus' heliocentric universe. Martin Luther himself described Copernicus as the 'fool who wanted to turn the art of astronomy on its head'.[38] The real theological controversy, however, in which both Catholics and Protestants took part, did not start until the seventeenth century in the wake of Galileo and Kepler.

Among astronomers, Copernicus' work, which had been eagerly anticipated, was recognized as the first astronomical treatise since Ptolemy to rival the *Almagest* in depth and scope. But the motion of the Earth (which was the central thesis of *On the Revolutions*) was, with one or two notable exceptions, dismissed. A typical reaction was that of the English astronomer Thomas Blundeville, who in 1594 wrote:

> Copernicus ... affirmeth that the earth turneth about and that the sun standeth still in the midst of the heavens, by help of which false supposition he hath made truer demonstrations of the motions and revolutions of the celestial spheres, than ever were made before.[39]

The most important German university in the sixteenth century was in Wittenberg, and there a group of astronomers under the guidance of the Protestant reformer Philipp Melanchthon came to a consensus as to how *On the Revolutions* should be interpreted. In essence, this boiled down to treating Copernicus' work as a tool for computing the positions of celestial bodies and believing that mathematical astronomy had nothing to say about physical reality. One of these astronomers was Erasmus Reinhold, who was Professor of Astronomy at Wittenberg from 1536 until his death in 1553, and he was impressed deeply by Copernicus' removal of the equant from Ptolemy's theory. In his copy of *On the Revolutions*, he emphasized the importance he attached to Copernicus' achievement when he wrote: 'The axiom of astronomy: celestial motion is circular and uniform or made up of circular and uniform parts.'[40] Reinhold was an excellent computational astronomer, and went on to produce a new set of tables based on Copernicus' theory. These were the widely respected *Prussian Tables* (1551), named after Reinhold's patron, the Duke of Prussia, in which the author wrote:

---

[38] Quoted from Blair (1990).   [39] Quoted from Kuhn (1957), p. 186.
[40] Quoted from Gingerich (1973a).

All posterity will gratefully celebrate the name Copernicus. The science of the celestial motions was almost in ruins; the studies and works of this author have restored it.[41]

Through the work of Reinhold, and also that of his successor as Professor of Astronomy at Wittenberg, Melanchthon's son-in-law Caspar Peucer, Copernicus became known widely in astronomical circles as a reformer of Ptolemaic astronomy; he was the man who had removed the equant.

In the years immediately following the publication of *On the Revolutions*, one astronomer viewed Copernicus' theory in an altogether different light, and that was Rheticus, who had also studied at Wittenberg before travelling to Frombork in 1539. Rheticus was captivated by the aesthetic nature of the heliocentric scheme, perhaps even more so than Copernicus himself was, and argued strongly for the acceptance of both the geometrical and the physical parts of the theory. Despite Rheticus' support, most sixteenth-century astronomers restricted themselves to the problem of computing accurate ephemerides. Rheticus was one of the few astronomers who ever believed in the Copernican universe in both the physical and technical sense.

Another astronomer who played a significant role in subsequent developments was Michael Mästlin, Professor of Mathematics at the University of Tübingen from 1584 until his death in 1631. Mästlin is best known as the man who taught Kepler the Copernican theory, though this was by no means his only contribution. Although Mästlin eventually became convinced that Copernicus was right, and taught the Copernican hypothesis at Tübingen, it is interesting to note that in his successful astronomy textbook *Epitome of Astronomy* he concentrated almost entirely on Ptolemaic astronomy.

Many of the procedures that were required when using Copernicus' planetary theories were extremely tedious, but surprisingly little effort was expended on devising more efficient mathematical techniques suitable for the purpose. Much of the problem stemmed from the fact that many astronomers in the sixteenth century were simply not good enough mathematicians to improve on what by and large were the methods of Ptolemy. There was one notable exception though, and that was François Viète who was one of the leading mathematicians of his day and who made the first significant steps toward the development of modern algebraic techniques. Viète was scathing about Copernicus:

Certainly an unfortunate computer, Copernicus was a still more unfortunate geometer, and so failed to do what Ptolemy failed to do, but made even more mistakes.[42]

---

[41] Quoted from Westman (1975).
[42] Quoted from Swerdlow (1975), which contains details of Viète's planetary theories.

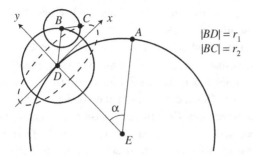

Fig. 5.9. The use of an ellipse in Viète's theory for a superior planet.

Viète's approach to astronomy was that of a mathematician. He was not concerned with representing the heavenly motions any more accurately than Copernicus or Ptolemy had, but in devising equivalent though more refined geometrical models. His theories were geocentric, but he was fully aware that geocentric and heliocentric systems could be equivalent geometrically and the question of which one represented reality was unimportant. Viète's influence on the development of astronomy was negligible, not least because his work on planetary theory was never completed, and by the time his manuscripts were discovered, Kepler had left Copernicus and Ptolemy far behind. However, given the importance of the ellipse in astronomy from Kepler onwards, it is interesting to note that Viète actually introduced the shape into his planetary theories – though not for the same purpose!

Copernicus' theory for the superior planets involved an epicycle on an eccentric circle and this is equivalent geometrically to a double epicycle. Figure 5.9 shows this double epicycle mechanism (compare Ibn al-Shāṭir's scheme in Figure 4.8, p. 106) in which the first epicycle, radius r_1, rotates so that as D revolves around E, BD remains parallel to the apsidal line EA, and the second epicycle rotates in the opposite sense so that $\angle DBC = 2\angle DEA$. This double epicycle can be replaced by an ellipse and the simplest way to show this is to use coordinates centred at D as shown in the figure, but it should, of course, be remembered that Viète did not have this method at his disposal, since the development of analytic geometry by Descartes and Fermat still lay in the future. The point C (which is actually the centre of a third epicycle in Ibn al-Shāṭir's theory) is at (x, y) where $x = (r_1 + r_2)\sin\alpha$ and $y = (r_1 - r_2)\cos\alpha$, from which it follows that

$$\frac{x^2}{(r_1 + r_2)^2} + \frac{y^2}{(r_1 - r_2)^2} = 1.$$

Thus, Viète replaced the two epicycles with a single ellipse having semi-major axis $r_1 + r_2$ and semi-minor axis $r_1 - r_2$.[43]

One thing that did lead to a minor improvement in Copernicus' system was the realization that things would be simpler conceptually if the axis of the Earth were considered to be fixed with respect to the stars (apart from precession of course). Toward the end of the sixteenth century, a number of astronomers recognized that Copernicus' insistence on three motions for the Earth was unnecessarily complicated. For example, the German astronomer Christoph Rothmann, who was a firm adherent of the Copernican theory, wrote:

> I know that in this point Copernicus is very obscure and difficult to understand. . . . This can be explained much more easily, and the triple motion of the earth is not necessary; the daily and the yearly motions suffice.[44]

Of the early followers of Copernicus in England – somewhat of a scientific backwater during the sixteenth century – most notable are John Dee and his pupil Thomas Digges.[45] Dee advocated openly the Copernican theory as the best method for computing future positions of celestial bodies, though whether he subscribed to the physical basis of the theory is unclear. Either way, Dee's acceptance of heliocentrism was significant, as he was a man of considerable influence in Elizabethan England. In 1576, Digges translated substantial parts of the first book of *On the Revolutions* into English and built into his vision of the Universe a radical suggestion. He replaced the idea of a sphere of fixed stars outside the sphere of Saturn with the concept of an infinite world with stars scattered through it. This idea was incorporated by Giordano Bruno into his pantheistic cosmology. Bruno advocated many unorthodox ideas, e.g. that the stars were all like our Sun, perhaps surrounded by planets and with other people inhabiting them.[46] He was burned as a heretic in Rome in 1600.

In the half century following Copernicus' death, most astronomers realized that *On the Revolutions* was of major significance, but the desire was there to

---

[43] It appears that Copernicus was aware of this method of generating an ellipse (Swerdlow and Neugebauer (1984), p. 135), but of course for him an ellipse was not an acceptable curve to incorporate into a planetary theory.

[44] From a letter to Tycho Brahe in 1590. Quoted from Pannekoek (1961).

[45] A discussion of early British Copernicanism can be found in Russell (1973).

[46] More detail on Bruno's cosmology can be found in Singer (1950) (which contains a translation of his *De l'infinito, universo e mondi*), and McMullin (1987), while the relationships between the ideas of Dee, Digges, and Bruno are discussed in Ronan (1967), pp. 27–39. A number of other radical thinkers saw the Copernican theory as an opportunity to break away from the scholastic tradition (see Boas Hall (1994), Chapter IV; also Johnson (1937)).

find a way of utilizing the attractive parts of Copernicus' theory without having to lose hold of the fundamental tenet of a stationary Earth. For many, this was simply too high a price to pay, even for the undoubted benefits of the Copernican system. Toward the end of the sixteenth century, an astronomer emerged who realized that it was possible to construct a world system that retained some of the elegance of Copernicus' heliocentric idea but which did not require a moving Earth. This was Tycho Brahe.

6

Tycho Brahe, Kepler, and the ellipse

Tycho Brahe

Tycho Brahe was born into a small privileged class of people who had controlled Denmark for over 200 years.[1] He was influenced by the Lutheran doctrine, prevalent in Denmark in the sixteenth century, that emphasized the importance of education, including the four mathematical sciences of the *quadrivium* – arithmetic, geometry, astronomy, and music – and at the age of 13 he went to the University of Copenhagen where he began to acquire his knowledge and love of astronomy.[2] He also studied astrology and cast horoscopes, believing that a better knowledge of the motions of the stars and planets would allow man to have greater power over his fate. The following passage from an oration delivered in 1574 gives us a clear idea of Tycho's views on the validity of astrology:

> To deny the forces and influence of the stars is to undervalue firstly the divine wisdom and providence and moreover to contradict evident experience. For what could be thought more unjust and foolish about God than that He should have made this large and admirable scenery of the skies and so many brilliant stars to no use or purpose – whereas no man makes even his least work without a certain aim. That we may measure our years and months and days by the sky as by a perpetual and indefatigable clock does not sufficiently explain the use and purpose of the celestial machine; for what it does for measuring the time depends solely on the course of the big luminaries, and on the daily rotation. What purpose, then, do these other five planets revolving in different orbits serve? . . . Has God made such a wonderful work of art, such an instrument, for no end or use? . . . If, therefore, the celestial

---

[1] An excellent biography is Thoren (1990).
[2] His education at Copenhagen is described in Christianson (1967). There was a solar eclipse on 21 August 1560 that was visible as a partial eclipse in Copenhagen and it seems likely that this event was instrumental in arousing Tycho's interest in observational astronomy.

bodies are placed by God in such a way as they stand in their signs, they must of necessity have a meaning, especially for mankind, on behalf of whom they have chiefly been created . . . [3]

Astrological predictions required accurate knowledge of planetary positions but by the time he was 20, Tycho had already satisfied himself that both the *Alfonsine Tables* and the *Prussian Tables* left a great deal to be desired. In 1563, he observed the conjunction between Jupiter and Saturn, an event that takes place every 20 years or so and which is especially significant for astrologers, and he noted the inaccuracies in the existing tables; those based on Copernicus were out by a few days and those based on Ptolemy by almost a month! Another important part of Tycho's development as an astronomer was his interest in the design and construction of observing instruments. Already in 1570 he had designed and built a giant quadrant measuring a massive 5.5 m in radius.

On 11 November 1572, Tycho noticed an unfamiliar star-like object in the sky near the three stars that comprise the right-hand half of the familiar W of Cassiopeia. It was brighter than the stars Sirius or Vega or, indeed, Venus, and lay outside the zodiac, so could not be a planet. It did not conform to other peoples' descriptions of comets (though Tycho had never seen a comet) and repeated observation over a number of nights showed that it did not move relative to the stars like a comet. When the new star appeared it was so bright that it was visible at noon, but by December had dimmed to the brightness of Jupiter, and by February/March 1573 was a bright star. Tycho followed the decay of the star until its final disappearance sometime around March 1574. We now know that what Tycho and other astronomers witnessed was a huge stellar explosion – a supernova.

This event had profound implications for the Aristotelian view of the Cosmos and was a turning point in Tycho's life. One of the reasons for the success of Aristotle's separation of the changeable sublunary world and the immutable heavenly spheres, was that essentially it was correct in that no changes were observed in the region beyond the Moon. The only serious problem was that posed by comets but, by insisting that comets were closer to the Earth than the Moon, Aristotle had ensured that few astronomers, at least until the fifteenth century, studied them carefully. Tycho was not a slavish follower of Aristotelian doctrine, but it must have had some influence on him. However, he was suffi-ciently independent-minded to measure the new star for diurnal parallax, and

---

[3] Quoted from Pannekoek (1961), p. 204. The oration was an inaugural address preceding a series of lectures that Tycho gave on advanced astronomy at the University of Copenhagen. This oration covered a wide variety of subjects, including Tycho's approach to biblical interpretation, a topic discussed by Howell (1998).

when he found none, to trust his observations. Since the diurnal parallax of the Moon, was easily detectable, he asserted categorically that the new star was further away than the Moon. Tycho regarded the new star as a special creation of God, and most people (probably including Tycho) were less concerned with arguments about its position than with its meaning. Michael Mästlin also observed the new star, and came to the same conclusions as Tycho about its position above the Moon.

As Kepler later remarked, the new star of 1572 heralded the birth of a new astronomer, for after this Tycho concentrated his efforts on astronomy, whereas previously his intellectual efforts had been directed more toward alchemy (he never gave up his researches into alchemy, though). In 1573, he published a short tract, *On a New Star*.[4] This publication, while not to everyone's liking, did establish Tycho as an authority on astronomical matters, and in 1574 he gave some lectures on advanced astronomy at the University of Copenhagen. Interestingly, Tycho based the technical part of his lectures on Copernican theory, though it is clear that he rejected any physical basis for the theory.

> In our time, however, Nicolaus Copernicus, who has justly been called a second Ptolemy, from his own observations found out that something was missing in Ptolemy. He judged that the hypotheses established by Ptolemy admitted something unsuitable and offending against mathematical axioms: he did not either find the Alphonsine computation meeting the heavenly motions. He therefore arranged, by the admirable skill of his genius, his own hypotheses in another manner and thus restored the science of the celestial motions in such a way that nobody before him has considered more accurately the course of the heavenly bodies. For although he devises certain features opposed to physical principles, e.g. that the sun rests at the centre of the universe, that the earth, the elements associated with it, and the moon move about the sun in a triple motion, and that the eighth sphere remains unmoved, he does not for all that admit anything absurd as far as mathematical axioms are concerned. But if we examine the matter thoroughly it appears right to blame the current Ptolemaic hypotheses in this regard. For they dispose the motions of the heavenly bodies in their epicycles and eccentrics as irregular with respect to the centres of these very circles. This is absurd, and by means of an irregularity they save unsuitably the regular motion of the heavenly bodies.[5]

Clearly, Tycho admired the mathematical aspects of the Copernican system and rejected the Ptolemaic equant, and it was to Copernicus' theory that he would turn when computing the positions of heavenly bodies. As a young man, Tycho may well have believed that in order to produce tables that were in agreement

---

[4] *De stella nova* . . . The full Latin title translates as *On a New Star, Not Previously Seen within the Memory of Any Age since the Beginning of the World*. An English translation can be found in Shapley and Howarth (1929).
[5] TYCHO BRAHE *An Oration on Mathematical Disciplines*. Quoted from Moesgaard (1973).

with observation, all that was required was to re-evaluate the parameters in the Copernican models. The accuracy of his subsequent observations, however, led to the realization that this alone was not sufficient.

In 1576, in order to persuade Tycho not to emigrate from Denmark, King Frederick II granted him the fiefdom of the island of Hven for the rest of his life. Tycho liked the isolation and decided to build a great house and observatory there that he named Uraniborg, after Urania, the Greek goddess of the heavens. It took over 4 years to build and on the walls he hung portraits of Timocharis, Hipparchus, Ptolemy, al-Biṭrūjī, Alfonso X, and Copernicus (as well as himself). Also, he had a workshop built in which a great number of ever more accurate observing instruments were produced.[6]

Tycho Brahe's first major contribution came in 1577 following his first sighting of a comet. He saw the comet first on 13 November and began immediately making careful observations, something he continued to do until the comet faded from view in January 1578. In November, its tail stretched 22° across the sky and the comet matched Venus for brilliance. By the end of December, Tycho concluded that the comet showed essentially no diurnal parallax and was convinced that it was further away than the Moon, at least 230 Earth radii away (about 4 times the mean Earth–Moon distance according to Copernicus). He noted that the tail of the comet always pointed away from the Sun[7] and came to the conclusion that it was, in fact, orbiting the Sun. Once again, Mästlin was in agreement.[8] Tycho's initial findings were written up in a report for the Danish Crown[9] and here we find that Tycho had already taken a fundamental step toward his own geoheliocentric world system. This was the assumption that Venus and Mercury orbit the Sun rather than the Earth.[10] According to this scheme, Tycho believed that the closest approach of Venus to the Earth came at a distance of 296 Earth radii, and so suggested that the comet had an orbit around the Sun outside that of Venus.[11]

His first public account of the comet did not appear until much later. In 1588, he published personally his *Concerning the More Recent Phenomena of*

---

[6] Christianson (2000) describes in detail the staff, structure and culture of Tycho's island and how Tycho Brahe influenced the lives of all those who worked there.
[7] He wasn't the first to make this discovery about comets. Fracastoro noted precisely the same thing in his *Homocentrica* of 1538.
[8] Mästlin's analysis of the comet is described in Westman (1973).
[9] A discussion of this report, which was only discovered in the twentieth century, together with an English translation, is given in Christianson (1979).
[10] Described by Copernicus as being due to Martianus Capella (see p. 130).
[11] Interestingly, he also contemplated the idea of an oval rather than a circular orbit, Boas Hall (1994), p. 118.

the Ethereal World in which he not only propounded his own theory of the comet at great length, but also reviewed and criticized corresponding analyses by nineteen other authors.[12] Tycho, along with most other people, believed that comets usually heralded bad news, and this one was no exception. In his original report he wrote:

> Likewise, on the evening when the comet first appeared after sunset, it was in the 8th house, which astrology ascribes to death. All this indicates that the comet augurs an exceptionally great mortality among mankind... the like of which has not occurred in many years. This will come about not only through gruesome pestilence and other deadly diseases... but it will also ensue from great wars and bloodsheddings, for the tail of the comet had a fiery, dark and martial appearance...[13]

The accuracy and systematic nature of Tycho's observations were unprecedented. Earlier astronomers had been able to achieve accuracies of about 10–15', whereas by 1581 Tycho's measurements were averaging errors of about 4' of arc. His goal was to reduce this to 1' – a full order of magnitude improvement in accuracy – and he began to achieve this around 1585.[14] One of Tycho's first major projects, which he began as early as 1576, was the accurate determination of Hven's latitude and he did this by two different methods. One was based on averaging the maximum and minimum altitudes of circumpolar stars, and the other on observations of the noon Sun. He found that the results from the solar method consistently were lower than those based on the stars, and by early 1584 he identified the reason: atmospheric refraction. Tycho's recognition of the serious effects of this phenomenon on astronomical observations was a major step forward.

However, Tycho's quantitative calculations of the effect of refraction were erroneous due to his corrections for solar parallax, which he took to have the excessive value found in Ptolemy. In order to get theory to match observation, Tycho's corrections for refraction had to account for the inaccurate solar parallax as well as for the phenomenon itself. Since solar parallax was not relevant to the fixed stars, the light from which also underwent refraction, Tycho was led

---

[12] The Latin title was *De mundi aetherei recentioribus phaenomenis*. A thorough discussion of all that was written about the 1577 comet is given in Hellman (1944). Gingerich and Westman (1988) have found that over half of the astronomical works published in the 1570s sought the meaning of the comet of 1577 or the new star of 1572. For a discussion of the role of comets in astronomical thought prior to Tycho, see Barker and Goldstein (1988) and references cited therein.

[13] Quoted from Christianson (1979).

[14] A study of the accuracy of Tycho's instruments can be found in Wesley (1978).

to believe that the effect of refraction on sunlight was different quantitatively from its effect on starlight.[15]

During the 1580s and early 1590s, Tycho derived improved parameters for the solar theory (still essentially Hipparchus'), and worked on the production of an extremely accurate catalogue of 777 stars. This was the first star catalogue available in Europe that was independent of Ptolemy's, and represented a decrease of an order of magnitude in the errors compared to the catalogue Copernicus had included in *On the Revolutions*. Because Tycho realized just how much effort went into obtaining accurate observations, he was extremely sceptical of ancient observations and distinctly unimpressed with the many complicated theories of precession that had been proposed based on them. His data suggested that the rate of precession was constant (he fixed it at 51″ per year) and that some of the old observations were erroneous, and discarded the popularly held belief that theories should be able to reproduce both ancient and modern observations.

The decrease in obliquity (which in the trepidation theory was produced by the same mechanism that generated precession) was real, however. Copernicus had modelled the variation in obliquity as due to a motion of the Earth, but apart from this it had been assumed previously to be due to a motion of the sphere of fixed stars. If this were the case, the ecliptic latitudes of the stars should have been constant over time, but their declinations (distances from the celestial equator) should have varied. On the other hand, if the variation in obliquity was due to a shift in the orbit of the Sun it would be the latitudes that had altered rather than the declinations. Tycho's observations enabled him to determine which of the alternatives was the true cause. He found that stellar latitudes had, indeed, changed, with those for stars near the summer solstice increasing and those near the winter solstice decreasing. Hence, the angle between the path of the Sun and the celestial equator gradually decreases, but the stars do not take part in this motion.[16]

Tycho's work required extensive and highly accurate computations and he was helped by the 4-month visit to Hven of the mathematician and astronomer

---

[15] A theory of refraction that was the same for the Sun and the stars was developed by Kepler in 1604, but astronomers had to wait until Descartes published what we now refer to as Snell's law in 1637 for the correct refraction law. The Dutchman Willebrord Snell, who visited Tycho during the winter of 1599–1600, discovered the law in 1621 but did not publish it. There is evidence that the English mathematician Thomas Harriot discovered the correct refraction law sometime before 1601. Harriot produced refraction tables for a number of substances and he sent them to Kepler in 1606 (Lohne (1975)).

[16] Tycho's views were not accepted universally. Indeed, in the eighteenth century, long after the invention of the telescope, distinguished astronomers were still putting forward evidence to support the idea of a constant obliquity (see Wilson (1980), p. 65).

Paul Wittich in 1580. Wittich brought with him the knowledge of the method of *prosthaphaeresis*, originally devised by Ibn Yūnus (see p. 96), by which multiplications could be reduced to a combination of addition, halving and the use of trigonometric tables. Tycho recognized immediately the benefit of the method and, together with Wittich, began to compile a manual of trigonometry to aid Tycho's assistants. This was never completed.[17]

Tycho's world system

Due largely to having provided the data from which Kepler discovered his laws of planetary motion, Tycho is often remembered as simply an observer (albeit a brilliant one) rather than as a theorist. In fact, he was one of the great cosmologists between Copernicus and Newton. He developed a world system that was designed to avoid what he viewed as the mathematical absurdity of Ptolemy's equant and the physical absurdity of Copernicus' moving Earth. Tycho was convinced that the Earth was at the centre of the Universe and 'not whirled about with an annual motion as Copernicus wished'.[18] The baseline for his cosmology was the *Almagest*. Although he admired Copernicus' achievement in removing the equant and was quite happy to use Copernicus' geometrical methods for computing planetary positions, he could not countenance the implications of the heliocentric system. Thus, even though the structure he ended up with is equivalent geometrically to that of Copernicus, almost certainly he derived it without ever departing from the hypothesis of a stationary Earth.

By 1577, Tycho had contemplated already the fact that Venus and Mercury might orbit the Sun rather than the Earth, but at that time he still thought that the superior planets orbited the Earth. When Wittich visited Hven in 1580, he carried with him a series of drawings containing various geometrical transformations of Copernicus' planetary theories.[19] One of these was a schematic diagram (very similar to that in Figure 6.1 though without the dashed lines) showing the Sun *S* orbiting the Earth *E*, with Mercury and Venus in orbit round the Sun. The Moon orbits the Earth, as do the three superior planets, which are all carried on Ptolemaic epicycles with the radius connecting each planet to the centre of its epicycle parallel to the Earth–Sun line. What was significant about Wittich's

---

[17] More details of this episode can be found in Thoren (1988).
[18] Quoted from Gingerich (1992), p. 92.
[19] The drawings were discovered in the Vatican library by O. Gingerich and some were published in Gingerich (1973a). Gingerich thought originally that they were made by Tycho but subsequent investigations revealed that they were by Wittich (see Gingerich and Westman (1981, 1988)).

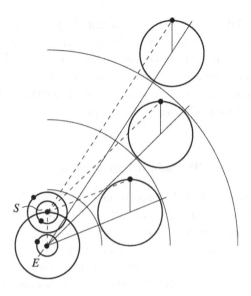

Fig. 6.1. Paul Wittich's diagram.

drawing was that all the epicycles were the same size as each other and also as the Sun's orbit around the Earth.

Of course, this is perfectly possible in a Ptolemaic universe, one is free to choose the radii of the epicycles as one pleases, but only at the expense of changing the sizes of the planetary orbits from the values established using the space filling spheres principle. However, in a Copernican system, with which Wittich was thoroughly familiar, the role of all these epicycles is replaced by a single circle – the orbit of the Earth round the Sun – and so perhaps this is why he drew the diagram as he did. In any case, it is a small step from Wittich's illustration to Tycho's geoheliocentric world system. One simply has to transform the epicyclic motions of the superior planets into eccentric circles centred on the Sun by constructing the parallelograms shown with dashed lines in Figure 6.1.

Tycho did not make this conceptual leap immediately – it took him a couple of years. For one thing, he was busy with other aspects of his observational programme. Another comet appeared in 1580 that diverted his attention; again, he concluded that it was far above the Moon. Then in the winter of 1582/83 he began a series of observations of Mars. Tycho realized that no one had as yet devised an observational technique for deciding between the Copernican and Ptolemaic universes, but perhaps Mars held the key. In Ptolemy's geocentric scheme, Mars was always further away from the Earth than the Sun, but in

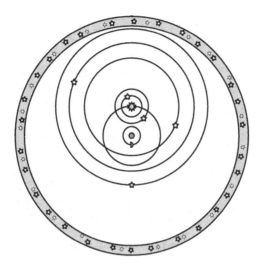

Fig. 6.2. The world system of Tycho Brahe.

Copernicus' heliocentric theory (and in the geoheliocentric theory suggested by Wittich's diagram) it was sometimes nearer than the Sun. Since, as far as Tycho was concerned, the Sun had a parallax of 3′, Mars would have a parallax in excess of this value when it was closer to the Earth than the Sun. If he could find such a parallax, he would have provided the first 'proof' of the falsity of Ptolemy's geocentric universe. Now, the Sun is about 20 times further away than Tycho thought, and the largest possible parallax angle for Mars is roughly 25″, which was beyond the limits of even Tycho's naked-eye astronomy. As a result, Tycho failed to find the phenomenon for which he was looking, but he wanted so desperately to find this parallax that eventually he deluded himself that he had done so.

By 1584, he had devised his geoheliocentric scheme (Figure 6.2) in which the Earth was immobile at the centre of the Universe, orbited by the Moon and the Sun, about which orbited the other five planets. The orbits of Venus and Mercury around the Sun were smaller than that of the Sun around the Earth, whereas the orbits of Mars, Jupiter and Saturn were larger. This ensured that Venus and Mercury always remained close to the Sun, while enabling the other three planets to be found anywhere in the zodiac.

As far as the determination of planetary longitudes is concerned, Tycho's geoheliocentric system is equivalent geometrically to both the geocentric universe of Ptolemy and the heliocentric world of Copernicus. This is illustrated for the superior planets in Figure 6.3. The solid circles represent the deferent and epicycle of Ptolemy's theory, with E the stationary Earth (placed at the

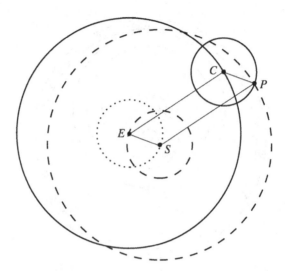

Fig. 6.3. The geometrical equivalence of Ptolemy's, Copernicus' and Tycho's system for the superior planets.

centre of the deferent for simplicity), S the Sun, C the centre of the epicycle and P the planet under consideration. The motion of the planet is linked to the Sun by virtue of the fact that $ESPC$ is always a parallelogram. In Copernicus' theory, S is at rest, with the Earth moving around it on the small dashed circle and the planet moving around the large dashed circle centred on the Sun. In Tycho's scheme, the Sun rotates around the Earth on the dotted circle and the planet again orbits on the large dashed circle. Thus, exactly the same planetary orbit can be thought of as resulting from three quite different physical origins. A similar equivalence can be obtained for the inferior planets. Tycho's system also retains the conceptual elegance of the Copernican universe, in that the orbit of the Sun round the Earth explains the second anomaly for all the planets.

However, there was a major problem. When Tycho actually computed all the technical details required to make his theory reproduce accurately the phenomena – including the sizes of the various orbits – he found that the orbit of Mars had to intersect that of the Sun, and he confessed that he initially

> could not bring myself to allow this ridiculous penetration of the orbs, so that for some time, this, my own discovery, was suspect to me.[20]

The problem with intersecting orbits was the Aristotelian belief that the celestial bodies were carried around on real, solid spheres, and Tycho, along with most

---

[20] Quoted from Thoren (1990), p. 254. There is a sense in which Tycho's problem is an illusion (see Margolis (2002), p. 49) but what is important here is that Tycho and his contemporaries believed that there was a problem.

others of his generation, subscribed to this view. However, it is not clear why he did not object to the fact that the spheres of Venus and Mercury also intersect that of the Sun.

Salvation for Tycho lay in his data on the comet of 1577 and that of his contemporary Michael Mästlin. Mästlin had, in addition to computing its orbit around the Sun, tabulated its daily distances from the Earth, and these ranged from about 155 to 1495 Earth radii. Tycho computed these distances from his own orbital data, with similar conclusions, and realized that if the results were correct, the comet would have had to pass through the supposed solid spheres. He began to doubt their existence and eventually was converted to this new viewpoint.[21] Tycho published finally his geoheliocentric scheme, which he always regarded as the most significant achievement of his life, in 1588.[22]

Tycho was sure that the stars were situated just beyond Saturn, which in his geoheliocentric universe meant they were about 14 000 Earth radii away. For why would God have created so much wasted space as was implied by the Copernican theory? Furthermore, if Copernicus was correct, the angular diameter of the stars (thought then to be up to $2'$)[23] would imply that the stars were of a comparable size to the Earth's orbit, whereas if they were just beyond Saturn, the implied sizes were about 4 times the diameter of the Earth, which, to Tycho, was much more reasonable. In fact, Tycho used this as further evidence for the superiority of his scheme over Ptolemy's, because in the universe described in the *Almagest*, the stars were 20 000 Earth radii away, which implied a proportionate increase in their size.

From a modern perspective, Tycho's system appears to be a step backwards. Copernicus had set the Earth in motion around the Sun and Tycho placed it firmly back in the centre of the Universe. Moreover, Tycho had reduced the size of the Copernican cosmos so that in his scheme it was only two-thirds of the size of the Ptolemaic universe. But to many sixteenth-century astronomers, it represented the best of both worlds, retaining all that was desirable from the Copernican theory, but without the motion of the Earth for which there was, after all, absolutely no evidence. The intersection of the spheres of Mars and the Sun was, however, a problem that troubled many astronomers since, although Tycho had ceased to believe in solid celestial spheres, this was by no means the general consensus. Tycho's world-view excited a great deal of debate in the seventeenth century and, as religious opposition to Copernicus

---

[21] Details of this conversion are given in Rosen (1985a).

[22] As an additional chapter in *Concerning the More Recent Phenomena of the Ethereal World*.

[23] In fact, all stars are essentially point-like objects; it is human vision that makes brighter stars appear larger than dimmer ones.

grew, his system became more and more influential. Ironically, at the time of Tycho's death in 1601, virtually no astronomers were believers in a heliocentric cosmology, but they all used Copernican techniques. After Kepler however, virtually all astronomers were to become Copernicans, but they had no need for his technical astronomy.[24]

Various disputes arose as to the original discovery of the geoheliocentric system. A German mathematician Nicholas Reymers Bär, otherwise known as Ursus, claimed to have discovered such a scheme in 1585 and that Tycho had stolen it from him. Ursus' scheme, published in his *Foundation of Astronomy*[25] in 1588, differed from Tycho's in that the orbit of Mars enclosed completely that of the Sun, and Tycho claimed that Ursus, on a visit to Tycho in 1584, must have looked surreptitiously at Tycho's papers and seen one of his diagrams that incorrectly had this feature. Helisaeus Röslin published what he claimed was his own world system in 1597, though it would appear to have been cribbed directly from Ursus, and Duncan Liddel, who had visited Hven in 1587 and 1588 and seen pre-publication details of Tycho's new theory, lectured in Germany on the 'Tychonic' system, claiming it as his own. Tycho was incensed. In 1598 he wrote:

> We have remedied [difficulties in the earlier systems] by means of a special hypothesis that we invented and worked out fourteen years ago, basing it on the phenomena. There are certain persons, of whom I know three with distinguished names, who have not been ashamed to appropriate this hypothesis and present it as their own invention. In due course I shall, God willing, point out the occasions on which they did it, and repudiate and refute their immense impudence, and I shall demonstrate that the fact of the matter is as I say, and I shall do that so clearly that it will be impossible for impartial men to doubt or contradict me.[26]

Ursus' scheme differed from Tycho's in another important respect: he assigned a daily rotation about its own axis to the Earth. He thus became one of the

---

[24] While the influence of Tycho's world system was short-lived in Europe, it had a much longer life in China. It was exported there by the Jesuits, many of whom were very knowledgeable in astronomy and for whom a moving Earth was anathema. The Tychonic system appears in the Imperial Encyclopedia of 1726. The heliocentric astronomy of the West did not really gain followers in China until the early nineteenth century. Part of the reason for this is the fact that the main concern of the Chinese with astronomical phenomena was in the production of calendars, and the principles by which these were made were arithmetical rather than geometrical in origin. Hence, the particular world-view to which one subscribed was not very important. In addition, there was very little transfer of ideas between China and the West in the eighteenth century (see Yoke (1983)).

[25] The *Foundation of Astronomy* is concerned mostly with mathematical methods used in practical astronomy, including a highly competent treatment of *prosthaphaeresis* based on the work of Jost Bürgi and Paul Wittich.

[26] Quoted from Gingerich and Westman (1988), p. 19.

first proponents of what became known as the semi-Tychonic world system.[27] Tycho did not believe in the daily rotation of the Earth, but did admit that it was a possibility. Commenting on Copernicus' triple motion of the Earth he wrote:

> But whether this third motion, that accounts for the daily revolution, belongs to the earth and nearby elements, is hard to say . . . . It is likely nonetheless that such a fast motion could not belong to the earth, a body very heavy and dense and opaque, but rather belongs to the sky itself whose form and subtle and constant matter are better suited to a perpetual motion, however fast.[28]

Tycho's lunar theory

Tycho began making observations of the Moon around 1582 and he continued to do so, though not always assiduously, for the next 15 years. He was well aware that both the *Alfonsine Tables* and *Prussian Tables* left considerable room for improvement, and also knew that the errors were worst at the octant points. As he put it in a letter to a Bohemian physician and part-time astronomer, Thaddeus Hayek:

> And even though [Copernicus] has arrived at a much more agreeable and probable theory for the moon than the Ptolemaic was, nevertheless not even this theory is sufficient to explain the lunar cycles in every case. In syzygies and quadratures he deserves praise, although not even in these places does he set forth everything with the necessary precision. But in the four places already mentioned, which are intermediate, he by no means saves the appearances . . . [29]

In a series of observations in 1594, Tycho discovered the first wholly new astronomical phenomenon since Ptolemy's time, and one that reduced the error from existing theories of the longitude of the Moon by almost 75 per cent. He observed that the Moon sped up as it approached the syzygies and slowed down near the quadratures. This phenomenon, which shows up as a displacement of about 40′ from the position at the octants predicted by previous theories, has ever since been known by the name Tycho gave to it: the 'variation'.[30]

In Tycho's original lunar theory, he accounted for the variation by modifying Copernicus' theory of the Moon (which is the same as that of Ibn al-Shāṭir, see Figure 4.7 p.105). He made the centre of the deferent revolve around the Earth on a small circle twice each synodic month, and by this mechanism

---

[27] A detailed description of the many Tychonic and semi-Tychonic planetary systems that were proposed is given in Schofield (1981).
[28] Quoted from Blair (1990).   [29] Quoted from Thoren (1990) p. 327.
[30] Tycho's discovery of the variation is described in detail in Thoren (1967b).

managed to produce displacements of about 45' in the position of the Moon
at the octants, while leaving the position of the Moon at the syzygies and
quadratures unchanged. He modified further the traditional theory by making
the centre of the larger epicycle (D in Figure 4.7) rotate around another small
circle with a period of 1 year. This mechanism produced a change in longitude
that oscillated between $\pm 11'$ over the period of 1 year. It is not known how
Tycho came to discover this annual variation in the Moon's motion, but it
was explained by Newton a century later as being due to the varying distance
between the Sun and the Earth–Moon system over the course of 1 year.[31]

However, this theory had the defect that the ratio of the greatest and least
distances of the Moon did not correspond to the observed changes in apparent
diameter. Ibn al-Shāṭir had managed to bring this ratio more into line with
reality from that inherent in Ptolemy's theory, but it was still too large. The
modification that was made to Tycho's model actually was performed by his
long-serving assistant, Longomontanus, and consisted essentially of replacing
the larger of Ibn al-Shāṭir's epicycles, of radius r_1, by two smaller ones of radii
$2r_1/3$ and $r_1/3$. The resulting set of epicycles had to be reshuffled so as to get
the scheme to work, but the final theory resulted in very accurate displacements
of about 40' 30'' in the position of the Moon at the octants, and the ratio of the
greatest to least distances between the Earth and the Moon had been reduced
from the Copernican value of 1.31 to the much more accurate 1.16 (1.141 being
the modern value).[32]

Tycho also considered the problem of the Moon's latitude, and observed
that the angle at which the Moon's orbit was inclined to the ecliptic was not
the constant 5° that everyone assumed.[33] Observations in 1587 had led him to
believe that the inclination of the orbit of the Moon actually was $5\frac{1}{4}^{\circ}$, but in
1595 his observations indicated that the orbital plane actually varied between
5 and $5\frac{1}{4}^{\circ}$ during a synodic month. He postulated a small nutation, reminiscent
of earlier theories of trepidation, with a period of half a synodic month; this is
illustrated (greatly exaggerated) in Figure 6.4. At a conjunction or opposition,
the pole of the orbit of the Moon is at P_1 and the inclination of the orbit is 5°,
whereas at quadrature this pole is at P_3 and the inclination is $5\frac{1}{4}^{\circ}$. This latitude
theory predicts an oscillation in the nodes of the lunar orbit: as the pole of the
orbit of the Moon moves around the small circle P_1, P_2, P_3, P_4 the ascending
node, which is B_1 at syzygy, oscillates through the points B_1, B_2, B_3, B_4 along

[31] The fact that Tycho arrived at exactly the 11' that modern theory predicts would appear to have
been somewhat fortuitous, especially as he was out by about 24° in the phase of the oscillation
(see Thoren (1990), Appendix 3).
[32] Details can be found in Jacobsen (1999).
[33] Full details of Tycho's lunar latitude theories are given in Thoren (1967a).

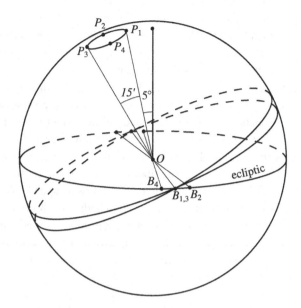

Fig. 6.4. The nutation of the pole of the orbit of the Moon in Tycho's theory.

the ecliptic. Tycho realized this and looked back at some previous observations that seemed to confirm this effect.

Ever since his patron King Frederick II had died in 1588, Tycho's status and influence had been on the decline and by the time the 19-year-old King Christian IV was crowned in 1596 he had had enough. He left Hven in April 1597, and after a few weeks in Copenhagen left Denmark altogether. He still harboured hopes that he might return to a position of power in Denmark, but when he realized that this was not to be he began to look around for another patron, and who better than the ruler of the Holy Roman Empire, Rudolph II, in Prague.

However, a problem arose in the form of his old adversary, Ursus, who since 1591 had been the Imperial Mathematician to the Emperor. Ursus found out that Tycho was planning to go to Prague and, perhaps thinking that his job was threatened, decided to counterattack. He published a book, *On Astronomical Hypotheses*,[34] in which he claimed that Tycho had stolen the idea for a geohelio-centric universe from him and remarked that Tycho should have stuck to what he was good at – observing – and left the important cosmological questions

---

[34] Jardine (1984) provides a survey together with a partial translation.

to those of higher intellect, like himself! He even descended to making fun of Tycho's nose, which had been disfigured severely in a brawl by an opponent's sword when Tycho was 20, and claiming that Tycho's wife and daughter had been available for the 'use' of visitors to Hven.

Tycho's reputation as an astronomer was much greater than that of Ursus and eventually he got the royal patronage he was after. The personal attacks that Ursus had launched in his book damaged what reputation he did have so much that he was forced to flee Prague soon after Tycho's arrival in 1599. Tycho set up his household in Benatky castle, just north of Prague, and gradually he began to turn the estate into a new Uraniborg so that he could continue with his observations. What Tycho badly needed, however, was manpower. On Hven he had had numerous assistants who could help him both with observing and with the mountain of complex astronomical calculations that were required, but he had left all of this behind. The man who did arrive in Prague in 1600 to be Tycho's assistant ended up completely revolutionizing astronomy.

Kepler

Of all the significant figures in the history of astronomy, Johannes Kepler is perhaps the most intriguing. His works display a mixture of rational brilliance and mystic flights of fancy, the two appearing to coexist in perfect harmony. He was born in 1571 in Weil der Stadt near Stuttgart.[35] In 1589, he became a student of theology at the University of Tübingen, intending to go on and become a Lutheran clergyman, and as part of his education in the Faculty of Arts he studied mathematics and astronomy. His teacher was Michael Mästlin, from whom he learned the Copernican theory[36] and it did not take long before Kepler was convinced that Copernicus was right. In 1596, in the preface to his first major work, he wrote:

> ... six years ago, when I was studying under the distinguished Master Michael
> Mästlin at Tübingen, I was disturbed by the many difficulties of the usual
> conception of the universe, and I was so delighted with Copernicus, whom
> Mr Mästlin often mentioned in his lectures, that I not only frequently defended his

---

[35] The standard biography of Kepler is the scholarly work of Caspar (1993); more concise accounts include those of Armitage (1966) and Voelkel (1999). A number of Kepler's letters are translated in Baumgardt (1952). Kepler's life and work is the subject of a novel, *Kepler*, by John Banville (Banville (1981)).

[36] It is probable that Mästlin taught the Copernican theory only to a few of the brighter students, including Kepler, while for most students he taught the easier parts of Ptolemaic theory (see Methuen (1996)).

opinions at the disputations of candidates in physics but even wrote out a thorough disputation on the first motion, arguing that it comes about by the earth's revolution.[37]

In 1594, he was recommended for the post of mathematics teacher at the Protestant seminary in Graz. Kepler had set his sights on entering the service of the Church but, significantly for astronomy and, indeed, science as a whole, he accepted the job. Here he could dedicate more of his time to astronomical matters and, in particular, he sought to explain why the number of planets, the sizes of their orbits and their speeds, were as they were. Here we see that already Kepler was asking new questions. Not since Philolaus' universe containing the sacred ten celestial spheres had the number of heavenly bodies been the subject of serious attention. Kepler's desire to find meanings in astronomical phenomena went hand in hand with his great interest in astrology, which he maintained throughout his life. His approach to this was rather different to that of most other astrologers, though. He believed that it was the continuous physical interaction between the heavens and the Earth that had a bearing on people's lives rather than the configuration of the heavens at any particular instant:

> How does the face of the sky affect the character of a man at the moment of his birth? It affects the human being as long as he lives in no other way than the knots which the peasant haphazardly puts around the pumpkin. They do not make the pumpkin grow but decide its shape. So does the sky; it does not give the human being morals, happiness, children, fortune, and wife, but it shapes everything in which the human being is engaged.[38]

In theoretical astronomy, Kepler came out openly in support of Copernicus. He knew that the tables constructed from the heliocentric theory were more accurate than those from Ptolemy, but it was the ability of the Copernican system to *explain* phenomena that was the source of Kepler's belief in its truth. Copernicus had provided the reason why the retrograde motion of the superior planets always occurred when they were in opposition, but that for Mercury and Venus it happened at conjunction. The heliocentric theory explained why the inferior planets were never seen in opposition and why their zodiacal periods were the same as that of the Sun. All these explanations followed from setting the Earth in motion round the Sun.

Kepler believed that this all pointed to the existence of a rational order behind the structure of the Copernican universe. He observed a 'wonderful resemblance' between the Holy Trinity and the Sun, stars and intervening space.

---

[37] KEPLER *Secret of the Universe*. Translations are from Kepler (1981).
[38] Quoted from Peterson (1993).

God the Father lay at the centre of the Universe, God the Son on the outer boundary and God the Holy Ghost filled the space in-between.[39] His attempts to understand the reasons behind the structure of the Universe formed the subject of Kepler's first famous work, the *Secret of the Universe*, published in Tübingen in 1596.[40]

The *Secret of the Universe*

Kepler never regretted his change in chosen career; in fact, he believed that the greatest service he could provide was to uncover the beauty in God's design. In a letter to Mästlin in 1595, after discovering what he believed to be the reason behind the spacings of the planets, he wrote:

> For a long time I wanted to become a theologian; for a long time I was restless. Now, however, behold how through my effort God is being celebrated through astronomy.[41]

As far as Kepler was concerned (since he was a Copernican) there were six planets, and he began by searching for numerical patterns in the dimensions of their orbits. This did not work, so he hypothesized the existence of new, as yet undiscovered, planets between Mars and Jupiter and between Venus and Mercury, and tried to fix their distances so that a pattern would emerge, but again he had no success. Kepler's next idea originated from his astrological work. We have noted already that conjunctions of Jupiter and Saturn were considered particularly significant in astrological circles. These events happen every 20 years or so and each successive conjunction occurs about 117° behind the previous one.[42] He plotted the positions of these conjunctions in the zodiac

---

[39] An in-depth study of Kepler's religious and philosophical beliefs and how they interacted with his scientific work is beyond the scope of this book. For a detailed account see, for example, Kozhamthadam (1994).

[40] The publication of this work, with the Latin title *Mysterium cosmographicum*, was supervised by Mästlin, who was pleased that Kepler was openly advocating the heliocentric theory. Since some of Kepler's treatment of Copernicus was rather brief, Mästlin decided to add what he believed to be the best exposition of the theory, Rheticus' *First Report*, to the publication (see Rosen (1985b)). Kepler's cosmological theories are analysed in Field ((1988)) and the philosophical ideas that underpinned his approach to astronomy are explored in Martens (2000).

[41] Quoted from Gingerich (1972).

[42] The rates of revolution of Jupiter and Saturn around the zodiac are 30.35 and 12.22° per year, respectively. The difference between these is 18.13° per year and so it takes $360/18.13 \approx 19.86$ years for the two planets to be seen at the same longitude again. In this time, Saturn has moved on $19.86 \times 12.22 \approx 242.7°$, which is $360 - 242.7 = 117.3°$ behind the previous conjunction.

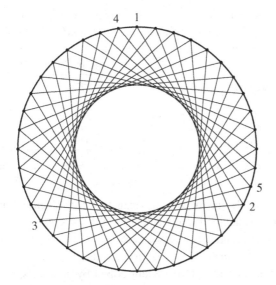

Fig. 6.5. Successive conjunctions of Jupiter and Saturn. Given a conjunction at 1, the next will be at 2, and so on.

on a circle and joined them up (see Figure 6.5) and noticed that the ratio of the radii of the outer circle to the inner circle that is formed by this procedure looked remarkably like the ratio of the sizes of the orbits of Saturn and Jupiter. The inner circle is very close to being inscribed inside an equilateral triangle and, if this were the case, the ratio of the radii would be 2, which was fairly close to the ratio of Saturn's distance to that of Jupiter (about 1.8 in the Copernican theory) and he tried to find similar relationships involving different polygons (squares, pentagons and so on) and match them to other planetary ratios. This did not work either, but 'the end of this useless attempt was also the beginning of the last, and successful one'.

In all Kepler's previous theorizing there was nothing to explain why there were precisely six planets,[43] but now he stumbled across an idea that both solved this riddle and gave an explanation of the planetary distances. Instead of circumscribing and inscribing circles around regular polygons, why not do the same with spheres and regular polyhedra? This has the advantage that it is a three-dimensional theory but, more importantly, since there are only five such

---

[43] Rheticus had suggested that there were six because six was the first perfect number (a perfect number is one that is the sum of its proper factors, $6 = 1 + 2 + 3$, $28 = 1 + 2 + 4 + 7 + 14$, etc.; these numbers had been the subject of study since Pythagorean times) but Kepler did not consider this a likely reason. Koyré (1973) finds it curious that neither Rheticus nor Kepler came up with the reason that the Universe was created in 6 days.

Table 6.1. *Ratios of planetary distances according to*
Kepler's theory of regular solids

		Kepler	Copernicus
Saturn/Jupiter	Cube	577	635
Jupiter/Mars	Tetrahedron	333	333
Mars/Earth	Dodecahedron	795	757
Earth/Venus	Icosahedron	795	794
Venus/Mercury	Octahedron	577	723
		707	

solids, there is exactly one for each interplanetary gap. When he found he could obtain reasonable quantitative agreement from this system, he was overjoyed; he felt that he had found the Creator's plan.

Starting from Mercury, Kepler envisioned its orbit as a sphere inscribed in an octahedron that was itself circumscribed by another sphere, the orbit of Venus. The orbits of the other planets (Earth, Mars, Jupiter, and Saturn) were each spheres separated similarly by an icosahedron, a dodecahedron, a tetrahedron, and a cube. In Table 6.1, the numbers in the third column are the values computed by Kepler for the radii of the inscribed sphere in the given solid, if the circumscribed sphere has a radius of 1000. The fourth column shows numbers Kepler calculated from Copernican theory taking into account the thickness of the planetary spheres, which are required to fill the space from the minimum to maximum distance of the planet from the Sun. In order to get satisfactory agreement for Venus and Mercury, Kepler found that he had to take the circle inscribed in the square formed by the middle edges of the octahedron (which gives the value 707 in table) rather than the inscribed sphere (which gives 577). In his final scheme, everything fits to within 5 per cent, apart from the Jupiter/Saturn ratio, which Kepler explained away by appealing to the fact that Jupiter and Saturn were so far away.

The numbers predicted by the new theory and those computed from Copernicus do not quite match, but Kepler was not deterred. The planetary theory in *On the Revolutions* is worked out with respect to the mean sun, rather than the Sun itself, but Kepler was interested in the reasons behind the physical structure of the Universe, and surely it would be the distances to the Sun that would be important, not the distances to some mathematically constructed point. He needed, therefore, to know the maximum and minimum distances of each planet from the Sun itself, based on Copernican theory, which could be computed by combining the theory of the Earth with that of the planets. The figures actually were provided by Mästlin, with whom Kepler kept up an active correspondence, but they do not lead to any improvement.[44] Kepler never lost faith in the idea that the number of planets was determined by the number of regular solids. He reminded readers of the *Harmony of the World* (1619) that he had made this discovery, though he did by then recognize that the theory as set out in the *Secret of the Universe* was too simplistic.[45]

The judgement of history relegates Kepler's polyhedral hypothesis for the planetary spacings, which fills most of the *Secret of the Universe*, to an interesting curiosity. But three of the twenty-three chapters are about a different problem – that of explaining the speeds of the planets – and here Kepler's treatment proved extremely significant. Kepler noted that as one moved out through the planets, the periods of their orbits increased at a greater rate than their distances from the Sun. Hence, the speeds of the planets decreased. Kepler's conclusion was that the driving force behind the Universe comes from the Sun (he referred to a 'moving soul' within the Sun) and that the effect of this 'force' weakens as one moves away from the Sun. This also provided added justification for the position of the Sun at the centre of the Universe.

Quite how the orbital periods depended on the distances from the Sun eluded Kepler for the time being – he simply did not have data accurate enough for him to find the correct law[46] – but in retrospect we know that he was asking the right questions. Kepler also realized that his physical hypothesis implied that the planets moved at varying speeds as they moved round their orbits, by virtue of the fact that the distance to the Sun varied. Again, Kepler did not have the information at his disposal that could point him to the correct relationship between speed and distance, but the desire to find a physical cause for the planetary motions had led him to a crucial realization: planets do not move at a constant speed. Of course, he did not yet have any proof for his hypothesis, but he had broken away from the 2000-year-old dogma of uniform circular motions.

Kepler realized that Ptolemy had achieved just such a change in speed by his introduction of the equant. When a planet is closest to the Earth in Ptolemy's theory and, hence, furthest from the equant, it must move more quickly than at

---

[44] A detailed analysis of the numerical accuracy of the polyhedral hypothesis, including corrections to Mästlin's calculations, is given in Brackenridge (1982).

[45] A second edition of the *Secret of the Universe* (with extensive annotations) was published in 1621, after Kepler had made the discoveries of his three laws of planetary motion. This suggests that Kepler still saw some value in his early work (see Field (1988)).

[46] It is often written that at this time Kepler suggested that the relationship between the orbital period T and the distance r was $T \propto r^2$, but this is not true. In the *Secret of the Universe*, Chapter XX, Kepler proposed the relationship $(T_2 - T_1)/T_1 = 2(r_2 - r_1)/r_1$, where r_1 and r_2 are the distances of successive planets ($r_2 > r_1$) and T_1 and T_2 are their respective periods. This law is equivalent to $T_2/T_1 = 2(r_2/r_1) - 1$ which is not of the form $T \propto r^\alpha$ for any α. Kepler did later hypothesize the law $T \propto r^2$ in Chapter 39 of the *New Astronomy* (1609).

other parts of its orbit so as to maintain its uniform angular motion with respect to the equant point. Copernicus had removed the equant and replaced it by an extra epicycle, but nevertheless, the effect was pretty much the same, except that now it was the distance to the Sun that was important. Kepler preferred Ptolemy's equant mechanism since it seemed related more directly to his physical idea of the Sun as the driving force in the Universe.

Throughout his life, Kepler sought to discover the harmonies that underlay God's Creation. Already in 1599 he was planning a sequel to the *Secret of the Universe* but, in order to make progress, Kepler needed more data. The one person who possessed large quantities of accurate observational data was Tycho Brahe and, as luck would have it, it was precisely this data with which Kepler was entrusted.

Tycho's assistant

The story of how Kepler came to be Tycho Brahe's assistant is a complicated one.[47] Before Kepler completed the *Secret of the Universe*, he wrote a letter about it to Ursus, asking for the Imperial Mathematician's opinion on his construction involving the Platonic solids. Unbeknown to Kepler, Ursus published this letter (which, unsurprisingly as it was from a young man to one of considerable status, was full of praise for Ursus) on the title page of his attack on Tycho, *On Astronomical Hypotheses.*

When it was finished in 1597, Kepler sent a copy of the *Secret of the Universe* to several astronomers, including Tycho, again asking for an opinion. Unfortunately, Kepler never received Tycho's reply and only heard through Mästlin that Tycho was upset about Kepler's letter having been published in Ursus' book. Kepler was mortified. He knew little of the feud between Tycho and Ursus but wanted to keep on the right side of Tycho so that he could gain access to his observational data. Kepler also knew that his days in Graz were numbered. Following the return of Archduke Ferdinand to the province of Styria in 1598, Graz was not a place for Protestants to live, and so Kepler decided that it was in his interests to apologize to Tycho, which he managed to do in a manner Tycho found acceptable. In a letter to Kepler in 1599, Tycho wrote:

> Even though I have not met you face to face, most learned sir, nevertheless I love you very dearly on account of the excellent qualities of your mind . . . [48]

---

[47] Described in detail in, for example, Jardine (1984) and Thoren (1990) (see also Ferguson (2002)).
[48] Quoted from Thoren (1990).

The fact that Ursus had been forced out of Prague was not sufficient for Tycho, and he was intent on ruining Ursus by dragging him through the courts on the ground that he had not submitted his offensive book to the official censor prior to publication. Tycho believed that evidence of Ursus' behaviour from Kepler would further help his cause and so, since clearly he showed promise as an astronomer as testified by the *Secret of the Universe*, Tycho invited him to become his assistant.

Kepler visited Tycho at Benatky in early 1600, but things did not work out immediately – the two men did not really get on with each other. Tycho could not manage to secure Kepler an official position, and Kepler returned to Graz to sort out his affairs there. Then two significant things happened. Longomontanus, Tycho's most senior assistant, after finishing his revisions of the lunar theory, left to return to Denmark,[49] and Ursus died. Tycho still wanted to vindicate his honour and formally establish that he was the originator of the geoheliocentric hypothesis, and decided that he would produce a properly reasoned demonstration of his claims. At the same time, life became unbearable in Graz for Kepler, and he moved back to Prague with his wife and daughter. He joined the Tycho household, still with no official position, and Tycho set him to work on what became *A Defence of Tycho against Ursus*, often referred to as the *Apologia*.[50]

In this short treatise – which is more than just a simple refutation of Ursus' claims to be the discoverer of the geoheliocentric hypothesis – Kepler made one of the first major efforts to trace the history of ancient astronomy and examined the question of the nature of astronomical hypotheses in general.[51] Throughout the history of astronomy, the debate had raged as to whether the mathematical theories used to describe celestial motions actually said anything about reality. Many, including Ursus, believed that the fact that a theory correctly predicted planetary longitudes did not provide any evidence that the planets actually moved as implied by the theory, since it was perfectly possible for correct conclusions to be drawn from false premises. However, Kepler realized that just because two geometrical theories gave the same predictions, it did not

---

[49] After Tycho's death, Longomontanus (Christian Sørenson) developed his own all-encompassing semi-Tychonic cosmology (see Moesgaard (1975, 1977)).

[50] This work was not published until 1858 and it is translated in Jardine (1984). Martens (2000) describes the *Apologia* as an early modern treatise on realism.

[51] Unlike most of his predecessors, Kepler often referred to the contributions of earlier philosophers and astronomers. In his last major work, the *Rudolphine Tables*, he included an account of the history of astronomy from its ancient origins. For a detailed discussion of Kepler's approach to history, see Grafton (1992).

mean that they had the same status as possible models of reality:

> Even if the conclusions of two hypotheses coincide in the geometrical realm, each
> hypothesis will have its own peculiar corollary in the physical realm.[52]

Clearly, Kepler believed that understanding the physical nature of the Universe should be the subject of astronomical study, in sharp contrast to those like Osiander – the author of the preface to *On the Revolutions* – who believed that astronomical hypotheses were unimportant provided they predicted the phenomena correctly.

In August 1601, Tycho obtained finally an official position for Kepler, under the condition that Tycho's planetary tables would be published as the *Rudolphine Tables*. Soon afterwards, in October, Tycho died and the emperor put Tycho's incomplete works in Kepler's hands[53] and promoted him to the position of Imperial Mathematician. Even though Rudolph died in 1612, Kepler still used his name in the title of the planetary tables he produced in 1627.

Over the next 5 years, Kepler worked extremely hard trying to understand the reasons behind the motions of the planets, and now he had the data he needed. One of the tasks Longomontanus had been involved with before he left Prague had been an analysis of the motion of Mars, and Kepler took this over after his departure.[54] This was particularly fortunate for Kepler – he later attributed it to Divine Providence – because only the orbit of Mars has an eccentricity large enough to make possible the discovery of the correct shape of its orbit from Tycho's data.[55] The culmination of Kepler's studies into the orbit of Mars was the *New Astronomy* (1609),[56] which he referred to as 'warfare' with Mars, itself the god of war.

---

[52] KEPLER *Apologia pro Tychone contra Ursum*. Quoted from Jardine (1984).

[53] Kepler supervised the first complete printing of the great work Tycho had been preparing ever since the 1580s, *Astronomiae instauratae progymnasmata* (*Introductory Exercises Toward a Restored Astronomy*) (1602).

[54] Kepler bet Longomontanus that he would solve the problem of the orbit of Mars within 8 days! (see Caspar 1993, p. 126).

[55] Even for Mars, the deviation from circularity is very small. The orbit is an ellipse with minor axis 0.996 times the major axis. The eccentricity of Mercury's orbit is more than twice that of Mars, but observational data for Mercury was scant and inaccurate.

[56] The full Latin title, which begins *Astronomia nova ...*, translates as *New Astronomy Based upon Causes, or Celestial Physics, Treated by Means of Commentaries on the Motions of the Star Mars from the Observations of Tycho Brahe*. The first English translation was published as recently as 1992 (Kepler (1992)) and all quotations are taken from this source. Some lengthy passages are translated in Koyré (1973) in which the work is discussed in some detail. Another book in which a thorough account is given of the contents of the *New Astronomy* – particularly the physical aspects of Kepler's astronomy – is Stephenson (1987) and the technical aspects of the work are described in, among others, Aiton (1969) and Whiteside (1974).

The *New Astronomy*, and the first two laws of planetary motion

The style of the *New Astronomy* was different totally from that of Ptolemy or Copernicus. Astronomers before Kepler had presented their final, more or less polished, theories, providing few clues as to how they were derived. Kepler, on the other hand, described his process of discovery in great detail, telling the reader about all his fruitless efforts as well as his successful ideas. He justified his approach by considering other great voyages of discovery, like those of Columbus and Magellan, which would have been much less enjoyable to read about if those describing them had only told us about their successes. Not only was Kepler's style novel, the astronomy that he presented really was new, with a completely different emphasis to that of his predecessors. Instead of trying to reproduce the celestial motions as accurately as possible with certain special geometrical constructions, Kepler was attempting to understand the basic structure of God's Universe. The appropriate geometrical tools were not to be decided beforehand, but to be discovered as part of the exploration process.

In the lengthy introduction, Kepler set out his arguments against Aristotelian physics and for the Copernican view of the Universe. He was attempting to develop a physics based on causes, and criticized Aristotle's notion that earthly bodies naturally move toward the centre of the Earth:

> A mathematical point, whether or not it is the centre of the world, can neither affect the motion of heavy bodies nor act as an object toward which they tend.[57]

Kepler then described his own theory of 'gravity', which contains the seeds of what would, in Newton's hands, become the universal theory of gravitation; he thought of gravity as a mutual attraction between bodies similar to the force of magnetism. The one aspect of Aristotelian physics that Kepler did not manage to shake off, and which hindered him throughout his attempts to find a physical theory of the Universe, was the principle of inertia. Kepler believed that unless a body was under the influence of some external 'force' it would remain motionless:

> Every corporeal substance . . . has been made so as to be suited to rest in every place in which it is put by itself, outside the sphere of influence of a kindred body.[58]

Kepler presented numerous arguments as to why the Copernican system was to be preferred to the Ptolemaic or to that of Tycho Brahe, and even described at

---

[57] KEPLER *New Astronomy*, Introduction.
[58] KEPLER *New Astronomy*, Introduction. Ironically, it was around this time that Galileo was coming to realize that Aristotle's principle of inertia was erroneous and that it was uniform motion that was the natural state for a body.

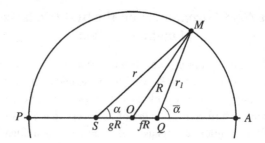

Fig. 6.6. Kepler's vicarious hypothesis.

considerable length how Copernicanism can be reconciled with the Scriptures. For those not sufficiently enlightened to accept his arguments he wrote:

> But whoever is too stupid to understand astronomical science, or too weak to believe Copernicus without affecting his faith, I will advise him that, having dismissed astronomical studies and having damned whatever philosophical opinions he pleases, he mind his own business and betake himself home to scratch his own dirt patch, abandoning this wandering about the world.[59]

After having left the reader in no doubt that his aim was to reform astronomy so that it both reproduced accurately the phenomena and was based on physical causes, Kepler began his description of his battle with Mars.

Kepler's first attack was extremely useful in some respects, but flawed, and he called it a 'vicarious hypothesis', i.e. a substitute for the real thing. When working on the *Secret of the Universe*, Kepler realized that the equant mechanism, incorporated into a heliocentric universe, had the effect of slowing a planet near aphelion (the point on the orbit furthest from the Sun) and increasing its speed near perihelion (the point closest to the Sun). Whereas Copernicus had taken great pains to remove the equant from planetary theory so as to restore astronomy to the fundamental Pythagorean principle of uniform circular motion, Kepler saw no place for physically meaningless epicycles, and so reinstated Ptolemy's device. Ptolemy had made the distance of the equant from the centre of the deferent exactly the same as the eccentricity of the orbit, and Kepler wondered if he could achieve a system that agreed better with observations by moving its position.

In Figure 6.6, M is Mars orbiting on an eccentric circle of radius R around the Sun S, which is a distance gR from the centre O. Because Kepler was convinced of the physical role played by the Sun, the point S was the actual Sun and not the mean sun which had formed the basis of Copernicus' planetary theories. The angle α is the true anomaly, which needs to be computed as a function of time. The point M is made to rotate with uniform angular speed with respect to an equant Q, which is placed a distance fR from O, as shown. The angle $\bar{\alpha}$, the mean anomaly, thus increases uniformly with time. We will begin by using mathematical techniques not devised until about 100 years after Kepler (in particular the binomial theorem and the series expansion for $\tan x$) to see what can be achieved.

From elementary geometry we have it that

$$r_1 \sin\bar{\alpha} = r \sin\alpha \quad \text{and} \quad r_1 \cos\bar{\alpha} + (f+g)R = r\cos\alpha,$$

which, on division, leads to

$$\tan\alpha = \left(1 + \frac{(f+g)R}{r_1 \cos\bar{\alpha}}\right)^{-1} \tan\bar{\alpha}. \tag{6.1}$$

The right-hand side can now be expanded in powers of the small quantities f and g. The cosine rule applied to triangle OQM gives $R^2 = r_1^2 + f^2 R^2 + 2fRr_1 \cos\bar{\alpha}$, from which

$$R/r_1 = 1 + f\cos\bar{\alpha} + O(f^2).$$

Substituting this back into Eqn (6.1) we find that

$$\tan\alpha = [1 - (f+g)(\sec\bar{\alpha} + f) + (f+g)^2 \sec^2\bar{\alpha}]\tan\bar{\alpha} + O(f^3),$$

the symbol $O(f^3)$ being used as shorthand for any third-order quantity. It follows, after some tedious algebra, that

$$\alpha = \bar{\alpha} - (f+g)\sin\bar{\alpha} + \tfrac{1}{2}g(f+g)\sin 2\bar{\alpha} + O(f^3). \tag{6.2}$$

Now, as Kepler eventually would discover, planets move on elliptical orbits, and the actual relationship between the true anomaly (the longitude as measured from the Sun) and the mean anomaly ($2\pi t/T$, where t is the time and T is the orbital period), is[60]

$$\alpha = \bar{\alpha} - 2e\sin\bar{\alpha} + \tfrac{5}{4}e^2 \sin 2\bar{\alpha} + O(e^3),$$

where e is the eccentricity of the ellipse. These two expressions agree up to second order if we take $f = 3e/4$ and $g = 5e/4$, i.e. $f/g = 3/5$.

---

[60] Note that in the seventeenth century it was customary to measure the anomalies from aphelion (A in Figure 6.6), though nowadays they are measured usually from perihelion so as to permit a unified treatment of planets and comets. Formulas can be transferred from one system to the other by simply replacing α and $\bar{\alpha}$ by $\pi + \alpha$ and $\pi + \bar{\alpha}$, respectively.

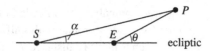

Fig. 6.7. Geocentric latitudes.

Kepler's procedure was nothing like the above, however. He used a complicated iterative geometrical procedure to work out from Tycho's observations
the best possible value for f/g. This clearly was incredibly tedious, since he
wrote:

> If this wearisome method has filled you with loathing, it should more properly fill
> you with compassion for me, as I have gone through it at least 70 times at the
> expense of a great deal of time . . .[61]

Kepler's calculations led him to conclude that, based on a radius of 100 000, he
should take $f = 7232$ and $g = 11\,332$, for which $f/g \approx 0.64$, which is very
close to the optimal value. The maximum deviation from the ideal elliptical
value is about $2'$ of arc, which represents a huge improvement over Copernicus
and is less than the errors introduced from other sources in Kepler's calculations.[62] As far as longitudes were concerned, Kepler's vicarious hypothesis
succeeded brilliantly, and he continued to use it as a method for determining
longitudes long after he had discarded it as a physical theory.

Next, Kepler turned his attention to latitudes. In Kepler's physically motivated astronomy there was no place for Copernicus' cumbersome theory in
which the planetary latitudes were linked to the motion of the Earth. As far as
Kepler was concerned, Mars was orbiting the Sun, so it followed that its motion
lay in a plane through the Sun, and he determined that this plane had an inclination of $1° 50'$ to the ecliptic.[63] In a heliocentric system, the latitude as viewed
from the Earth can be considerably larger than this orbital inclination, as the
Earth can be much nearer the planet than the Sun. Thus, when a planet P is at
opposition, we have the situation shown in Figure 6.7, in which α is the latitude
as viewed from the Sun, which, in the case of Mars, has a maximum value of
$1° 50'$, and θ is the geocentric latitude. By measuring θ and determining α from

---

[61] KEPLER *New Astronomy*, Chapter 16. The reason Kepler carried out so many iterations was
due largely to the lack of any theory for dealing with redundant observations (see Gingerich
(1973b)). Kepler's iterative scheme is described succinctly in Kozhamthadam (1994),
Chapter 6.

[62] See Whiteside (1974).

[63] Kepler actually used three different methods for computing the inclination, each yielding the
same answer. This was confirmation that the plane of the orbit did pass through the Sun (see
Jacobsen (1999)).

his simple latitude theory, Kepler then could solve the triangle SEP and, hence, determine the distance to the planet in terms of the Earth–Sun distance.

Before Kepler, astronomers had been satisfied with separate theories for latitude and longitude, but to Kepler, who for the first time was basing his astronomy on physical principles, both phenomena should have the same cause. Thus, he checked to see whether his hypothesis matched latitude observations. The angles involved are small but, nevertheless, the accuracy of Tycho's observations was sufficient to show Kepler that his vicarious hypothesis was wrong. For an ellipse with major axis of length $2R$ and with the Sun at one focus, the Sun–Planet distance is given in powers of the eccentricity by

$$r/R = 1 + e\cos\alpha - e^2\sin^2\alpha + O(e^3),$$

but the vicarious hypothesis has (from the cosine rule applied to the triangle OSM in Figure 6.6: $R^2 = r^2 + g^2R^2 - 2rgR\cos\alpha$):

$$r/R = 1 + g\cos\alpha - \tfrac{1}{2}g^2\sin^2\alpha + O(e^3).$$

Thus, to obtain agreement to first order, we need to take $g = e$, and Kepler found (using completely different methods, of course) that to construct a model based on the equant that predicted accurately planetary distances, he needed to place the equant and the Sun equidistant from the centre of the orbit, exactly as Ptolemy had done in his geocentric scheme. The longitudes are then given, from Eqn (6.2), by

$$\alpha = \bar{\alpha} - 2e\sin\bar{\alpha} + e^2\sin 2\bar{\alpha} + O(e^3),$$

which now differs from the true relation at second order, the error being $(e^2/4)\sin 2\bar{\alpha}$. The maximum error of about $8'$ occurs when $\bar{\alpha} = \pm 45°$, i.e. near the octants. There is no way that anybody working with observations made prior to Tycho's could have detected such an error, but Kepler was well aware of the accuracy of the data he was working with and realized there was a problem:

> Since the divine benevolence has vouchsafed us Tycho Brahe, a most diligent observer, from whose observations the $8'$ error in this Ptolemaic computation is shown, it is fitting that we with thankful mind both acknowledge and honour this benefit of God. For it is in this that we shall carry on, to find at length the true form of the celestial motions...[64]

Kepler now had perfectly good theories for both longitude and latitude, but they involved different geometrical constructions. His refusal to accept an error of $8'$ when he tried to explain everything from a single model led

---

[64] KEPLER *New Astronomy*, Chapter 19.

him onto a tortuous, but ultimately rewarding, path of discovery. Errors that previously would have been considered acceptable became the driving force behind the development of a radical, new astronomy. He turned his attention to the theory of the Earth's motion which he recognized as the key to a deeper understanding of planetary orbits. Observations necessarily are made from the Earth and then calculations performed to infer things about the orbit of a planet round the Sun. Because of the high degree of precision to which Kepler was working, this procedure could only work if the position of the Earth was known accurately. Now, in Kepler's physical conception of the Universe, there should be no difference between the theory for the Earth and for the other planets. But in *On the Revolutions*, all the planets except the Earth had an epicycle that played the role of a Ptolemaic equant. In essence, neglecting the variations in eccentricity that took place over a long period of time, Copernicus' theory for the orbit of the Earth was simply an eccentric circle, exactly like Hipparchus' solar theory devised over 1500 years previously. The reason that solar theory had lagged behind theories for the planets was that the methods developed for making accurate observations of planetary positions were no use for the Sun as it is never seen against the backdrop of fixed stars, and the one technique for obtaining very accurate solar positions was based around eclipse observations, which in turn depend on the highly complex motions of the Moon.

Kepler's method for obtaining accurate positions of the Earth was ingenious. He could determine the position of the Earth from observations of Mars, provided he could place Mars accurately. Martian longitudes were fine, as we have seen, but not distances. Kepler got round this by using observations of Mars separated by its zodiacal period (687 days) so that, however far Mars was from the Sun, it was the same each time. With this device he managed to show that the Earth's orbit was better represented by a Ptolemaic equant–eccentric mechanism, exactly like that of the other planets.[65]

Kepler appreciated the equivalence between Copernican and Ptolemaic planetary theory, and he knew that, in Ptolemy's scheme, the epicycle represented the orbit of the Earth round the Sun. In order to modify Ptolemy's astronomy so as to take advantage of the refined Earth orbit, the whole of Kepler's theory, complete with equant and bisected eccentricity, had to be attached to the planet's deferent. With his improvement in the theory of the Earth's motion, Kepler

[65] Koyré (1973) suggested that Kepler's calculations demonstrated that the Earth's eccentricity should be bisected as in Ptolemaic planetary theory, but Kepler's method was not accurate enough to show that the centre of the orbit of the Earth should be placed exactly half way between the Sun and the equant (Wilson (1968)). He assumed simply that this was the case because it fitted-in best with his physical theory in which speeds vary inversely with the distance from the Sun.

demonstrated conclusively the greater simplicity of the heliocentric structure over the geocentric one, and realized he had hammered another nail in the coffin of Ptolemaic astronomy:

> And finally . . . the sun itself . . . will melt all this Ptolemaic apparatus like butter, and will disperse the followers of Ptolemy, some to Copernicus' camp, and some to Brahe's.[66]

Kepler had suggested in the *Secret of the Universe* that the speed of an individual planet was inversely proportional to its distance from the Sun.[67] In the *New Astronomy*, he provides a geometrical demonstration of the fact that, at least at perihelion and aphelion, this is precisely what Ptolemy's equant mechanism achieved. He also appreciated that, at other points on the orbit, this distance law was only satisfied approximately, but he put this down to the equant mechanism being a geometrical hypothesis that gave only approximate agreement with the physical law. The latter he assumed to be true throughout the orbit.

What was the 'force' that made the planets move in this way? Kepler believed that it was similar to magnetism, which had been the subject of a very influential book, *On the Magnet*,[68] by the respected Englishman (and royal physician) William Gilbert in 1600. Gilbert concluded that the Earth was a giant magnet and Kepler reasoned that the 'motive virtue' present in the Sun which drove the planets on their orbits was also present, to a lesser extent, in the Earth, and that this was responsible for the Moon's motion. Kepler's force was not the same as magnetism, as it did not attract the planets to the Sun; instead, he envisioned a rotating Sun with rotating filaments emanating from it that drive the planets round. Kepler's erroneous conception of inertia meant that in order for the planets to be moving continually, they had to be subject constantly to a force in the direction of motion. This motive virtue spread out like light as one moved away from the Sun and, hence, its effect diminished with distance. But, unlike light – which spreads out spherically and, hence, varies inversely as the square of the distance – Kepler wanted to produce a motion that decayed in direct proportion to the distance, and so he believed that the effect of the

---

[66] KEPLER *New Astronomy*, Chapter 26.
[67] Kepler did not express the law in terms of the instantaneous speed of the planet, a concept with which he was not entirely at ease, but instead said that the time taken to traverse equal arcs was proportional to the distance from the Sun. This still leaves the concept of instantaneous distance, but Kepler seems to have been more comfortable with this, even if, understandably, he did not know how to treat it properly in his mathematics.
[68] A translation of *De magnete* was produced in the late nineteenth century and reprinted in the 1950s (Gilbert (1958)). Gilbert argued strongly in favour of the daily rotation of the Earth, but chose to sit on the fence regarding heliocentrism (Russell (1973)). Margolis (2002) argues that his outlook was clearly Copernican even if he never stated the fact explicitly.

filaments was in some way concentrated near the ecliptic plane in which the planets move.

This could not be the whole story, though, because otherwise the planets simply would orbit on circular paths centred on the Sun. So Kepler thought that each planet had to possess its own innate ability to steer itself along its orbit. He tried to work out what sort of additional motion a planet would have to have in order to turn a circular orbit centred on the Sun into an eccentric circular orbit, and he tried various devices including epicycles and librations (oscillations toward and away from the Sun). He concluded that the necessary variation in solar distance was extremely complicated and, thus, that in a physical astronomy, eccentric circular orbits were unnatural.

Nevertheless, he persevered with them and attempted to work out how a planet would move around such an orbit under the influence of his distance law. Since the distance was varying continuously, this was no easy task without the use of the calculus, and Kepler had to develop his own methods.[69] With the calculus, however, it is clear that, for a planet moving on a circular orbit, Kepler's distance law is equivalent to the statement

$$\frac{d\theta}{dt} \propto \frac{1}{r},$$

where θ is the angle between the line connecting the planet to the centre of the orbit and some fixed line, and r is the distance to the (eccentric) sun. In other words, $t \propto \int r \, d\theta$.

Kepler took the practical approach of dividing the circular orbit into 360 parts and in each part assuming that the distance to the Sun was constant; then $\int r \, d\theta$ can be treated as a sum. This procedure was 'mechanical and tedious' and Kepler looked for an alternative. He remembered that Archimedes had shown that the area of a circle could be obtained by dividing it into 'infinitely many' triangles, and in a similar, but less rigorous, vein, Kepler argued that the sum of all the (infinitely many) radii would be proportional to the area. If he could simply compute the area of the sector to calculate the time, this would represent a great saving of time. But it did not quite work, and the problem is illustrated in Figure 6.8.

On the left, we have two positions of the planet, P_1 and P_2, as it orbits the Sun S, which is displaced from the centre of the orbit O. As the planet moved from P_1 to P_2, Kepler wanted to sum up the lengths of all the lines connecting the planet to the Sun as being proportional to the area of the triangle $P_1 S P_2$. But

---

[69] The philosophical basis that underpins Kepler's treatment of infinitesimals and in particular his inspiration from the fifteenth century Platonist, Nicholas of Cusa, is described in Aiton (1973).

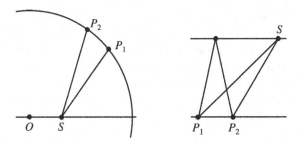

Fig. 6.8. The difference between the area law and the distance law.

he knew that this could not be the case, since these lines were not perpendicular to the base of the triangle. In the right-hand diagram there are two triangles, the areas of which clearly are identical, since they have the same base and height, but it is clear that the distances from the vertex to the base in the oblique triangle are greater. It was intuitively obvious to Kepler that his area law would be equivalent to his distance law only if the lines connecting the planet to the Sun always were perpendicular to the path of the planet, which, of course, only happens if the Sun is at the centre of the orbit. But the eccentricities involved are very small, so Kepler argued that the area law, being much simpler to work with, could be used as an approximation to the true distance law. The title to Chapter 40 of the *New Astronomy* was 'An imperfect method for computing the equations from the physical hypothesis, which nonetheless suffices for the theory of the sun or earth'.

Kepler then computed longitudes for Mars using this area law and found errors from those computed using the vicarious hypothesis of 8′ at the octants. Conscious of the fact that there were a number of possible causes for this error, he proceeded systematically to show that they were all, including that caused by using the area law instead of the distance law, far too small to account for the 8′ discrepancy. Finally, after a great deal of toil, Kepler realized that there was only one possibility left: the orbit was not a circle.[70]

This was, of course, a crucial realization, but it is not at all surprising that it took Kepler some time to arrive at it. Circles had been in the forefront of astronomical theories throughout history, and no other curves, other than the straight line, had ever been used in the explanation of physical phenomena.

---

[70] Of course, the planetary paths in the Ptolemaic and Copernican systems are not circles either, they are complicated curves built up from uniform circular motions; Kepler described them as having the shape of pretzels (*New Astronomy*, Chapter 1). To those before Kepler, however, the actual path was of secondary, or even negligible, importance, but in the new physical astronomy it was fundamental.

Once he had dropped boldly the notion of circularity, he was forced to step out into a largely untrodden mathematical world, and numerous new questions were raised. Most obvious was the question of the actual shape of the orbit. A circle is a well-defined geometrical object, but saying that something is not a circle does not narrow its form down very far. Kepler showed a great deal of faith in his whole strategy at this point, since it is by no means obvious that a solution to his problem existed that was both plausible physically and tractable mathematically.

To find what shape the orbit really was, Kepler calculated the position of the planet based on his eccentric circle theory and from observation. Since he now had accurate positions for the Earth based on observations of Mars, he could use the same procedure in reverse to obtain positions for Mars. He found that, away from the line of apsides, the planet was always closer to the Sun than the eccentric circle:

> ... the orbit of the planet is not a circle, but comes in gradually on both sides and returns again to the circle's distance at perigee. They are all accustomed to call the shape of this sort of path 'oval'.[71]

Kepler looked round for a physically plausible mechanism that could give rise to an oval orbit. In trying to devise a mechanism that converted a circular orbit into an eccentric circle, Kepler earlier had used epicycles, but had found that the required rate of rotation was non-uniform and, hence, physically unacceptable. However, if the epicycle turned at a uniform rate (something for which Kepler could come up with a plausible physical mechanism[72]) the orbit would not be a circle, but an oval, and this was precisely what Kepler was after. So, notwithstanding the fact that he had argued already that epicycles could not physically be real, he proceeded to use precisely such a device. Kepler's oval, constructed from a circle and a uniformly rotating epicycle, is illustrated in Figure 6.9 and using modern techniques we can judge its success.[73]

We begin with a circle of radius R centred on the Sun S, on which is mounted an epicycle of radius eR, centre C. Mars M rotates around the epicycle so that CM makes an angle t with the line SC extended. Here, t is the time in units of $T/2\pi$, where T is the orbital period of Mars or, in other words, t is the

[71] KEPLER *New Astronomy*, Chapter 44.
[72] In essence, Kepler hypothesized that the force from the Sun was sometimes attractive and sometimes repulsive, depending on the relative orientations of the Sun and planet. Such a phenomenon could be explained if, for example, we consider the Sun as a fixed magnet and a planet as a magnet, the axis of which retains a constant direction with respect to the stars (see Stephenson (1987)).
[73] Kepler's investigations into the oval orbit are discussed in Aiton (1978).

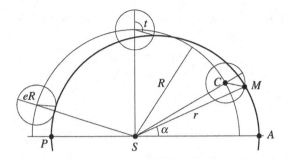

Fig. 6.9. Kepler's oval.

mean anomaly. The distance from the Sun to Mars is r, and $\angle MSA$ is the true anomaly, α. According to Kepler's area law approximation, α varies so that the area of the sector MSA increases at a uniform rate. The cosine rule on triangle SCM gives $r^2 = R^2(1 + 2e\cos t + e^2)$ and the area law is equivalent to (since the area of a sector with angle $\delta\theta$ is $r^2\delta\theta/2$),

$$\frac{d\alpha}{dt} = \frac{c}{r^2},$$

(6.3)

for some constant c. It follows that

$$\alpha = \frac{c}{R^2} \int (1 + 2e\cos t + e^2)^{-1} dt$$

(6.4)

$$= \frac{2c}{R^2(1 - e^2)} \tan^{-1}\left[\left(\frac{1-e}{1+e}\right)\tan \tfrac{1}{2}t\right],$$

where the constant of integration has been chosen so that $\alpha = 0$ when $t = 0$. We also require $\alpha = \pi$ when $t = \pi$, so we must choose $c = R^2(1 - e^2)$, and then

$$\tan \tfrac{1}{2}\alpha = \left(\frac{1-e}{1+e}\right)\tan \tfrac{1}{2}t.$$

(6.5)

If we expand the integrand in Eqn (6.4) in powers of e, and integrate term by term, we obtain:

$$\alpha = t - 2e\sin t + e^2 \sin 2t + O(e^3),$$

which is precisely the same as the result from the bisected eccentricity model and, thus, subject to the same errors of $8'$ near the octants.

Interestingly, Kepler's oval has a simple and exact polar equation.[74] It follows from Eqn (6.5) that

$$\cos t = \frac{(1 + e^2)\cos\alpha - 2e}{1 + e^2 - 2e\cos\alpha},$$

and, hence, that

$$\frac{r^2}{R^2} = 1 + 2e\cos t + e^2 = \frac{(1 - e^2)^2}{1 + e^2 - 2e\cos\alpha}.$$

If we expand this in powers of e, we obtain

$$\frac{r}{R} = 1 + e\cos\alpha - \tfrac{3}{2}e^2\sin^2\alpha + O(e^3).$$

Thus, the eccentric circle model with which Kepler began overpredicts the true distances and the oval underpredicts them by the same amount (at second order). The oval orbit, since it improves on neither longitudes nor distances, must therefore be considered a failure. However, it was a stepping stone toward the correct path because, in working with the oval and the area law, Kepler had to compute areas of its sectors – something he could not do – so he simplified his laborious calculations by assuming that it was an ellipse! It was not the ellipse that corresponds to the actual orbit – in particular the Sun was not at a focus – but it was an ellipse nevertheless. Thus, at this stage Kepler was using two 'approximations' to simplify his calculations – the area law and an elliptical orbit – both of which would turn out not to be approximations at all.[75]

Kepler had reached this point in his war with Mars by mid 1603, and he decided to set Mars aside for a while to concentrate on his researches into optics, including a study of atmospheric refraction.[76] He resumed battle in summer 1604, and sometime in early 1605 he had a flash of inspiration. He noticed that at quadrature (Q in Figure 6.10) the distance of Mars from the Sun was precisely half the distance between aphelion and perihelion, $|AP|$. He hypothesized that the distance r would vary with θ according to the equivalent of the formula $r = a(1 + e\cos\theta)$, for which he devised a complicated and unconvincing pseudo-magnetic physical theory, and referred to the resulting

[74] This was first derived as recently as 1962 by Kuno Fladt (Whiteside (1974)). Aiton (1978) has pointed out that since at this time Kepler believed the area law to be an approximation to the true distance law, this polar equation should be thought of only as an approximation to the actual oval Kepler had in mind.

[75] Strictly speaking, the area law and elliptical orbits *are* approximations due to the perturbing effects of the other planets.

[76] The *Astronomiae pars optica* (*The Optical Part of Astronomy*) was published in 1604. It contained influential work on pinhole images as well as the first essentially correct account of human vision.

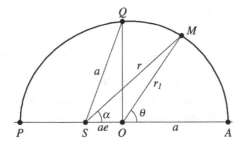

Fig. 6.10. Kepler's puffy-cheek orbit.

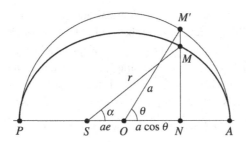

Fig. 6.11. The elliptical orbit.

orbit as the 'puffy-cheek' path because it was broader near aphelion than near perihelion.

From the cosine rule on triangle OSM and

$$r \cos \alpha = ae + r_1 \cos \theta = ae + \frac{r_1(r - a)}{ae},$$

we can derive the expansion for r in powers of e:

$$\frac{r}{a} = 1 + e \cos \alpha - e^2 \sin^2 \alpha + O(e^3),$$

exactly as for an ellipse. Kepler's calculations confirmed that this new orbit reproduced planetary distances accurately and he could have carried his calculations through for the 'puffy-cheek' path and, hence, find that the anomaly α was also predicted accurately by this orbit together with the area law, but, instead he went on, realizing that with a minor modification he could make his 'puffy-cheek' a perfect ellipse. Kepler kept the equation $r = a(1 + e \cos \theta)$ but, instead of making θ the angle between OM and OA, he took it to be the angle between OM' and OA, where M' is the point on the circle having diameter PA (known as the auxiliary circle) which is also on the perpendicular to PA through M (see Figure 6.11). This angle is known as the 'eccentric anomaly'.

It is clear that $|NM'| = a \sin \theta$ and a simple application of Pythagoras' theorem yields

$$|NM| = [r^2 - a^2(e + \cos \theta)^2]^{1/2} = a \sin \theta \sqrt{1 - e^2} = |NM'| \sqrt{1 - e^2},$$

a geometrical property that Kepler knew implied that AMP was an ellipse. It is straightforward to derive the standard polar equation for an ellipse from Kepler's equations.[77] Since $r \cos \alpha - ae = a \cos \theta$ and $a \cos \theta = (r - a)/e$, we immediately have

$$r = \frac{a(1 - e^2)}{1 - e \cos \alpha}.$$

Although Kepler had no good empirical evidence for throwing away the 'puffy-cheek' orbit in preference to the ellipse, he was convinced of the truth of the elliptical orbit.[78] Moreover, although Kepler had only derived this new orbital shape for Mars, he assumed simply that since the cause of the motion was the same for all the planets, all the planets must behave in the same way. Nowadays, Kepler's first law of planetary motion usually is stated as:

Planets move on elliptical orbits with the sun at one focus

but it is worth noting that nowhere in the main body of the *New Astronomy* is the word 'focus' mentioned, and it was only later in his *Epitome of Copernican Astronomy* that Kepler emphasized this aspect of planetary orbits. Kepler's physical mechanism by which the planets were forced to move in such a fashion was incorrect, but the idea that the study of celestial motions should be based on physical causes influenced all those who followed in his footsteps.[79]

Kepler derived a very elegant relation that connects the mean anomaly (i.e. the time t) to the eccentric anomaly, θ. It is obvious from Figure 6.11 that the areas of the sectors $SM'A$ and $OM'A$ differ by the area of the triangle $SM'O$, which is $(a^2 e/2) \sin \theta$ and, since $OM'A$ is a sector of a circle, its area is $a^2\theta/2$. Now, M rotates around its elliptical orbit so that the area of the sector SMA is proportional to the time t and, since $|NM'|/|NM|$ is a constant ratio, it follows

[77] First written down in 1664 by Nicholas Mercator.
[78] The justification that Kepler provided for the ellipse was, to say the least, misleading (Donahue (1988)).
[79] Most of Kepler's contemporaries remained unconvinced of the value of turning astronomy into a physical science. Even Michael Mästlin had his doubts. In a letter to Kepler written in 1616, he wrote: 'Concerning the motion of the moon you write you have traced all the inequalities to physical causes; I do not quite understand this. I think rather that here one should leave physical causes out of account, and should explain astronomical matters only according to astronomical method with the aid of astronomical, not physical, causes and hypotheses' (Holton (1988)).

that the area of the sector $SM'A$ is also proportional to the time. When $t = \pi$, the area of $SM'A$ is simply that of a semicircle, $\pi a^2/2$, and so the area of the sector $SM'A$ at an arbitrary value of t is $a^2 t/2$. We, thus, have what is now known as 'Kepler's equation':

$$t = \theta + e \sin \theta. \qquad (6.6)$$

It was the constant ratio $|NM|/|NM'|$ that enabled the area law to be applied easily and made the elliptical orbit so attractive. However, in order to determine θ for a given t from Eqn (6.6), Kepler had to use a tedious iterative procedure and challenged mathematicians to find a better method. Many tried, and Kepler's method was improved upon, but the procedures remained difficult and time-consuming.[80]

Throughout the *New Astronomy*, Kepler made extensive use of the area law, but initially he believed that it was only an approximation to the physically correct but mathematically inconvenient distance law. By the time he wrote the *Epitome of Copernican Astronomy* 10 years later, he believed in the truth of the area law and had changed the distance law to the equivalent of the statement that it was the component of the velocity perpendicular to the Sun–planet line that varied inversely with the distance from the Sun.[81] In other words,

$$r \frac{d\alpha}{dt} = \frac{c}{r},$$

which is precisely the same as the area law, Eqn (6.3). Not only does this revised form of the distance law fit in with the area law, it also fits in very well with Kepler's physical conception of magnetic filaments rotating in circles around

---

[80] Kepler's iterative procedure was as follows. Guess an approximate solution θ_0 for θ. Then $\theta_0 + e \sin \theta_0 = t + \varepsilon$ for some ε. Let $\theta_1 = \theta_0 - \varepsilon = t - e \sin \theta_0$. Then $\theta_1 + e \sin \theta_1 = t + O(\varepsilon^2) + O(e\varepsilon)$. Thus, provided both e and ε are small, θ_1 should be a better approximation. This procedure can then be repeated. Despite the fact that satisfactory methods of solution have been known for a long time, Kepler's equation has continued to be a source of interest right up until the present. For details, see, for example, Battin (1987), Chapter 5, and Colwell (1993), the latter containing an extensive list of references. Interestingly, Kepler was not the first to use this iterative procedure for the solution of the equation $t = \theta + e \sin \theta$. An equation of exactly the same form was derived by Islamic astronomers to aid the reduction of observations to the center of the Earth so as to account for parallax. Ḥabash al-Ḥāsib, a contemporary of al-Khwārizmī (ninth century), developed an iterative solution procedure that was equivalent to Kepler's (see Kennedy (1956a)). In the nineteenth century, attempts to find a series solution to Eqn (6.6) led Bessel to study the functions that now bear his name (see Watson (1944), Chapter 1). In fact, the solution of Eqn (6.6) can be written as $\theta = t + \sum_{n=1}^{\infty} 2(-1)^n n^{-1} J_n(ne) \sin nt$, where J_n is a Bessel function of the first kind.

[81] Kepler's treatment of the area law in the *Epitome of Copernican Astronomy* is described in Davis (2003).

the Sun, the influence of which diminishes in proportion to the distance from the Sun. Also, from Kepler's equation,

$$\frac{dt}{d\theta} = 1 + e\cos\theta = \frac{r}{a}$$

and, hence,

$$a\frac{d\theta}{dt} = \frac{a^2}{r},$$

so saying that M moves around an ellipse subject to the area law is equivalent to saying that the speed with which M' moves around the auxiliary circle varies inversely with the Sun–planet distance. Although the ideas developed in Kepler's mind over a number of years, it is stated nearly always that Kepler's second law,

> *Planets move so that the line connecting them to the sun sweeps out equal areas in equal times*

was given in the *New Astronomy*.

Kepler's first two laws of planetary motion truly were revolutionary. With them he overthrew the fundamental premise of all mathematical astronomy that preceded him, i.e. that celestial motions were built up from uniform circular motions.[82] The fact that Kepler discovered these laws is made all the more remarkable by the fact that he did so from observational data alone and without the help of Newtonian dynamics. Crucial to his discovery was his insistence that the motion of the planets was a *physical* process, with the Sun playing a predominant role. Although Kepler's physics was erroneous, it did enable him to ask some of the right questions and to look in the right place for the answers.

The work for the *New Astronomy* was pretty much complete by the end of 1605, but problems arose over its publication. Apart from some financial difficulties there was the problem that, since the work was based on Tycho's observational data, Tycho's heirs insisted on the right to censor it, and they were not happy that the work was based on a heliocentric universe. Eventually, Kepler managed to overcome these difficulties and the book appeared in 1609 – and was almost totally ignored! The reception of Kepler's work and its gradual assimilation into the mainstream of astronomical thought[83] will be discussed in the next chapter. It should be noted, however, that the *New Astronomy* was

---

[82] There was still a very deep sense in which Kepler believed in the perfection of circles and circularity, feelings that manifest themselves in his later work, *Harmony of the World* (see Brackenridge (1982)). Many other influential astronomers remained obsessed with circularity, notably Galileo.

[83] Described in detail in Russell (1964).

extremely difficult reading. While Kepler's descriptions of his dead ends and false leads may be fascinating to the modern reader interested in the question of how Kepler arrived at his conclusions, they would have been lengthy and unnecessary diversions at the time. On top of this, Kepler's mathematics was often clumsy and involved concepts with which most of his contemporaries were unfamiliar. Ellipses were little studied and their geometrical properties largely unknown, and until the development of the calculus by Newton and Leibniz, techniques for dealing with motion around elliptical orbits subject to an area law were hopelessly inadequate.

The *Harmony of the World* and the third law

The first 40 years of Kepler's life were by no means plain sailing, but he managed to overcome those difficulties that presented themselves and, by 1611, had established his reputation as one of the foremost astronomers of his day. Then tragedy struck. His wife became seriously ill toward the end of 1610, and died in July 1611. His three children all contracted smallpox in January 1611 and, although the youngest and eldest survived, his 6-year-old son Frederick died. On top of this, the political situation in Prague forced Kepler's patron, the Emperor Rudolph, to abdicate in May 1611 and he died the following year.

The patronage of the emperor, who was much more interested in artistic and scientific endeavour than in religious differences, was the main reason that Kepler, a Protestant, had managed to live without persecution in early seventeenth-century Prague. With Rudolph's death, Kepler had to leave Prague, and in 1612 he took up a new position as District Mathematician in Linz, a job that basically was created for him by influential people who wanted Kepler working in their city. This was rather a comedown for the Imperial Mathematician, but he at least received a regular salary and had the time to begin the laborious calculations of the orbital parameters required to fit the celestial bodies other than Mars to their elliptical orbits around the Sun. These calculations formed the basis of the *Rudolphine Tables*, which were published in 1627.

In 1613, as well as remarrying, Kepler became interested in the problem of determining the volumes of wine barrels of different-shapes. He realized that the ideas he had developed when determining areas of curved regions as part of his war with Mars could be brought to bear on this subject, and this led to the publication of his *New Solid Geometry of Wine Barrels* in 1615.[84] Kepler was

---

[84] *Nova steriometria doliorum vinariorum.* Kepler's method involved dividing volumes into infinitely many infinitesimal regions and represents a primitive integral calculus. It was expanded systematically by Bonaventura Cavalieri.

thus not only working at the forefront of astronomical research, but also paving the way for those who would later create the new mathematical methods of the eighteenth century.

Kepler also planned to produce a textbook that would combine the Copernican heliocentric theory with his own discoveries into a comprehensive whole. The first part of this work, the *Epitome of Copernican Astronomy*, went to press in 1617, but the book (described in the next chapter) was not completed until 1621. Kepler's life, never simple, took another downward turn in 1617 when his mother was accused of witchcraft and threatened with torture and execution. He expended considerable energy in fighting on her behalf through the complex legal process that ensued, and Frau Kepler eventually was cleared, though she died soon afterwards. On top of all this, three of Kepler's children died within a 6-month period in 1617–18.

In spite of the tragedies that befell him, Kepler managed to turn his attention back to the project he had abandoned in 1599, the sequel to the *Secret of the Universe*, which was to provide a comprehensive study of harmony as it occurred in geometry, arithmetic, music, astrology, and astronomy. It is tempting to think that the study of harmony rather than the computation of astronomical tables was just what Kepler needed at this trying time. To complete his theories of harmony in astronomy, he wanted to know the relationship between the speeds of the planets and their distances from the Sun. This he discovered on 15 May 1618 and a few days later he wrote:

> Now, because eighteen months ago the first dawn, three months ago the broad daylight, but a very few days ago the full sun of a most remarkable spectacle has risen, nothing holds me back. Indeed, I give myself up to a sacred frenzy.[85]

True to his word, he completed his work quickly – the *Harmony of the World* was finished by 27 May!, though the printing of the book took more than a year.[86] It is believed widely that of all Kepler's works, this is the one that gave him the most pleasure.

The *Harmony of the World* is a fascinating book, with mathematical discussions on such things as the constructibility (using straight edge and compass alone) of the regular polygons, tessellations of the plane and semi-regular

[85] Quoted from Caspar (1993), p. 267.
[86] Some modifications needed in the light of his new discovery were incorporated during 1619 (Field (1988), p. 143). The first complete English translation of the *Harmonice mundi* is that of Kepler (1997) which contains a lengthy introduction and extensive notes. Interestingly, Kepler dedicated the work to King James I of England, whom he believed had it in his power to reunify Protestants and Catholics.

polyhedra or Archimedean solids[87] (Kepler gave the first-known proof that there are exactly thirteen of them). Kepler also discussed the relative merits of long-established geometrical methods and newly developed algebraic techniques for the solution of various problems. For example, Kepler believed that regular polygons that were not constructible using a ruler and compass were 'unknowable', and played no role in God's design of the Universe.[88] When Jost Bürgi showed him that an algebraic relation could be established between the side of a regular heptagon and the radius of its circumscribing circle, Kepler insisted that this was of no use as it did not provide a geometrical construction.

The fifth and final book of the *Harmony of the World* attempts to explain the harmonic relationships that govern astronomical phenomena and contains Kepler's statement of his third law of planetary motion. Kepler had been attracted by the Pythagorean idea of the harmony of the spheres from his *Secret of the Universe* days. He believed that were the heavens filled with air, the planets would produce audible sounds, but that in the absence of air it was an intellectual harmony that existed. In his early work he had assigned speeds to the planets such that the ratios of these speeds resulted in musical consonances. He chose the relative speeds of Saturn, Jupiter, Mars, Earth, Venus, and Mercury as 3, 4, 8, 10, 12 and 16, respectively, so that, for example, Mars and Jupiter produced a ratio of 2, which corresponds to an octave, whereas Venus and Mars produce a ratio of 3 : 2, a fifth. From these speeds, together with the known orbital periods, Kepler could calculate the relative distances of the planets from the Sun, and he obtained a better agreement with Copernican theory than he had from his polyhedral hypothesis.

However, calculations based on Tycho's accurate observations showed Kepler that this theory based on simple arithmetic ratios was not correct, and instead he sought musical harmony in relations between constructible polygons. He asked then in what way the design of the Universe reflected this complex

---

[87] An Archimedean solid is a convex polyhedron which has faces that are all regular polygons (of at least two different types) and the vertices of which are all identical. No work of Archimedes describing these solids has survived, but Pappus of Alexandria (AD fourth century) credits him with their discovery (Boyer (1989), Section 8.12).

[88] In Kepler's time, the only known constructible regular polygons were those with three, four and five sides and those derivable from them, i.e. those with the number of sides of $2^n \, 3^m \, 5^l$ with n a non-negative integer and m, $l = 0$ or 1. In fact, Kepler includes a proof in Book I of the *Harmony of the World* that regular polygons with a prime number of sides greater than 5 are not constructible, but his proof is flawed. In 1796, when only 19, Gauss proved that a regular polygon with a prime number of sides p is constructible if, and only if, $p - 1$ is a power of 2. Gauss' result shows that it is possible to construct regular polygons with 3, 5, 17, 257 and 65 537 sides. It is not known if there are any more such primes p.

Table 6.2. *The harmonies implied by the daily motions of the planets near aphelion and perihelion.*
(A diesis is the difference between a whole minor tone and a semitone, i.e. $24/25 = 9/10 \div 15/16$.)

		Apparent daily motion	Musical harmony
Mercury	Aphelion	164′	5 : 12
	Perihelion	384′	Octave and minor third
Venus	Aphelion	94′ 50″	24 : 25
	Perihelion	97′ 37″	Diesis
Earth	Aphelion	57′ 3″	15 : 16
	Perihelion	61′ 18″	Semitone
Mars	Aphelion	26′ 14″	2 : 3
	Perihelion	38′ 1″	Fifth
Jupiter	Aphelion	4′ 30″	5 : 6
	Perihelion	5′ 30″	Minor third
Saturn	Aphelion	1′ 46″	4 : 5
	Perihelion	2′ 15″	Major third

musical harmony and found his answer in the apparent daily motions of the planets as viewed from the Sun:

> ... we should look not how high any particular planet is from the sun, nor what space it traverses in a single day—for that is rational and astronomical, not instinctive—but how large an angle the daily motion of each planet subtends at the actual body of the sun, or how large an arc on one common circle drawn about the sun, such as the ecliptic, it seems to complete on any particular day.[89]

Elliptical orbits with the area law imply that planets move faster near perihelion than near aphelion, and Kepler computed the daily motion in longitude of each planet at these points. The ratios of these motions he found to be approximated closely by harmonic ratios (see Table 6.2). He also went on to deduce similar harmonies between pairs of planets. These harmonic ratios were then *tempered*: tuned so that the planets would form a six-part harmony. For example, Saturn's ratio of 4 : 5 became 64 : 81 when 'tuned'.

Kepler thus had two different types of harmony in his celestial scheme. On the one hand he had the planetary spacings (which he believed still were guided by the regular polyhedra), and on the other he had the harmonies implied by the maximum and minimum angular speeds of the planets. He needed a harmonic relationship between the angular speeds of the planets and their distances from

---

[89] KEPLER *Harmony of the World*, Book V, Chapter IV.

the Sun in order to complete his theory, and he discovered this relationship in May 1618. Kepler's statement of his third law is equivalent to

The square of the period of a planet is proportional to the cube of its mean distance from the Sun.

The different aspects of the planetary motions then can all be tied together. At aphelion and perihelion, the area law is equivalent to Kepler's original distance law (speed varies inversely with distance) and so the ratio of angular speeds at these points satisfies

$$\omega_a/\omega_p = r_p^2/r_a^2,$$

where r_p and r_a are the perihelion and aphelion distances, respectively. Inserting the tempered values for ω_a/ω_p from his harmonic theory, Kepler could obtain the ratio r_p/r_a, from which the planet's eccentricity is determined easily:

$$e = \frac{r_a - r_p}{r_a + r_p} = \frac{1 - r_p/r_a}{1 + r_p/r_a}.$$

Next, Kepler calculates the mean angular speed of each planet; not the arithmetic mean of the extreme motions, $A = (\omega_a + \omega_p)/2$, and not the geometric mean $G = \sqrt{\omega_a\omega_p}$, but the quantity $\omega = G - (A - G)/2$ instead.[90] It is to this mean motion that Kepler then applies his newly discovered law to deduce the mean distance, and then the maximum and minimum distances are determined because the eccentricity is known. Kepler compared the results for perihelion and aphelion distances found by this elaborate procedure with those determined from Tycho's observational data, and the agreement is quite impressive.[91]

---

[90] Though Kepler does not derive this curious quantity, a simple analysis shows that he was calculating the angular speed ω that satisfies the relations $\omega/\omega_p = r_p^2/r^2$ and $\omega/\omega_a = r_a^2/r^2$, where $r = (r_a + r_p)/2$ is the mean distance. If we add and multiply these relations together we can derive

$$\frac{\omega A}{G^2} = 2 - \frac{r_p r_a}{r^2}$$

and

$$\frac{\omega}{G} = \frac{r_p r_a}{r^2},$$

and if we then combine these relations, we obtain $\omega = 2G^2/(A + G)$. On the assumption that A and G are close together, we have

$$\omega = 2G^2(2G + A - G)^{-1} = G\left(1 + \frac{A - G}{2G}\right)^{-1} \approx G - \tfrac{1}{2}(A - G).$$

[91] See Brackenridge (1982) for an analysis.

Table 6.3. *The accuracy of Kepler's third law based on Kepler's data.*

	Period T (years)	Relative distance a	T^2	a^3
Mercury	0.242	0.388	0.0584	0.0580
Venus	0.616	0.724	0.3795	0.3795
Earth	1.000	1.000	1.000	1.000
Mars	1.881	1.524	3.540	3.538
Jupiter	11.86	5.200	140.61	140.73
Saturn	29.33	9.510	860.08	867.69

The third law brought the whole of Copernican planetary theory together. While Copernicus had improved on Ptolemy in that the relative planetary distances could be determined from his theory by observation, there was no reason in Copernican theory for these distances being what they are. As far as Kepler was concerned, his third law provided that final link, and at the same time further justified his belief that the nature of the Universe could be described by simple mathematical relationships. The third, or harmonic, law was not given much emphasis in the *Harmony of the World*; in particular, Kepler did not publish a table of values showing the accuracy with which the third law relates the periods and distances of the planets. However, he must have performed these computations, and results are shown in Table 6.3.[92]

Kepler later (in the *Epitome of Copernican Astronomy*) developed a physical theory to 'explain' the third law. According to Kepler, there were four factors that influenced the period of a planet. First, there was the obvious fact that the further away a planet was from the Sun, the longer its orbital path, and this length is proportional to the radius of the orbit. Second, the force emanating from the Sun which pushed the planet around its orbit decayed in proportion to the distance. If all the planets were identical, this would imply that the period was proportional to the square of the radius. In order then to fit in with his newly discovered law, Kepler concluded that the planets must be different. In Kepler's physics, each planet resisted the force of the Sun, and this resistance was proportional to the quantity of matter in the planet; but to counter this, a larger planet would be able to absorb more of the solar virtue, and Kepler assumed that this would be proportional to its volume. Taken together, these last two factors cause an increase in the orbital period of a planet in proportion to its density. Kepler thus assumed that the volumes of the planets were proportional to their distances from the Sun and the quantity of matter they contained to the

[92] The numbers here are taken from Gingerich (1989).

square root of this distance, so that planetary densities varied inversely as the square root of the orbital radius, and in this *ad hoc* way he could reproduce his third law, $T^2 \propto a^3$, and absorb it into his physical astronomy.

There is a fundamental difference between Kepler's third law in modern astronomy and the law as appreciated by Kepler himself. In a Newtonian framework, the third law is true for any planet regardless of volume or mass, but for Kepler the law was only true for planets with a very precise relationship between density and orbital radius. In many ways the law as stated by Kepler is not a physical law at all, but simply a reflection of the design of the Universe.

7

Galileo, the telescope, and Keplerian astronomy

Galileo

Galileo Galilei, one of the founders of modern science, enrolled at the University of Pisa in 1581 as a student of medicine and philosophy but soon was drawn toward the study of mathematics, including mechanics and Ptolemaic astronomy.[1] In 1589, he became a professor of mathematics at Pisa with a keen interest in the study of motion and a desire to replace qualitative and, to him, implausible Aristotelian ideas, with laws that were both quantitative and more credible. He devoted much of the early part of his life to the study of dynamics or, more accurately, to the creation of a new mathematical science of moving bodies, and by 1604 had deduced that the distance travelled by a freely falling object was proportional to the square of the time taken.[2]

During the 1590s, Galileo developed an absorbing interest in the Copernican system and, although it made great physical sense to him, he (unlike Kepler) kept most of his ideas on the subject to himself.[3] Sometime around 1595 he constructed an argument for the cause of tides based on the daily rotation of the Earth, but he did not put it into print until 1616.[4] He argued that, since the

---

[1] There is a wealth of literature concerning Galileo. A good starting point is Machamer (1998).
[2] See Dugas (1988). That Galileo was not in total command of the dynamics of free fall at this time is clear from the fact that he also believed that the speed was proportional to the distance travelled rather than the time.
[3] Galileo's progression from rejection to acceptance of Copernicanism is described in Drake (1987). During his early years as a teacher, Galileo composed a *Treatise on the Sphere*, which basically followed Sacrobosco's thirteenth-century work. He was still teaching this elementary Ptolemaic material after he had converted to heliocentrism (see Drake (1978)).
[4] The *Discourse on the Tides* was written in the form of a letter to Cardinal Orsini (translated in Finocchiaro 1989). Prior to Galileo, most scholars (e.g. Kepler) attributed tidal motion correctly to the influence of the Moon on the oceans, based on the easily observed fact that certain tidal phenomena are related to the Moon's phases. The first detailed treatment of the relationship between the Moon and the tides is due to Posidonius (first century BC (see Darwin (1962),

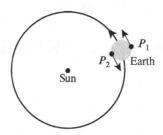

Fig. 7.1. Galileo's theory of the tides.

motion of a point on the surface of the Earth is made up of two components –
one due to the daily rotation of the Earth and the other to the annual motion
of the Earth around the Sun – the linear velocity of the point P_1 in Figure 7.1
(where it is midnight) is the sum of the linear velocity of the Earth in its orbit
around the Sun and the linear velocity due to the diurnal rotation. However, at P_2
(where it is noon) the linear velocity is the difference between these quantities.
Galileo put it like this:

> Thus I believe that it is clear how, though each part of the earth's surface moves
> with two very uniform motions, nevertheless within a period of twenty-four hours it
> moves sometimes very fast, sometimes slowly, and twice at intermediate speeds;
> and this change results from the combination of these two uniform motions, diurnal
> and annual.[5]

This constant speeding-up and slowing-down of a point on the Earth's surface
caused the oceans to slosh back and forth, resulting in the tides. This theory
is flawed,[6] but it is clear from his work on the subject that he recognized
the essentially localized nature of tidal phenomena. There is a periodic tide-
generating force (which Galileo got wrong) but it is the geometry of the ocean
basin that determines the actual behaviour of the fluid within it.

When Kepler sent him a copy of the *Secret of the Universe* in 1597, Galileo
thanked him and told him that he had been convinced of the Copernican hypothesis for some time but added that he was not prepared to publish anything
to that effect:

> Many years ago I came to agree with Copernicus, and from this position the causes
> of many natural effects have been found by me which doubtless cannot be

---

pp. 81–5). Galileo's theory is discussed at length in Palmieri (1998). The influence of
Copernicanism on Galileo's early investigations in dynamics is discussed in Naylor (2003).
[5] Quoted from Finocchiaro (1989).
[6] The argument confuses two different frames of reference. The motion of the Earth is
considered-relative to the Sun, but it is the motion of the water relative to the Earth that Galileo
was trying to explain.

explained by the ordinary supposition. I wrote down many reasons and arguments, and also refutations of opposite arguments, which, however, I did not venture until now to divulge, deterred by the fate of Copernicus himself, our master, who, although having won immortal fame with some few, to countless others appears . . . as an object of derision and contumely. Truly, I would venture to publish my views if more like you existed; since this is not so, I will abstain.[7]

Kepler encouraged Galileo to be less cautious. In his reply he wrote:

Be confident, Galilei, and proceed! If I am right, only a few of the chief mathematicians of Europe will keep aloof from us; such is the power of truth.[8]

In 1604, Galileo lectured on the supernova of that year and showed that the absence of diurnal parallax indicated that the new star occupied the region of the heavens above the Moon, supposedly unchanging according to Aristotle. By 1609, Galileo was what might be described as a cautious Copernican; in private he was prepared to advocate the motion of the Earth, but in public, for fear of ridicule, he was not. One of Galileo's reasons for not recoiling from the concept of a moving Earth was that his researches into the nature of motion had convinced him that Aristotle's theory was false. For example, Aristotle said that bodies fall with speeds proportional to their weights, but Galileo's experiments revealed that the speed of fall was independent of weight. More fundamental was Galileo's realization that motion could persist without any applied force. He did not quite formulate the principle of inertia that Newton arrived at a century or so later, because in Galileo's mind it was circular motion that was the natural state; thus, no force was necessary for the Earth to spin continually and, similarly, no force was required for the planets to orbit the Sun. This circular inertia was consistent with the ancient Pythagorean doctrine of uniform circular motion, but Galileo also applied his principle to terrestrial physics. In Galileo's famous work on projectiles, he assumed that the horizontal part of the motion (which was, in the absence of air resistance, not subject to change) actually was part of a circular motion, the radius of which was that of the Earth, and thus it was only approximately rectilinear.

While Galileo was Professor of Mathematics at Padua – perhaps the leading Italian university of the time – he learned of an invention that had the effect of making objects appear to be closer than they really were. Galileo was interested, and he had soon (by autumn 1609) built himself a telescope[9] with a magnification factor of about 8. He did not (as is sometimes said) invent the telescope,

---

[7] Quoted from Pannekoek (1961).   [8] Quoted from Pannekoek (1961).
[9] Galileo's preferred name was a perspicillum, and Kepler approved, so it is perhaps surprising that this did not last; for the explanation, see Rosen (1947).

and the original instruments he used probably were rather less effective than a modern pair of binoculars, but he was the first person to use such a device to make a systematic survey of the the heavens, with devastating consequences for man's view of the Universe.[10]

Not unnaturally, Galileo's first target was the Moon,[11] and when he looked at it he found it was not perfectly smooth and spherical as a heavenly body should be, according to Aristotle, but instead had shadows that moved as the angle of the Sun varied. He noticed that the border between the light and dark parts of the Moon was not perfectly straight. Near this border, in the dark portion, he observed light patches that grew larger and gradually merged with the light part of the surface as the Moon got fuller. He concluded that these were the tops of mountain ranges. Thus, the Moon was seen to be like the Earth in many ways, and not some pure crystalline object. He explained also the appearance of a slight illumination in the dark part of a thin crescent moon (sometimes referred to as 'the old moon in the new moon's lap') as arising from 'Earthshine', the reflection of sunlight onto the Moon by the Earth.[12] This offended Aristotelians for whom the Earth could not shine like a planet. Galileo observed the Milky Way through his telescope and found that it was made up of many, many faint stars. In fact, he soon realized that wherever he looked in the heavens, he could see more stars with his telescope than without it.

He also observed that, through a telescope, stars and planets looked different – the planets looked like discs, but the stars remained as twinkling points of light. He realized that both planets and stars were much smaller than naked-eye observations suggested; indeed, the telescope allowed the first accurate determination of the sizes of planets.[13] This removed one of the objections that Tycho Brahe had proposed concerning the vast distance to the fixed stars that was implied in *On the Revolutions*. Based on naked-eye observations, Tycho had calculated the sizes of stars that would be implied by this great distance, and found that they would have to have diameters that were of the same order of magnitude as the radius of the orbit of the Earth, something Tycho thought was ridiculous. Telescopic observations showed, however, that the apparent

---

[10] Another person to look at the heavens through a telescope around this time, maybe even earlier than Galileo, was Thomas Harriot (see, for example, Montgomery (1999), pp. 106–13). Harriot, who was one of the leading mathematicians in Elizabethan England, was responsible for the first recorded astronomical observation in North America, made during the attempted colonization of Virginia in 1585 (Yeomans (1977)).

[11] A century before Galileo, Leonardo da Vinci had written: 'Construct glasses to see the moon magnified' (Montgomery (1999), p. 97).

[12] Galileo was not the first to put forward this explanation (Ashbrook (1984), p. 198).

[13] For example, the traditional value for the angular diameter of Venus at apogee was about $3'$, but Galileo determined the much more accurate value of $\frac{1}{6}'$ (see van Helden (1989)).

diameters of the fixed stars were much smaller than had been thought previously. On the other hand, the advent of the telescope did not result in anybody being able to measure any stellar parallax, and so Copernicans had to keep revising upwards the distance to the fixed stars.

By early 1610, Galileo had built a telescope that could magnify about 20 times, and he turned his attention to Jupiter and saw four small star-like objects close to the planet. Observations over the next few nights showed that while these 'stars' moved with respect to each other and to Jupiter, they participated in Jupiter's motion with respect to the fixed stars, and he realized that they must be satellites of the planet. This discovery further undermined Aristotelianism, as it implied the existence of another centre of rotation. It also aided the Copernican cause because the Earth ceased to be the only planet with a satellite, thus lessening the force of any argument that was based on the special nature of the Earth. While Galileo's discoveries strengthened his Copernican convictions, he still chose, for the time being, not to publicize his beliefs.

He did, however, publish the results of his telescopic observations and he did this very quickly to ensure priority – within 2 months he had written up his findings in *The Starry Messenger*.[14] Since he had long wanted to secure an official position in the employ of the Medici family, he decided to call the new objects the Medicean Stars. His efforts were rewarded, and later in 1610 he moved to Florence to become Philosopher and Chief Mathematician to the Medici court.

Galileo's discoveries concerning Jupiter encouraged him him to point his telescope at other planets to see if they, too, had satellites. Later in 1610 he observed Saturn, but he was baffled by its curious shape:

> I have observed that Saturn is not a single star but three together, which always touch each other. They do not move in the least among themselves and have the following shape oOo, the middle being much larger than the lateral ones.[15]

---

[14] *Sidereus nuncius*, translated in Drake (1957). In fact, in 1614 Simon Mayr (Marius in Latinized form) published his *Mundus Jovialis* (*World of Jupiter*) in which he claimed to have observed the four 'stars' near Jupiter during December 1609 and to have reached the same conclusion as Galileo as to what they were. Since Mayr, who was a believer in a Tychonic arrangement of the planets, did not publish his observations there and then, it is generally Galileo who is given the credit for discovering the first four moons of Jupiter; a discussion of this controversy over precedence is given in Johnson (1931). In Mayr's work he suggested the names for these satellites that are in use today (Io, Europa, Ganymede, and Callisto) based on an idea of Kepler's.

[15] Quoted from Shea (1998). It was Christiaan Huygens in the late 1650s who first explained Saturn's shape as being due to a ring encircling the planet. Various theories were proposed during the first half of the seventeenth century, and these are described in van Helden (1974a, 1974b). Huygens was also the first to discover (in 1655) a satellite of Saturn (now called Titan) and he conjectured, on numerical grounds, that no more would be found: the number of known

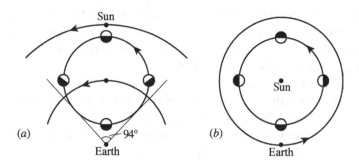

Fig. 7.2. The phases of Venus (half-shaded circles) in the (*a*) Ptolemaic, and (*b*) Copernican systems.

Galileo realized that telescopic observations of Venus might enable him to determine whether it orbits around the Earth below the Sun – as in the Ptolemaic system – or around the Sun – as in the Copernican and Tychonic systems. In Ptolemy's universe, Venus should never be observed as a full disc, since it was always closer to the Earth than the Sun and always within ± 47° of it; but if it orbited the Sun, then it should display a full range of phases, just like the Moon. Figure 7.2(*a*) shows Venus at four points in its passage around its Ptolemaic epicycle and, since more than half of the dark side of the planet always is visible from the Earth, the planet will be observed only in crescent phases. On the other hand, in a heliocentric universe (see Figure 7.2(*b*)) all possible phases can be observed. Galileo's observations showed that Venus did display the complete range of phases from crescent to full and back again and, hence, the traditional Ptolemaic view was false.[16]

Galileo communicated these and other findings to an excited Kepler, and the two engaged in a lively correspondence. Observations of sunspots revealed that the Sun was rotating, and Kepler naturally was thrilled that what he had supposed in his physical theory had been shown to be true. Galileo's discoveries were instrumental in maintaining Kepler's enthusiasm through the difficult times he

---

planets and satellites was then 12, which is 2 × 6 and 6 is the first perfect number. However, in 1671, G. D. Cassini, one of Europe's foremost observational astronomers, discovered another moon of Saturn (and ultimately three more) and this kind of speculation began to lose its appeal. Alexander (1962) describes in detail the history of Saturnian observations from ancient times to 1960.

[16] The same argument – which in the context of its time was not quite as conclusive as it might appear (see Ariew (1987)) – applies to Mercury, though the phases are much harder to observe. Kepler published details of Galileo's findings concerning the phases of Venus in his *Dioptrice* of 1611. In this mathematical work, Kepler developed his earlier work on optics and described the laws that govern the passage of light through lenses and also the design of a telescope that was superior to Galileo's and became the standard in astronomy from the 1630s onwards.

was experiencing and, on receiving a copy of *The Starry Messenger*, Kepler wrote quickly his *Conversation with the Starry Messenger*[17] in which he, unlike many other contemporaries, endorsed Galileo's findings and looked ahead to what other discoveries might lie in store for astronomers. Kepler pointed out that many of Galileo's conclusions were not new, but admitted that the idea of viewing the heavens through a telescope had not occurred to him; he had assumed that the thick blue air would have blocked out any details of the heavenly bodies. Now that Galileo had shown the space between the planets to be filled with a very thin substance, astronomers were free to speculate about what else might be observed. For example, Kepler hypothesized that Jupiter, like the Earth, must rotate so as to push its satellites around in their orbits and, sure enough, this was later observed to be the case, the period of rotation being established by G. D. Cassini in 1664 as 9 h 56 min.[18] From this point on, very few astronomers believed in the Ptolemaic structure of the Universe.

Perhaps the last serious defender of Ptolemy was Christoph Clavius, an early member of the Order of Jesuits and the founder of Jesuit mathematical and astronomical studies.[19] Clavius played a leading role in the reform of the calendar in 1582, and his opinions were extremely influential among those philosophers and theologians who later would criticize Galileo, most of whom would have learned their astronomy from Clavius' popular textbook, *Commentary on the 'Sphere' of Sacrobosco* (1570). This work was more than simply a description of Sacrobosco's little book. It brought together Aristotle's cosmology and Ptolemy's mathematical astronomy, including Peurbach's detailed physical interpretations. Clavius was well aware that there were many rival cosmologies to the one in which he believed. He was particularly critical of the sixteenth-century revival in theories based on homocentric spheres (e.g. those of Fracastoro and Amico), which seems to suggest that these cosmologies had a sizeable following. He also argued against Copernicus, using the usual arguments against a moving Earth, but clearly he admired the Polish astronomer's mathematical astronomy because he decided to incorporate the Copernican theory of precession into his geocentric astronomy in the 1593 edition of his textbook. This

---

[17] A translation of which can be found in Rosen (1965).
[18] See Hockey (1999) for details of early telescopic observations of Jupiter. It is perhaps also worth mentioning that not all telescopic discoveries were as illuminating. Cassini himself observed a satellite of Venus in 1686 and the same non-existent object was observed subsequently by no fewer than fifteen different observers during the seventeenth and eighteenth centuries (Ashbrook (1984), pp. 281–3). The arrival of the telescope did not signal the immediate end for naked-eye astronomy since, until the incorporation of cross hairs in the 1660s, it was not easy to make accurate positional observations with a telescope. The last great practitioner of naked-eye astronomy was Johannes Hevelius (Montgomery (1999), p. 174).
[19] The astronomy of Clavius is studied in detail in Lattis (1994).

also was done by the Italian astronomer Giovanni Magini who, in his *New Theories of the Celestial Orbs Conforming to the Observations of N. Copernicus* (1589), did away with the trepidation theory used in the *Alfonsine Tables* and replaced it with Copernicus' complicated alternative. In Copernicus' system, the variation in precession and obliquity was caused by a complex motion of the Earth, but when transformed into a Ptolemaic system the mechanism becomes a complicated motion of the fixed stars.

Galileo's telescopic discoveries were made when Clavius was over 70 years old. He was sceptical initially, but soon came to endorse Galileo's findings, though not his interpretations. He recognized that the astronomy he had defended all his life was in need of reexamination:

> I do not want to hide from the reader that not long ago a certain instrument was brought from Belgium. It has the form of a long tube in the bases of which are set two glasses, or rather lenses, by which objects far away from us appear very much closer, and indeed considerably larger, than the things themselves are. This instrument shows many more stars in the firmament than can be seen in any way without it, especially in the Pleiades, around the nebulas of Cancer and Orion, in the Milky Way, and other places . . . and when the moon is a crescent or half full, it appears so remarkably fractured and rough that I cannot marvel enough that there is such unevenness in the lunar body. Consult the reliable little book by Galileo Galilei, printed in Venice in 1610 and called *Sidereal Messenger*, which describes various observations of the stars first made by him.
>
> Far from the least important of the things seen with this instrument is that Venus receives its light from the sun as does the moon, so that sometimes it appears to be more like a crescent, sometimes less, according to its distance from the sun. At Rome I have observed this in the presence of others more than once. Saturn has joined to it two smaller stars, one on the east, the other on the west. Finally, Jupiter has four roving stars, which vary their places in a remarkable way both among themselves and with respect to Jupiter – as Galileo Galilei carefully and accurately describes.
>
> Since things are thus, astronomers ought to consider how the celestial orbs may be arranged in order to save these phenomena.[20]

As to quite how the phenomena were to be saved within a Ptolemaic framework, he left no clue.

Galileo's discoveries with the telescope soon were confirmed by others, and he became arguably the most celebrated scientist in the whole of Europe. His revealing observations forced people completely to reassess the nature of the Universe and man's place within it. One person whose attitude changed was Galileo himself. Before 1610, he had kept his Copernican views suppressed

---

[20] From the final (1611) edition of Clavius' *Commentary on the 'Sphere' of Sacrobosco.* Translation from Lattis (1994), p. 198.

(at least in public) but his telescopic observations began to change all that. He became Copernicus' most ardent supporter in Italy and, perhaps more significantly, also an extremely hostile (and eloquent) critic of the whole of Aristotelian physics, which he realized needed to be overthrown. Galileo, a devout Catholic, went to Rome in 1611 to argue for the Copernican theory but, although he did manage to convince people of the truth of the recent telescopic discoveries, he had less success persuading others to believe in his interpretation of the results.

Whereas his observations of Venus had demonstrated that the Ptolemaic model of the heavens could not be upheld, it did not distinguish between the heliocentric theory of Copernicus and Tycho Brahe's geoheliocentric scheme. The main reason for the invention of the Tychonic system was the reluctance to accept a moving Earth, but Galileo had no such reluctance; indeed, he thought he had a proof of the Earth's motion via the tides. He thus chose completely to ignore the Tychonic system and, instead, to contrast Copernicus' scheme with that of Ptolemy.[21]

One of the key figures who opposed Galileo was Cardinal Bellarmine, the chief theologian to the Roman Catholic Church and the man who had sent Giordano Bruno to the stake. He was the principal advocate for Pope Paul V, and his skills at arguing his case were respected highly. Bellarmine was sufficiently open-minded to accept that, if conclusive proof of the motion of the Earth were found, a reinterpretation of the Scriptures would be necessary, but he did not believe that any of Galileo's telescopic discoveries amounted to such a proof because the Tychonic system could explain them equally well, but with a stationary Earth.

In the battle with the theologians that followed Galileo's new discoveries,[22] the debate concerned not only the Copernican system but also Galileo's scientific method. Galileo believed that mathematics was the language of nature and that problems should be formulated mathematically. The truth of the results obtained from such theoretical considerations could then be verified by experiment. The more tests they passed, the more confidence one could have

---

[21] Other planetary systems were in the air around this time. Many believed that there was no proof that Mars, Jupiter, and Saturn revolve around the Sun, though Galileo's telescopic observations had confirmed this for Mercury and Venus. Thus, men like Francis Bacon, Joseph Blancan, and Charles Malapert supported a Capellan system in which Mercury and Venus orbit the Sun but all the other planets the Earth. The widely respected Giambattista Riccioli suggested that, since Jupiter and Saturn had been shown to have satellites, they were primary planets like the Sun and, hence, orbit the Earth, whereas the others would orbit the Sun.

[22] This battle is described in detail by de Santillana (1961) and all the significant documents which pertain to the affair have been translated in Finocchiaro (1989), a work that also provides a detailed chronology of events.

that one was dealing with a 'true' model of reality. As Galileo wrote in 1615:

> I think that in discussions of physical problems we ought to begin not from the
> authority of scriptural passages, but from sense-experiences and necessary
> demonstrations.[23]

Thus, Galileo was providing a way to the truth that was an alternative to the Bible, and this was what many members of the Church found so offensive. Galileo is reported to have said that the Bible teaches how to go to heaven, not how the heavens go.

Galileo was not without powerful friends and, although he was told privately to desist from teaching his opinions, he believed still that he could persuade the Catholic leaders that Copernicus' heliocentric universe was not at odds with Scripture. Pope Paul V was having none of it, however, and in 1616 ordered Galileo to be silenced. This gagging order was relaxed 8 years later after the election in 1623 of a more liberal Pope, Urban VIII, who as a cardinal had shown a friendly interest in Galileo's work. Under the new arrangement, Galileo was permitted to discuss the Copernican theory, provided he treated it as an astronomical hypothesis and not a model of reality.

The question of what to do with Copernicus' *On the Revolutions* was a difficult one for the Church. On the one hand, they would have liked to ban it for its advocacy of a moving Earth, but against this was its usefulness to astronomers. Eventually, they decided, not to proscribe it, but to censor it:

> If certain of Copernicus' passages on the motion of the earth are not hypothetical,
> make them hypothetical; then they will not be against either the truth or the Holy
> Writ, in a certain sense they will be in agreement with them, on account of the false
> nature of suppositions, which the study of astronomy is accustomed to use as its
> special right.[24]

A list of required corrections was issued in 1620. These were fairly small in number and merely toned down some of the passages where Copernicus seemed to be arguing for the truth of his system.[25]

Following the easing of relations between Galileo and the Church after the election of Pope Urban VIII, Galileo felt sufficiently encouraged to embark on his epoch-making *Dialogue Concerning the Two Chief World Systems, the Ptolemaic and the Copernican*, which was published in 1632.[26] There are three

[23] From his *Letter to the Grand Duchess Christina*, translated in Drake (1957).
[24] The instructions for censorship were drafted by Cardinal Caetani (Gingerich (1992), p. 113).
[25] The majority of the copies in Italy were censored, but it would appear that the decree had little effect in other countries (Hine (1973), Gingerich (1992), p. 79).
[26] The *Dialogue* was written in Italian with the title *Dialogo sopra I due massimi sistemi del mondo Toelmaico e Copernico*. It was translated into English by Thomas Salusbury as long ago as 1661 and this translation was revised by Giorgio de Santillana (Galilei (1953)). A more modern translation is that of Stillman Drake (Galilei (1962)).

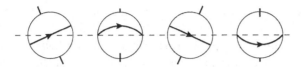

Fig. 7.3. The motion of sunspots. The dashed line represents the ecliptic.

speakers, Simplicio (a traditionalist named after the sixth-century commentator on Aristotle, but whose name carried an obvious double meaning), Salviati (the spokesman for Copernicanism), and Sagredo (an open-minded man who is persuaded largely by Salviati) and the discussion takes place over a period of 4 days at Sagredo's palace. The first day consists of a systematic demolition of Aristotelian natural philosophy, showing, on the one hand, that Aristotle's physics does not stand up to logical analysis and also demonstrating how recent telescopic discoveries had undermined completely some of its basic principles. In particular, Galileo seeks to dismantle the Aristotelian distinction between the terrestrial and supralunar regions. On the second day, the discussion turns to the motion of the Earth. Aristotle's theory of motion is picked apart and Galileo refutes all the arguments usually put forward to prove that the Earth was stationary.

The advantages of the Copernican theory were discussed on the third day. This included both the arguments that Copernicus had used himself and more recent ones such as the discovery of the full range of phases of Venus and the phenomenon of sunspots. Back in his days of telescopic discoveries, Galileo had observed dark patches on the Sun which moved and, after a period of continued observation, concluded that they were on the surface of the Sun and that the Sun rotated with a period of about 1 month.[27] The behaviour of these spots can be used to make a very convincing case for the motion of the Earth, though Galileo's argument in the *Dialogue* is uncharacteristically obscure.[28] Sunspots are concentrated near the solar equator, and Figure 7.3 shows the typical motion of a sunspot as seen from the Earth at four equally spaced times of the year. When observing a sunspot from the Earth, the daily rotation of the Earth plays a negligible role (except, of course, that we can only make observations during the day!) and the observed motions are consistent with a Sun that rotates on an

---

[27] Galileo was not the first to make telescopic observations of sunspots; this was Thomas Harriot in London in 1610; Harriot's observations are described in detail in North (1974). Many people, notably the Jesuit Christoph Scheiner, thought that these spots were planets orbiting the Sun much closer to it than Mercury, but Galileo argued that they moved too slowly and were often too large for this explanation to be correct (see, for example, Shea (1970)).

[28] The argument given here is based on that given in Smith (1985) and clarified in Hutchison (1990).

axis inclined to the ecliptic but maintaining a fixed direction with respect to the stars.[29]

To reproduce the same phenomena with a stationary Earth is much more difficult, because the diurnal motion is now a motion of the Sun. In this case, we would have to attribute an extra daily rotation to the Sun about a rotation axis perpendicular to the celestial equator so as to keep the same part of the Sun facing the Earth during each day, on which the monthly rotation of the sunspots could be observed as they are. This clearly is much more complicated and hence, based on the principle that if two theories predict the same phenomena then the simpler is the most likely, the observed paths of sunspots provide powerful evidence for the motion of the Earth; in fact, the Earth and Sun must each be rotating. Thus, the idea of a stationary Earth is untenable and the Ptolemaic system must be rejected.

Sagredo, after having been informed of the telescopic discoveries that confirm the Copernican theory, says: 'What pleasure the telescope would have given Copernicus', but Salviati observes that our admiration for Copernicus is enhanced by the fact that he proposed a heliocentric universe without these benefits. The last day is used to discuss Galileo's 'proof' of the motion of the Earth based on the tides.[30] In all the conversations between the three men, Galileo achieves the maximum effect by building up the traditional arguments through Simplicio and then, just when they appear invincible, destroying them.

The *Dialogue* had a huge influence. It was written in a style that brought home to many people the overwhelming weight of evidence in favour of a heliocentric universe. This is how John Playfair, Professor of Natural Philosophy at Edinburgh, described it in 1819:

> His dialogues contained a full exposition of the evidence of the earth's motion, and set forth the errors of the old, as well as the discoveries of the new philosophy, with

---

[29] To Galileo this was a natural motion. He rejected Copernicus' three motions for the Earth, instead arguing that the Earth's axis would tend naturally to remain fixed with respect to its orbit and, hence, he took the more modern view that (apart from precession) the motion of the Earth was made up of just two rotations.

[30] Described previously (Figure 7.1). Salviati states that 'it is impossible to explain the movements perceived in the waters and at the same time maintain the immovability of the vessel which contains them'. In the *Dialogue*, Galileo expanded on his earlier theory, attempting to account for the monthly and annual inequalities exhibited by the tides. The English mathematician John Wallis tried to improve on Galileo's theory in 1666 by treating the Earth–Moon system as a single body, the centre of gravity of which described an orbit round the Sun. This idea was a good one, but Wallis' theory was still based on the same fundamental misconceptions as Galileo's (Aiton (1954)). The question of whether the tides could possibly provide a proof of the Earth's motion, even in the context of Newtonian physics, was addressed by Burstyn (1962) and Aiton (1965), the former believing that tides do offer such a proof, and the latter arguing to the contrary. There is no simple answer to this question (see Palmieri (1998)).

great force of reasoning, and with the charms of the most lively eloquence. They are written, indeed, with such singular felicity, that one reads them at the present day, when the truths contained in them are known and admitted, with all the delight of novelty, and feels one's self carried back to the period when the telescope was first directed to the heavens, and when the earth's motion, with all its train of consequences, was proved for the first time.[31]

In some ways, Galileo was misrepresenting the current state of affairs among astronomers, virtually none of whom seriously entertained the world system of Ptolemy any more. As early as 1601 Kepler had written:

Thus today there is practically no one who would doubt what is common to the Copernican and Tychonic hypotheses, namely, that the sun is the centre of motion of the five planets, and that this is the way things are in the heavens themselves—though in the meantime there is doubt from all sides about the motion or stability of the sun.[32]

The real debate in astronomical circles was between the Copernican and the Tychonic systems, so it would appear somewhat surprising that Galileo chose not to mention the geoheliocentric scheme at all.[33] Also noteworthy is Galileo's silence on Kepler's astronomy based on elliptical orbits. Instead, he defends the heliocentric system based on uniform circular motions.

The *Dialogue* was received enthusiastically by Galileo's friends and by more forward-thinking scholars. However, the meaning behind the words was clear enough for the Church authorities to be outraged, and the Pope agreed that Galileo had overstepped the mark. The Inquisition was unleashed. In 1633, Galileo was threatened (probably only half-heartedly) with torture and made to renounce his belief in the Copernican system; he was forbidden to write anything about the mobility of the Earth, and the *Dialogue* was banned.[34] Galileo was put under house arrest, where he remained until his death in 1642. Much has been made of Galileo's imprisonment, a lot of it rather exaggerated. Compared with what happened to many others who chose to oppose the Church, he was treated rather well. It was during this period that Galileo wrote up his life's researches

---

[31] Quoted from Playfair (1822).   [32] KEPLER *Apologia*. Translation from Jardine (1984).

[33] Margolis (1991) has argued that Galileo paid particular attention to those arguments for Copernicanism that are the most awkward for a proponent of the Tychonic system, but that he refrained from any explicit mention of Tycho's scheme because he believed that the Pope's tolerance of his work was based on an understanding that while he would attack the Ptolemaic hypothesis, he would leave open the possibility of a stationary Earth through the Tychonic theory.

[34] Printing of a censored version of the *Dialogue* was allowed in 1744, but the ban was not lifted totally until 1822.

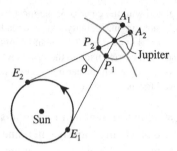

Fig. 7.4. The effect of the annual rotation of the Earth on occultations and transits of the moons of Jupiter.

into the theory of dynamics, *Discourses and Mathematical Demonstrations Concerning Two New Sciences* (1638).[35]

Galileo's recantation of Copernicanism perhaps marks the pinnacle of the war against the heliocentric system. Prior to 1610, arguments for Copernicus' thesis had been largely aesthetic and unconvincing. But by 1633, the situation had changed. Telescopic observations had shown Ptolemy's world system to be false, and Kepler's new astronomy had transformed the accuracy of planetary tables. From this point onwards, anti-Copernicans were fighting a losing battle.

One of Galileo's most significant contributions to mathematical astronomy was his work on the periods of the moons of Jupiter. During the 2 years following their discovery, Galileo spent a great deal of time on this project, soon realizing that it was non-trivial. Part of the problem was the difficulty in finding a suitable reference time at which a satellite's position could be fixed, and he settled eventually on the instances when a moon disappeared from view behind Jupiter (the apogee as he called it, or occultation of the orbit of the satellite) and those when it disappeared in front of the planet (perigee or transit). These two events could be distinguished by the direction of travel of the satellite in question. Galileo's results, computed from a large number of observations, were inconsistent, but he was able to correct matters when he realized that the annual motion of the Earth round the Sun had a significant effect.

In Figure 7.4, E_1 and E_2 are two different positions the Earth might occupy in its orbit round the Sun when viewing Jupiter and its moons. Crucial to Galileo's determination of the periods of the Jovian satellites was accurate observation of the apogees and perigees. But depending on the position of the Earth, the

---

[35] The Italian title was *Discorsi e dimonstrazioni matematiche intorno a due scienze attinetti alla mechanica ed i movimenti locali.*

Table 7.1. *Mean orbital periods of the moons of Jupiter.*

	Galileo		Modern theory
	Published	Computed	
Io	$1\,\mathrm{d}\,18\frac{1}{2}\,\mathrm{h}$	1 d 18; 28, 26 h	1 d 18; 28, 34 h
Europa	$3\,\mathrm{d}\,13\frac{1}{3}\,\mathrm{h}$	3 d 13; 20, 51 h	3 d 13; 17, 42 h
Ganymede	7 d 4 h	7 d 3; 55, 14 h	7 d 3; 58, 48 h
Callisto	16 d 18 h	16 d 17; 56, 14 h	16 d 18; 0, 0 h

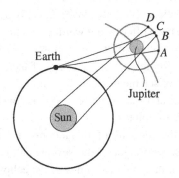

Fig. 7.5. Eclipses of the moons of Jupiter.

apogee (perigee) would be observed at A_1 (P_1) or A_2 (P_2) and the angular difference θ can be as large as 23°.

After incorporating the corrections implied by this realization, Galileo published (in his *Discourse on Floating Bodies* in 1612) his computed periods for the four moons he had observed (shown in Table 7.1). The first column gives the values Galileo published, and these were rounded from those he computed (shown in the second column). For comparison, the third column lists values computed from modern theory for the period 1610–14.[36] Galileo's calculations clearly represent a significant achievement.

Galileo also noticed, when timing the invisibility of a satellite during an occultation, that sometimes the moon would remain invisible for rather longer than his calculations suggested it should, and he soon realized the cause. The satellite was in the shadow of Jupiter, just as our Moon goes into the shadow of the Earth during a lunar eclipse. The situation is illustrated schematically in Figure 7.5. One might expect the satellite to be invisible from the Earth as it

---

[36] Galileo's values are taken from Swerdlow (1998b); whereas the numbers in the final column are from Johnson (1931).

revolves around Jupiter on the orbit $ABCD$, during the times between A and C when it is hidden from view behind the planet. But between B and D it is in the shadow of the Sun and so, in total the satellite will remain invisible from A to D.

An eclipse, unlike an occultation or a transit, is a phenomenon that does not depend on the position of the Earth in its orbit. Galileo realized because of this that eclipses of the Jovian satellites could be used to help solve one of the great problems of navigation, i.e. the determination of longitude. The local times of eclipses could be tabulated for some reference point (e.g. Florence) and then the differences in the local time at which these were observed would correspond to a difference in longitude from Florence. In 1612, Galileo sent a proposal to the king of Spain, who had been offering a reward for anyone who could 'discover the longitude', but nothing came of it. He proposed later the same thing to the Dutch in 1632, again with no success. The use of the satellites of Jupiter to determine longitude exercised the minds of astronomers and navigators for the next 200 years, however, and was very successful as a method for determining accurately the longitude of points on land.

Investigations into the motions of the moons of Jupiter led to another major astronomical discovery later in the seventeenth century. A number of astronomers noticed that the predicted times of the eclipses of Io did not correspond to the actual times, with the errors being greatest when Jupiter was at conjunction, and least at opposition. The Danish astronomer Ole Christenson Römer concluded that this must be due to the as yet undetected finite speed of light, with light taking about 22 min to cross the orbit of the Earth (the modern value is 16 min 38 s).[37]

The *Epitome of Copernican Astronomy*

While Galileo was advocating Copernicanism noisily in Rome, Kepler was adding his weight to the cause through his new theoretical astronomy. The longest and most systematic of all Kepler's works is the *Epitome of Copernican*

---

[37] See Boyer (1941). Following Aristotle, the orthodox view was that light was transmitted instantaneously. Galileo, and later Descartes, both tried unsuccessfully to measure the speed of light. This is another phenomenon for which priority of discovery is difficult to establish with certainty (see Débarbat and Wilson (1989)). Römer's conclusion generally was not accepted at the time but was confirmed early in the eighteenth century by the discovery of the aberration of starlight (see p. 307).

Astronomy, published in parts between 1618 and 1621.[38] In the *Epitome*, Kepler committed himself unreservedly to heliocentrism, and the book was placed immediately in the Church's Index of Prohibited Books. The work did not contain the lengthy reports of observations and parameter derivations that are to be found in Kepler's other writings, and was intended for a more general audience. It is written in the form of questions and answers, a style fairly typical of sixteenth-century textbooks, and was read widely.

Although the title suggests that it is a summary of Copernican theories, the book is actually a detailed textbook of heliocentric astronomy covering everything from elementary spherical astronomy to the laws of planetary motion that Kepler himself had discovered. Right from the start, Kepler emphasized that his astronomy was based on physical principles, and large parts of the *Epitome* are descriptions of his celestial physics. Here we have for the first time a complete theory of the Solar System without the complex geometrical devices of epicycles and the like that had dominated astronomical thought since Apollonius. Kepler has left both Ptolemy and Copernicus far behind although he describes his own work as Copernican because the whole of his physical theory originates from the choice of the Sun as the centre of the Universe.

Many of the arguments that Kepler used are based on his ideas on the overall harmony of the design of the Universe. For example, when he came to estimate the Earth–Sun distance, he assumed simply that the ratio of the Earth's radius to the distance between the Earth and the Sun is the same as the ratio of the volumes of the Earth and the Sun, since 'nothing is more in accord with the correct, elegant and ordained order'.[39] Kepler quoted Aristarchus' value of $1/2°$ for the angular diameter of the Sun and, hence, the Earth–Sun distance is $\cot 1/4° \approx 229$ solar radii. The assumed proportion implies that the Earth–Sun distance then can be expressed in Earth radii as $(\cot 1/4°)^{3/2}$, and Kepler obtained a figure of $3469\frac{1}{3}$, about 3 times larger than the Copernican value (though still over 6 times too small) and corresponding to a solar parallax of $1'$. He then justified his new value by noting that parallax observations of Mars indicated that, even when that planet is at its closest to the Earth, it is still more than 1200 Earth radii away and so the Sun must be a great deal further away than this. Of course, by placing the Sun 3 times further away from the Earth than Copernicus had done, he had to argue that the fixed stars were even further

[38] *Epitome astronomiae Copernicanae.* No complete English translation appears to be available, but books IV and V have been translated in Kepler (1995) and extracts are translated in Koyré (1973).
[39] Quoted from van Helden (1985), p. 83.

away than indicated by Copernican theory. The result of Kepler's calculations was a distance to the fixed stars of 60 000 000 Earth radii, and he argued against disbelieving this by saying:

> ... it is much more probable that the sphere of the fixed stars should be 2,000 or 1,000 times wider than the ancients said than that it should be 24,000 times faster than Copernicus said.[40]

By far and away the most novel aspect of the *Epitome of Copernican Astronomy* is Kepler's lunar theory.[41] As we have seen, the Moon was subject to some irregularities that did not feature in the motions of the planets. First, there was evection, discovered by Ptolemy, that linked the motion of the Moon to that of the Sun. Second there was the variation, discovered by Tycho Brahe, that manifested itself as a speeding up of the Moon near the syzygies and a corresponding slow down near the quadratures. Finally, there was the annual equation, incorporated by Tycho into his modified lunar theory (and independently discovered by Kepler, sometime before 1599) that has its origins in the change in the Earth–Sun distance over the course of a year.

In order to fit the Moon into his physical theory, the Earth would have to have physical properties similar to those of the Sun, or else how would the Moon be driven round the Earth in its orbit? Kepler invoked numerous justifications for attributing physical causes to the Earth. One of these involved applying his third law to the four newly discovered moons of Jupiter, the data for which he obtained from Simon Mayr's *World of Jupiter*. He quoted their relative distances from Jupiter, a, as 3, 5, 8, and 13, and their periods, T, as 1 day $18\frac{1}{2}$ h, 3 days $13\frac{1}{3}$ h, 7 days 2 h and 16 days 18 h, respectively. From these we obtain values of T^2/a^3 of 0.12, 0.10, 0.10 and 0.13. Not exactly constant but, given the crudity of the data, not bad at all. Kepler thus argued that the Jovian system was like a mini Solar System, with the planet rotating on its axis, carrying its moons with it, exactly as the Sun does. If this is the case for Jupiter, then why not the Earth? A spinning Earth thus not only accounted for the diurnal rotation of the heavens but also provided one of the primary causes for the motion of the Moon.

If it was the spinning Earth that was pushing the Moon around its orbit it might seem natural for the Moon to orbit near to the Earth's celestial equator. Kepler inferred from the fact that actually it moves near to the ecliptic that it was the motive virtue of the Sun that was dominant in determining the motion of the Moon, and this was reasonable, since in fact the Moon is in orbit around the Sun with the motion around the Earth just making this orbit rather erratic.

---

[40] KEPLER *Epitome of Copernican Astronomy*, IV. Translation from Kepler (1995).
[41] Described in detail by Stephenson (1987).

The intricacies of the lunar motion showed that the Moon had to be moved by a complex interplay between the effects of the Sun and the Earth, and Kepler hit upon the idea that the irregularities in the motion of the Moon were due to the effect of sunlight.

Kepler needed to quantify the effect of sunlight for use in his physical theory and here again he fell back on an argument based on design. What could be more natural than a year of 360 days made up of 12 synodic months of 30 days each? This was surely the perfect state of affairs, and so the deviations from this would be due to the effect of sunlight. Thus, sunlight caused the Earth to rotate slightly faster than the natural rate so that it completed $5\frac{1}{4}$ extra rotations during each orbit round the Sun. The Moon, on the other hand, completes 12 synodic months each year and undergoes an extra motion in longitude of just over 132°, so Kepler argued that the increase in the Earth's rotation rate due to sunlight caused an increase in the speed of the Moon, which resulted in an 'extra' motion of just under 11° per synodic month. He then argued that this basic strength would be modified by factors such as the distance between the Sun and the Earth, and the angle between the Sun and Moon as seen from the Earth. These variations in the efficacy of sunlight were then incorporated into a fairly elaborate geometrical theory that accounted for evection and variation.

The so-called 'annual equation' was, in Kepler's view, simply a manifestation of the fact that the effect of sunlight on the rotation rate of the Earth increased as the Earth moved closer to the Sun. Consequently, the Earth rotated faster when it was near perihelion but slowed down near aphelion, and this changed the length of the apparent day (the time between successive crossings of the meridian by the Sun). The length of the apparent day varies for other reasons (the fact that the axis of the Earth's rotation is not perpendicular to the ecliptic plane and the varying speed of the Earth as it orbits the Sun) and these irregularities are measured by the equation of time.[42] Since astronomical observations in the seventeenth century were made with reference to the actual position of the Sun, they had always to be corrected using the equation of time, and this was particularly important for the Moon, because it moves relatively rapidly, and an error in 30 min in the time of observation corresponds to an error in longitude of about 16′, easily detectable with the naked eye. Kepler thus believed that the annual equation of the Moon was, in fact, a modification to the equation of time rather than an intrinsic property of the motion of the Moon.

---

[42] The equation of time is the difference between the local apparent time (based on the actual position of the Sun) and the local mean time (based on the mean solar day of 24 h). The first detailed mathematical treatment of the equation of time was given by Ptolemy in the *Almagest* (see p. 71) and it was in common use in ancient and medieval times (see Kennedy (1988)), but the first reliable table of the equation was not published until 1673 (see Kollerstrom (2000), p. 23).

Kepler's physical explanations for the lunar motion were ingenious and he was very proud of them, but they were essentially *ad hoc*, being derived to fit the observed phenomena. They did have one great advantage over geometrical theories, though, since changes in speed did not have to be accompanied by changes in distance as they did in systems based around epicycles. Thus, Kepler's theory did not predict absurdly large changes in the apparent diameter of the Moon over the course of a month, as Ptolemy's and, to a lesser extent, Ibn al-Shāṭir's had done. Kepler's lunar theory found little support among his contemporaries and is little known today. It did, however, provide the inspiration for the later work of Jeremiah Horrocks, which in turn underpinned that of Isaac Newton.

All three of Kepler's laws of planetary motion were expressed clearly in the *Epitome*. These ideas did not catch on immediately, but between 1630 and 1650 the *Epitome of Copernican Astronomy* probably was the most widely read astronomical work in Western Europe.[43] Elliptical orbits generally were accepted by 1655 but, largely because of the difficulty in application, the second law was used virtually always in a simpler (and only approximate) form. The third law was of little use in practical astronomy and had no satisfactory theoretical basis; as a consequence, it did not attract much attention in the mid seventeenth century.[44]

The *Rudolphine Tables* and their impact

The main reason that Kepler's laws were accepted by astronomers was not the *Epitome*, though – it was the publication of the *Rudolphine Tables* in 1627. These long-awaited tables, from which planetary positions could be computed based on a heliocentric Universe with elliptic planetary orbits, were delayed for a number of reasons, one of which was simply the huge quantity of work that went into them. Kepler described this effort, with a justifiable lack of modesty, as the

unexpected transfer of the whole of astronomy from artificial circles to natural causes that were most profound to investigate, difficult to explain and difficult to calculate, my attempt being the first . . .[45]

[43] Russell (1964).
[44] One might think that Kepler would have used his third law to compute the mean solar distances of the planets from their sidereal periods, which were determined very accurately, but he did not. There are at least two possible reasons for this. First, he may have anticipated criticism for the use of a law that was grounded in his speculations about harmony rather than derived from systematic observation, and second, he may well have believed that, as with other essentially correct laws, the third law was subject to small variations.
[45] From the preface to the *Rudolphine Tables*, which is translated in Gingerich (1972).

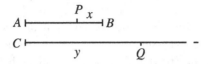

Fig. 7.6. Napier's geometrical definition of logarithms.

Another source of delay was Kepler's discovery of logarithms, which he described as a 'happy calamity', and his subsequent investigations of them. The invention of logarithms simplified greatly many of the tedious arithmetical calculations required by astronomers and, unlike many other mathematical advances, was an almost instant success. Laplace later remarked that:

> it is an admirable contrivance ... which, by reducing to a few days the labour of many months, we may almost say doubles the life of astronomers, and spares them the errors and disgusts inseparable from long calculations.[46]

Logarithms were developed in the 1590s by the Scotsman, John Napier, and independently by the Swiss instrument maker, Jost Bürgi.[47] However, Bürgi did not publish his work until 1620, whereas Napier's first account of his theory appeared in print in 1614, and so Napier usually is credited with the invention.[48]

The basic idea that lies behind logarithms is the correspondence between geometric progressions

$$r^1, r^2, r^3, \ldots$$

and the arithmetic progressions formed by the exponents:

$$1, 2, 3, \ldots.$$

Multiplying two numbers in the geometric series together is equivalent to adding the corresponding exponents. The closer r is taken to 1 the closer adjacent terms in the geometric progression are to each other, and the geometric progression can be made to include as many numbers as one requires within a given interval.

From this basic idea, Napier developed a geometrical definition of logarithms, illustrated in Figure 7.6. The upper line AB is of fixed finite length, and

[46] Laplace *L'exposition du système du monde* (1796), Book V, Chapter IV. Translation from Laplace (1809).

[47] Bürgi has been mentioned already for his work on *prosthaphaeresis*, a technique that was used prior to the invention of logarithms for simplifying certain arithmetic calculations (see note 25 on p. 164). Bürgi spent some time with Kepler in Prague, but Kepler did not appreciate fully the value of logarithms until reading Napier's work (see, for example, Armitage (1966), p. 167).

[48] Napier's original account *Mirifici logarithmorum canonis descriptio* (*Description of the Wonderful Canon of Logarithms*) contained tables and a brief explanation of how they were to be used. The theory behind the construction of the tables appeared posthumously in 1619.

the point P moves along it in such a way that its speed is proportional to the distance from B. If we consider a succession of small time intervals Δt, with P being at P_1, P_2, P_3, ..., and assume that the speed is constant during each time interval (and equal to the speed of P at the beginning of the interval), then the distance $|P_1 P_2|$ is given from the definition of the motion by $k|P_1 B|\Delta t$, where k is the constant of proportionality. Hence,

$$|P_2 B| = |P_1 B| - |P_1 P_2| = |P_1 B|(1 - k\Delta t).$$

It follows that the lengths $|P_1 B|$, $|P_2 B|$, $|P_3 B|$, ... form a decreasing geometric progression, and the length x indicated in the figure corresponds to this geometric progression in the limit as $\Delta t \to 0$. The associated arithmetic progression is derived from the lower line, on which the point Q moves with uniform speed equal to the initial speed of P, P and Q starting from A and C at the same time. If we denote the distance $|CQ|$ by y, then y is what Napier termed the logarithm of x, which we can write as $y = \mathrm{NLog}(x)$.

It was the laborious calculations required in spherical astronomy that led Napier, who was not a professional mathematician, to invent logarithms (he sent some preliminary results to Tycho Brahe for approval[49]) and actually he dealt with logarithms of sines. Regiomontanus' tables of sines were based on a circle with radius 10^7 and, because of this, Napier chose 10^7 for the length $|AB|$, this being the largest number for which he needed the logarithm. To see how Napier's logarithms are related to functions used today, we can interpret Napier's definitions using the modern language of the calculus. The geometrical definitions are equivalent to the differential equations

$$\frac{dx}{dt} = -x, \qquad \frac{dy}{dt} = 10^7,$$

with the condition that $y = 0$ when $x = 10^7$. Eliminating t leads to

$$y = 10^7 \ln(10^7/x)$$

in terms that we now call the 'natural logarithm'. Napier's logarithms thus have the property that

$$\mathrm{NLog}(x_1 x_2) = \mathrm{NLog}(x_1) + \mathrm{NLog}(x_2) - 10^7 \ln 10^7 \qquad (7.1)$$

and so, with the help of a table of these logarithms, any multiplication can be reduced to two additions, a huge saving in time when the numbers to be multiplied had large numbers of digits.[50]

---

[49] Kline (1972), p. 256.
[50] Katz (1998), p. 418 gives details of how Napier actually applied his logarithms to trigonometric problems.

Napier's invention greatly impressed Henry Briggs, who was at the time the Professor of Geometry at Gresham College, London. The two of them agreed that the system should be modified to make it easier to use. Briggs tells the story as follows:

> I myself ... remarked that it would be more convenient that 0 be kept for the logarithm of the whole sine [as it was in Napier's original work], but that the logarithm of the tenth part of the whole sine, that is to say 5° 44′ 21″ [i.e. the angle whose sine is $\frac{1}{10}$] should be 10^{10}. And concerning that matter I wrote immediately to the author himself; and ... I journeyed to Edinburgh, where being most hospitably received by him, I lingered for a whole month. But as we talked over the change in the logarithms he said that he had been for some time of the same opinion and had wished to accomplish it; he had, however, never published those he had already prepared, until he could construct more convenient ones if his affairs and his health would permit of it. But he was of the opinion the change should be effected in this manner, that 0 should be the logarithm of unity, and 10^{10} that of the whole sine, which I could not but admit was by far the most convenient.[51]

These modified logarithms are, apart from the position of the decimal point, just logarithms to the base 10 ($\log_{10}$) or, as they became known, common logarithms. Briggs went on, after Napier's death, to produce tables of common logarithms accurate to fourteen decimal places, and these formed the bases of tables of logarithms right up until the arrival of the electronic calculator. As an aid to computation, common logarithms are simpler than Napier's original logarithms because setting $\log 1 = 0$ ensures that there is no term equivalent to the $10^7 \ln 10^7$ in Eqn (7.1) to subtract, i.e. $\log_{10}(x_1 x_2) = \log_{10}(x_1) + \log_{10}(x_2)$.

Napier's work contained a discussion on the use of decimal fractions and his notation was the same as that which we use today. Because logarithms were welcomed widely throughout the scientific community, the publication of Napier's work and then Briggs' tables had the effect of spreading the use of decimal fractions across Europe.

Kepler carried out a study of Napier's logarithms during the winter of 1621/2 and wrote a short book on the subject.[52] He constructed his own version of logarithms, following Napier's approach rather than Briggs', which are related to our modern natural logarithms through

$$\text{Keplerian } \log x = 10^5 \ln(10^5/x).$$

The *Rudolphine Tables* contains tables of these logarithms and was the first book to require the use of logarithms in a scientific application. Kepler was thrilled with this new device, though his former teacher Michael Mästlin commented,

---

[51] Quoted from Coolidge (1949).   [52] *Chilias logarithmorum*, eventually published in 1624.

'it is not seemly for a professor of mathematics to be childishly pleased about any shortening of the calculations'.[53]

The great success of the *Rudolphine Tables* in predicting planetary positions is well illustrated by the reaction of Peter Crüger, Professor of Mathematics at Danzig, who until their appearance had been unimpressed by Kepler's work. Responding to Philip Müller, Professor of Mathematics at Leipzig, on the subject of improving the planetary tables produced by Longomontanus, he wrote (in 1629):

> But I should have thought that it would be a waste of time now that the *Rudolphine Tables* have been published, since all astronomers will undoubtably use these ...
> For myself, so far as other less liberal occupations allow, I am wholly occupied with trying to understand the foundations upon which the Rudolphine rules and tables are based, and I am using for this purpose the *Epitome of Copernican Astronomy* previously published by Kepler as an introduction to the tables. This epitome which previously I had read so many times and so little understood and so many times thrown aside, I now take up again and study with rather more success seeing that it was intended for use with the tables and is itself clarified by them ... I am no longer repelled by the elliptical form of the planetary orbits; Kepler's proofs in his *New Astronomy* have convinced me.[54]

We know now that the errors in Kepler's tables were about 30 times smaller than those in previous astronomical tables, though this would not have been apparent when they were first published, and not all astronomers were convinced immediately.[55]

Whereas Kepler's planetary theory, in the form of the *Rudolphine Tables*, was a success, the same could not be said for his physics. One of the most significant astronomical treatises published between those of Kepler and Newton was that written by the Parisian librarian Ismael Boulliau in 1645.[56] Boulliau supported strongly the idea of elliptical orbits, but was not prepared to accept Kepler's physical theory. In the introduction he wrote:

> After I had long considered Kepler's *Commentaries on Mars* and his *Epitome of Copernican Astronomy*, and had seen that his elliptical hypothesis represents the observed celestial motions more exactly than all the others, I did not cease praising and commending his felicitous ability and cleverness. At last I determined to seek the truth of a hypothesis so apt and appropriate, and to confirm the truth thus found

---

[53] Quoted from Caspar (1993), p. 309.   [54] Quoted from Russell (1964).
[55] The Polish Jesuit Michael Boym introduced the *Rudolphine Tables* to the Far East, declaring in 1646 that they were 'of inestimable value in calculating partial and complete solar eclipses, together with celestial movements' (Szczesniak (1949)).
[56] *Astronomia Philolaica*, the reference to the Pythagorean Philolaus reinforcing the fact that the Earth is not immobile.

by reasons. But I saw that in the explanation of the hypothesis that man had left many roughnesses, had enunciated many things obscurely, and had thrust abstruse physical causes upon us in place of demonstrations; also, he had not demonstrated certain things that required demonstration.[57]

Boulliau took particular exception to Kepler's supposition that the effect of the motive virtue in the Sun decays inversely with distance in the ecliptic plane, and argued instead that, if such a physical force existed, it would have to decay in inverse proportion to the square of the distance, just as the intensity of light does. However, like many of his contemporaries, Boulliau did not believe any such force did exist; instead, his view was that we should look to geometry for the causes of celestial motion, i.e., planets moved as they did because the geometrical form of an ellipse represented the natural motion of such bodies.

Kepler's second law had its roots in his physical theory and, having discarded the physics, Boulliau could discard the troublesome second law, too. In the mid seventeenth century, application of the second law involved the solution of the equation $t = \theta + e \sin \theta$ for θ, which involved messy and non-geometrical trial-and-error-type approaches and thus was considered unsatisfactory by those who looked for elegance in planetary motion. Boulliau replaced the second law with one that was, to him, more appropriate. As he was still fixated with the aesthetic beauty of uniform circular motion, he proposed a scheme in which a planet always was moving instantaneously on a circular path at constant angular speed, but in which the circles were of continually changing radius. These circles, taken together, made up a surface in three-dimensional space, and he took this surface to be a cone, and then the planetary path was the intersection of this cone with the plane of the orbit, i.e. an ellipse!

In 1653, Seth Ward, Savilian Professor of Astronomy at Oxford, demonstrated that Boulliau's hypothesis was equivalent to treating the empty focus of the elliptical orbit as an equant point, about which a planet would move with uniform angular speed (something Boulliau had never suspected and that Kepler had examined and discarded in the *Epitome of Copernican Astronomy*[58]). Boulliau's theory – which gave meaning to the otherwise purposeless empty focus of the elliptical orbit – was followed widely, but Boulliau realized eventually that his replacement law did not predict planetary positions satisfactorily.[59]

---

[57] Quoted from Wilson (1970).
[58] In Book V, Part I, 5. In the *Rudolphine Tables*, Kepler claimed that one of his friends, the Jesuit Albert Curz (Curtius in Latinized form) used the empty focus as an equant.
[59] Another attempt to make the empty focus an equant point was made around 1690 by G. D. Cassini. He proposed that planetary orbits took the form of ovals defined by the condition that the product of the distances of a point on the curve from the two foci is constant. Such curves are now referred to as 'Cassini ovals', or 'cassinoids'.

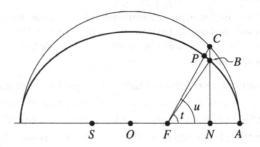

Fig. 7.7. Boulliau's modified version of Kepler's second law.

In 1657, he produced a modified theory that provided answers in much better agreement with the true area law, though still not equivalent to it.

This modified theory is illustrated in Figure 7.7, in which the heavy curve represents the elliptical orbit, and the lighter curve the circumscribing circle. The Sun is at S, and the empty focus of the ellipse is F. In Boulliau's original theory, the planet was at B, where the angle AFB is the mean anomaly, t. In the modified theory, we construct the line BN, perpendicular to the axis of the ellipse which, when extended, cuts the circumscribing circle at C. Finally, we join C to the focus F, and the planet P is situated where this line cuts the ellipse. Based on the Tychonic data of the time, this version of the second law actually is as accurate as Kepler's law.[60]

If we denote the eccentricity of the orbit by e, then we know from the basic geometry of an ellipse that $|BN|/|CN| = \sqrt{1 - e^2}$. Thus, writing u for the angle subtended by the planet at the empty focus, we have

$$\tan t = \sqrt{1 - e^2} \tan u,$$

from which

$$u = t + \tfrac{1}{4}e^2 \sin 2t + O(e^4).$$

As Newton would later demonstrate (see p. 268), this is correct up to second order in e.

Notwithstanding the success of Keplerian astronomy, many influential astronomers were still not prepared to accept the idea of a heliocentric universe. One such was Giambattista Riccioli, a well-respected professor at the Jesuit College in Bologna, who published his *New Almagest* in 1651. Riccioli was a serious astronomer and knew that Ptolemy's universe could no longer be upheld,

---

[60] A comparison of the accuracy of the two methods is given in Wilson (1970).

but his religious beliefs forced him to argue against the Copernican hypothesis:

> ... all Catholics are obliged by prudence and obedience ... not to teach
> categorically the opposite of what the decree lays down.[61]

In the *New Almagest*, he produced forty-nine arguments that were in favour of heliocentrism, and seventy-seven that were against, and thus the weight of the argument[62] favoured an Earth-centred cosmology! However, even Riccioli recognized that in terms of predictive power, astronomy owed much to Kepler, particularly with regard to the motion of Mars.

As far as Jupiter and Saturn were concerned, the *Rudolphine Tables* were not really an improvement on other competing planetary theories. In his *Reformed Astronomy* of 1665, Riccioli listed seventy-one observations of Saturn, both ancient and modern, and compared them with various planetary tables. He concluded that the tables he himself had computed and those of Boulliau were the best, while those of Kepler and Longomontanus were not far behind.[63] John Flamsteed, the first Astronomer Royal, wrote in 1674:

> ... the places of the planet Jupiter have been, for these last two years, some 13 or
> 14 minutes forwarder in the heavens, than Kepler's numbers represent; and ... his
> motions are not much better solved by any others ... .[64]

In other words, all the planetary tables were similarly inaccurate. One problem here was the accuracy of the orbital parameters that Kepler had computed, and Flamsteed did manage to improve on Kepler's theory for Jupiter by recalculating these parameters. The real problem remained hidden, though. In the eighteenth century, Laplace showed that the gravitational effect of these two large planets on each other causes deviations in their orbits from perfect Keplerian ellipses, that are of far greater significance than the errors introduced by using eccentric circles and equants rather than elliptical orbits.

---

[61] Quoted from Russell (1989). According to Russell, Riccioli asserted that Catholics were under no obligation to believe that the Copernican system was heresy (as the Holy Office decree of 1633 had stated), but that out of respect for the Church they ought not to maintain its truth in public. There were, in fact, no serious attempts by the Holy Office in Rome to discipline those of a Copernican persuasion and outside Italy astronomers were pretty much free to write what they believed.

[62] Almost literally. The frontispiece of Riccioli's *New Almagest* shows his own world system (described briefly in note 21 on p. 209) being weighed against that of Copernicus while Ptolemy's system lies discarded on the ground.

[63] Riccioli's observations and calculations contain numerous errors (Wilson (1970)).

[64] Quoted from Wilson (1970).

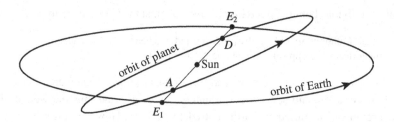

Fig. 7.8. Transits of inferior planets.

Transits of Mercury and Venus

One of the reasons that many astronomers were impressed by Kepler's tables was their success at predicting the transit of Mercury, which took place in 1631. Transits of Mercury or Venus take place when the planet lies directly between the Earth and the Sun, and so a necessary condition for a transit is that we have an inferior conjunction. The orbits of the inferior planets do not lie in the ecliptic plane but are instead inclined slightly to it, so in order for a transit to occur the planet must either be at its ascending node (A in Figure 7.8) with the Earth at E_1, or at its descending node D with the Earth at E_2.

In 807, an observed spot on the Sun was interpreted as a transit of Mercury even though it lasted for 8 days and there were a number of other medieval reports of transits of Mercury or Venus, though none has been authenticated.[65] Kepler claimed also to have observed a transit in 1607 but later, when he realized there could not have been one on the day of his observation, he acknowledged his error. In 1629, he did, however, predict a transit that 2 years later (after his death) became the first to be observed. Transits of Mercury last at most about 4 h and, since pre-Keplerian theories of Mercury were in error by several days in the times of predicted inferior conjunctions, Kepler's successful prediction (even though he was out by about 5 h) was extremely impressive.[66]

Apart from providing strong evidence for the accuracy of the *Rudolphine Tables*, this transit was important for two reasons. First, it helped greatly in the accurate determination of Mercury's orbit (since the same error in angular measurement results in a much greater uncertainty in the planet's position in its

---

[65] Goldstein (1969) discusses the transit reports of al-Kindī, ninth century, Ibn Sīnā (Avicenna) eleventh century, Ibn Rushd (Averroës) and Ibn Bājja (Avempace) twelfth century. The only one that could possibly have been a transit is that of Ibn Sīnā.
[66] In terms of the predicted longitude, Kepler's error amounts to $14' 24''$. By comparison, the error in the set of tables published by the Dutch astronomer Philip Lansberg in 1631–32 was $1° 21'$, and that in the Prussian Tables based on Copernicus's theory was about $5°$ (Wilson (1970)).

orbit near elongation than when directly in front of the Sun[67]) and, second, it provided the first indisputable quantitative measure of the apparent magnitude of a planetary disc. Pierre Gassendi, who was one of the three people who are known to have observed the transit of 1631, was very surprised at Mercury's small size (he measured its angular diameter at about 20″, whereas Tycho Brahe had estimated its apparent diameter at mean distance as 2′ 10″). Johannes Hevelius made accurate observations of the transit of 1661, and discovered that Mercury was even smaller than Gassendi had thought. Hevelius also found that only those astronomical tables based on Kepler's theory of elliptical orbits predicted a transit on the correct day.

The British astronomer Vincent Wing used a modified version of the second law rather than Kepler's area law, but clearly he believed that Kepler's successful predictions of Mercury transits represented very strong empirical evidence for his elliptical orbits, since he wrote in the posthumously published *Astronomia Britanica* of 1669:

> But this is proved especially by the planet Mercury, which on 28 Oct 1631, and again on 23 Oct 1651 and 23 April 1661 was interposed between our vision and some part of the body of the sun; on each occasion the Keplerian tables, conforming to the Copernican hypothesis, best agreed with the truth, while the tables of Longomontanus and Argolus, conforming to the Tychonic system, contained errors of many days.[68]

Transits of Venus are much rarer than those of Mercury. Whereas there were fourteen transits of Mercury in the twentieth century, only five transits of Venus have ever been observed – those in 1639, 1761, 1769, 1874 and 1882 (though the sixth is on 7 June 2004). The 8-year gap between the two transits in the eighteenth century and again between the two transits in the nineteenth century is due to the fact that thirteen sidereal periods of Venus is about 2921 days, which is very nearly 8 years. Thus, if the Sun, Venus and the Earth are aligned at a given time, they will be again 8 years later. The orbit of Venus is inclined at a little over 3° to that of the Earth and the discrepancy between the two periods leads to a change in ecliptic latitude of Venus between the two alignments of about 22′, which is less than the angular diameter of the solar disc (32′), and so, provided the first transit is not too close to the centre of the solar disc, there will be another after an 8-year gap. However, there cannot be another 8 years after that.

---

[67] Transits of Mercury actually occur about thirteen times per century, though each one is visible only from certain parts of the globe. They became less significant when Halley became able to observe the planet in daylight to within 15° of the Sun.
[68] Quoted from Wilson (1973). The dates are Old Style (see note 3 on p. 120).

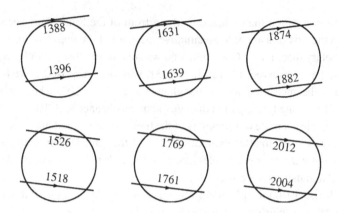

Fig. 7.9. Transits of Venus between 1396 and 2012.

Figure 7.9 shows the paths of Venus across the Sun for the eleven transits between 1396 and 2012. There was only one transit in the fourteenth century, with Venus passing just above the solar disc in 1388, but since then all transits have occurred in pairs. Venus was near its ascending node during the transits in the fourteenth, seventeenth and nineteenth centuries, but near its descending node during the transits in the sixteenth and eighteenth centuries, as it will be also in the twenty-first century.

Kepler predicted the transit of Venus in 1631, but this turned out to be visible only in America. He failed to predict the transit of 1639 which was visible from Europe, however, since the *Rudolphine Tables* implied that Venus would pass below the Sun. As a result, astronomers were not ready with their telescopes to observe the spectacle. The transit did not go unobserved though, thanks to the efforts of the young and largely self-taught Englishman, Jeremiah Horrocks.

Horrocks obtained a copy of the *Rudolphine Tables* in 1637 and comparisons of observations with Kepler's and other contemporary tables soon convinced him of the superiority of Keplerian astronomy (though he did not subscribe to the physical side of this new science). He set about trying to improve the tables by making more accurate observations,[69] beginning by refining the elements of the orbit of the Earth, reducing the eccentricity from Kepler's value of 0.018 to 0.0173,[70] which he derived by combining Tycho's solar theory (in which Tycho had assumed a solar parallax of 3′) with Kepler's value of 1′ for the solar parallax. This change, which reduced the errors in Tycho's theory by more than

---

[69] Horrocks' observational procedures were quite sophisticated and often involved the design and construction of his own instruments (Chapman (1990)).
[70] Still about 3 per cent too large, Wilson (1980), p. 67.

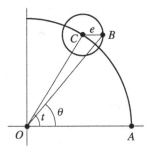

Fig. 7.10. Horrocks' approximation to the area law.

50 per cent, then had a knock-on effect for all the planetary theories. Horrocks directed most of his efforts on Venus and set about trying to determine its orbital parameters with great accuracy. He concluded from his observations and calculations that the mean solar distance (in terms of the radius of the orbit of the Earth) was 0.7233, exactly as predicted by Kepler's third law, rather than Kepler's value of 0.7241. He wrote in his *Venus Visible on the Sun* (1662), published over 20 years after the author's untimely death aged only 22:

> ... the proportion that obtains between the periods of the motions of the planets and the semi-diameters of their orbits is most exact, as Kepler, its discoverer, rightly states, and as I by repeated and most certain observation have found; indeed there is not an error of even a minute ... [71]

So Horrocks was a believer in Kepler's first and third laws but, as with many other astronomers, he had problems in applying the second. As we have seen, Kepler had reduced the application of the second law to the solution of the equation $t = \theta + e \sin \theta$ for the eccentric anomaly θ. In the *Rudolphine Tables*, Kepler tabulated t (in fact the logarithm of t) for equally spaced values of θ, from which the user of the tables had to interpolate between the non-uniformly spaced values of t. Horrocks instead derived a geometrical approximation, which is illustrated in Figure 7.10, and clearly has its origins in the use of epicycles. Given the mean anomaly t, we locate a point C on the unit circle and then construct a circle of radius e, where e is the eccentricity of the planetary orbit. If we denote the angle AOB by θ, an application of the sine rule (Horrocks actually used the law of tangents) to the triangle OBC, BC being parallel to OA, gives

$$\frac{1}{\sin \theta} = \frac{e}{\sin(t - \theta)}$$

[71] HORROCKS *Venus in sole visa*. Quoted from Wilson (1978).

Fig. 7.11. Determining the size of Venus as viewed from the Sun.

from which $e \sin \theta = \sin(t - \theta) \approx t - \theta$ if t and θ are close together. Horrocks knew that he was only approximating the area law, but he underestimated the magnitude of the error.[72]

With his new orbital parameters for Venus, Horrocks predicted that there would be a transit in 1639 (a supreme test of the accuracy of his theory) and he wrote to his friend William Crabtree asking him to attempt to observe the predicted transit, so as to reduce the chance that bad weather might interfere with the observation. As far as we know, Horrocks and Crabtree were the only two people to observe the event.[73] Horrocks waited patiently through the predicted day of the transit and probably was resigning himself to failure when, not long before sunset,

> ... the clouds, as if by *Divine Interposition*, were entirely dispersed ... and I then beheld a most agreeable sight, a spot, which had been the object of my most sanguine wishes, of an unusual size, and of a perfectly circular shape, just wholly entered upon the sun's disc ... I was immediately sensible that this round spot was the planet *Venus*, and applied myself with the utmost care to prosecute my observations.[74]

Horrocks measured the apparent diameter of Venus at $1' \, 16''$ (about 10 times smaller than contemporary wisdom suggested) and used this, together with some Keplerian style speculation, to estimate the Earth–Sun distance. He first calculated the apparent diameter of Venus as it would be observed from the Sun using the method illustrated in Figure 7.11. If we denote the radius of Venus by r, we then have

$$\tfrac{1}{2}\alpha \approx \tan \tfrac{1}{2}\alpha = r/d_2, \qquad \tfrac{1}{2}\beta \approx \tan \tfrac{1}{2}\beta = r/d_1,$$

so β can be determined in terms of α provided the ratio d_2/d_1 is known.

---

[72] Similar geometric approximations were developed around the same time by Bonaventura Cavalieri (Wilson (1989b)).

[73] Given the rarity of the event, Horrocks was extremely fortuitous in his prediction. The idea that there might be a transit did not occur to him until the end of October 1639 (Old Style) and the transit took place on November 24! (Chapman (1990)). According to Fernie (1996), Horrocks and Crabtree never met but communicated entirely through letters, though this seems unlikely (Applebaum (1975)). Maor (2000) relates that the two met at Cambridge, but after leaving the University maintained their friendship through correspondence only.

[74] HORROCKS *Venus in sole visa*. Quoted from Fernie (1996).

Horrocks accomplished the determination of this ratio by using Kepler's third law. Since $d_1 + d_2$ is the Earth–Sun distance, we have

$$\frac{1}{T_V^2} = \left(\frac{d_1 + d_2}{d_1}\right)^3 = \left(1 + \frac{d_2}{d_1}\right)^3 ,$$

where T_V is the orbital period of Venus (in years). Thus, using $\alpha = 76''$, Horrocks obtained the result $\beta \approx 28''$. Then he did the same for Mercury based on Gassendi's measurement of $\alpha \approx 20''$, and again arrived at $\beta \approx 28''$. In order to provide an explanation for his third law, Kepler had assumed that planetary volumes were directly proportional to solar distances, but Horrocks showed that this would have implied an apparent diameter for Mercury of over $2'$. So instead, Horrocks hypothesized that the *diameters* of the planets were proportional to their orbital radii, which in turn would imply that the apparent diameter of each planet as seen from the Sun was the same, in agreement with his results for Venus and Mercury. He applied this idea to the Earth and concluded that the solar parallax was tiny – only $14''$, corresponding to an Earth–Sun distance of about 15 000 Earth radii. Although the method clearly was dubious, the result was much closer than previous estimates, and demonstrates that some astronomers at least were coming to terms with the fact that the Solar System was much larger than people had previously been prepared to contemplate.[75]

Horrocks' faith in Kepler's harmonic law influenced Thomas Streete, one of the leading astronomers working in England at the time. In his *Astronomia Carolina* (1661) he used the law to calculate the mean solar distances of the planets from their sidereal periods, leading to improved accuracy for Mercury, Venus, and Mars. He made use also of the modified version of Kepler's second law due to Boulliau (though wrongly attributed by Streete). It was from this work that Isaac Newton became aware of Kepler's first and third laws.[76] Streete also utilized Horrocks' values for the solar parallax and the eccentricity of the Earth, and as a result the planetary tables he computed were rather better as a predictive tool than those of his contemporaries, Boulliau and Wing. They were reprinted many times, well into the eighteenth century.

---

[75] Another astronomer who realized that the accepted value for solar parallax was far too large was the Belgian, Gottfried Wendelin. Based on observations of planetary diameters, he concluded that the parallax of the Sun was no more than $15''$. Toward the end of the seventeenth century, G. D. Cassini devised an indirect method by which he managed to put an upper bound of $12''$ on the solar parallax (van Helden (1989)). Accurate direct measurements had to wait until the transits of Venus in 1761 and 1769 (see p. 315).

[76] See Whiteside (1964) and Wilson (1978).

Horrocks' lunar theory

Kepler's lunar theory was, as a predictive tool, far behind his planetary theories, and he acknowledged this. In the preface to the *Rudolphine Tables* he wrote that there was

> clear evidence . . . that the motions of the sun, moon and [precession] are not up to mathematical precision, but have rather slight physical increases and decreases in an irregular way.[77]

In the mid seventeenth century, Boulliau, Wing, and Streete all produced lunar theories[78] though none were as accurate or as conceptually simple as that of Horrocks – developed between 1637 and his death in 1641, but not published until 1672 (by the Astronomer Royal, Flamsteed).

It was Flamsteed's advocacy of Horrocks' theory that led to it becoming well known. For example, William Whiston, in his *Astronomical Lectures* (1728) referred to 'Mr Horrox's Lunar Hypothesis, as cultivated and explained by Mr Flamsteed'.[79] Flamsteed noted that, for a particular date in 1672, Horrocks' theory was accurate to within $2'$ of arc whereas those of Boulliau and Streete were in error by roughly $15'$. Horrocks' theory also seemed to predict distances better, since the observed decrease in the apparent diameter of the Moon over the course of 4 months at the end of 1671 and the beginning of 1672 was also predicted by Horrocks' theory, but not by other contemporary theories.

Kepler had used elliptical orbits within his lunar theory, but he had not made much of this construction because it did not work very well; it was Horrocks who first developed a lunar theory based properly on elliptical motion. Horrocks' ideas grew out of a careful study of the *Epitome of Copernican Astronomy*. By comparing the predictions of Kepler's lunar theory with eclipse observations he eventually was led to a theory in which the Moon moves on an elliptical orbit, the eccentricity and major axis of which oscillate (see Figure 7.12).[80] At any instant, the Moon is considered to be moving on an ellipse, the centre of which is C with a focus at E, the Earth. The point C, however, moves on a circle, the centre D of which lies on the mean apsidal line for the Moon's orbit, $E\bar{A}$. Thus, the eccentricity $|EC|$ and the apsidal line EA oscillate about their mean value and position, respectively.[81] This oscillation is linked to the motion of the Sun S by the relation $2\angle SE\bar{A} = \angle CD\bar{A}$, and the whole mechanism is superimposed

---

[77] Quoted from Gingerich (1972).   [78] All are described in Wilson (1989b).
[79] Quoted from Kollerstrom (2000), p. 81.
[80] The origin of Horrocks' lunar theory is described in detail in Wilson (1987b).
[81] In Flamsteed's published account, in which $|ED| = 0.055\,05$ and $|DC| = 0.011\,731$
(the length of the semi-major axis of the elliptical orbit being fixed at 1), he actually took the

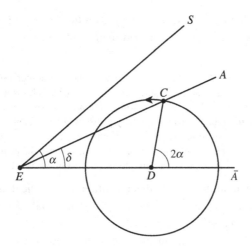

Fig. 7.12. Horrocks' mechanism for the Moon.

on a steady motion of the mean apsidal line with respect to the stars, which is implied by the differing lengths of the sidereal and anomalistic months.[82]

The idea of considering an orbit as a Keplerian ellipse, the defining parameters of which vary with time, was introduced to astronomers via Horrocks' lunar theory and would re-emerge in the latter half of the eighteenth century as an extremely powerful technique for determining the perturbing gravitational effects of one planet upon another.

The contributions of Kepler and Galileo

The period between Copernicus and Newton was a transitional one for mathematical astronomy. In the late sixteenth century, descriptions of the heavens were, as they had been since the time of the ancient Greeks, purely geometrical, but through the work of Kepler this began to change. Kepler's attitudes

---

eccentricity to be the length of the projection of EC onto the line $E\bar{A}$, i.e. $|EC| \cos \delta$. It is not clear whether Flamsteed misunderstood Horrocks' mechanism or whether Horrocks himself had actually done the same. Certainly, Flamsteed's form for the eccentricity leads to simpler calculations. For more on this issue, see Gaythorpe (1957) and Kollerstrom (2000), Chapter 7. The idea of an apse oscillating about its mean position was not new; it was implicit in many of the lunar theories that preceded Horrocks'. Ptolemy's use of prosneusis had just such an effect, for example, though the period of the oscillation was only half a synodic month (see p. 74).

[82] In one anomalistic month, the Moon advances (see Table 1.1, p. 7)
$(27.555/27.322) \times 360° \approx 363° \, 4'$ with respect to the stars and, hence, the mean apogee advances by just over $3°$ per lunar revolution.

to astronomical hypotheses were made clear in the *Apologia* (see p. 175) but this was not read widely. He reiterated them in the *Epitome of Copernican Astronomy*:

> ... physics is popularly deemed unnecessary for the astronomer, but truly it is in the highest degree relevant to the purpose of this branch of philosophy, and cannot, indeed, be dispensed with by the astronomer. For astronomers should not have absolute freedom to think up anything they please without reason; on the contrary, you should give *causus probabiles* for your hypotheses which you propose as the true cause of the appearances, and thus establish in advance the principles of your astronomy in a higher science, namely physics or metaphysics ... [83]

The major obstacle in the way of this transition to a physical astronomy was Aristotelian physics, which had become, since the Middle Ages, intertwined with Christian cosmology.

Although many people were involved in shaping the changes that took place during the seventeenth century, Kepler and Galileo stand out, but their contributions were very different. From a mathematical point of view it is Kepler's achievements that are the most impressive, but it is certainly true that of the two, Galileo is, and was, the more famous. Kepler's talents were not recognizable so easily to his contemporaries – his great discoveries were published alongside his more fanciful speculations, and in order to use his theories you had to be a highly competent mathematician. Also, when we look back on the events of this period, we do so with the knowledge that Kepler was right in some things but wrong in others, and we tend to ignore things such as his erroneous physical theory or his mysticism, choosing instead to remember the phenomenal achievement of his three laws of planetary motion. This filter was not available to his contemporaries and, although Kepler's mathematical astronomy clearly had to be taken seriously, it was not at all obvious that he had made the fundamental breakthrough we now know he achieved.

Galileo, on the other hand, was a competent mathematician, but not one of the great mathematical astronomers. His strength was his ability to use experimental results to argue persuasively against traditional prejudices. During the seventeenth century, there was a change in attitude in Europe toward the inductive sciences, and Galileo was at the forefront of this movement. His work laid the foundations on which Newton later would erect the subject of mathematical physics, and the fact that he clashed so publicly with the Roman Catholic Church and was forced to recant his views has helped to elevate him to the status of a martyr to science.

---

[83] KEPLER *Epitome of Copernican Astronomy*. Quoted from Jardine (1984).

Both men helped pave the way for what was to follow. Galileo's attacks on Aristotelianism demonstrated that the old physics was not sustainable, and many then attempted to determine the form the new physics should take. The increased accuracy of Kepler's planetary astronomy helped to shift the debate away from the Greek ideal of uniform circular motion and enabled others to ask themselves what shape the planetary orbits actually are and how the planets move round them.

8

The universal theory of gravitation

The Cartesian vortex theory

When it comes to our Solar System, pretty much the whole of modern-day predictive astronomy is based on the law of universal gravitation, introduced by Isaac Newton. Newton's work was the crowning achievement of a century of investigations into the subject of mechanics, which began with the work of Galileo. A detailed history of the development of dynamical ideas in the seventeenth century is beyond the scope of this book, but we will discuss briefly the main ideas and attitudes that shaped astronomical thought.[1]

By the mid seventeenth century, it was clear that the Ptolemaic universe, with the Moon, Sun and planets orbiting a stationary Earth, did not correspond to reality. Evidence suggested that the Earth was of a similar nature to the Moon and the other planets, and Aristotelian explanations of planetary phenomena in terms of natural motions were no longer acceptable. What was now required was an explanation in terms of terrestrial physics, a subject that, with Aristotle gone, was wide open. One such attempt was Kepler's, and his ultimate conclusions in the form of his laws of planetary motion were of immense significance, but little attention was paid to the precise nature of his physics. Another bold attempt was that of René Descartes, whose vortex theory was published in 1644.

Whereas Kepler had been concerned with the technical details of planetary astronomy, Descartes was interested in cosmology in a much broader sense. Kepler was led to his discoveries by a study of the small irregularities in planetary motions, but Descartes was led to his vortex theory by a consideration of the structure of the Universe as a whole. He accepted the Copernican system, since it explained the phenomena with the fewest assumptions, though he realized

---

[1] For a thorough discussion of this topic, see Westfall (1971).

that the same phenomena could be explained by a geoheliocentric universe. As a predictive tool, the vortex theory was way behind Keplerian astronomy, but it was immensely influential, particularly in The Netherlands and France, and played a key role in the subsequent development of ideas about the mechanics of the heavens.

The nature of the scientific method was a keen subject of debate in the early seventeenth century. Galileo had set the ball rolling (so to speak) with his insistence that mathematical theory was backed up by experiment, and in 1620, Francis Bacon, one of the most distinguished philosophers of science during the Renaissance, published the *New Organon* in which he espoused a general scientific philosophy that was based solely on experiment and research.[2] Descartes' method, on the other hand, was rational rather than empirical. For him, the relationship between science and metaphysics was crucial, and he based his physics on metaphysical principles. He sought a primary self-evident axiom from which everything else would follow by the application of mathematical laws, and he found this in the famous proposition: 'I think, therefore I am.' For example, he used his method to prove the existence of God: since he (Descartes) had ideas about perfection and infinity, and since the causes of these ideas must be perfect and infinite, a perfect infinite being must exist. He stated his opposition to the Baconian view very clearly in his *Philosophical Principles* (1644): 'I will explain the results by their causes, and not the causes by their results.'[3]

In Cartesian physics, matter was equated with its geometrical form – in other words, Descartes identified matter with the space it occupied. Neither could exist in isolation, and so a vacuum (space without matter) was impossible. There thus had to be some substance that filled the heavens. Another consequence of Descartes' views on the nature of matter was that all phenomena should be explicable in terms of geometry and motion. Apart from the supposed interaction between mind and body, the only possible causal mechanism in the Cartesian universe was impact; motion was the result of the action of matter on matter.

Descartes worked on his system of the world between 1629 and 1633, but the condemnation of Galileo deterred him from publishing. In his *Method* (1637)[4] he included a summary of his work on the mechanics of the heavens, but the

---

[2] An organon is a system of logical or scientific rules, particularly that of Aristotle. The title thus indicates that the new scientific method was to replace the old Aristotelian one. Bacon, who rejected Copernicanism and underrated the importance of mathematics in science, was far less influential than Galileo (see, for example, Cushing (1998), pp. 22–4).

[3] Quoted from Pannekoek (1961). The Latin title was *Principia philosophiae*.

[4] The full title of this work, a classic in philosophy, was *Le discours de la méthod pour bien conduire sa raison, et chercher la vérité dans les sciences*, and its publication brought the

definitive version of the vortex theory appeared in his *Philosophical Principles*. His theory was an attempt to provide a general explanation of planetary motion. As he put it:

> I have not described in my *Principles* all the motions of each planet, but I have supposed in general all those that the observers have found and I have attempted to explain the causes.[5]

The fundamental idea of his theory was that the space between the planets was filled with fluid matter containing a number of rotating vortices that carried the planets around in their orbits. These vortices were of differing strengths, e.g. a very powerful vortex was centred on the Sun, and this was responsible primarily for the motion of the six planets, but then there were lesser vortices centred on the Earth and Jupiter that accounted for the motions of their satellites. These vortices exerted pressures on each other, resulting in slight distortions, and these deviations from perfect circularity were the causes of the inequalities in planetary motions. Descartes used the analogy of the motion of water to show how such a mechanism could explain all the observed phenomena:

> Let us suppose that the matter of the sky where the planets are, turns without ceasing, as also a vortex at the centre of which is the sun, and that its parts which are nearer the sun move more quickly than those which are further away, up to [the distance of Saturn] and that all the planets (in whose number we henceforth include the earth) remain always suspended between the same parts of this matter of the sky; for by this alone and without employing other machines, we shall easily understand all the things that we notice in them. For as in the winding of rivers where the water folds back on itself, turning in circles, if some straws or other very light bodies float amidst the water, we can see that it carries them and moves them in circles with it; and even among these straws one can notice that there are often some which also turn about their own centre; and that those which are closer to the centre of the vortex which contains them make their revolution more quickly than those which are more distant; and lastly that, although these vortices of water design always to rotate in rings, they almost never describe entirely perfect circles, and extend themselves sometimes more in length and sometimes more in width, so that all parts of the circumference which they describe are not equally distant from the centre; thus one can easily imagine that all the same things apply to the planets; and it needs only this to explain all their phenomena.[6]

---

author instant fame. It contained three appendices, each containing an application of Descartes' method. One of these was entitled *Geometry* and contained his ideas on coordinate geometry and algebra. This was the only work Descartes ever published on mathematics, but it had a huge influence. The association of algebraic equations with curves and surfaces was an essential step in the invention of the calculus later in the seventeenth century. Another of the appendices was on dioptrics, and in this, Descartes extended Kepler's laws on lenses and published, for the first time, the correct law of refraction – Snell's law.
[5] Quoted from Aiton (1972).   [6] Quoted from Aiton (1972)

Though he was not entirely consistent on the matter, Descartes thought that the axial motion of the Earth and of the other heavenly bodies was simply a consequence of the way they were created. Once set in motion, they continued in this rotating state and, although they would be slowed down by their interaction with the surrounding fluid, this retardation would be imperceptibly small. Descartes used the analogy of a child's top to illustrate his point:

> ... without any noticeable diminution, because the larger a body is, the longer also can it retain the agitation which has been impressed on it, and the duration of five or six thousand years that the world has existed, compared with the size of a planet, is not so much as a minute compared with the smallness of a top.[7]

To Descartes, uniform motion in a straight line was a state in the same way that rest was a state. A cause was required to change a state and, hence, he was led to the concept of rectilinear inertia: unless something acts to change it, a body will remain at rest or continue in uniform motion in a straight line. Formally, this is equivalent to what has become known as Newton's first law of motion, but it meant something rather different to Descartes. In his system, inertial motion did not happen, since objects were continually subject to the impacts of other bodies, a void being impossible. Thus, this inertia represented a *tendency* to move in a straight line. A crucial corollary was that an object in circular motion was forever trying to fly off at a tangent (this tendency was known as the centrifugal force) and so some external force was required to pull it toward the centre.

Following Galileo's pioneering work, the concept of inertia had been studied actively in the early seventeenth century by men such as Evangelista Torricelli. Descartes' ideas about inertia corresponded, by and large, with experience, and when he published his statement of rectilinear inertia in 1644, it was a commonly held belief. Once one accepts that uniform motion in a straight line persists unless some external force acts, it seems only a small step to the realization that the effect of a force is to produce a change in velocity – an acceleration – but this conceptual breakthrough was slow in coming.

Another fundamental principle in Cartesian mechanics, although this time an erroneous one, was the idea of the conservation of motion. Descartes perceived motion as possessing its own independent existence so that it could pass from one object to another during an impact. Motion was caused by God, and since God was immutable, the quantity of motion should remain unchanged. Unfortunately for Descartes, he quantified motion in terms of the product of size and speed and, hence, in terms of a scalar, rather than a vector, quantity. Since the Cartesian universe was driven by the continual impacts between objects,

---

[7] Quoted from Aiton (1989).

Descartes applied his laws of motion to derive a theory of collisions, but his incorrect conservation law led to an unsatisfactory theory.[8]

The vortex theory, as developed by Descartes, was qualitative in nature; it had to be, since the mathematical methods required to study fluid vortices had not been invented yet. The theory provided a psychologically satisfying explanation of celestial phenomena and certainly was ingenious. However, it was riddled with inconsistencies and could not (and did not pretend to) compete with contemporary quantitative mathematical astronomy. Although one of the system's weaknesses was the fact that fluid motion was understood poorly in Descartes' time, this also had the beneficial effect that it was hard for anybody else to refute Cartesian ideas. For half a century, the mechanical philosophy of Descartes replaced the old Aristotelian physics as the accepted wisdom of the vast majority. It was the awesome predictive power of the Newtonian theory of gravitation that would lead eventually to its demise, and Descartes' theory was later much ridiculed, particularly in England. In his famous *History of Physical Astronomy*, Robert Grant wrote in 1852:

> No doubt, we think, can exist that this celebrated fiction exercised a most pernicious influence in retarding the progress of sound mechanical ideas relative to celestial physics. Like the theory of solid orbs, it at length utterly disappeared before the advancing light of true science, after continuing for nearly a century to indulge its adherents with the miserable delusion that it revealed to them the whole secret of the mechanism of the universe.[9]

Pre-Newtonian conceptions of gravity

During the seventeenth century, science began to organize itself. In particular, scientific endeavour in England and France began to revolve around the activities of two influential groups: the Royal Society in England and the Academy of Science in France. The Royal Society was founded in 1662, and soon after its formation Charles II asked about the possibility of using astronomical observations to aid navigation, a subject of vital importance to a seafaring nation like Britain. The Royal Society recommended that an observatory be created, and the purpose-built Royal Observatory was founded at Greenwich in 1675 with John Flamsteed as the first Astronomer Royal. While this sounds rather grand, actually the whole thing was done on the cheap, and Flamsteed had great difficulty in acquiring the necessary observing instruments. In 1666, the Academy of Science was formed in France and the Paris Observatory founded a year later.

---

[8] See, for example, Clarke (1977).   [9] Grant (1966), p. 19.

One of the topics that occupied the minds of many of the members of these learned societies was the nature of gravity. At the beginning of the seventeenth century, many natural philosophers looked to the magnetical philosophy of William Gilbert for the source of a new universal physics with which to describe the Copernican heliocentric universe;[10] Gilbert's ideas influenced Kepler and Galileo as well as many other astronomers. By about 1650, however, the popularity of magnetic astronomy had declined in favour of a mechanistic approach, though in England a residual influence remained. Christopher Wren, for example, gave a lecture in 1657 in which he held Gilbert up as an example of a positive influence on astronomy, as opposed to the negative approach of Galileo!

One of the leading proponents of Gilbert's philosophy in England was John Wilkins who, in his *Discovery of a World in the Moon* (1640), described gravity as

> ... a respective mutual desire of union, whereby condensed bodies, when they come within the sphere of their own vigour, do naturally apply themselves, one to another by attraction or coition.[11]

Following Wilkins, both Wren and Robert Hooke[12] utilized ideas that were a mixture of the magnetic philosophy and the increasingly popular mechanistic interpretations of the Cartesians. These ideas were interwoven in the discussions between the two men concerning the comet of 1664. In his *Cometa* of 1678, Hooke wrote:

> ... I suppose the gravitating power of the Sun in the centre of this part of the Heaven in which we are, hath an attractive power upon all the bodies of the Planets, and of the Earth that move about it, and that each of those again have a respect answerable, whereby they may be said to attract that sun in the same manner as the Load-stone hath to Iron, and the Iron hath to the Load-stone.[13]

But just as a magnet has no effect on certain materials, Hooke believed that some bodies would not be influenced by gravitational attraction.

In Europe, Descartes' influence was very strong; his popularity snowballed and his ideas came to dominate seventeenth-century scientific thought. Followers included Henricus Regius, Jacques Rohault, Pierre Sylvain Régis, and Nicolas Malebranche. Regius tried to separate the physics from the metaphysics in Descartes' system, and in so doing was disowned by Descartes for

---

[10] Ideas based on magnetism were used also to support the geocentric world view (see Baldwin (1985)).
[11] Quoted from Bennett (1989).
[12] Hooke was appointed by the Royal Society (on a meagre wage) to demonstrate a new and interesting experiment at every session.
[13] Quoted from Bennett (1989).

whom this linkage was fundamental. Others followed Regius' lead, however, notably Rohault, whose *Le traité de physique* (1671) became the standard work on Cartesian physics. Régis also tended to view physics as an empirical science and advocated the achievement of consistency in physical theories. Whereas others argued that a particular theory could be developed for a specific problem and that all these theories could be considered in isolation, Régis argued that all the different theories should be consistent with one another.

The first person to try to advance Cartesian physics beyond the level achieved by Descartes was Malebranche. He did not believe that motion could be transferred from one object to another through a collision, but instead fell back on the doctrine of occasionalism in which the impact was simply an *occasion* for God to make the change. This difference was purely one of metaphysics, but he effected also a change in Cartesian physics. In Descartes' cosmology there were three elements: luminous, transparent, and opaque. Luminous matter consisted of fine particles that made up the stars, whereas the coarse particles that made up the body of the Earth and planets were opaque. The space between the stars and planets – the so-called subtle matter – was made up of globules, or boules, of transparent material. Malebranche replaced these boules by small elastic vortices, and through this device managed to provide an explanation for the phenomena of heat and light.

The Cartesians spent a great deal of time discussing the nature and cause of terrestrial gravitation. Both Copernicus and Galileo adhered to the Platonic viewpoint that gravity was a natural inclination of bodies to move toward each other. For example, Copernicus believed that the spherical shape of the heavenly bodies was due to the mutual attraction of their parts:

> For my part I believe that gravity is nothing but a certain natural desire, which the divine providence of the Creator of all things has implanted in parts, to gather as a unity and a whole by combining in the form of a globe. This impulse is present, we may suppose, also in the sun, the moon, and the other brilliant planets, so that through its operation they remain in that spherical shape which they display.[14]

To Kepler, gravity – which he compared to magnetism – was a tendency for like bodies to approach each other. He took the tides as evidence that the Moon exerted an attractive force on the Earth, but this gravity was not related to orbital motion. The Sun did not attract the planets – it pushed them around.

It was Descartes who turned Kepler's ideas into a theory of universal gravitation, something that was a natural consequence of the collisions between bodies. In the Cartesian system, gravity was perceived as being analogous to the tendency of floating bodies to move toward the centre of whirlpools, and

---

[14] COPERNICUS *On the Revolutions*, I, 9.

Descartes described a mechanism by which this could occur. This mechanism had to account for the fact that bodies were attracted toward the centre of the Earth, whereas the whirlpool analogy suggested that they should move toward the axis of the vortex and, like much of the Cartesian theory, it contained inconsistencies. Descartes assumed that terrestrial gravity diminished as one moved away from the Earth sufficiently rapidly so that it did not affect the Moon or planets.

The fact that the physical cause for the motion of the planets lay in the Sun had been Kepler's idea, and he had also assumed, after the discovery of Jupiter's moons, that the force responsible for their motion was inherent in Jupiter. Similarly, Descartes attributed the motion of planetary satellites to small vortices centred on those planets, and others were of a like mind. John Wallis in 1666 wrote:

> As the sun by its motion about its own axis, is with good reason judged to be the *physical* cause of the *primary* planets moving about it; so there is the like reason to believe, that *Jupiter* and *Saturn* moving about their axes, are the physical cause of their *satellites* moving about them . . .[15]

In a work on the moons of Jupiter (also published in 1666), Giovanni Borelli[16] suggested that the orbits of the satellites were caused by a balance between the centrifugal force of the Moons (i.e. the tendency to move away from Jupiter) and some force pulling it toward the planet. Added to this was a force acting tangentially that pushed the planets around in their orbits. Borelli believed that by studying the Jovian system, general conclusions about the Solar System could be drawn.

> We shall assume, therefore, that the planet tends to approach the sun, while at the same time it acquires the *impetus* to move away from the solar centre through the *impetus* of circular motion: then, so long as the opposing forces remain equal (the one is in fact compensated by the other), [the planet] cannot come closer to, nor move further away from, the sun, and must remain within a certain, fixed space; consequently, the planet will appear to be in equilibrium and floating.[17]

Of course, if this were an exact balance, circular motion would result – which was not observed – but Borelli assumed that, in fact, there was a continuous disequilibrium in which the two opposing forces alternately dominated each other, resulting in elliptical orbits. The force pulling the moons of Jupiter toward the

---

[15] Quoted from Wilson (1970).
[16] Borelli was one of the most influential members of the Accademia del Cimento, formed in Florence in 1657 by a group of Galileo's followers. The Academy was forced to close only 10 years later under pressure from the Church. Details of Borelli's celestial mechanics can be found in Koyré (1973) and a modern view of Borelli's work is given in Meli (1998).
[17] Quoted from Koyré (1973).

central planet (and by analogy the force that pulled the planets toward the Sun) was inherent in the orbiting body and, hence, constant, and Borelli assumed that the centrifugal force increased as the planet approached the Sun and diminished as it moved further away. Hence, any disturbance to a perfect circular orbit would result in an oscillation about that orbit with the centrifugal tendency alternately being less than, and greater than, the gravitational attraction.

The underlying cause of planetary motion in Borelli's system was as it had been for Kepler, the rotation of the Sun. The light emanating from the Sun rotated with the Sun and pushed the planets around in their orbits. Borelli differed from Kepler in that the effect of the rays was cumulative and he believed that there was a tendency for a planet to retain the speed imparted by the solar rays. It was not clear what limited the planet's speed in Borelli's system, nor why the continued action of the rays was necessary.[18] Borelli's ideas necessarily were vague because he did not know how the centrifugal force varied as a function of the speed of the satellite and the distance from the planet; this was discovered by Christiaan Huygens.[19]

Cartesian physics – and in particular the treatment of gravitation – advanced to a new level with the work of Huygens, who incorporated a quantitative element into the theory. Crucial to his success was his correct appreciation of centrifugal force. Huygens' solution was first published in his great work on pendulum clocks, the *Horologium Oscillatorium* (1673), but in 1669 Huygens had sent his result in the form of an anagram to Henry Oldenburg, the secretary of the Royal Society, to ensure his claim to priority:[20]

> If a body revolves in a circle with the time that a pendulum of equal length with the radius of the circle makes a double swing, then the centrifugal force of the body will be equal to its weight.[21]

Since, in modern terminology, the period of a pendulum of length r is $2\pi\sqrt{r/g}$ (g being the acceleration due to gravity), a body moving so that it went round a circle of radius r in this time would have speed $v = \sqrt{gr}$. Huygens' assertion is

---

[18] The fact that rays of light from the Sun played a role in the motion of the planets suggests that Borelli originally developed his ideas in the context of planetary motion and transferred them subsequently to the Jovian system (where he had to assume the presence of rays emanating from the planet) so as to conform to the ban on the teaching of Copernicanism (see Armitage (1950)).

[19] Huygens was one of the great scientists of the seventeenth century and wrote on many subjects including mathematics and astronomy. He invented the pendulum clock in 1656, and around the same time discovered the nature of Saturn's curious shape. His contributions to mechanics, which are less well known than his contributions in other fields, are discussed in Gabbey (1980).

[20] In fact, Newton independently arrived at the same result in 1665 (see p. 253 below).

[21] Quoted from Hall and Boas Hall (1969).

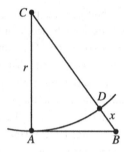

Fig. 8.1. Huygens' derivation of the form of the centrifugal force.

then equivalent to saying that the centrifugal force $F(v, r)$ satisfies the equation $F(\sqrt{gr}, r) = mg$, i.e. $F = mv^2/r$. Of course, Huygens could not express his result like this because he did not have the concept of mass as opposed to weight (this was one of Newton's many contributions). A simplified version of Huygens' derivation is illustrated in Figure 8.1.[22] Imagine that a particle and an observer travel around a circle, centred at C with radius r, and that at some instant they are at A, moving in the direction AB with speed v. In a time t, the particle moves around the circle to D, where the length of the arc $AD = vt$, but if allowed to move freely, the particle would move along the tangent to B where $|AB| = vt$. The points C, D and B lie on an approximate straight line, the approximation being better the smaller the value of t that we take. Thus, the particle would move away from the observer by a distance $|DB| = x$ and Huygens used this distance to measure the strength of the tendency to move off on a tangent. Now $|BC|^2 = |AB|^2 + |AC|^2$ and so, in modern notation,

$$r + x = \sqrt{r^2 + v^2 t^2} = r \left(1 + \frac{v^2 t^2}{r^2} \right)^{\frac{1}{2}} \approx r \left(1 + \frac{v^2 t^2}{2r^2} \right),$$

from which we obtain $x = v^2 t^2/2r$. Galileo had shown previously that a uniformly accelerated body moves according to the formula $s = at^2/2$, where s is the distance travelled and a is the acceleration, and so Huygens could conclude that the centrifugal acceleration – though he does not use the word 'acceleration' – is v^2/r.

Huygens' theory of centrifugal force was extremely significant. It was the final nail in the coffin of the idea of circular inertia that had dominated thought for so long. It made clear that uniform circular motion required the constant action of a force to balance the centrifugal force. Huygens was a Cartesian,

[22] This derivation did not become known until after Huygens' death (see Dijksterhuis (1961) and Jammer (1957) for more details).

and so any such force had to be caused by a mechanism based on impact, and he devised just such a mechanism to explain gravity. Note that Huygens was trying to explain gravity via circular motion; Newton later would explain circular orbits from a correct understanding of gravity.

The basis for Huygens' idea, which had already been espoused by Descartes, was that the centrifugal tendency of the subtle matter that was rotating around the Earth caused the heavier opaque matter to be forced inwards toward the centre of the vortex. One problem with this idea – which has already been noted – was that gravity acted toward the centre of the Earth, whereas the rotation of the fluid vortex was around the axis of rotation of the Earth. Huygens thus supposed that there was actually a spherical vortex, with particles of the subtle matter moving about the Earth in all possible directions. He supposed that heavy bodies did not rotate around the Earth like the fluid making up the vortex of the Earth because there was insufficient time for the many tiny impulsions that they received from the fluid matter to cause any significant horizontal motion. Heavy bodies thus possessed no centrifugal force relative to the Earth, unlike the subtle matter. In Huygens' theory, the weight of a body was equal to the centrifugal force of a quantity of the surrounding fluid, equal in volume to that of the body, and applying his quantitative theory he came to the conclusion that the fluid matter must be moving seventeen times faster than the speed of a point on the equator of the Earth due to the diurnal rotation.

A consequence of Huygens' theory was that the daily rotation of the Earth must reduce the effect of gravity. Since the centrifugal force was proportional to the square of the speed, his calculation indicated that the effect of gravity at the equator should be reduced by a factor of $1/17^2 = 1/289$. He then used this value to show that a pendulum clock that worked accurately at one of the poles would run slow by just over $2\frac{1}{2}$ min per day if used at the equator. Around this time, men were experimenting by taking pendulum clocks to different parts of the world, and the results of their experiments did suggest that the gravitational force was not constant, the data showing partial agreement with Huygens' result.[23]

Another consequence of the daily rotation of the Earth in Huygens' theory was a deviation of a plumb line from the vertical. At the latitude of Paris, this should have amounted to about 6' of arc, but such a deviation was not observed. This led Huygens to believe that the Earth was not a perfect sphere.

---

[23] In particular, Jean Richer travelled to Cayenne which is on the coast of South America, near the equator, to make observations that would help in the determination of the solar parallax. He took with him a pendulum that was designed to beat seconds in Paris and found that he had to shorten it by about 28 mm in order for it to keep time in Cayenne (Chapin (1995)). Richer's trip is described in detail by Olmsted (1942).

This deviation is contrary to what has always been supposed to be a very certain truth, namely, that the cord stretched by the plumb is directed straight toward the centre of the earth . . . Therefore, looking northward, should not the level line visibly descend below the horizon? This, however, has never been perceived and surely does not take place. And the reason for this, which is another paradox, is that the earth is not a sphere at all but is flattened at the two poles, nearly as an ellipse turning about its smaller axis would produce. This is due to the daily motion of the earth and is a necessary consequence of the deviation of the plumb line mentioned above. Because bodies by their weight descend parallel to the direction of this line, the surface of a fluid must put itself perpendicular to the plumb line, since else it would stream farther downward.[24]

Since gravity was caused by the vortices of fluid outside the Earth, Huygens reasoned that within the Earth its effect was constant. He computed the ratio of the polar and equatorial diameters to be $577 : 578$ on this assumption.

Isaac Newton and the *Principia*

It often is remarked that Galileo died and Newton was born in 1642, but this is not strictly true. At the time, Italy used the Gregorian calendar, but England was still using the Julian calendar and so, when Newton was born on Christmas Day 1642 in England, it was already 1643 in Italy. Of course, this takes nothing away from the interesting proximity between the death of the founder of the science of mechanics and the birth of the man who would turn Galileo's groundwork into the most successful scientific theory of all time. The lives of these two great intellectual giants spanned the years 1564–1727, a period of profound change in scientific attitudes often referred to as the 'Scientific Revolution'.[25]

Newton had an unremarkable education as a boy. He went to Cambridge in 1661 and came into contact with the mathematics and natural philosophy of the day. He read widely, learning mathematics from the works of Descartes and Wallis, and studied books on a wide range of subjects, mixing a flair for physics and optics with more unorthodox interests in alchemy and biblical chronology.

---

[24] HUYGENS *Le discours de la cause de la pesanteur* (*Discourse on the Cause of Gravity*), 1690 (translation from Pannekoek (1961), p. 269). This work was published as an appendix to his *Le traité de la lumière* (*Treatise on Light*), but most of its contents had been presented to the Academy of Science in Paris as early as 1669; for more details, see Snelders (1980). However, the quantitative analysis of the shape of the Earth was performed after the publication of Newton's *Principia* in 1687 (see Todhunter (1962)).

[25] A great deal has been written about Newton since his death in 1727 (Hall (1999) lists thirty-six major biographies). The definitive modern biography of Newton is Westfall (1980) which also contains an essay detailing other bibliographic information. A shortened version, with much of the technical content of the full biography omitted and therefore accessible to a wider audience, is Westfall (1993).

The rapid pace at which he acquired knowledge (and he was primarily self-taught) is indicated by the fact that he succeeded Isaac Barrow as Lucasian Professor of Mathematics at Cambridge in 1669.

Newton began to think about celestial mechanics in his early twenties, possibly prompted by the appearance of the comet of 1664. It seems likely that he learned of the Ptolemaic and Copernican world-views from Galileo's *Dialogue* and from works by Gassendi. He also familiarized himself thoroughly with the *Philosophical Principles* of Descartes, from which he learned the principle of inertia.[26] Entries in a student notebook begun in 1661 show that he read Streete's *Astronomia Carolina* and that he was aware of Kepler's first and third laws, but there is no evidence that he ever read Kepler's *New Astronomy*. He seems to have had confidence in the accuracy of the third law, but not the first.[27] As to the second law, Newton fell in with his contemporaries. Streete's work did not contain any direct statement of the area law but, instead, discussed the methods of Boulliau and Ward for working out the true anomaly as a function of time. Newton worked with these and also produced a few alternatives of his own – there is no mention of Kepler's second law in any of his writings until 1684.

It did not take long before Newton was making profound and original contributions of his own. He realized early on that Aristotle's attempt at understanding motion was inadequate, and began to develop his own ideas based on those of Galileo and Descartes. In his early work on collisions, he realized that Descartes' idea of conservation of motion was flawed, because it failed to take into account the direction of motion. Newton thus was led to the idea of the conservation of momentum – where momentum is mass × velocity and, hence, a vector quantity – though he did not express it this way since he had not at that time achieved a satisfactory understanding of the concept of mass, and the language of vectors was not to be invented until the nineteenth century. By the use of this modified version of Descartes' conservation law, he managed to derive the correct rules for both elastic and inelastic collisions.

As far as planetary motion was concerned, Newton again took the lead from Descartes. In the second half of the seventeenth century, the vortex theory was the only widely supported physical explanation of celestial motion, and Newton seems to have adhered to it, just as did most of his contemporaries. For example, in notes on the endpapers of Wing's *Astronomia Britanica*, he

---

[26] A discussion of all the works that influenced Newton's early thoughts on astronomy can be found in Whiteside (1970a). Newton's very early astronomical studies are described in McGuire and Tamny (1985).

[27] Whiteside (1964).

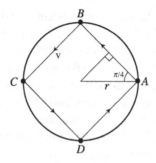

Fig. 8.2. Newton's derivation of the form of the centrifugal force.

attributed the physical cause of the irregularities in the motion of the Moon around the Earth to the pressure of the vortex of the Sun on that of the Earth.

The traditional story that Newton first thought of universal gravitation in 1666 after observing an apple falling from a tree while at home in Lincolnshire to escape the plague that had descended on Cambridge is, to say the least, misleading. There is no evidence that Newton considered the mutual attraction of heavenly bodies before 1679, or that the actual law of universal gravitation occurred to him before 1684. The theory of gravity that Newton published in 1687 was not simply a flash of inspiration – it was the result of a great deal of intellectual effort spent grappling with the fundamental principles of dynamics. He did, however, consider the possibility that the gravity of the Earth might extend as far as the Moon, and this may well have been inspired by the fall of an apple, but the terrestrial gravity he was reflecting upon was not gravity as eventually he understood it.[28]

The key ingredient in the development of Newton's ideas that allowed him to consider the effect of gravity on the Moon came from his study of uniform circular motion. Following Descartes, he assumed that any body moving in such a manner had a centrifugal tendency away from the centre of motion that caused the body continually to try to move off on a tangent and, like Huygens, he looked for a quantitative measure of this 'force'. To get a handle on things, Newton considered a body that was moving around a circular path of radius r on the inside of a spherical surface. Then he approximated the situation by assuming that the body moved around the inscribed square $ABCD$ shown in Figure 8.2 at a constant speed v, bouncing off the sphere at each of the vertices. At each impact, the component of the motion perpendicular to the tangent is

---

[28] For an account of the origin of the anecdote concerning the apple, see Westfall (1980), p. 154. A detailed analysis of the claim that Newton did not believe in universal gravitation prior to 1684 is given in Wilson (1970).

reversed, which implies a change in the quantity of motion of $2mv \cos \pi/4$. The total change in the quantity of motion over the four impacts, divided by the quantity of motion the body possesses (i.e. mv), is thus $8 \cos \pi/4$.[29] The length of the perimeter of the square $ABCD$ is clearly $8r \cos \pi/4$, and so we have the result that

$$\frac{\text{total change in quantity of motion}}{\text{quantity of motion of body}} = \frac{\text{perimeter of square}}{\text{radius of circle}}.$$

Newton realized that this result would hold for any inscribed regular polygon, and we can demonstrate this as follows. Half the interior angle of a polygon with n sides is $\pi/2 - \pi/n$ (this plays the role of the marked angle in the diagram) and so at each impact the change in the component of motion perpendicular to the tangent is $2mv \cos(\pi/2 - \pi/n)$, which must be multiplied by n to get the total change as the body moves around its polygonal path. The perimeter of the polygon is $2nr \cos(\pi/2 - \pi/n)$ and so the result stated at the end of the previous paragraph extends to this new situation. If it is true for any regular polygon, Newton could proceed to the limit of a polygon with an infinite number of sides, i.e. the circle itself. In this way, he established that the total change in the quantity of motion of the body over the course of its circular obit is $2\pi mv$. Then he argued that the magnitude of the instantaneous force acting on the body, f, must be this total change divided by the period, and in this way arrived at the correct form for the centrifugal force,

$$f = \frac{2\pi mv}{2\pi r/v} = \frac{mv^2}{r}.$$

The first use to which he put his new formula was the accurate determination of the acceleration due to gravity using a conical pendulum. His result was very accurate and roughly twice as large as the value determined by Galileo and given in the *Dialogue*.

Newton had read Borelli's work in which it was suggested that the centrifugal force of a planet was balanced by an attraction toward the planet and, in 1666, Newton made the crucial realization that Kepler's third law was consistent with an inverse square law for the centrifugal tendencies. Assuming circular orbits, Kepler's third law relates the periods T of the planets to their orbital radii r through the equation $T^2 \propto r^3$. Since $T \propto r/v$, where v is the speed of the planet, we can conclude that $v^2 \propto 1/r$. The centrifugal force of the planet is

---

[29] This is a meaningless physical quantity, since we have added the magnitudes of four vector quantities, each of which has a different direction (indeed, the vector sum is zero). Newton knew what he was doing, though, and his subsequent argument was sound (see Herivel (1960), Brackenridge (1995), pp. 42–54).

proportional to v^2/r, and thus to $1/r^2$. Hence, gravity, which balances this, must vary as the inverse square of the distance. Since the Earth–Moon distance is about 60 Earth radii, the pull of gravity on the Moon must be about 3600 times smaller than the gravitational force at the surface of the Earth.

To test this, Newton conducted his so-called 'Moon-test'. Basing his observations on an inaccurate value for the diameter of the Earth, he calculated that the effect of gravity on the Moon was actually a little over 4000 times greater than the gravitational pull at the surface of the Earth.[30] Many years later, Newton wrote that he had found the comparison to 'answer pretty nearly', but it would appear that at the time he was slightly disappointed in the lack of quantitative agreement, and so turned his mind to other studies, in particular his researches into optics. It should also be emphasized that at this time Newton still adhered to the Cartesian vortex theory, so what he had actually demonstrated was that, in the whirlpool of the terrestrial vortex, the endeavour of bodies to move out of their circular orbits varied, at least approximately, as the inverse square of the distance from the Earth. Thus, while this statement clearly is important, Newton was still a long way from any concept of universal gravitation and had not yet formulated his dynamics in terms of central forces and tangential inertia.

The year 1679 was a crucial one in the development of Newton's thought on planetary motion. By this time Newton had come to appreciate the many inadequacies in the vortex theory and had freed himself of much of the restrictive Cartesian baggage. He was unwilling still (like most other astronomers) to accept Kepler's second law as being true, since it had no firm mechanical foundation. The area law was, however, becoming more well known, particularly through the works of Nicholas Mercator – a Danish mathematician who lived in London – and it was clear that whatever the correct law, it had to give results close to those produced by the area rule. Newton recognized that planetary orbits certainly were approximately elliptical, but did not believe there was sufficient empirical evidence to be certain of their exact elliptical nature. The catalyst for the fundamental change in Newton's thought was an exchange of letters with Robert Hooke.

During the latter half of the seventeenth century, there was much discussion in England concerning the precise nature of planetary orbits and the relation

---

[30] It has been suggested that Newton assumed that there were 60 miles per degree of latitude on the Earth, a mile being 5000 ft (whereas a more accurate value is $69\frac{1}{2}$ miles, each of 5280 ft). This would lead to a factor of about 4332 (see Chandrasekar (1995)). Newton would later do the calculations again with the more accurate data (see p. 271). Accounts of the so-called 'Moon-test' by William Whiston, Henry Pemberton and Newton himself are given in Herivel (1965a).

between celestial motion and mechanics, three of the key figures being Christopher Wren, John Wallis, and Robert Hooke. In a lecture in 1674 – *Attempt to Prove the Motion of the Earth* – Hooke had suggested that *all* heavenly bodies attract things toward their own centres, this being the cause of their spherical nature and also the motion of bodies within their sphere of influence.

> This depends upon three suppositions. First, that all the celestial bodies whatsoever, have an attraction or gravitating power toward their own centres, whereby they attract not only their own parts, and keep them from flying from them, as we may observe the earth to do, but that they do also attract all other celestial bodies that are within the sphere of their activity ... The second supposition is this, that all bodies whatsoever that are put into a direct and simple motion, will so continue to move forward in a straight line, till they are by some other effectual powers deflected and bent into a motion, describing a circle, ellipse, or some other more compounded curve line. The third supposition is, that these attractive powers are so much the more powerful in operating, by how much the nearer the body wrought upon is to their own centres.[31]

Hooke's remarks show that he was not only moving away from the restricted view that gravity was a property of an individual planet, but also getting to grips with the dynamics, in that there is no mention of centrifugal endeavour. The orbit is due to the continual deflection of the body from its tangential path, toward the centre.

In November 1679, Hooke wrote to Newton in an attempt to re-establish a working relationship that had suffered due to serious disputes between the two men on the subject of the nature of light. He included the request:

> And particularly if you will let me know your thoughts of that of compounding the celestial motions of the planets of a direct motion by the tangent and an attractive motion toward the central body ...[32]

There is no evidence that, before the receipt of this letter, Newton ever had understood orbital motion in this way, but instead had always treated it as a balance between the opposing centrifugal and central forces. Thought of as a state of equilibrium between equal and opposite forces, circular motion appeared to be an argument against the principle of inertia and represented an obstacle to the correct understanding of the principles of dynamics. In his reply to Hooke, Newton declined to give his thoughts on the matter, instead choosing to discuss an experiment to prove the daily rotation of the Earth, which had been the main subject of Hooke's 1674 lecture.

---

[31] Quoted from Westfall (1980), p. 382, (with changes to reflect modern spelling). Hooke's lecture was republished in 1679 in his *Lectiones Cutlerianae (Cutler Lectures)*.
[32] Quoted from Whiteside (1964).

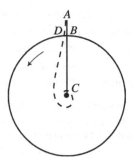

Fig. 8.3. Newton's spiral path for a freely falling body on a rotating Earth.

The objection to diurnal rotation had, for a long time, been that an object dropped from a tower should be left behind as the Earth turned beneath it, thus landing slightly to the west of the point from which it was released, something that was not observed. Newton suggested that the object actually should land to the east because the tangential velocity at the top of the tower was greater than that at the bottom; hence, the falling body should rotate faster than the Earth.[33] He drew a sketch indicating the path that the body would take, which he thought would be a sort of spiral. In Figure 8.3, which is drawn from the point of view of an observer rotating with the Earth, the path of fall of an object dropped from the top of a tower at A is indicated by the dashed line, and we can see that the body strikes the Earth at D, which is ahead of the bottom of the tower, B.

Newton's reply had been hasty, and his spiral path was a serious mistake. By drawing the path continuing through the Earth, he had converted the problem into one of orbital motion about the centre of attraction, and Hooke was quick to realize this. He wrote to Newton correcting the error and stating that he believed the correct path for such an orbit would resemble an ellipse, and he referred to his own theory of circular motions made up of a direct tangential motion and an attractive one toward a centre. Newton's next reply was that of an irritated man; nevertheless, he acknowledged his error and went on to demonstrate to Hooke that he did have the expertise to handle orbital motion correctly, but at the same time he said that he did not consider the matter particularly important. Hooke did not give up. He wrote two further letters in which he stated that he

---

[33] This problem has a long and fascinating history (see, for example, Armitage (1947), Koyré (1955), Drake (1988)). The first person to provide a satisfactory quantitative estimate of the eastward displacement of a falling body was Gauss. A mean displacement of 0.149 cm was observed in experiments by E. H. Hall at Harvard in 1902 for objects dropped from a height of 23 m. Gauss' theory (neglecting air resistance) predicts 0.179 cm (Armitage (1947)).

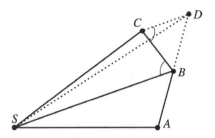

Fig. 8.4. Newton's derivation of the area law.

believed gravity to be an inverse square force, and asked Newton what shape the correct orbit for such a force law was. Newton did not reply.[34]

The problem did serve to revive Newton's interest in dynamics, though, and he took up the challenge of determining the force law that would cause a body to revolve around a point of attraction in an elliptical orbit. With Hooke's ideas about the dynamics of circular motion, Newton quickly mastered his own dynamical principles, and took a huge step forward when he managed to show that any body moving subject to a central force would satisfy Kepler's law of areas. Elliptical motion was clearly non-inertial, and so a body moving in such an orbit must be continually subject to a force, but it is not obvious at all what the direction and magnitude of such a force should be. Newton may well have believed that the force varied inversely as the square of the distance from the Sun and was directed toward the Sun, but he wanted to provide a mathematical demonstration of the fact. These demonstrations were written up in a short tract usually referred to as *De motu* (*On Motion*) that Newton sent to Edmond Halley in 1684.[35]

Newton's proof that any body moving under the influence of a force that always acts in the direction of a fixed point obeys Kepler's area law is illustrated in Figure 8.4. His method was to replace the continuously varying force with a series of equally spaced impulses. Suppose that the body at *A* moves inertially (i.e. in a straight line), but is acted upon at equal intervals of time by an impulse directed toward the fixed point *S*. In the first interval of time, the object moves from *A* to *B* and would continue to *D* in the second interval if left undisturbed. However, the action of the force toward *S* is to deflect the body to *C* where

---

[34] The correspondence between Newton and Hooke on gravitation has been the subject of much discussion (see, for example, Patterson (1949), (1950), Lohne (1960), Westfall (1967), Wilson (1989a)).

[35] An English translation is given in Herivel (1965a). Just how quickly Newton managed to link Kepler's first and second laws to an inverse square law of attraction is still the subject of debate (see, for example, Kollerstrom (1999)).

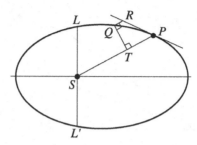

Fig. 8.5. The consistency of the inverse square law and elliptical orbits.

DC is parallel to BS. Now, since the area of SAD is twice the area of SAB, it follows that the triangles SAB and SBD have the same area. Also, since DC is parallel to BS, the areas of SBC and SBD are the same. Hence, area $SAB =$ area SBC. Finally, the intervals of time are made shorter and shorter, a process Newton described as follows.

> Now let the number of triangles be increased, and their width decreased indefinitely, and their ultimate perimeter [ABC] will ... be a curved line; and thus the centripetal force by which the body is continually drawn back from the tangent of this curve will act uninterruptedly, while any areas described, [SABS and SACS], which are always proportional to the times of descriptions, will be proportional to those in this case.[36]

Newton also proved the converse result: that any body moving according to the area law was under the influence of a net force that remained directed toward a fixed point. This general relation between motion subject to Kepler's second law and central forces was the basis of all Newton's subsequent work on orbital motion.

The logical next step was to investigate the motion of a body in an elliptical path under the action of a force directed toward one of the focuses of its orbit, and this is precisely what Newton did. His method[37] is illustrated in Figure 8.5, in which we can think of S as representing the Sun, at one focus of the elliptical orbit, and P a planet. If the planet were moving inertially, it would do so along the tangent of the ellipse to R in a short time Δt, but the force acting along PS results in the planet actually moving to Q, where RQ is parallel to PS. Newton's argument then proceeds as follows. The magnitude of the force F is proportional to $|RQ|$ and for a given force (from Galileo's results on uniformly accelerated motion) $|RQ|$ is proportional to $(\Delta t)^2$. Thus,

$$|RQ| \propto F(\Delta t)^2.$$

---

[36] NEWTON, *Principia*, Book I, Proposition 1.   [37] Described in Herivel (1965b).

Next, we can invoke the area law which says that Δt is proportional to the sector SPQ or, since Δt is small, to the area of the triangle SPQ. Combining this with the above proportion gives[38]

$$F \propto \frac{|RQ|}{(|SP|\cdot|QT|)^2},$$

where T is the foot of the perpendicular from Q to SP. So far, the argument is independent of the actual shape of the orbit. Newton then showed that for an ellipse, in the limit as Q approaches P, the ratio $|RQ|/|QT|^2$ tends to $1/|LL'|$ which is constant and, hence, that $F \propto 1/|SP|^2$. In this way, he managed to demonstrate that elliptical orbits with focal attraction were consistent with an inverse square law.

Another hint at the universality of gravitation came from analysis of cometary orbits. The appearance of some particularly bright comets during the period 1680–82 excited a great deal of interest in England, as had the comets of 1664 before them. The generally accepted view was that cometary motion was distinct from planetary motion. Wallis, for example, expounded Horrocks' cometary theory (straight lines modified by the magnetic action of the Sun), whereas Wren followed Kepler in believing that comets travelled in straight lines at constant speed.[39] The comet of November 1680 was travelling toward the Sun, whereas that which appeared 2 weeks later was travelling away from it. Flamsteed suggested that perhaps they were one and the same object, which had reversed its motion in the vicinity of the Sun. (Actually, this comet passed extremely close to the Sun, within 1 solar radius, and so the reversal of direction was extremely rapid.) Flamsteed corresponded with Newton, who became interested in the orbits of comets. He denied initially that the two comets could possibly be the same object, but later changed his opinion. So perhaps comets were subject to the same forces of attraction as planets.[40]

In 1684, at the Royal Society, Wren, Halley, and Hooke discussed the problem of motion under an inverse square central force. Hooke claimed that he could derive the laws of planetary motion from the inverse square relation, but Wren and Halley were sceptical, particularly as Hooke refused to divulge his methods. Halley decided to consult Newton. Newton, to Halley's amazement, claimed already to have solved the problem using mathematical techniques he

---

[38] Brackenridge (1995) discusses in detail the processes by which Newton arrived at this fundamental dynamical result.

[39] Various cometary theories that were in use during the sixteenth and seventeenth centuries are described in Ruffner (1971).

[40] A discussion of the role of comets in the development of Newton's ideas is given in Bork (1987), Kollerstrom (1999) and Ruffner (2000).

had devised himself, the answer being an ellipse.[41] A few months later, Newton sent Halley the short tract *De motu*, containing his derivations of the relationships between central forces and the area law and between elliptical orbits and inverse square forces. Halley immediately recognized that Newton's results represented a huge step forward in celestial dynamics, and for the next 3 years encouraged and cajoled Newton to publish his results.

What Halley interpreted as a reluctance to publish the results he had obtained already was, in fact, something rather different. Newton had become utterly absorbed in the problems of dynamics and the development of his theory of gravitation. For nearly 3 years he shut himself off from the rest of the world and toiled night and day, sometimes forgetting to eat, perfecting his ideas. On numerous occasions he requested Flamsteed to send him accurate observational data so that he could confirm his theoretical predictions e.g. data on Jupiter's satellites that Newton used to test the accuracy of Kepler's third law, and observations of Jupiter and Saturn near their point of closest approach with which he could investigate the effect of one planet upon the other.[42]

There was a great deal to do. Much was required to turn his demonstrations of the consistency between Kepler's laws and central inverse-square forces into a theory of universal gravitation capable of explaining planetary motion. We will not attempt to trace Newton's progress during this period leading up to the appearance of his masterpiece, but instead turn our attention to the finished

[41] Hooke later (in 1686) claimed priority over the inverse square law, but Newton pointed out that to Hooke this was simply a hypothesis, whereas he had proved it. As we have seen, Hooke had suggested to Newton (in a letter of 6 January 1680) that the force of gravity varied inversely as the square of the distance from the Earth, but his derivation of this result was confused and fundamentally flawed, based as it was on Kepler's erroneous distance law. Furthermore, Newton believed that he himself was the first to establish that planets move in elliptical orbits. In a letter to Halley in 1686 he wrote: 'Kepler knew ye Orb to be not circular but oval, and guest it to be elliptical' (Turnbull (1961), Volume II). According to Lohne (1960), Newton was the only one who had the technical expertise to solve the problems of celestial mechanics, but that Hooke sketched out the programme of research that underlies the *Principia*. Westfall (1967) argues that this gives Hooke rather more than his due, the temptation to do so probably resulting from the shabby way that Newton treated Hooke. Hooke was more concerned with the inverse square nature of gravity than its universality, but it was the latter which was fundamental to the enormous power of the final theory. Newton never forgave Hooke for the charge of plagiary, and the latter remained Newton's enemy until his death in 1703.

[42] Newton needed evidence to support his idea that *all* celestial bodies exerted forces on each other, but Flamsteed's observations were inconclusive. The relationship between Newton and Flamsteed deteriorated over the course of the next 20 years. Westfall (1980), p. 586 describes the relations between the two men as oscillating 'between Newton's unremitting contempt for Flamsteed and Flamsteed's unquenchable need to win Newton's respect'. Newton took the view that the Astronomer Royal's observations were public property, but Flamsteed, who financed much of his work himself, thought they were his own. A large part of Flamsteed's work was published in 1712 without the author's approval; this included a catalogue of about 3000 stars. His life's work was later published in the form he himself wanted under the title *Historia Britannica coelestis*, but not until 1725, 6 years after his death.

product, the *Mathematical Principles of Natural Philosophy*, known simply as the *Principia*, published in three editions in 1687, 1713, and 1726.[43] The fact that the work was published at all is testament to the hard work and encouragement of Halley, who was clerk to the Royal Society at the time and, although by no means wealthy, paid for it to be printed.

The main body of the *Principia* is divided into three books, the first two on the motion of bodies in free and resisting media, respectively, and the third, *The System of the World*, on the mechanics of the heavens. Books I and II concern geometry applied to motion. It was this rational mechanics that was required to unite Earthly and celestial motions. The subject of Book II is the motion of fluids and of particles in resistive media (i.e. fluids) which was all part of the need to discredit the Cartesian theory. Newton's refutation of the vortex theory was not particularly sound because he did not have a satisfactory theory of fluid motion, but nevertheless he was confident enough to finish the second Book with the remarks:

> Therefore the hypothesis of vortices can in no way be reconciled with astronomical phenomena and serves less to clarify the celestial motions than to obscure them. But how these motions are performed in free spaces without vortices can be understood from book 1 and will now be shown more fully in book 3 on the system of the world.[44]

Book III contains Newton's statement of the law of universal gravitation and his explanation of its consequences for the structure of the Solar System. The treatment is unashamedly mathematical; this was to make the arguments as convincing as possible to the qualified reader and also to put off the ignorant from being critical. At the beginning of the third book he wrote.

> On this subject I composed an earlier version of book 3 in popular form, so that it might be more widely read. But those that have not sufficiently grasped the principles set down here will certainly not perceive the force of the conclusions, nor will they lay aside the preconceptions to which they have become accustomed over many years; and therefore, to avoid lengthy disputations, I have translated the substance of the earlier version into propositions in the mathematical style, so that they may be read only by those who have first mastered the principles.

[43] The title of the work, *Philosophiae naturalis principia mathematica*, was a direct challenge to Descartes' *Principia philosophiae*. Newton chose to emphasize that he was restricting his attention to natural philosophy and that the underlying principles on which his work was based were mathematical. The second edition was edited by Roger Cotes and the third by Henry Pemberton. Until recently, the standard English translation was that of the third edition made by Andrew Motte in 1729, revised by Florian Cajori in 1934 (Newton (1934)). The source of all the quotations reproduced here is the translation of the third edition by I. Bernard Cohen and Anne Whitman (Newton (1999)).
[44] NEWTON *Principia*, Book II, Scholium to Proposition 53 (see, for example, Aiton (1972) for a discussion of Newton's criticism of the vortex theory).

The *Principia* marks the end of one era and the beginning of another, and its publication is one of the most significant events in the history of science. Virtually all celestial phenomena were explained in terms of one fundamental principle: universal gravitation. As far as Newton and his followers were concerned, Kepler's laws changed from speculation to fact. The history of mathematical astronomy from Newton's day to the present is more or less the study of the programme for future research in astronomy provided by the law of gravitation. Newton was well aware of the power of the principle that he had discovered, as is clear from the following passage from his preface to the first edition of the *Principia*:

> And therefore our present work sets forth mathematical principles of natural philosophy. For the basic problem of philosophy seems to be to discover the forces of nature from the phenomena of motions and then to demonstrate the other phenomena from these forces. It is to these ends that the general propositions in books 1 and 2 are directed, while in book 3 our explanation of the system of the world illustrates these propositions. For in book 3, by means of propositions demonstrated mathematically in books 1 and 2, we derive from celestial phenomena the gravitational forces by which bodies tend toward the sun and toward the individual planets. Then the motions of the planets, the comets, the moon, and the sea are deduced from these forces by propositions which are also mathematical.

With regard to the propositions in Books I and II, Newton could not be criticized for failing to represent reality, because no such claim was made; he simply was explaining the mathematical consequences of certain suppositions. It was only after defining, mathematically, a whole new dynamics that he then applied it to the physical world in Book III and managed to explain a whole host of different celestial phenomena. As time progressed and mathematical techniques became more sophisticated, those gaps that did exist in Newton's theory gradually were filled and explanations for more and more phenomena succumbed to the power of universal gravitation.

Although Newton's work was developed mathematically, its philosophical implications were enormous. The prevailing view was that space and matter were one and the same thing with motion caused by the interactions between different pieces of matter; this was the starting point for Descartes' physics. Yet the *Principia* suggested that space mostly was empty with isolated bodies moving within it, each influencing the motion of all the others through some mysterious gravitational attraction.

Newton's laws and orbital motion

After some preliminary definitions, the *Principia* begins with a statement of Newton's three laws of motion:

1. *Every body perseveres in its state of being at rest or of moving uniformly straight forward, except insofar as it is compelled to change its state by forces impressed.*
2. *A change in motion is proportional to the motive force impressed and takes place along the straight line in which that force is impressed.*
3. *To any action there is always an opposite and equal reaction; in other words, the actions of two bodies upon each other are always equal and always opposite in direction.*

The first law is, of course, just the principle of rectilinear inertia. It defines qualitatively the concept of force in Newtonian dynamics: a force is something that causes a body to deviate from uniform motion in a straight line (or from rest).[45] In modern language, we would say that a force is something that causes a change in velocity, i.e. an acceleration. To Newton, absolute space and time were consequences of the omnipresence and eternity of God,[46] and so Newton regarded this velocity as being relative to absolute space but knew that there was no way actually of determining whether a body was at rest in this fixed frame of reference:

> Moreover, absolute and relative rest and motion are distinguished from each other by their properties, causes, and effects. It is a property of rest that bodies truly at rest are at rest in relation to one another. And therefore, since it is possible that some body in the regions of the fixed stars or far beyond is absolutely at rest, and yet it cannot be known from the position of bodies in relation to one another in our regions whether or not any of these maintains a given position with relation to that distant body, true rest cannot be defined on the basis of the position of bodies in relation to one another.[47]

The second law quantifies the action of a force. Newton was the first person to draw a clear distinction between the concepts of weight and mass. His thoughts on gravity had told him that the weight of a body varies with its position, but there had to be some related property that was intrinsic to the body, that he called the 'quantity of matter' and which arose from a combination of a body's density and its volume. Of course, this begs the question 'What is density?', which Newton did not attempt to answer. The quantity of motion is the product of the mass with the velocity, and is a vector – the momentum. Newton formulated the second law in terms of impulsive forces, but applied it to the case of continually acting forces (which he thought of as the limit of a sequence of impulses) in

---

[45] This is a modern interpretation. Newton's own views on just what was meant by the concept of force were not necessarily so simple (see for example, Jammer (1957)).
[46] For a discussion of Newton's concepts of space and time, see McGuire (1978).
[47] NEWTON *Principia*, Book I, Scholium to Definitions.

264 The universal theory of gravitation

which case (in the modern language of the calculus) the second law is

$$\mathbf{F} = \frac{d}{dt}(m\mathbf{v}), \qquad (8.1)$$

and this really serves as Newton's definition of mass.[48] It measures a body's resistance to a change in motion, i.e. its inertia.

The third law, when combined with the second, gives Newton's version of Descartes' law of the conservation of motion. If two bodies A and B interact, with A exerting a force $\mathbf{F}_B$ on B and B exerting a force $\mathbf{F}_A$ on A, Newton's third law says that, in the absence of any other forces,

$$\mathbf{F}_A = -\mathbf{F}_B.$$

Combining this with the second law, we obtain the law of conservation of momentum,

$$m_A\mathbf{v}_A + m_B\mathbf{v}_B = \mathbf{c},$$

where $\mathbf{c}$ is a constant vector.

Much of the *Principia* is concerned with the motion of bodies under the influence of central forces or, as Newton called them, 'centripetal forces'. In this we see that Newton had realized crucially that it was much simpler to consider things from a frame of reference in which the point of attraction was fixed rather than from the point of view of the body in motion. In this way, centrifugal forces – which were not forces at all in Newton's new dynamics – were replaced by forces that acted continually toward a fixed point.

After stating the laws on which his dynamics was based, Newton went on to prove eleven lemmas that showed how to apply methods from the calculus to geometrical problems involving curved lines – his so-called method of 'first and last ratios'.[49] Here, we find the clearest statement Newton ever gave of the concept of the limit process:

> Those ultimate ratios with which quantities vanish are not actually ratios of ultimate quantities, but limits which the ratios of quantities decreasing without limit are continually approaching, and which they can approach so closely that their difference is less than any given quantity, but which they can never exceed and can never reach before the quantities are decreased indefinitely.[50]

---

[48] The conceptual struggle involved in replacing Newton's impulsive definition with Eqn (8.1) is discussed in Hankins (1967).

[49] This was the method that Newton had developed to remove the troublesome infinitesimal quantities from his calculus (see for example, Kline (1972)). There is much of mathematical interest in these lemmas. Interested readers should consult, for example, Chandrasekar (1995) or Densmore (1996).

[50] NEWTON *Principia*, Book I, Scholium to Lemma 11. While the definition of a limit given by Newton may seem a bit vague by modern standards, his understanding of the concept

This aspect of Newton's work should not be overlooked. It was not sufficient just to have a great insight into the problems of natural philosophy; to reform astronomy in the way Newton did required a hitherto unknown level of mathematical expertise. The question arises naturally as to why Newton, who had developed his method of fluxions based on analytic geometry, chose to demonstrate all the results in the *Principia* using synthetic geometry. Looking back on the events of the 1680s, Newton later wrote (anonymously):

> By the help of the new *analysis* Mr. Newton found out most of the propositions in his *Principia Philosophiae*: but because the ancients for making things certain admitted nothing into geometry before it was demonstrated synthetically, he demonstrated the propositions synthetically, that the system of the heavens might be founded on good geometry. And this makes it now difficult for unskilful men to see the analysis by which those propositions were found out.[51]

However, while there is a certain logical simplicity in this explanation, very few papers of Newton's containing these earlier analytic demonstrations have been found. Certainly, it is true that the ideas of the calculus lie behind many of the derivations in the *Principia*, but it may well be that Newton himself believed that a demonstration in terms of synthetic geometry carried more weight, and his method of first and last ratios represented the natural way of extending classical Greek geometry to the study of motion along curved lines.[52]

Another obvious reason for choosing to present his ground-breaking theory in this way was that it was more likely to be intelligible to his contemporaries.[53] Exactly the opposite is true nowadays, though, the techniques of synthetic geometry being largely unknown to the scientists of today. We will thus, except where Newton's geometrical demonstrations are particularly simple and elegant, describe the main results of the *Principia* that are relevant to celestial mechanics using (essentially eighteenth-century) analytic tools.[54]

(as judged by how he applied it) would appear to have been better than he is often given credit for (see Pourciau (2001)).

[51] Quoted from Westfall (1980), pp. 423–4.

[52] Whiteside (1970b) shows the direct correspondence between some of Newton's synthetic demonstrations and the equivalent analytic ones. It should also be remembered that when Newton made these claims he was engaged in a bitter dispute with Leibniz over the invention of the calculus, and it was thus in his interest to claim that he was using the new analysis before Leibniz. A recent assessment of the mathematical techniques actually employed by Newton is that given by Guicciardini (1998, 1999).

[53] Actually, very few of Newton's contemporaries understood fully the *Principia*'s technical content. The concepts introduced in the work were of necessity difficult, but this was compounded by the fact that the *Principia* was written in a hurry and was not organized in a way that made understanding easy. A famous anecdote records that as Newton walked down a Cambridge street, an undergraduate said under his breath, 'There goes the man that writt a book that neither he nor anybody else understands', (Whiteside (1970b)).

[54] Readers wishing to study Newton's geometric derivations can consult Densmore (1996).

Immediately following his exposition of the method of first and last ratios, Newton tackled the general problem of orbital motion subject to a central force, and proved the results relating such forces to the area law and elliptical motion to inverse square attraction that were discussed earlier. He also showed that if a body were moving due to the action of a central force along a path given by any conic section (i.e. an ellipse, parabola or hyperbola) then that force had to be of inverse square type. A modern formulation of these results might proceed as follows. A central force (where we assume that the magnitude of the force varies only with the distance r from the point of attraction and not with the direction of the force) can be represented as $\mathbf{F} = -f(r)\mathbf{e}_r$, where $\mathbf{e}_r$ is the unit radial vector, a unit vector the direction of which is from the centre of attraction to the body, which for convenience we will refer to as a planet. (The minus sign clearly is superfluous, but we include it so that for a force of attraction $f(r)$ will be positive.) If we then use Newton's second law, we obtain the second order differential equation[55]

$$-f(r)\mathbf{e}_r = m\ddot{\mathbf{r}} = m(\ddot{r} - r\dot{\theta}^2)\mathbf{e}_r + m(2\dot{r}\dot{\theta} + r\ddot{\theta})\mathbf{e}_\theta. \qquad (8.2)$$

The tangential component of this equation is simply

$$2\dot{r}\dot{\theta} + r\ddot{\theta} \equiv \frac{1}{r}\frac{d}{dt}(r^2\dot{\theta}) = 0,$$

from which we deduce immediately that $r^2\dot{\theta}$ is a constant, usually given the symbol h. Now, in a small time δt, a planet sweeps out a sector of area δA which is approximately $r^2\delta\theta/2$. If we divide by δt and take the limit as $t \to 0$ we arrive at the result

$$\frac{dA}{dt} = \tfrac{1}{2}r^2\dot{\theta} = \tfrac{1}{2}h \qquad (8.3)$$

and, hence, the rate of change of area is constant, which is Kepler's second law. Up to this point we have used only the fact that the force has no tangential component; the magnitude of the force only enters the analysis when we treat the radial component of Eqn (8.2). Substituting for $\dot{\theta}$ in terms of h we can write

$$m(\ddot{r} - h^2r^{-3}) = -f(r) \qquad (8.4)$$

and, in principle, if we know the polar equation of the path of the planet, we

---

[55] The second equality comes from differentiating the relation $\mathbf{r} = r\mathbf{e}_r$ twice, noting that $\dot{\mathbf{e}}_r = \dot{\theta}\mathbf{e}_\theta$ and $\dot{\mathbf{e}}_\theta = -\dot{\theta}\mathbf{e}_r$. Here $\mathbf{e}_\theta$ is the unit tangential vector, the unit vector perpendicular to $\mathbf{e}_r$ in the direction of increasing θ. This equation (in component form) was first worked out in the 1740s, though the mathematical techniques needed to derive it – elementary calculus and polar coordinates – were available at the end of the seventeenth century (Guicciardini (1996)).

can use this to determine f. For example, if the orbit is a conic section with the centre of attraction at a focus, the path has the form

$$\frac{\ell}{r} = 1 - e\cos\theta,$$

where ℓ (the semilatus rectum, sometimes referred to simply as the 'parameter') and e (the eccentricity) are constants defining the precise nature of the path;[56] and from this, using the relation $\dot\theta = hr^{-2}$, we have $\ddot{r} = -(eh^2/\ell r^2)\cos\theta$ and, hence,

$$f(r) = -m(\ddot{r} - h^2 r^{-3}) = \frac{mh^2}{\ell r^2}\left(e\cos\theta + \frac{\ell}{r}\right) = \frac{mh^2}{\ell r^2}, \quad (8.5)$$

thus revealing the inverse square nature of the force.

What this derivation has shown is that if a body moves along a conic section under the influence of a central force toward a fixed point, then the magnitude of that force must vary inversely as the square of the distance. But what has not been shown, and what Newton wanted to infer, is that an inverse square central force implies motion along conic sections. In the second edition of the *Principia*, Newton added the following 'proof' of this converse result:

> From the last three propositions it follows that if any body P departs from the place P along any straight line PR with any velocity whatever and is at the same time acted upon by a centripetal force that is inversely proportional to the square of the distance of places from the centre, this body will move in some one of the conics having a focus in the centre of forces; and conversely. For if the focus and the point of contact and the position of the tangent are given, a conic can be described that will have a given curvature at that point. But the curvature is given from the given centripetal force and velocity of the body; and two different orbits touching each other cannot be described with the same centripetal force and the same velocity.[57]

The essence of this proof is that conic sections exhaust all the possible orbits in an inverse square force field. The final statement that it is impossible to have two distinct orbits tangential at a point, with the bodies having the same velocity there, requires justification, and many of Newton's contemporaries were not entirely convinced.[58]

---

[56] For example, take the intersection of the right circular cone $x^2 + y^2 = z^2$ and the plane $z = ex + \ell$. With polar coordinates such that $x = r\cos\theta$ and $x^2 + y^2 = r^2$, we have $r = er\cos\theta + \ell$ from which the result follows.

[57] NEWTON *Principia*, Book I, Corollary I to Proposition 13.

[58] Whether or not Newton's corollary represents an outline of a valid logical proof has been the subject of recent debate. Weinstock (1982) made the claim that Newton's argument is flawed logically, but Pourciau (1991, 1997) demonstrated subsequently that, although there is an omission in the logical procedure that Newton described, this gap is bridged easily (and Newton would merely have had to refer the reader to other results in the *Principia* in order to complete the proof).

The discovery that the area law was a natural consequence of motion subject to a central force did not, of course, make its practical application any easier. In the *Principia*, Newton gave a geometrical solution (due to Wren, though Newton does not mention the fact) based on the cycloid curve. But, to use Newton's words,

> ... the description of this curve is difficult; hence it is preferable to use a solution that is approximately true.[59]

The approximate solution that Newton gives is an iterative procedure, equivalent to what we now call 'Newton's method' for finding roots of equations. In essence, we use the fact that for x close to x_0, we have

$$f(x) \approx f(x_0) + (x - x_0)f'(x_0),$$

and so if we have an approximate solution θ_0 to the equation $t = f(\theta)$, where $f(\theta) = \theta + e \sin \theta$, a better one will be

$$\theta_1 = \theta_0 + \frac{t - f(\theta_0)}{f'(\theta_0)} = \theta_0 + \frac{t - \theta_0 - e \sin \theta_0}{1 + e \cos \theta_0}.$$

This process can then be repeated until the desired accuracy is reached.[60]

For the case of planetary motion, where e is very small, Newton suggests a more straightforward alternative – one clearly having its origins in Boulliau's empty focus equant hypothesis.[61] From Kepler's first and second laws, Newton deduced that the angle subtended by the planet at the empty focus, u, is related to the mean anomaly t via

$$u = t + \tfrac{1}{4}e^2 \sin 2t + \tfrac{2}{3}e^3 \sin^3 t + O(e^4),$$

and that, even for Mars, 'the error will hardly exceed one second'. Once u has been found, θ follows from elementary geometry.

One of Newton's early breakthroughs had been the discovery that Kepler's third law was consistent with an inverse square law for gravitation, at least as

---

[59] NEWTON *Principia*, Book I, Scholium I to Proposition 31. In the language of the time, the cycloid solution was considered 'mechanical' rather than 'geometrical'. Descartes described curves as 'mechanical' if they could not be represented in finite powers of x and y. James Gregory, inspired by Wren's solution, produced an analytic solution (in effect utilizing the Taylor expansion of the equation of the cycloid) in the 1660s (see Wilson (1970)).

[60] It is not at all obvious that Newton's geometrical procedure given in the Scholium to Proposition 31 of Book I is equivalent to the procedure described here. According to Cajori (Newton (1934)), this was first pointed out by the astronomer and mathematician John Couch Adams in 1882.

[61] See p. 226 and Whiteside (1964).

far as circular orbits were concerned. Now he could show that this was still true for elliptical orbits. In terms of the semimajor axis a and the eccentricity e, the area of an ellipse is $\pi a^2 \sqrt{1 - e^2}$ and the semilatus rectum ℓ is given by $a(1 - e^2)$. If a number of planets are moving in elliptical orbits, all subject to the same central force given by $f(r) = m\mu/r^2$, where μ is a constant, then for each body we have from Eqn (8.5) that $\mu = h^2/\ell$. It then follows, making use of Eqn (8.3), that

$$\frac{\text{area of ellipse}}{\text{period of revolution}} = \frac{\pi a^2 \sqrt{1 - e^2}}{T} = \frac{dA}{dt} = \tfrac{1}{2}h = \tfrac{1}{2}\sqrt{\mu a(1 - e^2)}, \tag{8.6}$$

from which we obtain Kepler's third law in the form

$$T^2 = 4\pi^2 \mu^{-1} a^3. \tag{8.7}$$

The mean motion is given by $n = 2\pi/T = \sqrt{\mu/a^3}$.

Newton thus demonstrated that Kepler's laws of planetary motion were consistent with an attractive force of gravity that pulled the planets toward the Sun and was inversely proportional to the square of the distance from the attracting body. Now Newton was aware that, while the planets appeared to move on ellipses that were fixed in space, the Moon did not, with successive lunar apogees observed to advance by about 3°. His correspondence with Hooke concerning the problem of free fall had led Newton to the conclusion that an orbiting body moving in a force field that deviated from an inverse square law would move in such a way that its apsidal line (the major axis of its elliptical orbit) would rotate. He thus went on to consider the type of force field that would result in a rotating ellipse. He showed in Proposition 45 of Book I that, assuming the orbit to be nearly circular, the angle between successive apses would be $180°/\sqrt{3 - m}$ if the magnitude of the attractive central force varied like r^{-m}. The fact that the apsidal lines of the planets seemed almost to be stationary with respect to the stars was more convincing evidence for the inverse square law of gravity ($m = 2$), since this implied an angle of 180° between successive aphelia and perihelia. For the Moon, however, his formula implied that m was very nearly $2\frac{4}{243}$. Newton did not discard the inverse square law, though. He realized that an advance in the apsidal line could be caused also by the perturbing effect of a third body (in this case the Sun) though, as we shall see, he never managed successfully to deduce the motion of the lunar apse from the theory of gravitation.

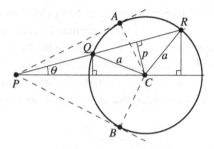

Fig. 8.6. The attraction of a spherical shell on an external particle.

Universal gravitation

A number of phenomena had caused Newton to consider that gravity was a property of all celestial bodies. In order to make the leap to *universal* gravitation, an understanding was required of how a large body like the Earth would attract an external object if its pull was the result of attractions from all its constituent parts. Newton answered these questions by proving some results concerning the attraction of thin spherical shells. His approach was hard-going; the description below is a modern adaptation.[62]

The particle is at P in Figure 8.6, and simple symmetry arguments show that the force on P due to the shell of radius a centred at C must be along the line PC. In order to work out the magnitude of the force, we draw a line from P that intersects the sphere at Q and R. Each of these points can be thought of as being part of a ring of points on the sphere, the centre of which lies on the axis PC and it is not difficult to show that the attractive effect of this circle of points is given by

$$\frac{2\pi a\rho p \, \mathrm{d}p}{d^2\sqrt{a^2 - p^2}},$$

where $d = |PC|$ is the distance between the particle and the centre of the shell, $p = d\sin\theta$ is the perpendicular distance from C to the line PQR, and ρ is the mass per unit area of the shell. It follows that the ring containing Q has the same attractive effect as that containing R and, in consequence, that a circle through AB (PA and PB being tangents to the shell) with its centre on the

---

[62] NEWTON *Principia*, Book I, Propositions 70 and 71. Here we follow Todhunter (1962), p. 2; an alternative approach is described in Littlewood (1948). Littlewood remarks that Newton's geometric proof 'must have left its readers in helpless wonder'. The geometric proof is also examined in detail in Weinstock (1984).

axis PC, divides the spherical shell into two parts, which attract the particle P equally, the total effect being

$$2 \times \frac{2\pi a\rho}{d^2} \int_0^a \frac{p}{\sqrt{a^2 - p^2}}\, \mathrm{d}p = \frac{4\pi a^2 \rho}{d^2}.$$

Thus, a spherical shell attracts an external particle as if all its mass were concentrated at its centre. In fact, Newton did not evaluate any equivalent of this integral, but demonstrated merely that the gravitational field exterior to a shell is inversely proportional to the square of the distance from the centre of the shell; the constant of proportionality was not determined. As no one had any idea about the densities of heavenly bodies, this is unimportant.

Newton then could conclude that the same result would be true for any sphere, the density of which was a function of radial distance alone. This implied that if the force that kept the Moon in its orbit and the force that made an apple fall to the Earth were caused solely by the gravitational pull of all the particles that make up the Earth, the calculation he had performed in 1666 should have given better agreement. Newton did the calculation again, this time using the more accurate data on the size of the Earth due to Jean Picard, and found excellent agreement.[63] This demonstration was crucial to Newton's confidence in his idea of universal gravitation.

Newton realized that the combined effect of the gravitational attraction of the particles that make up the Earth and the rotation of the Earth would lead to a flattening of the Earth at its poles.[64] He assumed that the Earth originally had been made up of fluid matter of constant density, and that this implied the actual shape was an oblate spheroid.[65] In other words, any cross-section through the Earth which includes the polar axis is an ellipse, which we can suppose has a ratio of major to minor axis of $1 + \varepsilon$, ε being the ellipticity or oblateness.[66] Newton showed that, within the body of a homogeneous Earth, the force of gravity would be directly proportional to the distance from the centre (unlike

---

[63] This calculation is outlined in Proposition 4 of Book III, and a more elaborate version, in which the Earth and Moon are treated as a two-body system, is given in Proposition 37. This latter calculation almost certainly was fiddled so as to give the impression of great precision, though, since Newton could not compute the centre of gravity of the Earth–Moon system accurately (see Westfall (1973), Kollerstrom (1991)).

[64] Newton discussed the oblateness of the Earth in Proposition 19 of Book III. His method is described in modern form in Todhunter (1962).

[65] An oblate spheroid is the shape formed by rotating an ellipse about its minor axis, and in the case of the Earth this axis corresponds to the polar diameter. Newton chose this shape primarily because he knew how to work with it.

[66] The term 'ellipticity' was introduced in the eighteenth century and is sometimes defined so that the major and minor axes have lengths in the ratio $1 + \varepsilon : 1$ as here, but one also finds this ratio as $1 : 1 - \varepsilon$. Of course, to first order in ε this makes no difference, since $(1 - \varepsilon)^{-1} = 1 + \varepsilon + O(\varepsilon^2)$.

Huygens, who thought that gravity was constant in the interior of the Earth), and if we assume that ε is small, the attraction at the pole due to the combined effect of all the particles that make up the Earth divided by the attraction at a point on the equator can be shown to be $1 + \frac{1}{5}\varepsilon$. Newton considered two columns of fluid extending from the centre of the Earth, one to the pole and one to the equator (the lengths of which are thus in the ratio $1 : 1 + \varepsilon$), and showed that the force of attraction on the polar column would be greater than that on the equatorial column by the factor $(1 + \varepsilon)/(1 + \frac{1}{5}\varepsilon) \approx 1 + \frac{4}{5}\varepsilon$. But, Newton argued, the two columns must be in equilibrium, so the different forces must be balanced by the effect of rotation which, as we have seen from the work of Huygens, was to reduce the effect of gravity at the equator by a factor of $1/289$.[67] Putting these results together, Newton concluded that

$$1 + \frac{4}{5}\varepsilon = \frac{289}{288},$$

and thus that ε is about $1/230$. Hence, the ratio of the polar to equatorial axes is $229 : 230$, in contrast to Huygens' value of $577 : 578$. Actually, the oblateness is closer to $1/300$, the discrepancy being due to the fact that the Earth is not homogeneous as Newton supposed.

Newton then applied his idea of balanced fluid columns to investigate the variation in the weight of a body as a function of latitude, and showed that, since the oblateness of the Earth was small, the change in weight varied with the square of the sine of the latitude. Using this result, he calculated that a seconds pendulum would have to be shortened by about 23 mm when taken from Paris to the equator. A comparison of his result with the data he had available led Newton to believe that the oblateness was, in fact, slightly greater than predicted with his theory, and he suggested that this was due to the Earth being denser toward its centre than near its surface (this would have the opposite effect to the one he was trying to create).

Looking back, we can see in Newton's work on the shape of the Earth the beginnings of a satisfactory theory. However, his explanations were difficult to follow, and relied on numerous assumptions that were not stated clearly and not at all obvious. To most readers of the *Principia*, Newton's theory of the shape of the Earth was pretty incomprehensible.

There was one final element needed to characterize the nature of gravitation. Newton's third law implied that whatever force one body exerted on another, an equal and opposite force was exerted on the former by the latter. The same had to be true for gravitation. This was explained by Cotes in the preface to the second edition of the *Principia*, as follows:

[67] Newton calculated the same value by a different method (Greenberg (1995), p. 2).

Furthermore, just as all bodies universally gravitate toward the earth, so the earth in turn gravitates equally toward the bodies; for the action of gravity is mutual and is equal in both directions. This is shown as follows. Let the whole body of the earth be divided into any two parts, whether equal or in any way unequal; now, if the weights of the parts toward each other were not equal, the lesser weight would yield to the greater, and the parts, joined together, would proceed to move straight on without limit in the direction toward which the greater weight tends, entirely contrary to experience.

The observational data showed Newton that the planets, the satellites of Jupiter and Saturn, and the Moon in its orbit around the Earth, all obeyed (as far as could be ascertained) Kepler's laws, and thus their motion was consistent with an inverse square law for gravitation. Moreover, spherical objects (like planets) would attract (as they appeared to do) as if their mass were all concentrated at their centres, if their gravitational pull was due to the pull from all the individual particles of which they were constituted. This knowledge led Newton to his law of universal gravitation:

Gravity exists in all bodies universally and is proportional to the quantity of matter in each.[68]

Together with the inverse square nature of the force, this means that the magnitude of the force of attraction between two particles (or spherical bodies) of mass m_1 and m_2, separated by a distance r, is Gm_1m_2/r^2, where G is the same for all bodies in the Universe (known as the 'universal constant of gravitation'). The fact that the (gravitational) masses in this expression are the same as the (inertial) masses that appear in the second law of motion (Eqn (8.1)) is, in Newton's system, simply an experimentally observed fact. If it were not the case, pendulums made of different materials but of the same length would have different periods. The equivalence of these quantities lies at the heart of Einstein's general theory of relativity (see Chapter 12).

In the case of a single planet orbiting the Sun, it follows that we should not treat the Sun as a fixed object because it is attracted by the planet. Newton thus was led to the examination of what we now call the two-body problem.[69] In Figure 8.7, let S represent the Sun (of mass M) and P a planet (of mass m), and let the position vectors of these bodies relative to some origin fixed in space be $\mathbf{r}_1$ and $\mathbf{r}_2$, respectively. Newton's law of gravitation, taken together with his second law of motion, shows that

$$M\ddot{\mathbf{r}}_1 = \frac{GMm}{r^3}\mathbf{r}, \qquad m\ddot{\mathbf{r}}_2 = -\frac{GMm}{r^3}\mathbf{r},$$

[68] NEWTON *Principia*, Book III, Proposition 7.
[69] NEWTON *Principia*, Book I, Propositions 57–63.

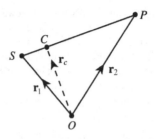

Fig. 8.7. The two-body problem.

where $\mathbf{r} = \mathbf{r}_2 - \mathbf{r}_1$ and $r = |\mathbf{r}|$ is the distance between P and S. Adding these equations reveals that the centre of gravity C of the two bodies (whose position vector is $\mathbf{r}_c = (M\mathbf{r}_1 + m\mathbf{r}_2)/(M + m)$) must move with constant velocity.[70] If we divide the first equation by M, the second by m, and subtract the results, we obtain, with $\mu = G(M + m)$,

$$\ddot{\mathbf{r}} = -\frac{\mu}{r^3}\,\mathbf{r}. \tag{8.8}$$

This is exactly the same equation as we get from assuming the Sun to be fixed, except in that case we have $\mu = GM$. It follows that in this two-body system, the planet still moves according to Kepler's first two laws (since it is subject to an inverse square central force) but the third law must be modified slightly, since μ is no longer the same for each planet. From Eqn (8.7) we have that $T^2\mu \propto a^3$, and so for two planets with masses m_1 and m_2, periods T_1 and T_2, and semimajor axes a_1 and a_2, we have

$$\frac{T_1^2(M + m_1)}{T_2^2(M + m_2)} = \frac{a_1^3}{a_2^3}. \tag{8.9}$$

Newton hoped to be able to use this result to make an accurate calculation of the semimajor axes of Jupiter and Saturn, but unfortunately there were still considerable uncertainties over the periods of these planets which were not resolved for another century.

On the assumption that a planet's mass (m) is much smaller than that of the Sun (M), Kepler's third law takes the form $T^2GM = 4\pi^2 R^3$ (T is the planet's orbital period and R its semimajor axis). Similarly, for a satellite of the planet, we have $t^2Gm = 4\pi^2 r^3$ (t being the satellite's orbital period and r its semimajor

---

[70] The precise form of the force is not important for this result, of course. It follows directly from Newton's three laws of motion and was stated by Newton in Corollary 4 to the Laws of Motion.

axis). Combining these results gives

$$\frac{m}{M} = \frac{r^3 T^2}{R^3 t^2},$$

and Newton used this formula to calculate the masses (as proportions of the Sun's mass) of those planets with satellites. Over the period of nearly 40 years between the publication of the first and third editions of the *Principia*, the data to which Newton had access got progressively more accurate, and his values for the masses of the Earth, Jupiter, and Saturn varied considerably. In the third edition, we find the values $1/169\,282$ for the mass of the Earth divided by that of the Sun (which is almost double the correct value, the error being due to Newton's inaccurate value for the solar parallax),[71] $1/1067$ for Jupiter (which is within 2 per cent of the modern value), and $1/3021$ for Saturn (which is about 16 per cent too big). This technique did not provide any hint as to the masses of Mercury, Venus, and Mars, however, as there were no known satellites orbiting these planets, but Newton felt confident that the masses (and consequently the gravitational effects) of these other planets were very small.

The seventeenth-century Solar System was a system consisting of the Sun, six planets, and numerous satellites, and Newton's theory implied that they each attracted each other. The treatment of the full dynamical problem was way out of reach, but Newton was able to make some progress by considering the case of three bodies and, in particular, for the case of the Sun's effect on the motion of the Moon.

Lunar theory and perturbations

Newton's lunar theory was developed in the latter half of the 1690s and first appeared as his *Theory of the Moon's Motion* in 1702.[72] The work appeared in revised form in the second and third editions of the *Principia*. What is striking about the original version is that it is not a 'theory' at all, but merely an algorithm for computing the Moon's position; it provides no justification based on the theory of gravitation.

---

[71] In the third edition, Newton took the solar parallax to be $10.5''$, but $1/169\,282$ is inconsistent with this and with other numbers given in Book III, Proposition 8. Garisto (1991) has shown that this was a hangover from a previous calculation, when Newton was using a parallax of $11''$. He should have written $1/194\,000$.

[72] The *Theory of the Moon's Motion* appeared as a pamphlet in English (now extremely rare) and a Latin version was published, also in (1702), in David Gregory's *Astronomiae physicae et geometricae elementa*, the first ever textbook on Newtonian astronomy. Numerous other editions and translations have been printed (see Cohen (1975)).

Unlike the material in the rest of the *Principia*, the lunar theory was designed as a predictive tool rather than a theoretical derivation of the consequences of gravitation. In the version published in the *Principia*, Newton did attempt to show how gravity could cause the various corrections he applied to the lunar motion, but his real aim, as the preface to the original version made clear, was to produce a method that would determine the position of the Moon sufficiently accurately to be of use in finding longitude at sea. Two major British naval disasters in the 1690s had been caused by the inability to find longitude accurately, and the Longitude Act, 1714 offered a huge reward for anyone who could solve the problem.[73]

When Newton had begun his work on the lunar theory, he had hoped that he would have been able to explain quantitatively the motion of the Moon using the theory of gravitation. He became increasingly frustrated, however, with his inability to produce an accurate mathematical theory. Eventually he was persuaded, largely by Flamsteed, that the best approach was to build on Horrocks' continually varying Keplerian ellipse, and this was precisely what he did.[74] To Horrocks' mechanism, Newton added a number of small sinusoidal oscillations so as better to fit the observations. He ended up with seven 'equations' of lunar motion; one of these accounted for the equation of centre and the evection known since Ptolemy, and the variation and the annual equation, which had been discovered by Tycho Brahe, corresponded to two of the others. The remaining four steps in Newton's algorithm were entirely new. In order to use the procedure, the position of the Moon had to be adjusted ('equated') seven times, and then, as Newton stated proudly, 'this is her place in her proper orbit'.[75]

The accuracy of Newton's seven-step procedure was the subject of debate then as now. Flamsteed claimed that it was no better than his own calculations based on Horrocks' theory, whereas Halley made great claims for the accuracy of Newton's approach. The truth would appear to lie somewhere in-between.

In the second edition of the *Principia*, Newton added a completely new section, containing a revised version of the *Theory of the Moon's Motion*, and claiming a justification for it based on universal gravitation:

---

[73] In fact, the problem eventually was solved by the development of an accurate chronometer by John Harrison, whose story is well documented in Sobel (1996). Chronometers were, however, prohibitively expensive and the use of lunar tables was the most common method for determining longitude until the nineteenth century. The first tables that were sufficiently accurate to be of practical use were those of Tobias Mayer (1753) (see p. 304).

[74] Horrocks' mechanism is described on p. 234. Flamsteed had used $|EC|\cos\delta$ for the eccentricity (see Figure 7.12, p. 235), but Newton, at Halley's suggestion, used $|EC|$. This leads to a significant improvement. Details of Newton's early struggles with a lunar theory can be found in Whiteside (1976).

[75] NEWTON *Theory of the Moon's Motion*. Quoted from Cohen (1975), p. 113. Details of the implementation and accuracy of Newton's procedure can be found in Kollerstrom (2000).

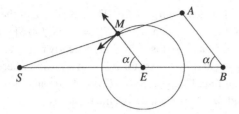

Fig. 8.8. The effect of the Sun on the Earth–Moon system.

I wished to show by these computations of the lunar motions that the lunar motions can be computed from their causes by the theory of gravity.[76]

In fact, Newton was not able to justify all the steps in his calculation procedure and had to resort to a certain amount of sleight of hand so as to get the numbers to agree, but the method he introduced for treating the motion of the Moon by considering the gravity of the Sun as a perturbing influence on the orbit of the Moon round the Earth was admired widely and paved the way for future developments.

Newton's attempt to deduce the perturbative effect of the Sun on the Earth–Moon system from universal gravitation was based on a diagram similar to that shown in Figure 8.8, in which E, S, and M represent the Earth, Sun, and Moon, respectively, and the Moon is assumed to move in a circular orbit around the Earth. Newton understood that the accelerative effect of the Sun on the Earth is very nearly the same as that on the Moon, since the two bodies are almost equidistant from the Sun. However, it is the small difference between these effects that leads to the irregularities in the orbit of the Moon. The Moon is attracted to the Earth by an inverse square force but is then subject to an extra force from the Sun, which can be thought of as having two components: one along the radius EM and one perpendicular to this, as indicated by the arrows in Figure 8.8. The idea is to try to relate the magnitude of these perturbing forces to the otherwise undisturbed Earth–Moon attraction.[77]

The magnitude of the acceleration of the Moon due to the attraction of the Sun can be written $\mu_S/|MS|^2$, where μ_S is the product of G and the Sun's mass, and that due to the Earth as $\mu_E/|ES|^2$, where μ_E is the product of G and the mass of the Earth. Newton began by assuming that the vector ES represents a measure of the acceleration of the Earth due to the Sun's attraction (and therefore also the average acceleration of the Moon over the period of one

[76] NEWTON *Principia*, Book III, Scholium to Proposition 35.

[77] The derivation below is a modified version of that given in Wilson (1989a).

orbit) and, hence, that $|ES| = \mu_S/|ES|^2$. The magnitude of the acceleration of the Moon due to the Sun is then $\mu_S/|MS|^2 = |ES|^3/|MS|^2$. In the diagram, $|MS|$ is less than $|ES|$, and so this acceleration clearly is greater than that of the Earth; we can represent it by the vector $AS = AB + BS$, where AB is parallel to the Earth–Moon line. Using $|AS| = |ES|^3/|MS|^2$ and the similarity of the triangles ABS and MES, we can derive the relations

$$|BS| = \frac{|ES|^3}{|MS|^3}|ES|, \qquad |AB| = \frac{|ES|^3}{|MS|^3}|ME|$$

and hence (neglecting quantities that are second-order in the small quantity $|ME|/|ES|)^{78}$ that

$$|BS| - |ES| \approx 3|ME|\cos\alpha, \qquad |AB| \approx |ME|.$$

The net perturbing effect of the Sun in the direction ES is $BS - ES$, and so the outward radial component of the perturbation due to the Sun is

$$3|ME|\cos^2\alpha - |ME| = \tfrac{1}{2}|ME|(1 + 3\cos 2\alpha),$$

the average value of which, over one orbit of the Moon, is $|ME|/2$. Thus, we have shown that the ratio of the mean radial perturbative acceleration to the acceleration of the Earth due to the Sun is $|ME|/2|ES|$. Finally, we use Kepler's third law (in the form of Eqn (8.7)), which implies that

$$\frac{\text{acceleration of Earth due to Sun}}{\text{acceleration of Moon due to Earth}} = \frac{\mu_S/|ES|^2}{\mu_E/|ME|^2} = \frac{T_M^2|ES|}{T_E^2|ME|},$$

where T_E and T_M are the periods of the Earth and Moon, respectively, and combining this with the previous result gives

$$\frac{\text{radial perturbative acceleration}}{\text{acceleration of Moon due to Earth}} = \frac{T_M^2}{2T_E^2} \approx \frac{1}{357.45},$$

the final figure being the one that Newton quotes in Book I, Proposition 45, Corollary 2. The effect of the perturbation is to decrease the pull on the Moon toward the Earth, and it is on the basis of this result that Newton showed that the mean effect of the Sun would be to produce an advance in the line of apsides of the orbit of the Moon and a retrogression of the nodes. His calculations predicted

---

[78] Using the cosine rule on triangle MES, we can derive

$$\frac{|ES|^3}{|MS|^3} = \left(1 - 2\frac{|MS|}{|ES|}\cos\alpha + \frac{|MS|^2}{|ES|^2}\right)^{-3/2} \approx 1 + \frac{3|MS|}{|ES|}\cos\alpha.$$

a rate of $19° 18' 1''$ per sidereal year for the retrogression of the nodes, which is only about $3'$ too small, but his value of $1° 31' 28''$ for the monthly advance of the apsidal line was only about half the measured value or, as Newton put it, 'the [advance of the] apsis of the moon is about twice as swift'.[79]

The magnitude of the tangential perturbing effect is $3|ME|\cos\alpha\sin\alpha$ and, hence, proportional to $\sin 2\alpha$. The average value of this effect over one lunar orbit is zero, and Newton assumed that he could neglect it. This turns out not to be the case, but it was not questioned by those who attempted initially to account for the observed motion of the lunar apse using the theory of gravitation. Instead, other possibilities were explored, e.g. a modification to the inverse square nature of the attracting force. The resolution of the problem came with the development of analytic techniques that allowed for a more systematic treatment of perturbations and which will be described in the next chapter.

Newton did make use of the tangential component of the perturbing force, however. He used it to explain the variation, discovered originally by Tycho Brahe. Since the effect is proportional to $\sin 2\alpha$, the perturbative acceleration of the Moon is positive between quadrature and syzygy and negative between syzygy and quadrature, implying that the Moon speeds up as it approaches syzygy and slows down as it nears quadrature, which is exactly the phenomenon Tycho had observed. Newton also realized that this perturbing effect, in conjunction with the radial component, would cause the curvature of the orbit of the Moon to be greater at the quadratures than at the syzygies and, hence, that the Moon would be further away at quadrature. He computed the ratio of the distances at quadrature and syzygy to be 70 : 69 and (in Book III, Proposition 29) went on to calculate the effect on the longitude of the Moon. He computed an inequality of the form $35' 10'' \sin 2\alpha$, the coefficient being slightly smaller than the actual value of about $39\frac{1}{2}'$ for the magnitude of the variation.

Other aspects of the motion of the Moon were also deduced as consequences of universal gravitation, e.g. the annual inequality (discovered independently by Tycho Brahe and Kepler) and the variation in the inclination of the lunar orbit (observed by Tycho). However, the largest correction to the mean motion of the Moon in Newton's theory came from the epicyclic style mechanism of Horrocks, which accounted for the evection, and nowhere were the numerical

---

[79] NEWTON *Principia*, Book I, Corollary 2 to Proposition 45. Waff (1976) gives a detailed discussion of Newton's work on the motion of the lunar apogee. The advance of the apsidal line and the retrogression of the nodes are simple consequences of the differing lengths of the sidereal, anomalistic, and draconitic months (T_s, T_a, and T_d, for example). We have $T_a/T_s - 1 \approx 0.008\,53$, so each sidereal month the perigee of the Moon advances by about $0.008\,53 \times 360° \approx 3° 4'$. However, $T_d/T_s - 1 \approx -0.004\,02$, so the nodes move backwards by about $1° 27'$ per sidereal month, or equivalently by about $19° 21'$ per sidereal year.

parameters that were used justified using the theory of gravity. Newton's lunar theory was thus far from perfect, and he was well aware of the fact. But in view of the immense difficulty of the problem he was considering, and the influence his ideas and methods had on those who followed in his footsteps, Newton's lunar theory should be considered as one of the great achievements of the *Principia*.

The lunar theory actually appears in the *Principia* in two quite different ways. In Book I, Section 11 (and notably in Proposition 66 and its twenty-two corollaries) Newton treats different aspects of the general problem of three mutually interacting bodies qualitatively. The Moon is not mentioned specifically, but it is clear from the choice of letters in his diagrams that it was the Moon for which the theory was intended. The general theory was then applied to the specific case of the Moon's motion in Book III, some of the calculations having been described above.

But Newton used also his perturbation theory to tackle three other problems, always starting from a diagram similar to that shown in Figure 8.8. By taking E to be the Sun, and S and M to be two different planets, Newton was able to consider the effect of one planet on another's orbit. The relative masses of the bodies that make up the Solar System showed that, apart from in the Sun–Earth–Moon system, the perturbing effects would be small, except perhaps in the case of Jupiter and Saturn. Newton did attempt to quantify this latter effect, but with little success.

Another application of the technique was to explain the tides. Let M be part of a belt of matter encircling the Earth E and then let this ring shrink until it is contiguous with the Earth. If the belt is fluid, it can be thought of as representing the seas, and Newton assumed that the motion of the fluid would be governed by the same laws as governed the motion of a solid particle. The theory then showed that the basic tidal phenomena were the combined effect of the perturbing effect of the Sun and the Moon. Notwithstanding the fact that the dynamical theory Newton developed was the first that could predict quantitatively the rise and fall of the oceans with tolerable accuracy, it is fundamentally wrong. Newton's analysis was based solely on the component of the perturbing force that was parallel to the gravitational attraction (the vertical component), but the dynamics of tidal motion are in fact determined mainly by the horizontal motion of the seas, which in turn is determined by the horizontal component of the tide-generating force.[80]

---

[80] See Palmieri (1998) for more details.

Finally, and most remarkably, Newton used his theory to explain the precession of the equinoxes, a phenomenon for which no one had previously suggested anything close to a correct explanation. He realized that the gravitational attraction of the Sun and Moon on the extra material around the equator of the Earth would result in its axis rotating in a small conical motion, like a top. He computed its effect as 50″ per year, agreeing with the accepted value of the time. Newton's solution to this problem was imperfect – hardly surprising, since no theory of rigid body dynamics existed[81] – but represented one of the more unexpected successes of gravitation theory. George Airy later said:

> . . . if at this time we might presume to select the part of the *Principia* which probably astonished and delighted and satisfied its readers more than any other, we should fix, without hesitation, on the explanation of the precession of the equinoxes.[82]

The basic idea of Newton's approach to the problem of precession was correct. He had shown that one of the perturbing effects of the Sun on the orbit of the Moon was to cause the nodes – the intersections of its orbit with the ecliptic plane – to regress. From his theory, Newton could calculate the rate of regression for a body orbiting the Earth once per day, i.e. at the same rate as the Earth rotates. Much as in the tidal theory, this body was then imagined to be a solid ring encircling the Earth and then finally attached to the Earth around the equator. Newton believed that the ring would be subject to the same regression of the points of intersection between it and the plane of the orbit around the perturbing body, which in this new situation corresponds to precession. Various technical results[83] led him to believe that the effect would amount to just over 9″ per year. But there was also the effect of the Moon on the equatorial bulge, and, based on measurements of the heights of tides at Bristol, Newton concluded that the attraction by the Moon of a particle on the surface of the Earth was about $4\frac{1}{2}$ times that of the Sun,[84] and thus that the effect of the Moon on precession would be about 41″, the total effect thus coming to 50″ per year.[85]

---

[81] A fully developed theory of rigid body dynamics first appeared when Euler published his *Theoria motus corporum solidorum seu rigidorum* in 1765.

[82] Quoted from Grant (1966), p. 38.

[83] For a recent analysis, see Dobson (1998, 2001).

[84] Westfall (1973) describes how Newton fiddled this number during the preparation of the second edition of the *Principia* so as to achieve the desired value for precession (see also Kollerstrom (1995)). Daniel Bernoulli later criticized Newton's value, instead deriving a figure of about $2\frac{1}{2}$: 1 based on tidal periods rather than heights. In fact, it is not possible to use the tides to give a reliable estimate for this ratio, which is now known to be about 2.18 : 1 (Wilson (1980), p. 88).

[85] The first person to publish a thorough analysis of Newton's approach to precession (pointing out all the flaws) was d'Alembert in 1749 (see note 43, p. 309).

The reception of Newton's theory

The acceptance of Newton's theory of gravitation following the publication of the *Principia* in 1687 was not rapid, though the work was anticipated eagerly after the appearance of a pre-publication review, written by Halley. Outside Britain, the work either was ignored or treated with hostility, but even in his own country the introduction of his theories into the academic curriculum was not immediate. David Gregory did introduce Newtonian philosophy to students at the University of Edinburgh before 1690 (though, for the sake of the students, not in any great depth), but Cambridge was, surprisingly given Newton's presence there, rather slower to introduce these new ideas, not beginning until 1699.[86] That being said, Newton was treated as an almost superhuman genius by his fellow countrymen. For example, Gregory, who became Savilian Professor of Astronomy at Oxford in 1691 (on Newton's recommendation), thanked Newton

> for having been at the pains to teach the world that which I never expected any man should have known.[87]

Adulation was not restricted to the scientific community. Many English poets saw Newton as the illuminator of the natural world and praise for him became a common theme, epitomized by Alexander Pope's famous *Epitaph*:

> Nature, and Nature's laws lay hid in night:
> God said, *Let Newton be!* and all was light.[88]

After its publication, three reviews of the *Principia* appeared, two of which were basically descriptions of its contents. The third, published in the *Le journal des sçavans*, presented a critical analysis, and this was not favourable.[89] The review drew a clear distinction between the mathematical achievements of Newton and his physical theory: 'The work of Mr Newton is the most perfect mechanics we can imagine' but is based on 'hypotheses which are generally arbitrary'. Then at the end: 'Mr Newton has only to give us a physics as exact as the mechanics'. This shows clearly the grounds on which the Newtonian

---

[86] A new translation of the influential Cartesian textbook *Treatise on Physics* by Rohault was prepared, with notes, by Samuel Clarke in 1697 for the use of Cambridge students, largely because it was felt that Newton's work was unintelligible. Clarke made no criticism of the vortex theory but, in a later edition of 1702, he did describe the attraction theory of Newton. In the 1710 edition, Clarke described the vortices as fictions and, in later editions, the Newtonian arguments against the Cartesian theory were spelled out (Hoskin (1962)).

[87] Quoted from Westfall (1980), p. 470, (with changes to reflect modern spelling).

[88] For more on the relationship between Newton's work and eighteenth-century poetry, see Nicolson (1946).

[89] The reviews were all anonymous, but it is suspected that the one in the *Le journal des sçavans* was by Régis. Quotes from this review are from Aiton (1972).

system would be attacked by the Cartesians. Lacking any firm foundation in a mechanistic physics, the conclusions of the abstract mathematics could, at best, be treated only as a guide to any description of reality.

Many French astronomers were sceptical in the extreme of Newton's theory; particularly conspicuous in their opposition were Huygens and Leibniz, both of whom were sent copies of the *Principia* by Newton. Huygens, though he clearly admired Newton, could not accept universal gravitation:

> I cannot agree with the principle which he supposes in this computation and elsewhere, viz. that all the small particles, which we can imagine in two or many different bodies, attract and try to approach one another. This I cannot admit because I think I see clearly that the cause of such an attraction cannot be explained by any principle of mechanics or by the rules of motion.[90]

He did, however, accept that the cause of planetary motion was a gravitational force that varied inversely with the square of the distance; it was the mechanism that generated this gravity with which he took issue. Central to this was the question of what filled the space between the planets, which in Newton's case was very little. Huygens, however, required a more substantial ether for his theory of light. Leibniz, in a letter to Huygens in 1690, wrote:

> I do not understand how he conceives gravity or attraction; it seems that to him it is only a certain immaterial and inexplicable virtue, whereas you explain it very plausibly through the laws of mechanics.[91]

Neither Huygens or Leibniz could accept action at a distance. Indeed, Newton himself was bothered by this:

> That gravity should be innate, inherent and essential to matter so that one body may act upon another at a distance through a vacuum without the mediation of anything else by and through which their action or force may be conveyed from one to another is to me so great an absurdity that I believe no man who has in philosophical matters any competent faculty of thinking can ever fall into it. Gravity must be caused by an agent acting constantly according to certain laws, but whether this agent be material or immaterial is a question I have left to the consideration of my readers.[92]

Although he could offer no explanation for the cause of gravity, Newton's position was that the attraction itself was clear from observations. To most

---

[90] HUYGENS *Discourse on the Cause of Gravity*, 1690. Translation from Pannekoek (1961).
[91] Quoted from Pannekoek (1961). The correspondence between Huygens and Leibniz during this period was dominated by discussions of Newton's work, which shows that they both recognized its significance whether or not they agreed with its central tenets.
[92] From a letter to Richard Bentley in 1693, quoted from Turnbull (1961), Volume III. Newton's correspondence with Bentley, later to become Bishop of Worcester, is the primary source for information concerning Newton's views on the physical problem of the structure and creation of the Universe (see for example, Kerszberg (1986)).

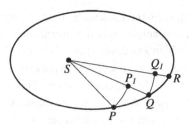

Fig. 8.9. The idea behind Leibniz's harmonic vortex.

Cartesians, this was less than convincing – without a satisfactory causal mechanism, Newton's theory was on very shaky foundations.

In 1689, Leibniz published *An Essay on the Causes of Celestial Motions.*[93] Leibniz's aim was to combine a mathematical description of heavenly phenomena with explanations based on physics, the absence of the latter being a fundamental flaw in Newton's work. Here, Leibniz was following Kepler in insisting that hypotheses should not only lead to predictions that correspond to observation, but also should be underpinned by a sound physical theory. As far as Leibniz was concerned, Newton not only had failed to follow Kepler's programme, but, by removing the vortices of celestial matter, was trying to explain the natural world through 'occult qualities' like attraction at a distance.[94]

In general structure, Leibniz's system was similar to that of Kepler's, since the planetary motions were thought of as being made up of a circular motion around the Sun and a libratory motion along the radius vector. The circular motion was due to the solar vortex, and the speed of rotation was inversely proportional to the distance from the Sun (equivalent to the area law), while the radial motion back and forth was due to the combination of gravity and centrifugal force (in Kepler's case this libration was due to alternate magnetic attraction and repulsion). Leibniz's vortex theory is illustrated in Figure 8.9, in which the positions P, Q, and R on the orbit are to be thought of as being separated by equal, infinitesimal intervals of time. At P, the motion is the result of a circular motion PP_1, centred on the Sun S, and a radial motion P_1Q. As Kepler had shown in the *Epitome of Copernican Astronomy*, the area law is equivalent to having the component of velocity perpendicular to the radius vector vary inversely with the solar distance, and so Leibniz's assumptions

[93] *Tentamen de motuum coelestium causis.* A translation is given in Meli (1993).
[94] Leibniz claimed that when he wrote the *Tentamen* he had not seen Newton's *Principia* but only the review of it in the *Acta eruditorum*. According to Meli (1991), however, Leibniz almost certainly had seen Newton's work, but by claiming his ideas were formulated before Newton's, he hoped to elevate their status. This theme is expanded upon in Meli (1993).

about the nature of the solar vortex guaranteed that the planet would move in accordance with Kepler's second law.

For Leibniz, the centrifugal endeavour was a real force caused by the circulation of the body, the magnitude of which, for a body orbiting in a non-circular orbit, was v^2/r, where v was the component of the velocity perpendicular to the radius vector. In his harmonic vortex – where this component varied inversely with r – Leibniz was led to the conclusion that the centrifugal force varied inversely with the cube of the distance from the Sun, and, utilizing Kepler's area law, he deduced that the magnitude of this force was h^2/r^3, where $h = r^2\dot\theta$ is a constant. Hence, in his theory, the force on the planet in the radial direction was h^2/r^3 minus the effect of gravitation. Leibniz then considered elliptical motion subject to the area law and showed, using his differential calculus, that

$$\frac{\mathrm{d}^2 r}{\mathrm{d}t^2} = \frac{h^2}{r^3} - \frac{h^2}{\ell r^2}$$

and, hence, that gravitation must vary inversely with the square of the distance.

The solar vortex provided the basic mechanism for the system, but Leibniz did not manage to develop a theory free of inconsistencies. For example, the fluid matter that made up the vortex was supposed to carry the planet in the transradial direction but not to interfere at all with the radial motion. In order to explain the radial motion, Leibniz introduced the concept of an ether that was distinct from the coarser matter that made up the vortex; this ether moved about in all directions, much as in Huygens' account of terrestrial gravity. Huygens, however, did not see the need for the harmonic fluid vortex that Leibniz had introduced; he realized that all that was needed to explain Kepler's laws was a satisfactory mechanism that led to inverse square attraction. Thus, he wrote to Leibniz:

> It is clear that the gravity of the planets being supposed in the reciprocal proportion of the squares of their distances from the Sun, this, together with the centrifugal force gives the eccentric ellipses of Kepler. But how, in substituting your harmonic circulation, and retaining the same proportion of the gravities, you deduce the same ellipses, it is this that I have never understood ... since the said proportion of the gravities, with the centrifugal force, alone produce the Keplerian ellipses according to the demonstration of Mr Newton.[95]

Of course, the harmonic vortex was superfluous, but Leibniz could not see that the circulation according to his harmonic law (i.e. the component of velocity perpendicular to the radius decreasing in proportion to distance from the Sun) followed from the law of attraction.

---

[95] Quoted from Aiton (1972).

There were other problems. Comets did not fit easily into Leibniz's framework and, perhaps more significantly, the fact that different planets obeyed Kepler's third law did not fit in with the harmonic circulation. Leibniz got around the latter problem by hypothesizing that the harmonic law existed only in narrow bands around each planet, but that the fluid in the large spaces between the planets did not rotate according to the same law. While this saved the phenomena, it was hardly convincing, and Newton found it easy to find fault with Leibniz's approach. Newton also accused Leibniz of making errors in his mathematical analysis, but in most cases these criticisms were misplaced.[96]

Another factor that increased French opposition to Newtonian ideas was the measurement of arcs of the surface of the Earth by Jacques Cassini (son of G. D. Cassini). If the Earth were an oblate spheroid, the length of 1° of latitude (corresponding to a change of 1° in the altitude of the celestial pole) would increase as one moved from the equator to the pole, but Cassini's measurements suggested that the Earth bulged at the poles rather than at the equator. Many people were happy to believe this result, since it cast doubt over Newton's theory, but it was by then an observed fact that gravity decreased toward the equator, and this did not appear to fit in with Cassini's data. As a result, the staunch defender of Cartesianism, J. J. d'Ortous de Mairan, invented an *ad hoc* law of attraction that (at least qualitatively) reproduced this behaviour.[97]

In a treatise on the shape of celestial bodies in 1732, Pierre-Louis-Moreau de Maupertius, one of the few supporters of Newton in France, compared the theories of Descartes and Newton and found in favour of the latter. However, his calculations on the shape of the Earth produced a very different value from that of Newton, and he concluded that in order to settle the question, more accurate measurements were needed. Accordingly, two groups set out to measure degrees of latitude and longitude; one travelled to Peru (the northern part, now Ecuador) in 1735, and another, headed by Maupertius himself, went north to Lapland in 1736. Maupertius' group returned in 1737 with data that showed that the degree

---

[96] Newton and Leibniz were the main protagonists in the most notorious priority dispute in the history of science – that over the invention of the calculus. This is described in great detail by Westfall (1980), Chapter 14. The facts of the dispute are not in doubt: Newton invented his fluxional calculus in around 1665 and Leibniz invented his differential calculus independently about 10 years later. However, Newton kept his discovery to himself, whereas, as Westfall writes, 'the second inventor published his calculus and thereby raised Western mathematics to a new level of endeavor. Newton realized as much in the end, and a good half of his fury was flung at Leibniz as a surrogate for his former self who had buried such a jewel in the earth'.

[97] Mairan's memoir was written in 1720 and its technical content is described in Todhunter (1962).

of latitude in Lapland was greater than that previously measured in France.[98] The measurements of Jacques Cassini then were performed again in 1740 (this time by another member of the Cassini family, César-François Cassini de Thury), and the results reversed the previous findings. There was now little doubt that the Earth was flattened at the poles. The return of the Peruvian expedition was then awaited to confirm these conclusions and to aid in an accurate determination of the ellipticity. For various reasons, no members of that group returned until 1744 and, while the data from Peru confirmed the oblate shape, different calculations resulted in values of the ellipticity ranging from 1/179 to 1/266.

The person who contributed most to the general acceptance of Newtonian principles throughout France was Voltaire, who spent the years 1726–9 in England. In his *Letters Concerning the English Nation* (1733) he brought Newton's theory to the attention of far more people than had previously been aware of it, and told the story of the falling apple, which he had learned from Newton's niece, Catherine Conduitt. Later, in 1738, Voltaire wrote a widely accessible summary of Newton's most important discoveries in optics and astronomy.[99] The first complete translation of the *Principia* into French was made by Voltaire's mistress, the Marquise du Châtelet, published posthumously in 1759.[100]

From France, Newton's ideas spread across Europe. A notable contribution was made by the Croatian Jesuit Rudjer Bošković, who published *A Theory of Natural Philosophy* in 1758, in which all physical processes came under the dominion of a force law between material points which was an inverse square attraction provided the particles were far enough apart, but which became infinitely large and repulsive as the particles moved closer and closer together.[101] In this way, Newtonian gravitation was, for the first time, combined with a theory designed to reproduce small-scale phenomena; in particular, Bošković saw in his approach an explanation for why material points could not occupy the same position in space.

Newton's achievement with the law of universal gravitation was enormous. He managed to unify the formally opposed principles of Bacon and Descartes in that he deduced rules from observation, but these rules were so powerful that

---

[98] Extracts from Maupertius' account of his expedition can be found in Fauvel and Gray (1987), p. 455 (see also Terrall (1992, 2003)).

[99] *L'élémens de la philosophie de Newton.* This work was translated immediately into English (Voltaire (1967)). For more details on Voltaire's advocacy of Newtonianism and the *Elements* in particular, see for example, Teske (1972).

[100] Madame du Châtelet's achievements in natural philosophy and mechanics are described in Iltis (1977).

[101] *Theoria philosophiae naturalis* (1758) was published in Vienna. An English translation of the 1763 edition published in Venice can be found in Boscovich (1966).

they allowed the theoretical determination (based on mathematics) of all sorts of apparently unrelated phenomena. Kepler's three laws had been shown to follow from a single general principle, and the same principle could be used to explain phenomena as diverse as the tides and the precession of the equinoxes. This was a state of affairs not even dreamed of before Newton. In the words of Voltaire: 'Before Kepler, all men were blind. Kepler had one eye, Newton had two.'[102]

One thing that the *Principia* did not bring about was an immediate increase in the accuracy of planetary tables. Well into the eighteenth century, such tables were essentially Keplerian in nature, with improvements due to the more accurate value of solar parallax and the eccentricity of the Earth that Horrocks had introduced, and more sophisticated observation techniques, including better corrections for refraction. Universal gravitation held the key to a step change in the accuracy of theoretical predictions, but astronomers had to wait until mathematicians had devised methods for determining quantitatively the effect of one planet on another's orbit before such a change could be realized.

There were many imperfections in Newton's demonstrations, some of which have been mentioned above, and other people were quick to latch on to them in order to help discredit the overall theory. In another way, these imperfections aided the advance of Newtonianism in that, as they were removed one by one by improved techniques, the impression was gained that there was nothing that universal gravitation could not explain, given sufficient mathematical ingenuity. One of the prime examples of this was the lunar theory, which, as we have seen, gave an erroneous value for the advance of the apsidal line and which was considered a serious objection to accepting the attraction theory. It took half a century before universal gravitation was reconciled with the motion of the lunar apogee, but with that achievement the Newtonian framework became virtually unassailable.

The demise of the vortex theory

As far as Cartesian physics is concerned, the period between the publication of the *Principia* and Newton's death in 1727 was characterized by attempts to explain the quantitative details of planetary theory, rather than simply being satisfied with an overall qualitative world view.

One year before the publication of the *Principia* in 1687, Bernard Fontanelle published his *L'entretiens sur la pluralité des mondes*, which described the

---

[102] Quoted from Brackenridge (1995), p. 10.

vortex theory in an accessible manner.[103] Another influential work was the *Search for Truth* (*La recherche de la vérité*) by Malebranche, the leading Cartesian among Newton's contemporaries. The sixth and definitive edition was published in 1713, a year after the second edition of the *Principia* had appeared, and it completely omitted Newton's theory of gravitation. Malebranche made his views on the role of mathematics abundantly clear: mathematical abstractions should not be confused with physical reality.

Leibniz was the first person to try to explain the details of planetary motion using Cartesian ideas and, as we have seen, managed to quantify the vortex theory in a way that produced inverse square gravity and elliptical orbits. Another attempt was made by Philippe Villemot, who published his *Le nouveau système ou nouvelle explication du mouvement des planètes* in 1707, a work Fontanelle praised for its originality but which Leibniz judged unfavourably. Villemot claimed that in order for the fluid making up the vortex to be in equilibrium, the total centrifugal force in each spherical layer had to be constant, which, expressed mathematically, is equivalent to saying that $r^2 \times v^2/r$ was constant, from which it follows that $v \propto r^{-1/2}$. With this arbitrary assumption, one then can conclude that the period of rotation of a planet being carried by the vortex, which is proportional to r/v, is proportional to $r^{3/2}$. Villemot claimed that he had for the first time provided a satisfactory derivation of Kepler's third law and attempted explanations of many other celestial phenomena (e.g. precession), but they were all equally superficial.

From about 1730 until the vortex theory finally was abandoned around 1755 – when Euler derived his equations of fluid motion – there were a number of attempts to reconcile the Cartesian and Newtonian viewpoints, notable contributers being Joseph Privat de Molierès, Johann Bernoulli, his son Daniel, and Jacques Cassini. Over this period, many advocates of Cartesian physics became gradually more tolerant of Newtonian ideas. An example is Pierre Bouguer[104] whose entry for the 1732 Academy prize was that of a convinced Cartesian. A greatly revised and extended version of his essay on the cause of the inclinations of the planets was published in 1748, and here Bouguer allied himself with the Newtonians, remarking that the force of attraction had to be admitted as a fact, the discovery of its cause being still awaited.

---

[103] Aiton (1989) describes this work as the first literary masterpiece having for its aim the popularization of science.

[104] Bouguer was a member of the group who travelled to Peru in 1735 to measure degrees of latitude and longitude (see p. 286).

9

Celestial mechanics

Analytic developments

The seventeenth century witnessed a complete transformation in astronomy. The Ptolemaic universe of uniform circular motions disappeared and was replaced by a system based on mechanical principles and Keplerian orbits. At the beginning of the eighteenth century, the methods being used to analyse the motions of the heavens were rooted still in the geometry of the ancient Greeks, but by the end of the century this, too, had changed, with dynamics reduced to the solution of differential equations: 'physical astronomy' became 'celestial mechanics'.

The great advance in the theory and applications of the new calculus contained in the works of Leibniz and his followers was instrumental in the rapid development of physical astronomy in Europe following the publication of the *Principia* and, ironically, the mathematical researches of adherents to Descartes' philosophy paved the way for the triumph of Newtonianism. Leibniz's differential calculus had been introduced in two papers, in 1684 and 1686, and studied with enthusiasm by Johann Bernoulli and his brother Jakob. Johann taught the new calculus to the Marquis de l'Hospital and the latter published the first textbook on the subject, *Infinitesimal Analysis* (1696).[1]

Pierre Varignon, who had also been taught the differential calculus by Johann Bernoulli, used this new mathematical language to describe many of Newton's geometrical propositions concerning central forces and, as a result, the essence of Newton's theory then could be read and understood by far more people, since knowledge of the differential calculus spread rapidly throughout Europe.[2] The

---

[1] The full title was *L'analyse des infiniment petits pour l'intelligence des lignes courbes*. There is an enormous literature concerning the early history of the calculus (see, for example, Boyer (1959), Kline (1972) and the works cited therein).
[2] Varignon also brought the calculus to bear on theories that utilized alternatives to Kepler's second law, e.g. those of Boulliau and Ward.

new calculus tempted mathematicians to try and provide analytical solutions to more and more challenging problems in mechanics and, as we shall see, the difficulties these problems generated led to fundamental improvements in the mathematical procedures themselves.

The calculus, as it was developed on the mainland of Europe, followed the notation introduced by Leibniz, whereas in England, Newton's symbols were used. Newton regarded variable quantities as being generated by motion, and he designated the rate at which a quantity changed (called a *fluxion*) by using so-called 'pricked letters', which consisted of placing a dot over the symbol representing the variable (which he called a *fluent*, i.e. something that flows). Thus, if x and y are fluents, their fluxions are denoted by $\dot{x}$ and $\dot{y}$. Fluxions of fluxions were written $\ddot{x}$ etc. Leibniz, on the other hand, used the symbols dx and dy to represent infinitesimal changes in the quantities x and y, and the ratio dy/dx to represent the rate at which y varied as x varied. This has the great practical advantage that the independent variable[3] is written explicitly, and changes of variable can be performed with relative ease because the notation suggests the correct rules for doing this:

$$\frac{dy}{dx} = \frac{dy}{du}\frac{du}{dx}.$$

Continental mathematicians were able to develop methods based on changes of variable more easily than their English counterparts, and these could be used to simplify the differential equations resulting from Newton's laws. At the risk of oversimplifying, we can say that Newton's approach was the more rigorous, in that he avoided the use of infinitesimal quantities – thinking instead in terms of limits of ratios – but Leibniz's procedures were easier to implement.

Clairaut, Euler, and d'Alembert

The enormous progress in mathematical astronomy made during the first half of the eighteenth century was due largely to the efforts of three men: Alexis-Claude Clairaut, Leonhard Euler, and Jean le Rond d'Alembert. Clairaut was an extraordinary child prodigy who read a paper on geometry to the Paris Academy at the tender age of 13, and a special dispensation allowed him to become a member when only 18. He became a close friend of Maupertius and Voltaire, both supporters of Newtonianism, and he aided the Marquise du

---

[3] In the eighteenth century, the differentials dx and dy usually were written separately so that if, for example, $y = x^2$, then one would have $dy = 2x\,dx$. The independent variable was distinguished by saying that its differential was constant, so that if x is the independent variable in the above, then $d^2y = 2(dx)^2$, where d^2y is shorthand for $d(dy)$, i.e. an infinitesimal change in dy.

Châtelet in her translation of the *Principia* into French.[4] Clairaut was a pioneer in the methods of the new calculus, and it is his applications of these ideas to celestial mechanics – in particular to the shape of the Earth (he was a member of the expedition led by Maupertius to Lapland in 1736 to measure the length of an arc of a meridian) and the motion of the Moon – for which he is best known.

Whereas Clairaut was born and died in Paris, Euler was born in Basel, Switzerland, where he studied under Johann Bernoulli, but he worked in St Petersburg from 1727 to 1741 and from 1766 until his death, with the intervening period in Berlin. Euler's achievements in mathematics were phenomenal, both in terms of quantity and quality, and his work on astronomy, although extremely significant, forms a minor part of his total contribution to science.[5] Within the field of astronomy, Euler contributed works on practical subjects such as the determination of orbits, methods for calculating the solar parallax, and the theory of refraction, as well as his highly theoretical work on celestial mechanics.

Like Clairaut, d'Alembert was a Parisian[6] and was admitted to the Academy at a young age (in d'Alembert's case, 24). He was an outstanding mathematician and wrote major works on many topics including partial differential equations and fluid mechanics, though perhaps the most famous of his scientific works is his *Treatise on Dynamics* (1743).[7] With Denis Diderot, d'Alembert edited the monumental *L'encyclopédie*,[8] which aimed to set down the basic principles of all the arts and sciences, and became perhaps the most influential French scientist in the mid eighteenth century.

One perceived deficiency of the *Principia* was Newton's failure to prove that an inverse square central force necessarily resulted in an orbit in the form of

[4] The published translation contained a lot of new material as well as Newton's text. Many of the additions were due to Clairaut (see Zinsser (2001)).

[5] Some of Euler's contributions to mathematics are described by Dunham (1999). Topics include number theory, logarithms, infinite series, algebra, and geometry, but celestial mechanics does not get a mention. Euler wrote and published more than any other mathematician – about 560 books and articles, and numerous unpublished memoirs. His achievements are made all the more remarkable by the fact that his eyesight deteriorated from the 1730s onwards, and he was totally blind by 1771. Details of his first period of work in St Petersburg, including early work on mechanics, are given in Calinger (1996).

[6] The single most famous incident in d'Alembert's life is his birth. He was found, abandoned by his mother, on the steps of the church of St Jean le Rond, from which his name is derived. Grimsley (1963) provides a non-mathematical biography, but for a description of his scientific activities, including his work in mathematics, see Hankins (1970).

[7] As well as its technical content, discussed in Fraser (1985a), *Le traité de dynamique* contains important observations on the underlying philosophy of mechanics. D'Alembert recognized that attitudes to mechanical principles had undergone considerable change since pre-Newtonian days, but bemoaned the lack of a logical foundation for the subject.

[8] The *L'encyclopédie ou dictionnaire raisonné des sciences, des arts, et des métiers* was published in twenty-eight volumes between 1751 and 1780. Most of the mathematical and scientific articles in *L'encyclopédie* were written by d'Alembert.

a conic section; rather, he had proved the converse: that such orbits imply an inverse square attraction. This was soon cleared up using the calculus, independent solutions being produced by Jakob Hermann and Johann Bernoulli in 1710.[9] One way to prove this result begins by making the substitution $u = 1/r$ in Eqn (8.4) (p. 266) which transforms that nonlinear differential equation into a linear one with θ as the independent variable:

$$\frac{d^2u}{d\theta^2} + u = \frac{f(1/u)}{mh^2u^2},$$

where $h = r^2\dot{\theta}$ is a constant. This clever substitution originated in the work of d'Alembert and Clairaut in the 1740s and 1750s.[10] If we then substitute $f(1/u) = m\mu u^2$, which corresponds to an inverse square law of attraction, we obtain the differential equation of the orbit

$$\frac{d^2u}{d\theta^2} + u = \frac{\mu}{h^2}. \tag{9.1}$$

In the 1730s, Euler developed a systematic procedure for the solution of linear ordinary differential equations with constant coefficients, which made the solution of this equation straightforward.[11] In the process, Euler created a unified theory of trigonometric and exponential functions and brought them all under the umbrella of the new calculus.

All solutions to Eqn (9.1) are of the form $u = A\cos(\theta - \theta_0) + \mu/h^2$, where A and θ_0 are arbitrary constants. If we write $\ell = h^2/\mu$ and $e = \ell A$, then this can be rearranged to give

$$r = \frac{\ell}{1 + e\cos(\theta - \theta_0)},$$

which is the polar equation of a conic section. The constant e is the eccentricity, and the point of closest approach always occurs when $\theta = \theta_0$. If $e = 0$, the orbit

---

[9] Hermann was a pupil of Jakob Bernoulli. His main work is his *Phoronomia* (1716), in which many of the problems from Books I and II of the *Principia* are treated. Hermann's approach was a sort of half-way house between the geometry of Newton and the fully fledged analytical mechanics that would later be created by Euler. A discussion of the interplay between calculus and geometry in Hermann's work is given in Guicciardini (1996) and Bernoulli's proof is discussed in Speiser (1996).

[10] The key step is to note that, since $h = r^2\dot{\theta}$ is a constant,

$$\frac{d^2r}{dt^2} = \frac{d}{dt}\frac{dr}{dt} = \dot{\theta}\frac{d}{d\theta}\left[\dot{\theta}\frac{d}{d\theta}\left(\frac{1}{u}\right)\right] = \dot{\theta}\frac{d}{d\theta}\left(-h\frac{du}{d\theta}\right) = -h^2u^2\frac{d^2u}{d\theta^2}.$$

[11] The details of Euler's procedure are contained in correspondence between Euler and Johann Bernoulli in 1739 (Fauvel and Gray (1987)), pp. 446–9).

is a circle, values of e in the range $0 < e < 1$ correspond to elliptical orbits, $e = 1$ gives a parabola, and hyperbolic orbits result when $e > 1$.

As Newton had been at pains to emphasize, universal gravitation implied that the motion of each celestial body was influenced by all the others. The *Principia* contains a number of attempts to quantify these effects; in some cases Newton was successful, in others not. The challenge facing the practitioners of the new calculus was to determine this influence theoretically. Because of the dominating effect of the Sun (it possesses nearly 99.9 per cent of the entire mass of the Solar System) each planet's orbit is very nearly elliptical, and the perturbations due to each of the other planets are small and could, with justification, be considered separately. Moreover, of the remaining mass, Jupiter makes up about 70 per cent, so its influence is the most significant. The theoretical study of the Solar System, over and above the basic Keplerian orbits, thus became the analysis of a large number of distinct three-body problems.

The transition from Newton's geometrical methods to an analytical approach to mechanics became firmly established with the publication in 1736 of Euler's two-volume *Mechanica*. In this work, which brought the author to the attention of the Parisian scientific community, Euler treated many of the problems that could be found in the *Principia*, but the three-body problem was conspicuously absent. The case of the mutual interaction of two bodies had been solved geometrically by Newton, and an analytical solution was contained in an essay by Daniel Bernoulli, who won a prize for it from the Paris Academy in 1734.[12] When three mutually attracting bodies were considered, progress appeared to be much more difficult. Clairaut wrote:

> I have deduced the equations given here at the first moment, but I only applied few efforts to their solution, since they appeared to me little tractable. Perhaps they are more promising to others. I have given them up and have taken to using the method of approximations.[13]

and, looking back on his efforts at finding a solution, Euler wrote:

> As often as I have tried these forty years to derive the theory and motion of the moon from the principles of gravitation, there always arose so many difficulties that I am compelled to break off my work and latest researches. The problem reduces to three differential equations of the second degree, which not only cannot be

---

[12] Kline (1972), p. 492. Daniel's father, Johann, also entered the prize contest, and the two entries were declared joint winners. Johann was distinctly unimpressed by the fact that his son had been judged his equal, and the episode led to a breakdown in the relationship between them. The contest actually asked for an essay on the physical cause of the inclinations of the planetary orbits to the plane of the equator, and Daniel's contribution included the first application of probability theory to astronomy (see p. 344).

[13] Quoted from Pannekoek (1961) p. 299.

integrated in any way but which also put the greatest difficulties in the way of the approximations with which we must here content ourselves; so that I do not see how, by means of theory alone, this research can be completed, nay, not even solely adapted to any useful purpose.[14]

Clairaut and Euler were pioneers in what become one of the most important areas of mathematical research in the nineteenth century. Between 1750 and the beginning of the twentieth century, over 800 papers relating to the three-body problem were published.[15]

The basic differential equations that describe the motion of three mutually interacting particles, as derived from Newton's second law, are, in vector notation,

$$\ddot{\mathbf{r}}_i = \sum_{j=1}^{3}{}' \frac{Gm_j}{r_{ij}^3}(\mathbf{r}_j - \mathbf{r}_i), \qquad i = 1, 2, 3, \tag{9.2}$$

where m_i and $\mathbf{r}_i$ are, respectively, the mass and position vector of body i, r_{ij} is the distance between bodies i and j, and the prime on the summation sign indicates that the term $j = i$ should be omitted. One approach, which might sensibly be pursued for some problems, is to consider the three-body problem as a perturbation of the two-body problem. Thus, when considering the motion of the Moon due to the interactions of the Sun and Earth, we might consider the Sun as a perturbing influence on the solution to the Earth–Moon two-body problem. Similarly, the effect of Saturn on the orbit of Jupiter could be considered as a perturbation of a known solution. It is appropriate for such problems to generate the equations of motion relative to one of the three bodies, which we will label as body 1. The equation corresponding to $i = 1$ in Eqn (9.2) can be subtracted from that corresponding to $i = 2$ to give

$$\ddot{\mathbf{r}}_{21} + \frac{\mu_{12}}{r_{12}^3}\mathbf{r}_{21} = Gm_3\left(\frac{\mathbf{r}_{31} - \mathbf{r}_{21}}{r_{23}^3} - \frac{\mathbf{r}_{31}}{r_{13}^3}\right), \tag{9.3}$$

where $\mu_{ij} = G(m_i + m_j)$ and $\mathbf{r}_{ij} = \mathbf{r}_i - \mathbf{r}_j$. An equation for $\mathbf{r}_{31}$ can be obtained similarly. If the right-hand side were zero, then the above equation is identical to Eqn (8.8) (p. 274) and we would simply have the two-body problem with its resultant elliptic orbits. The two terms on the right-hand side thus represent the perturbing influence of body 3 on this Keplerian motion; the first represents the effects of body 3 on body 2, while the second represents the effects of body 3 on body 1.

[14] From the preface to Euler's prize essay on lunar theory (1772). Quoted from Pannekoek (1961) pp. 299–300.
[15] Whittaker (1937), Chapter XIII.

In the case of body 1 being the Sun and bodies 2 and 3 are planets, G_3 is much smaller than μ_{12}, and so the right-hand side can be considered small. For the case of bodies 1, 2 and 3 being the Earth, Moon, and Sun, respectively, the right-hand side is again small, but for a different reason. In this case, it is because the perturbation comes from the difference between the effect of the Sun on the Moon and the effect of the Sun on the Earth, these two effects being very nearly equal.

Euler introduced one of the key mathematical ideas that allowed progress to be made in the solution of Eqn (9.3) when the right-hand side was small, namely the approximation of functions by trigonometric series. This was a huge step forward, as it allowed the integrations which needed to be performed to solve these second-order differential equations to be carried out relatively simply. In the expressions for the components of the perturbing forces in the coordinate directions that appear in the differential equations, the distance between bodies 2 and 3 appears as an inverse cube. Now, if we ignore the eccentricities in the orbits of the bodies so that r_{12} and r_{13} are constant,[16]

$$r_{23}^{-3} = r_{13}^{-3}(1 - 2\alpha \cos \theta + \alpha^2)^{-3/2} = r_{13}^{-3}(1 + \alpha^2)^{-3/2}(1 - g \cos \theta)^{-3/2},$$

where $\alpha = r_{12}/r_{13}$, θ is the difference between the longitudes of bodies 2 and 3 referred to body 1, and $g = 2\alpha/(1 + \alpha^2)$. Using the binomial theorem[17] we thus have

$$r_{23}^{-3} = r_{13}^{-3}(1 + \alpha^2)^{-3/2}\left(1 + \tfrac{3}{2}g \cos \theta + \tfrac{15}{8}g^2 \cos^2 \theta + \cdots\right). \quad (9.4)$$

If r_{12} and r_{13} are very different in magnitude (as they are for the lunar problem in which r_{12} is the Earth–Moon distance and r_{13} is the Earth–Sun distance) then g is much smaller than 1, and only a few terms in this series will yield accurate results. However, in the case of Jupiter and Saturn, g turns out to be about 0.84 and the series converges very slowly. In the case of the Earth and Venus, $g \approx 0.95$ and the situation is even worse. Questions of the actual convergence of infinite series like Eqn (9.4), as distinct from the use of such series as approximations, were not addressed seriously in the eighteenth century, though Euler did become concerned about this in his later years. He also suggested that perhaps it might be better to integrate the inverse cube directly using numerical quadrature. His colleague in St Petersburg – Anders Lexell – attempted this, but with little success.[18]

---

[16] Euler also treated the case of non-zero eccentricity.

[17] Discovered by Newton, and independently by James Gregory, in around 1670, though not provided with a rigorous proof until that of Abel in 1826 (Hairer and Wanner (1996), p. 251).

[18] In the 1760s, Euler suggested that for some purposes it might be better to approach the whole problem of perturbations by integrating numerically the differential equations using a

Another of Euler's significant contributions was his demonstration that an expansion in powers of $\cos\theta$ could be transformed into an expansion in terms of $\cos n\theta$,

$$r_{23}^{-3} = A + B\cos\theta + C\cos 2\theta + \cdots,$$

such series being much more straightforward to integrate. The coefficients A, B, C, ... were themselves expressed in terms of infinite series, but Euler showed that they satisfied a two-term recurrence relation, meaning that each coefficient depended on the values of the previous two. He thus only needed to calculate A and B from the series and then the others followed. This new representation could easily be integrated term by term and, moreover, the two integrations that were required improved the convergence of the series, since $\cos n\theta$ when integrated twice gives $-n^{-2}\cos n\theta$. While Euler's procedure still required considerable effort to be implemented, it made the three-body problem appear tractable for the first time.[19]

The motion of the lunar apogee

The Achilles heel of Newtonian celestial dynamics appeared to be the lunar theory. The larger of the lunar inequalities depends to some extent on the motion of the lunar apogee, the magnitude of which still lacked a theoretical explanation. The Moon was seen as the best test of the theory of gravitation, and in the 1740s this led Euler, Clairaut and d'Alembert to try and determine if the Sun's influence on the Earth–Moon system could explain the apsidal motion. All three used analytical approaches to the problem based around approximate methods of solution for Eqn (9.3), and all three were led to the conclusion Newton had reached, i.e. that the predicted motion of the lunar apogee was only half the

[19] time-stepping procedure. This is precisely what Clairaut did when studying the perturbations of Halley's comet (see p. 304).
Euler's method for expanding the inverse cube in terms of $\cos n\theta$ subsequently was improved by Lagrange. Lagrange made use of a mathematical concept that Euler himself had introduced, the complex exponential $\exp(i\theta)$, i being the square root of -1. Since $\exp(i\theta) = \cos\theta + i\sin\theta$, it follows that

$$(1 - 2\alpha\cos\theta + \alpha^2)^{-q} = (1 - \alpha e^{i\theta})^{-q}(1 - \alpha e^{-i\theta})^{-q}.$$

By expanding the factors on the right-hand side using the binomial theorem and multiplying the resulting series together, Lagrange simplified greatly the procedure by which the coefficients of $\cos n\theta$ in the desired expansion were calculated.

observed figure.[20] There was a strong feeling that the theory of gravitation was at fault but considerable disagreement as to how it should be rectified.

Euler suggested a number of possibilities as to why universal inverse square gravity appeared to yield an inaccurate prediction. The mass of the Sun might not be distributed uniformly, or the effect of the other planets might not be negligible, but in his view the most likely answer was that the inverse square law was not exact. This view was reinforced by the small, but perceptible, motion in the apsides of the planets, which implied that they were subject to forces in addition to just an inverse square attraction to the Sun, and also by the fact that Euler was unable to resolve the large irregularities in the motions of Jupiter and Saturn using the Newtonian law. He was influenced also by his belief that forces had to be imparted via contact; in the case of gravity, this involved contemplating the reintroduction of vortices, and he thought it implausible that in such a situation the law of gravity, would take such a simple, exact, and universal form.

D'Alembert, on the other hand, was reluctant to discard the inverse square law and hypothesized instead that the Earth–Moon system was subject to some additional force that affected the motion of the apse. He suggested magnetism as a possible source. Clairaut had his own doubts about the inverse square law and was delighted to find that Euler shared his misgivings. He suggested that the force law should be modified by the addition of a term that decayed more rapidly than the inverse square contribution – perhaps a term proportional to the fourth power of the distance – so that it would affect the Moon but become irrelevant for planetary motions:

> ... without doubt the moon obeys another law of attraction than the square of the distances, but on the contrary do not the principal planets demand this law in consequence of the observation of Kepler's laws? However, it is easy to reply to this difficulty by noticing that there are an infinity of laws of attraction which differ quite markedly from the law of squares for small distances and which disagree very little for great ones; one can imagine these laws easily with the familiar analysis. One can take for example an analytic quantity which expresses the relation of attraction to distance as composed of two terms, one having the square of distance as divisor, the other the fourth power ...[21]

---

[20] Actually, there was considerable disagreement as to what Newton had achieved in the *Principia*, due largely to the confusing way in which he reported his calculations. Some thought that Newton believed the inverse square law did not give the correct value for the motion of the lunar apse, but others (including Clairaut) thought that he believed the law to be sufficient to predict the true magnitude of the irregularity (see Waff (1976, 1995a)).

[21] Clairaut on the system of the world according to the principles of universal gravitation (1747). Quoted from Fauvel and Gray (1987).

Apart from the obvious problem that such a new law would have to give the same quantitative results in all cases where the inverse square law did agree with observation, this type of approach was criticized strongly on metaphysical grounds, the leading opponent being Georges, Comte de Buffon. Buffon thought it ridiculous to modify a physical law on the basis that it failed to account for one phenomenon, when it manifestly succeeded in predicting many others. Surely it would be more appropriate to look for a particular cause to explain this single phenomenon. Buffon suggested a number of different possibilities, including the magnetism of the Earth. He objected also to a physical law being represented by a two-term expression, since this destroyed the necessary simplicity that a universal law should possess. Clairaut replied to the first point that metaphysics alone was a poor guide to finding the truth, and, in response to the second, he pointed out that the need for more than one term was a consequence of the language of algebra and said nothing about the underlying physical simplicity.

In order to establish the veracity of a new force law, Clairaut needed more data on the motion of the apogee of satellites orbiting close to their parent body. Unfortunately, the apsidal motion of the moons of Jupiter and Saturn were as yet unknown, and those of Mercury and Venus were not determined with sufficient accuracy.

In 1748, Clairaut reviewed the theoretical calculations he had performed on the lunar perturbation problem. These calculations were based on a number of approximations, and he decided to push his method further, though he did not believe that this would have much effect on the final results. Buffon claimed subsequently that it was his criticisms that led Clairaut to go back to his original theory and see if it could be improved. A bitter argument between the two men ensued.[22]

The first simplification that Clairaut utilized in his attack on the lunar problem was to assume that the motion of the Moon, even when perturbed by the Sun, remained in a plane. Thus, right from the outset, he ignored any deviations in latitude. This then allowed him to reduce the three scalar equations given by Eqn (9.3) to two, which, in polar coordinates, can be written (compare Eqn (8.2)):

$$(\ddot{r} - r\dot{\theta}^2)\mathbf{e}_r + (2\dot{r}\dot{\theta} + r\ddot{\theta})\mathbf{e}_\theta = (-P - \mu r^{-2})\mathbf{e}_r + Q\mathbf{e}_\theta, \quad (9.5)$$

where the subscripts have been dropped so that $\mathbf{r} = r\mathbf{e}_r$ is the position vector of the Moon relative to the Earth, and P and Q represent, respectively, the inward radial and tangential components of the perturbing forces. The tangential

---

[22] The debate and subsequent dispute is described in Chandler (1975).

component of this equation is

$$\frac{1}{r}\frac{\mathrm{d}}{\mathrm{d}t}(r^2\dot\theta) = Q,$$

and so

$$r^2\dot\theta = \int Qr\,\mathrm{d}t + h, \tag{9.6}$$

where h is a constant of integration.

Clairaut's next step was ingenious – it greatly impressed Euler – and enabled him to obtain a differential equation for the orbit in which t was absent, analogous to Eqn (9.1) for the unperturbed case. He multiplied Eqn (9.6) by Qr, and integrated the resulting equation with respect to t, giving

$$\int Qr^3\,\mathrm{d}\theta = \tfrac{1}{2}\left(\int Qr\,\mathrm{d}t\right)^2 + h\int Qr\,\mathrm{d}t = \tfrac{1}{2}\left(\int Qr\,\mathrm{d}t + h\right)^2 - \tfrac{1}{2}h^2.$$

This can then be solved for $\int Qr\,\mathrm{d}t$ and the result substituted back into Eqn (9.6) to show that

$$r^2\dot\theta = \left(2\int Qr^3\,\mathrm{d}\theta + h^2\right)^{1/2},$$

which allows us to eliminate t from the radial component of Eqn (9.5). Introducing $u = 1/r$ and writing $I = 2h^{-2}\int Qu^{-3}\,\mathrm{d}\theta$ for convenience, we obtain

$$\frac{\mathrm{d}^2u}{\mathrm{d}\theta^2} + u = \frac{\mu}{h^2}(1+\Omega), \tag{9.7}$$

where the disturbing forces are represented by

$$\omega = (I+1)^{-1}\left(\frac{P}{\mu u^2} - \frac{Q}{\mu u^3}\frac{\mathrm{d}u}{\mathrm{d}\theta} - I\right),$$

which is zero if $P = Q = 0$.

Of course, Eqn (9.7) represents a much more difficult challenge than the equivalent equation with ω absent (Eqn (9.1)). However, we can write (reverting to the variable r):

$$\frac{h^2}{\mu r} = 1 - c\cos(\theta - \alpha) + \sin\theta\int\omega\cos\theta\,\mathrm{d}\theta - \cos\theta\int\omega\sin\theta\,\mathrm{d}\theta, \tag{9.8}$$

where c and α are arbitrary constants, which can be shown by direct differentiation to satisfy Eqn (9.7). This is still an equation for r (since ω depends on r), but one that lends itself well to approximation, since ω is relatively small.

Clairaut derived expressions for P and Q under the assumption that the Earth–Sun distance is constant, and then he substituted an approximate form for r into the right-hand side of Eqn (9.8). He originally chose, based on the empirical evidence,

$$\frac{k}{r} = 1 - e \cos q\theta,$$

where k, e, and q are undetermined constants. Thus, he used an approximation that represents a rotating ellipse and in which q should turn out to be slightly less than 1, so that θ must traverse through slightly more then 360° for the Moon to go from one perigee to the next. This approximation can be justified if the extra terms that are introduced by this process are sufficiently small. In performing the calculations, Clairaut made use of Euler's ideas on the expansion of functions in trigonometric series and had to make numerous other approximations,[23] but finally he obtained the equivalent of the expression

$$\frac{k}{r} = 1 - e \cos q\theta + A \cos 2n\theta + B \cos (2n - q)\theta + C \cos (2n + q)\theta,$$

$$(9.9)$$

where $n = (n_M - n_S)/n_M$, n_M being the mean motion of the Moon and n_S that of the Sun. In Eqn (9.9), the constants k, e, q, A, B, and C are all determined, with A, B, and C small, as required.[24] The value predicted for q was $0.995\,803\,6$, which implies a motion for the lunar apogee of $(1 - q) \times 360° \approx 1° \, 30' \, 39''$ per revolution – very close to Newton's value of $1° \, 31' \, 28''$ and about half that observed.

Clairaut realized he would get a more accurate value for q if, instead of using the approximate form $k/r = 1 - e \cos q\theta$, he used an expression of the form of Eqn (9.9), in which k, e, q, A, B, and C are all undetermined, but he did not anticipate that this would lead to a significant change. The calculations required for this more refined approach were extremely laborious and many more terms had to be kept in the equations. Nevertheless, Clairaut persevered and found that the extra terms in Eqn (9.9) *did* have an important effect on the calculated value of q and the implied motion of the lunar apse was nearly doubled, exactly what was required! Clairaut announced his sensational result to the scientific world in 1749.[25] One of the implications of Clairaut's analysis – that had not been suspected before – was that it was not simply the radial component of the perturbing force that affected the apsidal motion, but that the tangential component also had a significant effect.

---

[23] Detailed in Wilson (1980).   [24] In fact, max $(|A|, |B|, |C|) < 0.01$.
[25] It eventually was published in his *La théorie de la lune* (1752).

The procedure that Clairaut had used was by no means above criticism, and d'Alembert was quick to point out the flaws. A major fault with Clairaut's approach was that he had assumed the form of the solution based on observations rather than deducing it from the differential equation. D'Alembert showed how this could be avoided by developing an iterative solution procedure for Eqn (9.7). He also carried the algebraic development of the theory much further than Clairaut (who tended to substitute in numerical values appropriate to the lunar problem at every stage) so that the relationships between the derived quantities and the underlying parameters were identified more easily.[26]

Euler and d'Alembert both subsequently managed to improve their own methods for treating the lunar problem, and found that the value for the motion of the lunar apogee predicted by theory was in accord with observations. If we denote the sidereal, anomalistic, and draconitic months by T_s, T_a, and T_d, respectively, then d'Alembert showed that the effect of the Sun on the orbit of the Moon implied that

$$\frac{T_a}{T_s} - 1 = \tfrac{3}{4}m^2 + \tfrac{225}{32}m^3 + \cdots ,$$

$$\frac{T_d}{T_s} - 1 = -\tfrac{3}{4}m^2 + \tfrac{9}{32}m^3 + \cdots ,$$

where $m \approx 0.0748 = n_S/n_M$ (or equivalently the length of the sidereal month divided by the sidereal year). Newton's approach, and Clairaut's first approximation, are equivalent to neglecting the cubic terms in the above expressions. This leads to a fairly accurate answer for T_d and, hence, for the retrogression of the nodes, but because of the large coefficient of the m^3 term in the expansion for T_a, the first term provides a very poor approximation. Inclusion of the higher-order term results in a prediction for the advance of the apsidal line which is in accord with observation. Thus, something that had appeared to threaten the validity of Newtonian gravitation eventually became an extremely powerful argument in its favour. Euler was quick to recognize the implications:

> For it is very certain that it is only since this discovery that one can regard the law of attraction reciprocally proportional to the squares of the distances as solidly established; and on this depends the entire theory of astronomy.[27]

---

[26] See Wilson (1995a) for more details. D'Alembert's contributions to celestial mechanics possessed more mathematical rigour than those of Clairaut, but the latter's work usually was more easily utilized by astronomers.

[27] From a letter to Clairaut in June 1751. Quoted from Wilson (1980), p. 143. Euler still believed that the ether would introduce frictional forces which, over a long period of time, would result in changes to the planetary orbits.

An accurate lunar theory was not just of theoretical interest – it was important also in aiding navigation at sea, and Clairaut, Euler, and d'Alembert all produced lunar tables. These were more accurate than the positions calculated from Newton's *Theory of the Moon's Motion*, but still produced errors of up to 5′ of arc. The first tables that were of sufficient accuracy for practical use were those of Tobias Mayer, based on Euler's theory. Mayer not only understood the analytical theory, but also possessed great observational skills and appreciated the subtleties involved in the interpretation of observational data. His tables, which were generally reliable to within $1\frac{1}{4}'$, were first published in 1753, but became much better known following their revision and posthumous publication in 1770.[28]

Halley's comet

Clairaut was also involved in another important test of Newton's theory, and one that brought it a very public success. Universal gravitation implied that comets were subject to the same inverse square attraction by the Sun and so, unless they were moving sufficiently quickly, they would travel on, albeit very elongated, elliptical paths round the Sun. Hence, there was a possibility that they would return at regular intervals. Halley appreciated this and looked through previous records to see if there were sightings of comets separated by similar time intervals.

The comet of 1682 appeared to provide just such an example. Many aspects of its orbit appeared to be similar to those of the comets that had appeared in 1607 and 1531, the most notable of which was the fact that all these comets had travelled around the Sun in the opposite direction to the planets. Halley was convinced that these three comets were, in fact, one and the same, and he announced this to the Royal Society in 1696. His belief was reinforced when he discovered subsequently that comets had been observed in 1456, 1380, and 1305. The time intervals between sightings alternated between 75 and 76 years, and this led some astronomers to believe that there were two comets travelling

---

[28] The tables were edited by the fifth Astronomer Royal, Nevil Maskelyne, and published both in Latin (*Tabulae motuum solis et lunae novae et correctae*) and in English (*New and Correct Tables of the Motions of the Sun and Moon*). Mayer's heirs were awarded £3000 by the Board of Longitude as a recognition for his contribution to navigation, and Euler was given £300 for the theoretical work that formed the basis of the tables (see Forbes (1970), Forbes and Wilson (1995) for more details of Mayer's work). The thirty-one letters exchanged between Euler and Mayer have been translated in Forbes (1971a).

in the same orbit with identical periods of 151 years, with one near its aphelion when the other was at perihelion.[29]

Halley, however, believed that the variations in the period could be accounted for by the gravitational effect of the other planets – particularly Jupiter – on the comet. He wrote up his cometary investigations in 1705[30] and predicted that the comet would return sometime around the end of 1758, or possibly at the beginning of 1759, but he provided no mathematics to back up his prediction. To make a more accurate prediction required detailed and laborious calculations of the effects of the pull of Jupiter and Saturn on the comet's orbit, and these were carried out (by, in effect, performing the first ever large-scale numerical integration) by Clairaut and his associates, Joseph-Jérôme de Lalande and Madame Nicole-Reine Lepaute,[31] beginning in 1757.

At the outset, Clairaut believed that the only significant perturbations on the comet would be those due to close encounters with Jupiter, but he soon realized that the period of the comet was extremely sensitive to changes in speed near perihelion and, as a result, not only did the effects of Saturn on the comet have to be included, he also had to consider the effect of Jupiter and Saturn on the position of the Sun.

The calculations involved were horrendous, and many approximations had to be made to keep the number of numerical quadratures to a minimum.[32] In fact, Clairaut had to announce his results before the analysis was complete; otherwise, the comet might have appeared before the prediction of its return. In order to determine the accuracy of the procedure, it was tested on the data for the 1531 and 1607 comets, and the results compared with the actual events of 1607 and 1682. Based on the errors found, Clairaut predicted that the comet would reach its perihelion in April 1759, give or take a month.

In fact, the comet passed closest to the Sun on 13 March, having been first spotted on Christmas Day in 1758 by Johann Palitzsch, an amateur astronomer living near Dresden. The return of the comet had been anticipated by sections of the general public as well as by astronomers, and the success of the predictions based on the theory of gravitation reinforced the belief in the truth of the theory. As far as the majority were concerned, Halley had been right, and

---

[29] Details of these two-comet theories can be found in Waff (1995b). In fact, the comets of 1305 and 1380 were not Halley's comet at all, which appeared in 1301 and 1378 (Broughton (1985)).

[30] These investigations appeared in many editions and translations over the following 50 years (see Broughton (1985) for a list), the English title being *A Synopsis of the Astronomy of Comets*.

[31] Clairaut never acknowledged Mme Lepaute, the wife of the royal clockmaker, in any of his published work, apparently due to pressure from another woman (Broughton (1985)). Lepaute assisted Lalande in the production of ephemerides for over 20 years.

[32] Some details of the calculation procedure are given in Wilson (1993).

Newton's theory had triumphed yet again.[33] In his *Astronomical Tables*, Halley had written, referring to the comets of 1531, 1607, and 1682:

> You see therefore an agreement of all the elements in these three, which would be next to a miracle if they were three different comets . . . . Wherefore if according to what we have already said it should return again about the year 1758, candid posterity will not refuse to acknowledge that this was first discovered by an Englishman.[34]

Nicolas Louis de Lacaille referred to 'Halley's comet' in a session of the French Academy of Sciences in 1759, and it has borne Halley's name ever since.

Aberration and nutation

In the *Principia*, Newton had asserted that 'the orbit of the earth is sensibly perturbed by the moon',[35] but he did not provide any quantitative analysis. Newton claimed that the centre of gravity of the Earth–Moon system revolves around the Sun in an elliptical orbit (which is true to a high degree of approximation) and as the Earth and Moon take part in a monthly rotation about this point, the longitude of the Sun varies from its Keplerian value. In 1744, Euler calculated the inequality as $15'' \sin\theta$, where θ is the angular distance between the Sun and Moon, but this value is significantly in error due to the inaccurate values that he used for the ratio of terrestrial and lunar mass, and for the solar parallax – had he used the correct values he would have obtained a maximum inequality of nearer $6''$.[36]

The work of Euler, Clairaut, and d'Alembert on the three-body problem made it possible to perform a more systematic study of the orbit of the Earth, rather than simply assuming Keplerian motion for the centre of gravity. The first derivation of the perturbations of the orbit of the Earth due to the gravitational

---

[33] See Waff (1986) for a discussion of the public awareness of the comet's return in England during the 1750s. There was considerable controversy in astronomical circles regarding the accuracy of Clairaut's calculation, with d'Alembert pointing out that the error of 33 days was quite large when compared with the differences between successive periods of the comets. Clairaut later refined his calculations and eventually produced a figure he claimed was in error by just 19 days. Comparison with a modern analysis, due to Kiang (1971), shows that six of these days can be accounted for by Clairaut's neglect of the four innermost planets, and a further six due to the effects of the as yet undiscovered planets beyond Saturn (Broughton (1985)).
[34] Quoted from Gingerich (1992), pp. 150–1. The *Astronomical Tables* were first printed in 1719, but not published until 1749, 7 years after Halley's death.
[35] NEWTON *Principia*, Book III, Proposition 13. [36] Wilson (1980), p. 91

attraction of the other planets was by Euler, who thereby won the prize from
the Paris Academy in 1756. However, the first solar tables based on perturbation theory that were used widely – those of Lacaille (1758) – were based on
Clairaut's calculations. Clairaut's method, which was based on the same principles that he had brought to bear on the lunar problem, was easier to apply than
Euler's. Lacaille's tables were not free from either theoretical or observational
errors, but represented a marked improvement over the contemporary tables of
Halley and Jacques Cassini. As well as including perturbations due to the Moon,
Venus, and Jupiter, they also took account of the effects of two new astronomical
phenomena, both discovered by the English astronomer, James Bradley.[37]

In a series of observations in the late 1720s, Bradley noticed that stellar
positions underwent an annual oscillation which, because of its nature, could
not be the, as yet, undetected effects of stellar parallax.[38] If parallax had been the
cause, the observed displacements would have depended solely on the position
of the Earth in its orbit and been directed back toward the centre of the orbit
of the Earth. What Bradley found, however, were displacements which were in
the direction opposite to that in which the Earth was moving. The deviation was
small, the maximum amount being slightly less than 20.5″, but quite clearly
detectable using Bradley's instruments.[39]

For stars near the pole of the ecliptic – for which the velocity of the Earth is
always perpendicular to the Earth–star line – the effect of this so-called stellar
aberration is a deviation from the star's mean position of constant magnitude,
with the direction changing as the Earth goes round its orbit. Thus, such stars
trace out small circles of radius 20.5″. For stars in the ecliptic plane, the displacement vanishes when the Earth is moving directly toward or away from
the star, and is a maximum when the Sun, Earth and star are collinear. Thus,
such stars oscillate backward and forward about their mean position, the maximum displacement being 20.5″. Stars between these extremes move round
ellipses, the eccentricity of which depends on the ecliptic latitude of the star,
but with semi-major axes always of 20.5″, a quantity known as the 'constant of
aberration'.

In 1728, Bradley determined that the phenomenon was due to a combination
of the motion of the Earth relative to the 'fixed' stars, and the fact that light is

---

[37] Bradley was appointed to the Savilian Chair of Astronomy at Oxford in 1721 and succeeded
Halley as Astronomer Royal in 1742, the latter having taken over the post from Flamsteed in
1720.

[38] Hooke, Flamsteed and G. D. Cassini had all claimed (mistakenly) to have observed stellar
parallax, and it was in order to try to confirm Hooke's observations of 1669 that Bradley began
his project, initially in collaboration with the wealthy amateur telescope maker Samuel
Molyneux (see Hirshfeld (2001)).

[39] Bradley found 20.25″.

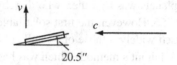

Fig. 9.1. The cause of stellar aberration.

not transmitted instantaneously. Thus, Römer's hypothesis of a finite speed for light, based on observations of eclipses of the moons of Jupiter, had been correct. Stellar aberration is a purely optical effect, the cause of which is illustrated in Figure 9.1 for a situation in which the Earth is moving at right angles to the light arriving from a star. In order for the light, which travels at speed c, to remain on the axis of the observer's telescope, which is moving with speed v at 90° to the light rays, the telescope must make an angle with the rays, the tangent of which is v/c. Bradley's measurements thus enabled him to calculate the ratio v/c and, given that he knew v with tolerable accuracy, he could then determine c.[40]

As well as its implications for observational astronomy, in which the positions of celestial bodies were measured against the background of 'fixed' stars, the discovery of aberration provided the first experimental confirmation of the motion of the Earth. Of course, the Copernican hypothesis was already an established scientific 'fact' by 1728, but its experimental verification still was an extremely significant event the history of astronomy. Bradley's observations also showed that stellar parallax was certainly less than 1″, and so the distance to the stars had to be revised upwards yet again.

The observations also suggested that stellar positions underwent a second oscillation, this time with a much longer period. He continued his observations through a full cycle – 18 years and 7 months – before announcing his explanation to the Royal Society in 1748.[41] The period of the oscillation was the same as the time it took for the nodes of the orbit of the Moon to complete a revolution around the ecliptic, and Bradley surmised correctly that it was due to the effect of the Moon on the oblate shape of the Earth, causing the axis of the Earth to wobble:

> I suspected that the moon's action upon the equatorial parts of the earth might produce these effects: for if the precession of the equinox be, according to Sir Isaac Newton's principles, caused by the actions of the sun and moon upon those parts,

---

[40] Bradley calculated that it takes light 8 min 13 s to reach the Earth from the Sun, very close to the modern value of 8 min 19 s (Débarbat and Wilson (1989)). A theoretical justification for aberration was provided by Clairaut in 1737 in his memoir *De l'aberration apparente des étoiles causée par le mouvement progressif de la lumière*. There is also an effect due to the daily rotation of the Earth, but the magnitude of this diurnal aberration is always less than 0.5″ and was of no concern to eighteenth-century astronomers.

[41] Bradley sent a preliminary account of his findings to the Paris Academy in 1737.

the plane of the moon's orbit being at one time above ten degrees more inclined to
the plane of the equator than another, it was reasonable to conclude that the part of
the whole annual precession which arises from her action would in different years
be varied in its quantity; whereas the plane of the ecliptic wherein the sun appears,
keeping always nearly the same inclination to the equator, that part of the
precession which is owing to the sun's action may be the same every year: and from
hence it would follow, that although the mean annual precession, proceeding from
the joint action of the sun and moon, were 50″, yet the apparent annual precession
might sometimes exceed, and sometimes fall short of that mean quantity, according
to the various situations of the nodes of the moon's orbit ....[42]

As well as identifying the cause, Bradley also determined the effect. He
suggested that the axis of the Earth underwent a small conical motion, tracing
out an ellipse with axes measuring 18 and 16″. This oscillation, which is superimposed on the slow conical motion of precession, is known as the 'nutation'
of the axis of the Earth and leads, as a consequence, to a variation in the obliquity of the ecliptic by $\pm 9''$ over an 18.6-year period. A quantitative theoretical
derivation of nutation from the equations for the Earth–Sun–Moon system, by
d'Alembert, followed soon afterwards.[43]

The discovery of aberration and nutation paved the way for a step change
in the accuracy of observational astronomy, with stellar positions determined
in terms of seconds, rather than minutes, of arc. It meant also that virtually
the whole of the vast body of existing observational data required revision.
Of course, an increase in accuracy of observations implied that much more
stringent tests could be applied to any theoretical predictions.

Two remaining problems

The anomalous motion of the lunar apse was not the only significant phenomenon that failed initially to yield to Newton's theory of gravity. When this
finally was explained by Clairaut, and confirmed by Euler and d'Alembert,
there remained two major irregularities without theoretical justification: the
secular acceleration of the Moon, and the so-called great inequality of Jupiter
and Saturn. The first of these was a gradual increase in the mean motion of the
Moon over the preceding 2000 years, to which Halley first drew attention in
1693. Its value, based on an analysis of ancient eclipse data, was believed in
the mid eighteenth century to be about 20″ per century per century. This type

[42] Quoted from Abetti (1954).
[43] *Les recherches sur la précession des equinoxes et sur la nutation de l'axe de la terre* (1749).
Euler provided subsequently his own derivation, and over the following 20 years the two men
developed the theory of rigid body dynamics (see Wilson (1987a)).

of cumulative change was termed 'secular', though when later astronomers realized that the inequalities might be oscillations with very long periods, the name was retained. The characteristic that distinguishes secular inequalities is that they have periods much longer than the orbital periods of the bodies in question, so that they lead to changes in the orbital parameters over a period of time. On the other hand, the inequalities that have periods comparable with those of the orbits can be thought of as oscillations about a fixed orbit.

The second problem was the very sizeable change in the mean motions of Jupiter and Saturn over time, a phenomenon that was not disputed,[44] but the form of which was understood poorly. Because of Clairaut's success with the lunar theory, astronomers were confident that universal gravitation would be able to provide the answers, but the technical difficulties needing to be overcome were daunting, and none of the three great practitioners of celestial mechanics in the mid eighteenth century lived to see a satisfactory theoretical explanation of either of these phenomena.

In order to understand the problems eighteenth-century astronomers faced when examining the evidence of the errant behaviour of Jupiter and Saturn, it will be helpful to look at the phenomenon with the benefit of more recent knowledge. Based on the nineteenth-century theory of the American mathematician and astronomer G. W. Hill, the angle by which Saturn is displaced from its mean position by Jupiter is (retaining terms whose amplitude is greater than 2')[45]

$$48'.49 \sin(0°.38492769t + 294°.63039)$$
$$-11'.40 \sin(11°.81346178t - 40°.19776)$$
$$-7'.04 \sin(5°.9067311t - 242°.82986),$$

in which t is measured in years and $t = 0$ corresponds to 1850. The periods of these three oscillations are approximately 935, 30.5, and 60.9 years, respectively. The effect of Saturn on Jupiter is given by the formula

$$19'.94 \sin(0°.38493647t + 114°.66699)$$
$$-3'.24 \sin(36°.25579742t - 250°.2646378)$$
$$-2'.68 \sin(29°.94136059t - 75°.94948)$$
$$-2'.05 \sin(5°.90673089t - 62°.81372498),$$

in which the terms have periods of roughly 935, 9.93, 12.0 and 60.9 years, respectively.

---

[44] It was noted by the Giacomo Maraldi (G. D. Cassini's nephew) in 1718 (Suzuki (1996)).
[45] See Wilson (1985), p. 34.

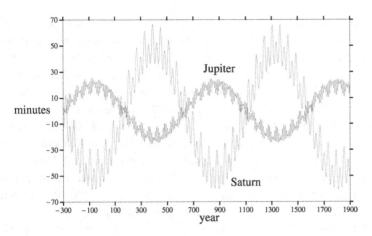

Fig. 9.2. The great inequality of Jupiter and Saturn.

A plausible reason for the inequality to have a period of about 900 years is as follows.[46] The greatest interaction between the two planets occurs when they are closest together (i.e. at conjunction) and as we have noted previously, these events occur every 19.86 years. The actual perturbative effect on the planets is influenced by their orbital parameters, (e.g. the position of the apsidal line) and thus the place in the zodiac at which the conjunction happens will influence the resulting perturbation. As Figure 6.5 (p. 171) shows, successive conjunctions form an approximate equilateral triangle in the zodiac, which shifts eastward every three conjunctions by about 8.1°. Thus, after forty-five conjunctions, which represents a period of just under 900 years, this triangle reoccupies (approximately) an earlier position and the cycle repeats.

The inequalities given by the above expressions are plotted in Figure 9.2 for the period 300 BC to AD 1900. The curves demonstrate why observations had failed to detect the correct form for the inequality, and why astronomers in the early eighteenth century did not appreciate the form of the irregularity they were trying to explain. Observational data were not continuous over many centuries, but concentrated in certain periods, e.g. those of the Babylonians in the third century BC, and those of Tycho Brahe in the sixteenth century. The small amplitude of the 'high-frequency' oscillations just added to the confusion.

The first person to note that there was a problem in reconciling the motions of Jupiter and Saturn with a Keplerian elliptical orbit was Kepler himself, and he suggested an oscillatory variation in the mean motions of the planets

---

[46] Due to George Biddell Airy, Astronomer Royal from 1835 until 1881.

with amplitude $28' 15''$ and period 354.5 years,[47] but did not include this in the *Rudolphine Tables*, believing that observations over many more years were required before this inequality could be determined with accuracy. Horrocks tried in vain to discover the nature of the inequality, as did Flamsteed, who was requested by Newton to look at observations of the two planets and see if they showed marked discrepancies from Kepler's tables when they were near conjunction. Any proposed solution that appeared to work for one era seemed to fail when applied to observations from another. Halley moved away from the idea of a cyclical change to that of a unidirectional variation. He hypothesized a uniform acceleration in the mean motion of Jupiter and a corresponding deceleration for Saturn. This was consistent with his view that the Moon had been undergoing a constant acceleration since antiquity.

Many eighteenth-century astronomers were aware of the serious problem that Jupiter and Saturn posed. Jacques Cassini (in 1746) provided a qualitative explanation of how gravitation might cause the observed inequality. He realized that if the orbits of Jupiter and Saturn were perfect concentric circles, both lying in the same plane, then the effects before and after a conjunction would cancel each other out. However, the orbits do not have this simple geometrical form, and Cassini argued that the different directions of the apsidal lines (the aphelia of the two planets were at the time about $80°$ apart) could result in the observed effect. Cassini did not, however, have the mathematical expertise to make any accurate quantitative predictions. He conjectured correctly that the accelerations and decelerations of Jupiter and Saturn actually were part of a periodic phenomenon, but his calculated period of 84 000 years was not very close to the mark.[48]

The Paris Academy made the problem of Jupiter and Saturn the subject of a prize contest in 1748, which resulted in the first major work on perturbation theory by Euler. Euler, in his essay – that won him the prize – did not achieve the desired goal of matching theory and observation, but did provide the first major inroads into the analytical treatment of planetary perturbations. The prize commission chose the same topic for the 1750 and 1752 contests; no prize was awarded for the first of these, but Euler won the latter, though still without resolving the issue.[49] In Euler's work we see the emergence of mathematical ideas that later, in the hands of Lagrange and Laplace, would become powerful

[47] Wilson (1985), p. 38.
[48] See Wilson (1985), p. 56 for more details.
[49] Daniel Bernoulli and Rudjer Bošković were runners-up in the 1748 and 1752 contests, respectively, but neither of their approaches had any impact on future developments (see Wilson (1995a)).

tools in the solution of problems of celestial mechanics and which eventually would lead to a successful understanding of the mutual interactions between Jupiter and Saturn.

One innovation that Euler introduced led eventually to the method of variation of orbital parameters. In the absence of perturbations, a planet moves in a Keplerian orbit round the Sun which is defined in terms of six constants: the orbital elements. These constants determine the shape and orientation of the orbit in space and the position of the planet on the orbit at a given time. Since the perturbations being considered are small, Euler argued, it makes sense to consider the resulting planetary path in terms of its Keplerian orbit, except that the orbital elements must now be considered as variables, albeit ones that change very slowly, i.e. at each instant the planet is imagined as moving on an elliptical orbit but the ellipse continually is changing its shape and orientation. As well as providing a useful starting point for a mathematical development, this idea also made the theory correspond more closely to practice, since astronomers had for a long time introduced slow changes in the orbital parameters into their tables so as better to fit the observations. We will not elaborate on Euler's method here (we will later describe the mature theory as worked out by Lagrange), but concentrate on one specific feature that arose in his analysis.

Considering the motion of Saturn as perturbed by Jupiter, Euler formulated a differential equation for the difference between the Sun–Saturn distance and its mean value which, in its essentials, was of the form:

$$\frac{d^2x}{d\theta^2} + m^2 x = a_1 \sin p_1\theta + a_2 \sin p_2\theta + \cdots,$$

the independent variable being Jupiter's eccentric anomaly. Here, m, a_i, p_i are known constants, and the terms on the right-hand side result from the trigonometric series expansions that Euler used to treat the effect of Jupiter on Saturn's orbit. The solution is (assuming for convenience that $x = 0$ when $\theta = 0$):

$$x = A \sin m\theta + \frac{a_1 \sin p_1\theta}{m^2 - p_1^2} + \frac{a_2 \sin p_2\theta}{m^2 - p_2^2} + \cdots,$$

where A is an arbitrary constant. Euler would immediately have realized that there was a potential problem: what happens if one of the ps is close to the value of m? How could he neglect confidently terms from the right-hand side without first establishing that none of their coefficients is large due to the presence of a so-called small divisor? The small divisor problem is of fundamental importance in perturbation theory, and Euler, with the mathematics available to him, was in no position to resolve it:

> And it is here that one encounters the greatest difficulty in the question proposed, a
> difficulty such that, if one does not reflect on it as deeply as possible, it is
> absolutely impossible to succeed in this research.[50]

In fact, when Euler first tackled the problem by making some simplifying
assumptions about the orbit of Jupiter, he found that one of the ps was actually
equal to m. In this case, the term in the solution corresponding to the $a \sin m\theta$
on the right-hand side of the differential equation is

$$\frac{a\theta}{2m} \sin m\theta,$$

which contains a factor (the eccentric anomaly θ) which continually increases
with time and thus eventually renders the solution useless. Even though terms
like these (which became known as 'arcs of circles') were not present in the
more accurate analysis, Euler believed at first that they might hold the key to
understanding the, as yet, unexplained motion of Jupiter and Saturn.

Euler's work on planetary perturbations was ground-breaking, but ultimately
unsuccessful. The same was true of his work on the secular acceleration of the
Moon. In the 1770s, Euler worked on improving the lunar theory, introducing
the idea of describing the motion with respect to a set of axes rotating with
the mean speed of the Moon. He received prizes from the Paris Academy in
both 1770 and 1772 for his work, which he believed demonstrated conclusively
that the secular inequality was not caused by gravitational attraction between
celestial bodies. In 1770 he wrote:

> it now appears well established that the secular equation of the movement of the
> moon cannot be produced by the force of attraction . . . ,

and in 1772 he went further:

> there no longer remains any doubt that the observed secular equation is the effect of
> the resistance of the medium in which the planets move.[51]

Euler died in 1783, 2 years before Laplace explained successfully the great
inequality of Jupiter and Saturn from the theory of universal gravitation, and
4 years before he claimed to have done the same for the secular acceleration of
the Moon.

---

[50] From Euler's prize-winning essay of 1752, *Les recherches sur les irrégularités du mouvement de Jupiter et Saturne*. Quoted from Wilson (1985), p. 110.
[51] Both quotes are taken from Wilson (1985), p. 21. The nature of any interplanetary medium and its effect on lunar and planetary motion was the subject of great debate in the eighteenth century. One contributor was Euler's son, Johann Albrecht, whose work, which won an award of merit from the Paris Academy in 1762, is described in Suzuki (1996).

Transits of Venus and the solar parallax

One thing that Newton's theory could not predict, was the actual size of the Solar System. The sizes of the planetary orbits were given by Kepler's third law in terms of the Earth–Sun distance – the astronomical unit – but this could not be determined accurately in terms of any terrestrial measure. The transits of Venus in 1761 and 1769 provided astronomers with the first opportunity to rectify this.

As the accuracy of observing instruments and techniques increased from the time of Hipparchus, the accepted value of the solar parallax was reduced gradually, with a corresponding increase in the Earth–Sun distance. Hipparchus had taken a solar parallax of $7'$, which would imply that the Sun was 490 Earth radii away from the Earth. Ptolemy reduced the parallax to just less than $3'$ and there it remained (with minor variations) until the advent of the telescope. As telescopic observations became more sophisticated, precise knowledge of the orbit of the Earth became crucial to the accurate determination of all lunar and planetary positions. Kepler was forced to accept a parallax of $1'$ because any larger value led to discrepancies between calculations and observations that were greater than the observational error. Horrocks then reduced this figure to $14''$ (which put the Sun some 15 000 Earth radii away).

The first proper scientific attempt to determine the parallax of the Sun was made by G. D. Cassini in 1672, based on simultaneous observations of the parallax of Mars at Paris and on the island of Cayenne. He produced a figure of $9.5''$, which is equivalent to a solar distance of about 140 000 000 km and was within 1 arc second of the correct value, but the accuracy had less to do with the precision of the observations than a desire to produce a parallax which would eliminate errors from the theory of the orbit of the Earth.[52] Between 1700 and 1760, a number of attempts were made to measure the solar parallax using Cassini's method but, even though observational techniques improved, the accuracy of the results did not and, while then it was accepted that the correct value was less than $15''$, no one could lay claim to an accurate direct measurement.

The fact that transits of the inferior planets could be used to determine the solar parallax was pointed out by the mathematician James Gregory in 1663. He wrote:

> This problem has a very beautiful application, although perhaps laborious, in observations of Venus or Mercury when they obscure a small portion of the sun; for by means of such observations the parallax of the sun may be investigated.[53]

---

[52] Cassini's method is described in detail in van Helden (1985), Chapter 12.
[53] Quoted from Woolf (1959).

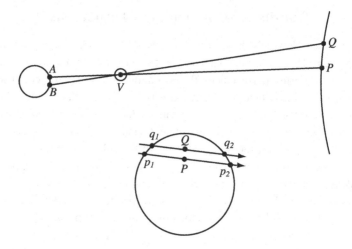

Fig. 9.3. Using transits of Venus to measure solar parallax.

The person responsible for the practical implementation of Gregory's idea was Edmond Halley, and the result was the first positive determination of the parallax of any heavenly body other than the Moon. Halley knew that he would not see the transit of 1761 (unless he lived to be 105!), but he submitted his proposal to the Royal Society in 1716. The essentials of the procedure are shown schematically in Figure 9.3.

In the figure, A and B represent two different places on the Earth from which the transit is observed, while P and Q represent the places on the surface of the Sun which are obscured by Venus, V. An observer at A would see Venus follow the path $p_1 p_2$ across the face of the Sun, whereas one at B would witness Venus moving from q_1 to q_2. The ratio $|AV|/|VP| = |BV|/|VQ|$ is known from Kepler's third law and, hence, the ratio $|AB|/|PQ|$ can be calculated. The distance $|PQ|$ as a fraction of the solar diameter can be determined from the times at which Venus is at p_1, p_2, q_1, and q_2, and then the diameter of the Sun can be calculated in terms of terrestrial units. The distance to the Sun then follows from the knowledge of its angular diameter. Exactly the same procedure can be applied to a transit of Mercury, but its proximity to the Sun means that the distance $|PQ|$ is too small to be determined accurately.

The transits of 1761 and 1769 gave rise to an international scientific collaboration on a scale not seen before and, in contrast to the transit of 1639, which was observed by two young men in northwest England, the transit of 1761 was scrutinized by over 120 professional astronomers viewing from at least sixty-two different places around the globe, with even more observing the

1769 transit.[54] One of those who witnessed the 1769 transit was Captain James Cook, whose first voyage to the South Seas in the *Endeavour* was made with the express purpose of viewing this astronomical event. He named the place on Tahiti from which he observed the planet's passage across the sun as Point Venus, and this name is still in use today.

Many calculations were made based on the 1761 transit, and numerous different values between 8.3 and 10.6″ were claimed for the solar parallax. The problems that had been encountered – some foreseen but others not – were then addressed, and the results of the 1769 transit were rather more successful, with values between 8.4 and 8.8″ being obtained. A thorough review of all the data in 1835 by Johann Encke, director of the Berlin Observatory, produced the figure of 8.57″, a figure that was accepted for many years. By the end of the nineteenth century, after two more Venus transits, astronomers had raised this figure to 8.80″, corresponding to a solar distance of 148 528 000 km. While the eighteenth-century transits of Venus did not produce a value for the solar parallax which was as accurate as Halley had thought it would be, they must be considered a great success. The size of the Solar System in terms of a terrestrial measure had been determined with a degree of precision that was far in excess of anything that had been possible beforehand.

Lagrange and Laplace

Celestial mechanics remained at the forefront of mathematical research throughout the eighteenth century. Many people contributed to the development of the subject, and many new mathematical techniques were devised in response to the difficulties that arose. Clairaut, d'Alembert, and Euler were the leaders in the field during the 1740s and 1750s and, when Clairaut died in 1765, the astronomical community could look back on a period of enormous and rapid progress in the theoretical branch of their subject. Some serious problems remained, however. As well as the discrepancies between theory and observation in the case of the secular acceleration of the Moon and the great inequality of Jupiter and Saturn, the most obvious failing was the lack of any general methods that were widely applicable. Over the following 60 years all this changed, due primarily to the efforts of two giants in the history of mathematics, Joseph Louis Lagrange and Pierre-Simon Laplace.[55]

---

[54] Woolf (1959) describes the events surrounding these transits in great detail. Another account is given by Maor (2000).
[55] A detailed scientific biography of Laplace can be found in Gillispie (1997).

Lagrange was born into a French-Italian family in Turin and discovered a passion for mathematics as a teenager. When only 18, he sent Euler a letter containing an outline of a new method for approaching problems in calculus. Euler had been working on similar methods but recognized immediately the superiority of Lagrange's approach, and named the subject the 'calculus of variations'.[56] Lagrange taught mathematics at the artillery school in his native city for many years, but through the influence of Euler and d'Alembert, was appointed to a much more prominent position by being named as Euler's successor at the Berlin Academy in 1766. D'Alembert also was instrumental in kick-starting Laplace's career, obtaining for him a post as Professor of Mathematics at L' École Militaire in Paris after receiving from Laplace a short mathematical essay in 1768, written while he was a student at the University of Caen in Normandy. Laplace became a member of the Paris Academy in 1773, and, in 1787, Lagrange left his post in Berlin also to become a member. Although competitors in many of their chosen areas of scientific endeavour, Lagrange and Laplace always maintained a deep respect for each other's contributions, and their rivalry, unlike that between many of their contemporaries, never turned into bitterness.

The two men had very different approaches. Lagrange was interested primarily in mathematics, and his work was always careful and clear. To him, elegance in mathematics was a virtue, and he always sought to derive the most general form of any result. Joseph Fourier said of him:

> The distinctive characteristic of his genius consisted in the unity and grandeur of view. He was attracted above all to a simple thought, just and very elevated . . . . All his mathematical compositions are remarkable for a singular elegance, by the symmetry of forms and generality of methods, and if one may speak thus, by the perfection of the analytical style.[57]

Laplace, on the other hand, was what might now be termed a 'mathematical physicist', who saw mathematics as a tool with which to study the physical world. He was 'always less enamoured with the beauty of mathematical speculation than he was anxious to unfold the system of the world'.[58] In contrast to the work of Lagrange, Laplace's technical writing was often hard to follow, with a considerable effort required on the part of the reader. However, neither man's achievements in celestial mechanics would have been accomplished without the

---

[56] The history of the calculus of variations, a subject that dominated Lagrange's work in mechanics, is charted in detail in Goldstine (1980) and Lagrange's contributions are examined in Fraser (1985b).
[57] From Fourier's *L'éloge* to Laplace, delivered in 1829, in which he also referred to the methods developed by Euler and Lagrange. Quoted from Fraser (1990).
[58] Quoted from Grant (1966), p. 111.

other. Throughout the latter part of the eighteenth century, essays of Lagrange stimulated Laplace toward new discoveries, and vice versa.

The libration of the Moon

Lagrange's interest in celestial mechanics was stimulated by the Paris Academy prize contest of 1764, for which it was asked

> ... whether it can be explained by any physical reason why the moon always presents the same face to us; and how, by observations and by theory, it can be determined whether the axis of this planet is subject to some proper movement similar to that which the axis of the earth is known to perform, producing precession and nutation.

Lagrange successfully resolved the first issue and won the prize.[59]

The physical cause of the equality between the length of a lunar day and a lunar month had been given by Newton,[60] who argued, by analogy with the tides on the Earth, that the Moon would not be spherical, but a spheroid, the greatest diameter of which, if produced, would pass through the centre of the Earth. The Moon could be in equilibrium only in this position, and any small displacement would result in slow oscillations about this configuration. Despite the efforts of people such as d'Alembert, no one had derived a theory capable of analysing such motions. What Lagrange accomplished was the development of a method by which the period of these so-called librations could be determined.

There are, in fact, two types of libration, the one under discussion being the physical libration. The part of the Moon visible from the Earth varies over the course of a month due simply to the orbit of the Moon being elliptical rather than circular. As the Moon orbits the Earth, its speed changes in accordance with Kepler's second law, but its rotation rate remains fixed, so that the part facing the Earth varies about a mean position. The rotation axis of the Moon is inclined to its orbital plane, and so the result of this oscillation – called the 'optical libration' – is an apparent motion from side to side and up and down during a lunar month.[61]

---

[59] The prize question is quoted from Itard (1975). The essay impressed Laplace, who, in 1783, wrote to Lagrange: 'The rigorous equality of the mean movement of the moon in its orbit with its mean movement in rotation being infinitely improbable, it is a very beautiful thing you have proved ... ' (Wilson (1980), p. 245).

[60] *Principia*, Book III, Proposition 38.

[61] The optical libration was described by Newton in the *Principia*, Book III, Proposition 17. About 59 per cent of the surface of the Moon is visible from the Earth over a period of time.

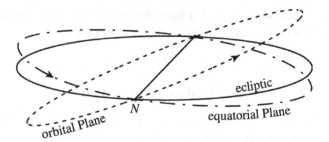

Fig. 9.4. The orientation of the rotation of the Moon. The point N is the ascending node of the orbit and the descending node of the equator.

Lagrange's work on the libration of the Moon was significant mainly because of the mathematical tools he used to study it, rather than for the results themselves. In his treatment of this particular problem from celestial mechanics, he derived general equations that have served as the basis for the study of dynamical systems ever since. His approach was founded on a combination of Maupertius' principle of rest and d'Alembert's principle, formulated in the latter's *Treatise on Dynamics*. The resulting method sometimes is referred to as the 'general principle of virtual velocities', and served as a cornerstone for all of Lagrange's subsequent work on dynamics.[62] As well as the details of the celestial mechanics, Lagrange's essay contained technical explanations of the procedures he was using, and in particular provided an outline of the derivation of what we now refer to as 'Lagrange's equations of motion'.

Lagrange derived the differential equations of motion for the nonspherical Moon from his new principle, and went on to show that the physical libration of the Moon should have a period of 119 months. He could not, however, explain another phenomenon – concerning the orientation of the axis of rotation of the Moon – discovered by G. D. Cassini and subsequently confirmed by Mayer. Cassini observed that the mean plane of the equator of the Moon made a constant angle with the ecliptic plane, that angle being 2° 30′.[63] Furthermore, the line of intersection of these two planes is also the line of nodes of the orbit of the Moon, with the ascending node of the orbit corresponding to the descending node of the equator (Figure 9.4). Lagrange could find no reason from his analysis why this should be so.

[62] Details can be found in Fraser (1983). More insight into the various dynamical principles in use in the mid eighteenth century can be gained from Lanczos (1986) and Dias (1999) (though a knowledge of French is required to understand the numerous quotations in the latter). A very informative, non-technical, summary of developments in mechanics during the eighteenth century can be found in Hankins (1985), which also contains a bibliographic essay citing numerous other relevant sources.
[63] Mayer revised this figure to 1° 29′. Roy (1978) gives the modern value as 1° 32′.

Lagrange gave a more complete solution in 1780, by which time he had improved and generalized his analytic methods considerably, and presented a formal derivation of his fundamental dynamical equations. Suppose we have a system of n particles with N degrees of freedom, so that the configuration of the particles in space is determined by N independent coordinates $q_1, \ldots, q_N$, for example. The position vectors of the particles are functions of these coordinates, so $\mathbf{r}_i = \mathbf{r}_i(q_1, \ldots, q_N)$, $i = 1, \ldots, n$, and, hence,

$$\dot{\mathbf{r}}_i = \sum_{\alpha = 1}^{N} \frac{\partial \mathbf{r}_i}{\partial q_\alpha} \dot{q}_\alpha. \tag{9.10}$$

The notation here, with vectors and partial derivatives, is very different to that used by Lagrange, but the concepts embodied in Eqn (9.10) are ones Lagrange understood fully. We can think of $\dot{\mathbf{r}}_i$ as being a function of the $2N$ quantities $q_\alpha, \dot{q}_\alpha$, $\alpha = 1, \ldots, N$, and it follows then that $\partial \dot{\mathbf{r}}_i / \partial \dot{q}_\alpha = \partial \mathbf{r}_i / \partial q_\alpha$, a result known as the 'cancellation of dots'. Moreover,

$$\frac{\mathrm{d}}{\mathrm{d}t} \frac{\partial \mathbf{r}_i}{\partial q_\beta} = \sum_{\alpha = 1}^{N} \frac{\partial^2 \mathbf{r}_i}{\partial q_\alpha \partial q_\beta} \dot{q}_\alpha = \frac{\partial \dot{\mathbf{r}}_i}{\partial q_\beta},$$

from Eqn (9.10). Hence, the operations $\mathrm{d}/\mathrm{d}t$ and $\partial/\partial q$ can be interchanged. If we define a function T by

$$T = \frac{1}{2} \sum_{i = 1}^{n} m_i \dot{\mathbf{r}}_i \cdot \dot{\mathbf{r}}_i \tag{9.11}$$

then, since $\dot{\mathbf{r}}_i$ is a function of $q_\alpha, \dot{q}_\alpha$, so is T, and using the cancellation of dots and the interchangeability of $\mathrm{d}/\mathrm{d}t$ and $\partial/\partial q$ we can show that

$$\frac{\mathrm{d}}{\mathrm{d}t} \frac{\partial T}{\partial \dot{q}_\alpha} - \frac{\partial T}{\partial q_\alpha} = \sum_{i = 1}^{n} m_i \ddot{\mathbf{r}}_i \cdot \frac{\partial \mathbf{r}_i}{\partial q_\alpha}, \qquad \alpha = 1, \ldots, N. \tag{9.12}$$

The quantity T is what we now call the 'kinetic energy' but in the eighteenth century, following the work of Leibniz, it was referred to as the *vis viva*, or living force, and its meaning was the subject of much debate.[64] Equations (9.12) are purely kinematic in character, arrived at by formal manipulations of various derivatives, but now we can incorporate the dynamics by utilizing the equations of motion.

Lagrange introduced what we would now call a 'potential function'. For any system of n mutually interacting particles, it is possible to determine a function, V, of the positions of the masses (the 'potential energy' as we would now call

---

[64] See, for example, Hankins (1985), pp. 30–6.

it, but called the 'perturbing function' by Laplace) such that the equations of
motion can be written (using modern notation for brevity) in the form

$$m_i \ddot{\mathbf{r}}_i = -\nabla_i V, \qquad i = 1, \ldots, n, \qquad (9.13)$$

in which $\nabla_i \equiv \left(\mathbf{i}\frac{\partial}{\partial x_i} + \mathbf{j}\frac{\partial}{\partial y_i} + \mathbf{k}\frac{\partial}{\partial z_i}\right)$. For the three-body problem in Eqn (9.2),

$$V = - \sum_{1 \le j < k \le 3} \frac{Gm_j m_k}{r_{jk}}.$$

Incorporating Eqn (9.13) into Eqn (9.12) yields

$$\frac{d}{dt}\frac{\partial T}{\partial \dot{q}_\alpha} - \frac{\partial T}{\partial q_\alpha} = -\sum_{i=1}^{n} \nabla_i V \cdot \frac{\partial \mathbf{r}_i}{\partial q_\alpha} = -\frac{\partial V}{\partial q_\alpha}, \qquad \alpha = 1, \ldots, N.$$

It is customary now to define a new function $L(q_\alpha, \dot{q}_\alpha)$ – the Lagrangian – by
$L = T - V.^{65}$ Then, since $\partial V/\partial \dot{q}_\alpha = 0$,

$$\frac{d}{dt}\frac{\partial L}{\partial \dot{q}_\alpha} - \frac{\partial L}{\partial q_\alpha} = 0, \qquad \alpha = 1, \ldots, N. \qquad (9.14)$$

These are Lagrange's equations of motion, and bear all the hallmarks of
Lagrange's mathematical style through their elegance of form and generality
of application. They reduce the derivation of the equations of motion to the
construction of a single scalar function, L, and could be applied to the n-body
problem (in which case $N = 3n$) whatever coordinates q_α with which one chose
to work. In his masterpiece *Le mécanique analytique* (Analytical Mechanics)
(1788), Lagrange illustrated the use of these equations with a number of applications and wrote:

> These different examples comprise nearly all the problems on the motion of a body
> or a system of bodies that the Geometers have solved; we have chosen them on
> purpose so that one may better judge the advantages of our method, by comparing
> our solutions with those found in the works of Messr. Euler, Clairaut, d'Alembert
> etc., in which one arrives at the differential equations only by reasonings,
> constructions and analyses often rather long and complicated. The uniformity and
> the swiftness of the course of [our] method are what should principally distinguish
> it from all others, and what we wished especially to show in these applications.[66]

By an expeditious choice of variables, Lagrange derived differential equations
that enabled him to show that the angular distance of the Moon from the mean
ascending node of its orbit had to be the same as the mean angular distance
of the prime meridian of the Moon from the node of the lunar equator. This is

[65] First done by Poisson in 1809 (Dugas (1988), p. 384).
[66] LAGRANGE *Analytical Mechanics*, Section 5, Part 2. Quoted from Fraser (1983).

precisely what Cassini had observed, and thus once more, what had presented itself as an apparently arbitrary fact turned out to be a consequence of universal gravitation.

Exact solutions to the three-body problem

As well as his important contributions to the development of perturbation theory, Euler (in the early 1760s) worked on an alternative technique for solving the three-body problem (Eqn (9.2)), which he hoped would shed light on the Sun–Earth–Moon system. If two of the bodies in a three-body system are much more massive than the third, one might assume that the small mass is affected by the other two bodies, but not the other way around. In such a situation, the two massive bodies – usually referred to as primaries – satisfy the equations for a two-body system, and thus move in Keplerian orbits. The problem then is to determine how the small body moves within the resulting gravitational field. Euler made significant advances in the study of this 'restricted three-body problem'; in particular, in 1767, he showed that exact solutions to the equations were possible, representing situations in which the three bodies remained forever in a straight line, the line itself rotating in space.[67]

The idea of finding exact solutions to the full three-body problem was then pursued by Lagrange. Although having declined to enter the prize contests of 1768[68] and 1770, Lagrange did enter the 1772 contest for which he shared the prize with Euler. He wrote to d'Alembert:

> I have considered the three-body problem in a new and general manner, not that I believe it is better than the one previously employed, but only to approach it [in another way].[69]

Apart from Euler's work discussed above, all attacks on the three-body problem had been based on perturbation theory. Lagrange tried a new approach and succeeded in finding a number of exact solutions.

The three-body problem is a system of nine second-order ordinary differential equations and is thus of order 18. Newton showed that the centre of gravity of any system of mutually interacting particles moves with constant velocity

[67] Boccaletti and Pucacco (1996), p. 216.
[68] The prize question of 1768 concerned the Moon. D'Alembert encouraged Lagrange to enter, but Lagrange replied: 'The king [of Prussia] would like me to compete for your prize, because he thinks Euler is working on it; that, it seems to me, is one more reason for me not to work on it' (Itard (1975)). Lagrange never met Euler, but the latter's influence on the former's work was perhaps greater than that of anyone else.
[69] Quoted from Itard (1975).

(what we would now call the 'conservation of linear momentum')[70] and so the system immediately can be reduced to one of order 12. This can be demonstrated from Eqn (9.2) by multiplying by the equation by m_i and summing over i. This results in

$$\sum_i m_i \ddot{\mathbf{r}}_i = 0,$$

since the terms on the right-hand side cancel in pairs. The centre of gravity has position vector $\mathbf{r}_c$, defined by $M\mathbf{r}_c = \sum_i m_i \mathbf{r}_i$, M being the sum of all the masses, and so it follows that $\ddot{\mathbf{r}}_c = 0$. This can be integrated twice to give

$$\mathbf{r}_c = \mathbf{a}t + \mathbf{b},$$

and the constant vectors $\mathbf{a}$ and $\mathbf{b}$ represent six constants of integration.

There are four more integrals that can be derived easily, all of which were known to Euler. If we take the vector cross product of Eqn (9.13) with $\mathbf{r}_i$ and then sum, we obtain $\sum_i m_i \mathbf{r}_i \times \ddot{\mathbf{r}}_i = 0$, which can be integrated once to give

$$\sum_i m_i \mathbf{r}_i \times \dot{\mathbf{r}}_i = \mathbf{h}, \tag{9.15}$$

where $\mathbf{h}$ is a constant vector. This result represents the conservation of angular momentum and contains three further constants of integration. Written out in component form as, of course, they had to be in the eighteenth century, these equations are termed the 'three integrals of areas' because they state that the sums of the products of the masses and the projections of the areas described by the line connecting the body m_i to the origin onto the coordinate planes, are proportional to time.[71] In vector form, we can see immediately that Eqn (9.15) defines a plane that remains fixed, however the masses move. If we take the centre of gravity as the origin, then any point in this plane has position vector $\mathbf{r}$ which satisfies $\mathbf{r} \cdot \mathbf{h} = 0$, $\mathbf{h}$ being the sum of the angular momenta of all the particles in the system. It was Laplace who made this realization and, for the Solar System, the plane is referred to as the 'invariable plane of Laplace'.[72]

---

[70] NEWTON *Principia*, Corollary 4 to the Laws of Motion.

[71] Laplace referred to these integrals as the 'principle of the Chevalier d'Arcy', after the French-Irish mathematician who was one of the first to derive them in analytical form. The appreciation that moments of vector quantities are vector quantities themselves (i.e. they add according to the parallelogram law) can be traced back to the work of Euler who wrote the equivalent of the equation $\dot{\mathbf{r}} = \omega \times \mathbf{r}$ for the rotation of a rigid body in 1765 (see Caparrini (2002)).

[72] See Caparrini (2002). The orientation of this plane is determined largely by Jupiter and Saturn since they contain most of the planetary mass. The ecliptic is inclined at about $1\frac{1}{2}°$ to this plane – which lies between the orbital planes of the two largest planets – though this figure varies over time as the orbit of the Earth is perturbed by the other planets.

The final integral of motion is the energy integral (historically called the 'vis viva integral' or the 'integral of forces viva'). If we dot Eqn (9.13) with $\dot{\mathbf{r}}_i$ and then sum over i we obtain

$$\sum_i m_i \dot{\mathbf{r}}_i \cdot \ddot{\mathbf{r}}_i = -\sum_i \dot{\mathbf{r}}_i \cdot \nabla_i V = -\frac{dV}{dt},$$

since V is simply a function of the coordinates of the masses. This can be integrated to give

$$T + V = E, \tag{9.16}$$

where T is the kinetic energy defined in Eqn (9.11). The constant E represents the total energy of the system. There are thus ten integrals for the n-body problem that can be shown to be independent, and for the three-body problem they reduce the order of the system of differential equations from 18 to 8.

By changing the independent variable from the time to one of the other dependent variables, Lagrange managed to reduce the three-body problem to a seventh-order system, and he found he could integrate the equations completely if he assumed that the ratios of the distances between the masses were constants. Joseph Alfred Serret, who edited the works of Lagrange (published in fourteen volumes between 1867 and 1892) commented on the 1772 essay on the three-body problem as follows:

> The first chapter deserves to be counted among Lagrange's most important works. The differential equations of the three-body problem ... constitute a system of the 12th order, and the solution required 12 integrations. The only known ones were those of the *force vive* and the three from the principle of areas. Eight remained to be discovered. In reducing the number to seven Lagrange made a considerable contribution to the question, one not surpassed until 1873 ... .[73]

One way to derive Lagrange's exact solutions is to seek solutions for which the resultant force on each body passes through the origin, which is taken to be the centre of gravity of the three bodies. If we cross Eqn (9.2) with $\mathbf{r}_i$ the above supposition implies that the left-hand side is zero and, hence,

$$\sum_{j=1}^{3}{}' \frac{m_j}{r_{ij}^3} \mathbf{r}_i \times \mathbf{r}_j = 0, \qquad i = 1, 2, 3.$$

Now, if we use the fact that $\sum_i m_i \mathbf{r}_i = 0$ – which expresses the fact that the origin is the centre of gravity – we can deduce the equation

$$m_2 \mathbf{r}_1 \times \mathbf{r}_2 \left(\frac{1}{r_{12}^3} - \frac{1}{r_{13}^3} \right) = 0,$$

---

[73] Quoted from Itard (1975).

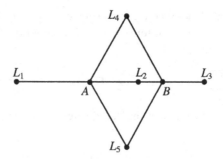

Fig. 9.5. The Lagrange points.

and two similar equations with the numbers 1, 2, and 3 permuted cyclically. There are two possibilities. Either $\mathbf{r}_1 \times \mathbf{r}_2 = \mathbf{r}_2 \times \mathbf{r}_3 = \mathbf{r}_3 \times \mathbf{r}_1 = 0$, in which case the masses lie in a straight line, or $r_{12} = r_{23} = r_{31}$, in which case the particles lie at the vertices of an equilateral triangle. In both cases, the equations can be integrated completely, but further work needs to be done to establish how the particles actually move.[74] In fact, each mass moves along a conic section with the centre of gravity of the system as a focus, the orbits all being similar to one another. If we take two of the masses as reference points, it turns out that there are five possible solutions for the position of a third particle. Two of these result in the equilateral-triangle solutions and the other three lead to straight-line solutions; collectively they are known as the 'Lagrange points'.

The simplest possibility is that the particles remain fixed relative to each other and rotate around their common centre of gravity at a constant rate. Figure 9.5, in which A and B are the positions of the two reference masses, illustrates the five Lagrange points (labelled L_1–L_5) for such a situation. The precise positions of L_1, L_2, and L_3 depend on the relative masses of the particles at A and B.

Lagrange believed that these solutions were just mathematical curiosities, having no relevance to the actual Universe. However, this is not the case. In 1906, an asteroid was discovered by Max Wolf (and named Achilles) oscillating about the point L_5 with respect to the Sun and Jupiter. Subsequently, more than twelve other asteroids were found orbiting close to the points L_4 and L_5. Collectively, they are known as 'the Trojans'. Other examples of asteroids in the Solar System oscillating about one of the points L_i with respect to two more massive bodies have subsequently been discovered.[75]

---

[74] See, for example, Roy (1978) Section 5.8.
[75] See, for example, Boccaletti and Pucacco (1996), p. 271.

Planetary and lunar perturbations

By the time Lagrange wrote his second memoir on celestial mechanics – with which the won the Paris Academy prize in 1766 – he was in full command of all the work in perturbation theory of his predecessors Clairaut, d'Alembert, and Euler. The topic of the 1766 contest concerned the mutual interactions of the moons of Jupiter, which the Swedish astronomer Pehr Wargentin had established did not move in uniform circular motion around their host planet, but instead were subject to small periodic inequalities, just like the planets themselves.

In his essay, Lagrange arrived at equations very similar in form to those that Euler had confronted when considering the Jupiter–Saturn problem, i.e.

$$\frac{d^2 x}{dt^2} + m^2 x = a_1 \sin p_1 t + a_2 \sin p_2 t + \ldots,$$

with the associated problem of possible small divisors in the solution. For the particular problem of the satellites of Jupiter, these equations turned out to be sufficient because, for each moon, one term on the right-hand side of the solution dominates all others. Thus, for the first moon, there was a term proportional to $\sin 2(\mu_2 - \mu_1)t$ in the equation, where μ_1 and μ_2 are the mean motions of the first and second satellites, Io and Europa, and m is μ_1 in this case. This leads to a term proportional to

$$\frac{\sin 2(\mu_2 - \mu_1)t}{\mu_1^2 - 4(\mu_2 - \mu_1)^2}$$

in the solution. Now $\mu_1 : \mu_2 \approx 2.008 : 1$ (this is computed easily from the data in Table 7.1, p. 215) and so the denominator is very small indeed. Consideration of this term alone allowed Lagrange to account for the observed irregularities in the motion of Io, and a similar analysis was successful for the other satellites.

So far, this was not much more than the reworking of the method of absolute perturbations – so-called because the perturbations were considered relative to fixed orbital parameters – to solve a new problem. But Lagrange did not stop there. He also considered what happened when the series solutions that were obtained based on a first approximation were substituted back into the differential equations to obtain a new, and hopefully more accurate, solution. Lagrange demonstrated that this method of successive approximations, that had been the basis of Clairaut's successful determination of the motion of the lunar apogee, was fraught with difficulty. In the first place, terms appeared in the second approximation that clearly were as large as those in the first (as Clairaut had found to his surprise and delight!) and, second, there was no way of avoiding the proliferation of 'arcs of circles'– terms in the solution that increased continually with time. Lagrange realized that these terms were a manifestation of the solution procedure rather than a representation of some real property of the solution itself, and their presence meant that any solution derived in using successive approximations in the method of absolute perturbations would be able to produce accurate predictions only for a finite time into the future. As an illustration, consider the function $\sin(1 + \lambda)t$ where λ is small. This is clearly bounded for all t, but an expansion in powers of λ begins

$$\sin(1 + \lambda)t = \sin t + \lambda t \cos t - \tfrac{1}{2}\lambda^2 t^2 \sin t + \ldots,$$

which shows clearly that the presence of a term proportional to $t \cos t$ in a perturbation expansion does not imply necessarily that the solution grows indefinitely.

The problems that Lagrange encountered with the method of absolute perturbations led him to develop his ideas on the variation of orbital parameters, a method that had been introduced by Euler. In this technique, a body was considered as moving on a Keplerian orbit defined by six orbital elements which were varying slowly. Differential equations for the elements need to be derived, and the procedures Lagrange employed for deriving them evolved over many years and appeared in many forms in his work. Lagrange's analyses formed the basis of Laplace's early investigations into celestial mechanics.

To give some idea of the principles involved, it will be helpful to examine briefly some of the main features of the method as it might now be presented. Consider two planets, of mass m_α and m_β, in orbit around the Sun. In the absence of any interaction between them, they would move in elliptical orbits defined by the elements $\alpha_1, \ldots, \alpha_6$ and $\beta_1, \ldots, \beta_6$, for example. When the mutual interactions are taken into account, these elements can be shown to vary according to differential equations of the form

$$\begin{aligned}
\dot{\alpha}_i &= m_\beta \phi_i(\alpha_1, \ldots, \alpha_6, \beta_1, \ldots, \beta_6; t) \\
\dot{\beta}_i &= m_\alpha \psi_i(\alpha_1, \ldots, \alpha_6, \beta_1, \ldots, \beta_6; t)
\end{aligned} \qquad i = 1, \ldots, 6. \quad (9.17)$$

Next, we expand α_i and β_i in powers of the masses m_α and m_β, which are small relative to that of the Sun, so that the elements vary slowly. Thus, we write

$$\alpha_i(t) = \sum_{j,k=0}^{\infty} \alpha_i^{jk}(t) m_\alpha^j m_\beta^k, \qquad \beta_i(t) = \sum_{j,k=0}^{\infty} \beta_i^{jk}(t) m_\alpha^j m_\beta^k,$$

where $\alpha_i^{jk}(t)$ and $\beta_i^{jk}(t)$ are functions of time to be determined. These expressions then can be substituted into Eqn (9.17), and the coefficients of like powers of m_α and m_β equated. We obtain

$$\dot{\alpha}_i^{j0} = \dot{\beta}_i^{0j} = 0, \qquad \text{for any } j,$$
$$\dot{\alpha}_i^{01} = \phi_i\left(\alpha_1^{00}, \ldots, \alpha_6^{00}, \beta_1^{00}, \ldots, \beta_6^{00}; t\right),$$
$$\dot{\beta}_i^{10} = \psi_i\left(\alpha_1^{00}, \ldots, \alpha_6^{00}, \beta_1^{00}, \ldots, \beta_6^{00}; t\right),$$

and so on. To carry the analysis up to the first order in the planetary masses, therefore, we can integrate simply the right-hand sides of the equations in (9.17), treating the elements as constants. Moreover, the problems for the two planets decouple, and when considering the perturbations of the planet m_α one only needs to consider the Keplerian orbit of the perturbing planet, m_β.

The six elements of a Keplerian orbit can be taken as

a	semi-major axis	Ω	longitude of ascending node
e	eccentricity	$\bar{\omega}$	longitude of perihelion
i	inclination	τ	time of perihelion passage

where a and e determine the shape of the orbit, i, Ω, and $\bar{\omega}$ fix the orientation of the orbit in space, and τ determines the position of a planet on its orbit given the actual time t. Before Laplace, it was customary to use the aphelion as the reference for the line of apsides, but Laplace realized that, in order to allow comets to fit into the same theory, it was more appropriate to use the perihelion. An alternative to τ is to use $\epsilon = \bar{\omega} - n\tau$ for the sixth element, where n is the mean motion. This is known as the 'mean longitude at epoch' and, since the mean longitude is simply $\bar{\omega} + n(t - \tau)$, ϵ is the mean longitude of the planet at $t = 0$.

Trigonometric series can be used to express the differential equation for an element of the planet m_α (to the first order of the planetary masses) in the form

$$\dot{\alpha} = \sum_{j,k} P_{jk} \cos(j n_\alpha t + k n_\beta t + \Lambda_{jk}),$$

in which j and k are any integers, the coefficients P_{jk} depend on the elements a, e, and i for both planets, and Λ_{jk} (a complicated function of the elements) is independent of t. The quantities n_α and n_β are the mean motions of the two planets (assumed constant).

When this is integrated, two types of term will result, as well as the constant of integration. First, if $j n_\alpha + k n_\beta = 0$ (and in particular if $j = k = 0$) the term in the sum is independent of t and we will get a term proportional to t in the solution. Otherwise, periodic terms will result and so the final solution will be of the form

$$\alpha = c_0 + c_1 t + \sum_{j,k} P_{jk} \frac{\sin(jn_\alpha t + kn_\beta t + \Lambda_{jk})}{jn_\alpha + kn_\beta}, \qquad (9.18)$$

where it is understood that pairs of integers for which the denominator in the final term vanishes are not included in the summation. The magnitude of the coefficients P_{jk} decreases rapidly as $|j + k|$ increases and, since n_α and n_β can never be known exactly anyway, it is reasonable to assume that $jn_\alpha + kn_\beta$ is never zero except when $j = k = 0$. The coefficient of the term proportional to t is thus simply P_{00}, and this term is what Lagrange and Laplace referred to as the secular variation in the element. Since the coefficients P_{jk} depend only on the elements a, e, and i for the two planets, the secular variation is a function of the orbits but not of the actual positions of the planets.

As far as Laplace was concerned, the successful determination of the secular inequalities affecting planetary motion was the key to progress, and the first major result he presented concerned the secular variations in the elements of a planet's orbit (attempts by Euler and Lagrange in this direction had led to contradictory results). Laplace focused his attention initially on the variation of the semi-major axis, which is related to the mean motion through Newton's modification of Kepler's third law Eqn (8.9). In a detailed analysis, he expanded the coefficient P_{00} up to third order in the eccentricities and inclinations of the orbits and showed that, to this degree of approximation, $P_{00} = 0$. Laplace thus concluded that the mutual interaction between two planets could not result in an acceleration in the mean motion of either of them which was independent of their relative position and, hence, the semi-major axes of the orbits could not be subject to a continual increase or decrease.[76] Laplace suggested that perhaps the observed anomalies in the mean motions of Jupiter and Saturn were due to interactions with comets. What he did not think of at this time was that they might be oscillations, dependent on the positions of the planets, with very long periods.

The situation with the other elements was different. It had been accepted widely since Kepler's *Rudolphine Tables* that the nodal and apsidal lines of planetary orbits were subject to secular variation, and Euler had concluded from his perturbation analysis that both these elements and the inclinations and eccentricities of the orbits changed over long periods of time. In other words, it was to be expected that the coefficient c_1 in the solution for these elements was nonzero. However, in 1774, Lagrange presented a memoir in which he

---

[76] Lagrange subsequently produced a proof that $c_1 = 0$, which was not restricted to small values of the eccentricities and inclinations (see Wilson (1985), pp. 198–205).

showed – much as he had done earlier for the problem of 'arcs of circles'– that the secular inequalities actually were periodic, the terms proportional to t in the solution simply representing developments of a periodic function in the eccentricities and inclinations of the orbits. As was so often the case, the key step involved choosing the most appropriate independent variables for the problem under consideration, and Lagrange managed to derive first-order linear differential equations for the secular variations of the nodes and inclinations. For example, in the case of Jupiter and Saturn, he found that the period of the oscillations in both the nodes and inclinations was 51 150 years. Both Lagrange and Laplace realized that the same approach could be brought to bear on the secular variations in eccentricity and aphelion, with the same conclusions. Thus, it appeared that all apparently secular phenomena caused by the mutual interactions of two planets were, in fact, slow oscillations.

The fact that mutual interaction between planets appeared to lead to oscillations in the semi-major axes, inclinations and eccentricities of their orbits, appeared to suggest that the Solar System was a stable entity, in that the positions of planets would remain within certain fixed limits for all time. However, actual calculations for specific planets required knowledge of their relative masses and, except for those planets with satellites, these were unknown. From the mid 1770s onwards, the search for a 'proof' of the stability of the Solar System became one of the main objectives of celestial mechanics. Lagrange continued to devise improvements to the techniques by which planetary perturbations could be determined, but he remained frustrated in his attempts to resolve the great inequality of Jupiter and Saturn. This resolution was achieved by Laplace who, following 10 years of silence on the matter, announced his findings in 1785.[77]

All the previous results on the Jupiter–Saturn problem suggested that the changes that were observed in the mean motions of these two planets were not caused by their mutual interaction. However, Laplace came to realize that, notwithstanding all the previous work, gravitational interaction had to be the cause. From the conservation of energy relation Eqn (9.16), and on the basis that the planetary masses are much smaller than that of the Sun, he derived the result

$$\frac{m}{a} + \frac{m'}{a'} + \frac{m''}{a''} + \cdots = f,$$

where f is a constant and m, m', m'', ... and a, a, a'', ... are, respectively, the masses and semi-major axes of the planets. The contributions from Jupiter and Saturn were much larger than from any of the other planets and thus a

---

[77] The full theory followed a few months later in *La théorie de Jupiter et de Saturne* (1786).

small increase in the mean motion of Jupiter, δn, would lead to a corresponding decrease in that of Saturn, $\delta n'$, and vice versa. Numerical calculations revealed that

$$\delta n' \approx -2.33 \, \delta n,$$

so that Saturn should be decelerating at about $\frac{7}{3}$ times the rate at which Jupiter was accelerating, and this appeared to agree with observation:

> It is therefore very probable that the observed variations in the movements of Jupiter and Saturn are an effect of their mutual action, and since it is established that this action can produce no inequality that either increases constantly or is of very long period and independent of the situation of the planets, and since it can only cause inequalities dependent on their mutual configuration, it is natural to think that there exists in their theory a considerable inequality of this kind, of which the period is very long.[78]

Being convinced of the cause was one thing, but finding the effect was another. Laplace took a lead from Lagrange's work in which he had shown that, in the interaction between two orbiting bodies, the effect of small divisors would lead to problems if the mean motions of the bodies, n and n' say, were commensurable, i.e. if $jn = kn'$ for some integers j and k. Now, in the case of Jupiter and Saturn it happens that $2n \approx 5n'$ and so Laplace looked for terms in the analysis that contained the factor $\sin(5n' - 2n)t$, even though he knew that these would be multiplied by cubes of small quantities (since $5 - 2 = 3$). With hindsight, knowing that the period of such an oscillation is $360°/(5n' - 2n) \approx$ 900 years, and that the sought-after inequality is periodic with a period of about 900 years, this is a fairly obvious thing to do, but Laplace had no prior knowledge of the form of what actually he was looking for. However, once he had hit upon the right idea it did not take him long to resolve the issue. Showing great technical ability in following a single aim through a mass of complex calculations, Laplace obtained $20'$ for the amplitude of the inequality of Jupiter in longitude and $46' \, 50''$ for that of Saturn, values he subsequently improved upon but which already compare well with the more accurate values given on p. 310 above.

The familiar pattern had repeated itself once more. An observed phenomenon had appeared out of line with Newton's theory of gravitation, but the fault had, as in all previous cases, turned out to be due to the incomplete understanding of the differential equations describing the motion rather than with the equations themselves. Laplace later wrote:

---

[78] LAPLACE *La mémoire sur les inégalités séculaires des planètes et des satellites* (1785). Quoted from Wilson (1985), p. 229.

The irregularities of the two planets appeared formerly to be inexplicable by the law of universal gravitation – they now form one of its most striking proofs. Such has been the fate of [Newton's] brilliant discovery, that each difficulty which has arisen has become for it a new triumph, a circumstance which is the surest characteristic of the true system of nature.[79]

Laplace's successful resolution of the cause of the great inequality of Jupiter and Saturn was a great success for perturbation theory, but it also served to highlight the method's intrinsic limitations. The only way to test whether enough terms had been taken in a perturbation expansion was to compare the results with observation. There was no way mathematically to bound the error, and one could not rule out the possibility that the near commensurability of the mean motions of two bodies might appear in the neglected terms and lead to a large effect.

One phenomenon now remained unexplained, and this was the secular acceleration of the Moon, which Euler had become convinced must be due to some form of ethereal resistance. Such an effect would cause the Moon to lose energy and, hence, fall toward the Sun. As a consequence, its period would decrease (due to Kepler's third law) and thus its mean motion would increase. Lagrange, for one, was unconvinced by this explanation, since most other planets did not appear to be accelerating and, in the case of Saturn in the early 1770s, the view was that the planet was decelerating.

As we have seen, Lagrange's analyses of planetary perturbations suggested that what appeared to be a unidirectional change might well be part of a very long period oscillation. Thus, instead of trying to isolate expressions for the longitude of the Moon proportional to $t + \mu t^2$, where μ is some small constant, perhaps one should be looking for terms like $\sin \mu t$. In a prize-winning memoir of 1774,[80] Lagrange showed that terms like this could not arise from the oblate shape of the Earth and, 6 years later,[81] also demonstrated that the non-spherical shape of the Moon could not lead to an acceleration of the desired magnitude. He dared even to suggest that the phenomenon might not exist, and this had the effect (which may well have been Lagrange's intention) of stirring astronomers into producing more hard evidence to back up their claims.

In his early twenties, Laplace had examined universal gravitation and asked himself why the theory failed to predict all observed phenomena. He did not doubt that a universal inverse square attractive force existed, but suggested

---

[79] Laplace *Celestial Mechanics*, Book IV, Chapter III. Quoted from Pannekoek (1961), p. 302.

[80] The Paris Academy prize contest for 1774 had as its subject the question of whether it was possible to explain the secular acceleration of the Moon by the attraction of all the celestial bodies or by the non-sphericity of the Earth and Moon?

[81] At the end of his second memoir on the libration of the moon (see p. 321).

that this force might not be propagated instantaneously and might depend in some way on the relative velocities of interacting particles.[82] He concluded that the observed secular acceleration of the Moon would result if gravity was propagated at a speed about 7 million times the speed of light. This was not sufficient evidence for him to advocate a finite speed of gravitational attraction, but if the acceleration of the Moon failed to yield to any other line of enquiry, this would offer a possible solution.

After his success with the great inequality of Jupiter and Saturn, Laplace turned his attention to the lunar problem. Again, he found the reason for the phenomenon, and demonstrated that universal gravitation was, indeed, able to cause the observed lunar acceleration, thus removing the need to hypothesize a resisting medium. The cause of the troublesome behaviour of the Moon turned out to be rather indirect. In 1787, Laplace showed[83] that the planets cause the eccentricity of the orbit of the Earth to undergo an oscillation of very long period, with the semi-major axis essentially constant. At the present time, the eccentricity is diminishing slowly, which in turn implies that the mean distance of the Earth–Moon system from the Sun is increasing. This reduces the perturbative effect of the Sun on the Moon, which causes the Moon to accelerate. The effect eventually will be reversed, and the Moon will begin to decelerate, but not for a few hundred thousand years! Pleased as he no doubt was with this result, Laplace was surprised at the size of the effect on the lunar longitude, since the more direct effect on the orbit of the Earth was an order of magnitude smaller.

Laplace's calculated value of just over $10''$ per century per century for the coefficient of t^2 in the expression for the longitude[84] was about 15 per cent too small, and so in his lunar theory he took the amplitude of this long-period variation from observational data. Nevertheless, for astronomers at the end of the eighteenth century, Laplace's achievement was an unmitigated success. However, in the 1850s, John Couch Adams extended Laplace's analysis by including higher-order terms and showed that actually this made the discrepancy between theory and observation greater. The effect of the change in the eccentricity of the Earth due to planetary action, in fact, only accounts for half of the observed acceleration and, in the 1860s, Charles Delaunay suggested that perhaps tidal friction was slowing the rotation of the Earth, and thus that the standard

---

[82] Daniel Bernoulli, in a prize-winning memoir on the tides (1740), previously had suggested that the attractive effect of the Moon might take as much as 2 days to reach the Earth.

[83] Laplace's memoir had the same title as Lagrange had used for his 1774 essay: *Sur l'équation séculaire de la lune*.

[84] Many authors quote this as the value for the secular acceleration though, more correctly, the acceleration is double this value (see Whittaker (1953), p. 147).

reference for time was a changing quantity, a hypothesis confirmed in the twentieth century with the advent of atomic clocks.[85]

Laplace's final lunar theory, which took account of secular changes in all its elements, not just the mean motion, brought the difference between theory and observation to less than 30 arc seconds. The motion of the Moon is affected by the non-spherical shape of the Earth, and Laplace showed how to work back from the observations to predict the oblateness of the Earth (he obtained 1/305). Another irregularity in the motion of the Moon (the parallactic inequality) depends on the Earth–Sun distance and, again working backwards from observation, Laplace determined a value of 8.6″ for the solar parallax. This use of theory to deduce fundamental parameters of celestial mechanics from standard observations was novel and unexpected:

> It is very remarkable that an astronomer, without leaving his observatory, by merely comparing his observations with analysis, may be enabled to determine with accuracy the magnitude and flattening of the earth, and its distance from the sun and moon, elements the knowledge of which has been the fruit of long and troublesome voyages in both hemispheres. The agreement between the results of the two methods is one of the most striking proofs of universal gravitation.[86]

Variation of orbital parameters

The theoretical side of the study of planetary perturbations in the eighteenth century was dominated by the method of variation of orbital parameters, which evolved over half a century beginning with Euler's initial tentative steps.[87] The final, polished, theory appeared in the second edition of Lagrange's *Analytical Mechanics*, which contained applications to celestial mechanics of the techniques the author had developed.[88] Lagrange's achievement was to reduce mechanics to the solution of differential equations. As he wrote in the

---

[85] Since friction is always a factor in tidal motion, tides act as a source of dissipation. In the case of the Earth–Moon system, the result is a gradual lengthening of both the day and the month. For a general description of this phenomenon, see Darwin (1962), Chapter 16, while for the mathematical theory, see Murray and Dermott (1999), Chapter 4.

[86] LAPLACE *The System of the World*, Book IV, Chapter V. Quoted from Pannekoek (1961), p. 306.

[87] Two of those who helped develop the technique were Euler's sons, Johann Albrecht and Charles, though the latter's work was ignored by or unknown to Lagrange (see Suzuki (1996) for details).

[88] The *Analytical Mechanics* was first published in 1788, though we know from a letter written to Laplace that it was essentially complete by 1782 (Fraser 1985b). As early as 1775, Lagrange had written to d'Alembert telling him that he had developed a complete theory of the variations of the elements of the planets resulting from their mutual action. Volume 1 of the second edition was published in 1811 and Volume 2 posthumously in 1816. The scope of the work is described in Dugas (1988), Part III, Chapter 11.

preface:

> One will find no diagrams in this work. The methods that I expound require neither geometrical nor mechanical constructions or reasonings, but only algebraic operations ordered in a regular and uniform development. Those who love analysis will be pleased to see mechanics become a new branch of it, and will be grateful to me for having thus extended its domain.[89]

The variation of orbital parameters as formulated by Lagrange was the starting point for much of the work done on celestial mechanics in the nineteenth century. Consider a perturbed two-body problem given by

$$\ddot{\mathbf{r}} + \frac{\mu}{r^3}\mathbf{r} = \nabla R. \tag{9.19}$$

If the right-hand side were zero, the solution would be known:

$$\mathbf{r} = \mathbf{q}(\alpha_1, \ldots, \alpha_6; t), \qquad \dot{\mathbf{r}} = \mathbf{p}(\alpha_1, \ldots, \alpha_6; t),$$

for example, where we have written explicitly the dependence of the solution on six orbital parameters, corresponding to the constants of integration for the problem. We then look for a solution to Eqn (9.19) in the form

$$\mathbf{r} = \mathbf{q}(\alpha_1(t), \ldots, \alpha_6(t); t), \qquad \dot{\mathbf{r}} = \mathbf{p}(\alpha_1(t), \ldots, \alpha_6(t); t),$$

from which

$$\dot{\mathbf{r}} = \frac{\partial \mathbf{q}}{\partial t} + \sum_{i=1}^{6} \frac{\partial \mathbf{q}}{\partial \alpha_i}\dot{\alpha}_i = \mathbf{p} + \sum_{i=1}^{6} \frac{\partial \mathbf{q}}{\partial \alpha_i}\dot{\alpha}_i. \tag{9.20}$$

Equation (9.19) is not sufficient to determine the six functions $\alpha_i(t)$ uniquely, and we thus choose to set

$$\sum_{i=1}^{6} \frac{\partial \mathbf{q}}{\partial \alpha_i}\dot{\alpha}_i = 0. \tag{9.21}$$

This particular choice for the three extra equations required has the advantage that, in the perturbed motion, both $\mathbf{r}$ and $\dot{\mathbf{r}}$ at time t will be given by the formulas for $\mathbf{r}$ and $\dot{\mathbf{r}}$ obtained from the two-body problem using the instantaneous values of the functions $\alpha_i(t)$.

If we differentiate Eqn (9.20) with respect to t, we obtain

$$\ddot{\mathbf{r}} = \dot{\mathbf{p}} = \frac{\partial^2 \mathbf{q}}{\partial t^2} + \sum_{i=1}^{6} \frac{\partial \mathbf{p}}{\partial \alpha_i}\dot{\alpha}_i = -\frac{\mu}{|\mathbf{q}|^3}\mathbf{q} + \nabla R$$

because of Eqn (9.19) and, since $\mathbf{q}$ is the solution to the two-body problem, it

[89] Quoted from Wilson (1995b).

follows that

$$\sum_{i=1}^{6} \frac{\partial \mathbf{p}}{\partial \alpha_i} \dot{\alpha}_i = \nabla R. \qquad (9.22)$$

Equations (9.21) and (9.22) represent six linear algebraic equations for the functions $\dot{\alpha}_i(t)$. Considered as six first-order differential equations for the position and velocity, they are equivalent to the three second-order Eqns (9.19).

In order to write these equations in a more elegant form, we dot Eqn (9.21) with $-\partial \mathbf{p}/\partial \alpha_j$, and Eqn (9.22) with $\partial \mathbf{q}/\partial \alpha_j$, and add the resulting equations. Since R is a function of position only, we obtain

$$\sum_{i=1}^{6} (\alpha_j, \alpha_i) \dot{\alpha}_i = \frac{\partial R}{\partial \alpha_j}, \qquad j = 1, \ldots 6, \qquad (9.23)$$

where

$$(\alpha_j, \alpha_i) = \frac{\partial \mathbf{q}}{\partial \alpha_j} \cdot \frac{\partial \mathbf{p}}{\partial \alpha_i} - \frac{\partial \mathbf{p}}{\partial \alpha_j} \cdot \frac{\partial \mathbf{q}}{\partial \alpha_i} = -(\alpha_i, \alpha_j).$$

The quantities (α_j, α_i) are known as 'Lagrange brackets' and Lagrange showed that

$$\frac{\partial}{\partial t}(\alpha_j, \alpha_i) = 0,$$

so that t does not enter explicitly into the expression for (α_j, α_i) but only appears in Eqn (9.23) through the terms on the right-hand side.[90] Finally, the six linear Eqns (9.23) can be inverted to produce equations for $\dot{\alpha}_i$ of the form of Eqn (9.17). With the six elements a, e, i, Ω, $\bar{\omega}$, and ϵ, these equations, known as 'Lagrange's planetary equations', are

$$\frac{da}{dt} = \frac{2}{na} \frac{\partial R}{\partial \epsilon}, \qquad \frac{de}{dt} = -\frac{q(1-q)}{na^2 e} \frac{\partial R}{\partial \epsilon} - \frac{q}{na^2 e} \frac{\partial R}{\partial \bar{\omega}},$$

$$\frac{di}{dt} = -\frac{\tan \frac{1}{2}i}{na^2 q} \left(\frac{\partial R}{\partial \epsilon} + \frac{\partial R}{\partial \bar{\omega}} \right) - \frac{1}{na^2 q \sin i} \frac{\partial R}{\partial \Omega},$$

$$\frac{d\Omega}{dt} = \frac{1}{na^2 q \sin i} \frac{\partial R}{\partial i}, \qquad \frac{d\bar{\omega}}{dt} = \frac{q}{na^2 e} \frac{\partial R}{\partial e} + \frac{\tan \frac{1}{2}i}{na^2 q} \frac{\partial R}{\partial i}, \qquad (9.24)$$

$$\frac{d\epsilon}{dt} = -\frac{2}{na} \frac{\partial R}{\partial a} + \frac{q(1-q)}{na^2 e} \frac{\partial R}{\partial e} + \frac{\tan \frac{1}{2}i}{na^2 q} \frac{\partial R}{\partial i},$$

in which $q = \sqrt{1 - e^2}$.

---

[90] The Lagrange brackets can be evaluated readily (see, for example, Brouwer and Clemence (1961), Chapter XI).

The shape of the Earth and its effect

During the decade 1775 to 1785, Laplace devoted little time to planetary perturbations. Instead, he worked on a mixture of purely mathematical topics, issues
related to chemical physics, and problems in mechanics. In the latter context,
he wrote on such things as the tides, the determination of the orbits of comets,
and the effect of the shape of the Earth on its gravitational attraction.

As we have seen, the shape of the Earth had been used to argue both for
and against universal gravitation in the early eighteenth century. Some measurements had suggested a sphere flattened at the poles in accordance with
Newton's theory, while others had indicated a prolate spheroidal shape. By
1740, it had been established that Newton was right, though the value of the
ellipticity of the Earth remained a subject of debate. Whereas Newton had assumed an oblate spheroidal shape, it was Colin Maclaurin in his *Treatise of
Fluxions* (1742) who first proved that this was an equilibrium shape for a rotating homogeneous fluid. In the second half of the eighteenth century, the shape
of the Earth was still of interest to astronomers, but for a different reason. The
nonsphericity of the Earth had consequences such as precession and nutation,
and it was important to be able to quantify these accurately. Attention thus
turned to the effects of the shape of the Earth on its gravitational attraction.[91]

The first major theoretical advance since the *Principia* in the study of the
effects of a nonspherical Earth was Clairaut's *La théorie de la figure de la terre*,
published in 1743.[92] Newton had assumed in his calculations that the Earth was
homogeneous, and had made erroneous claims about the effect that a variable
density would have. Clairaut showed that, for a shape built up from ellipsoidal
layers of different densities, the acceleration due to gravity g at a latitude β
would be related to that at the equator g_0 through the equation

$$g - g_0 = g_0 \left(\tfrac{5}{2}\lambda - \varepsilon\right) \sin^2 \beta,$$

where λ is the ratio of the centrifugal to the gravitational force at the equator,
and ε is the ellipticity (assumed small) of the Earth. This result is known as
'Clairaut's formula'.[93]

---

[91] The investigations into the attraction of non-spherical bodies from Newton to the mid
nineteenth century are described in Todhunter (1962).

[92] The transition from Newton's ideas, as enunciated in the *Principia*, to Clairaut's mature theory
is described in Greenberg (1995).

[93] G. G. Stokes showed in 1849 that this formula is correct whatever the internal constitution of
the Earth, provided it is an ellipsoid with small ellipticity (Rouse Ball (1908)). Clairaut's
derivation is described in Todhunter (1962), Chapter XI, and a derivation independent of the
density variation within the Earth can be found in Roy (1978), Section 10.2.2. If the Earth is
homogeneous, then we know from the work of Newton that $\lambda = 4\varepsilon/5$, and we obtain
$g = g_0(1 + \varepsilon \sin^2 \beta)$ in agreement with his *Principia*, Book III, Proposition 20.

Investigations into the shape of the Earth and the effect of its shape continued throughout the eighteenth century, with notable contributions from d'Alembert, but it was the combined efforts of Laplace and another great French mathematician – Adrien-Marie Legendre – that produced the next great advance. In 1784, Laplace published a treatise in which he developed Lagrange's idea of a potential to the attraction of a spheroid.[94]

Suppose we have a body V, the density of which at a point (x', y', z') with position vector $\mathbf{r}'$ is $\rho(\mathbf{r}')$. Then the gravitational acceleration induced at a point $\mathbf{r} = (x, y, z)$ (assumed to be outside the body) by a small volume element of the body $\delta V = \delta x' \delta y' \delta z'$, can be written

$$G\rho(\mathbf{r}')\frac{\mathbf{r}' - \mathbf{r}}{|\mathbf{r}' - \mathbf{r}|^3}\, \delta V = \nabla \left(\frac{1}{|\mathbf{r}' - \mathbf{r}|} \right) G\rho(\mathbf{r}')\, \delta V,$$

where $\nabla \equiv \left(\mathbf{i}\frac{\partial}{\partial x} + \mathbf{j}\frac{\partial}{\partial y} + \mathbf{k}\frac{\partial}{\partial z} \right)$. It follows that the gravitational acceleration due to the whole body is ∇U, where

$$U = G \int_V \frac{\rho(\mathbf{r}')}{|\mathbf{r}' - \mathbf{r}|}\, \mathrm{d}V. \tag{9.25}$$

Following the work of George Green[95] U is known as a 'potential function'. Laplace also was able to show that the same function U served as a potential function when the point (x, y, z) was inside the body, though in this case the argument is rather more subtle.

In 1785, Laplace went further and showed that, provided (x, y, z) is outside V, the function U satisfies

$$\nabla^2 U = 0,$$

where ∇^2 is what is now known as the 'Laplacian operator'. In Cartesian, coordinates, this has the form $\partial^2/\partial x^2 + \partial^2/\partial y^2 + \partial^2/\partial z^2$, although in Laplace's initial derivation of the equation that now bears his name, he used spherical polar coordinates.[96]

Knowledge of the potential function U is all that is required in order to be able to calculate the gravitational attraction at any point due to the body V,

---

[94] The first part of *La théorie du mouvement et de la figure elliptique des planètes* was a textbook of basic celestial mechanics; the second part was on the gravitational attraction of spheroids. For some reason, this book, his first, was not included in either of the two collections of his works (Gillispie (1997)).

[95] *An Essay on the Application of Mathematical Analysis to the Theories of Electricity and Magnetism* (1828), reprinted in Ferrers (1970). More on the history of the potential function can be found in Grattan-Guiness (1995).

[96] In 1813, Poisson showed that if (x, y, z) is inside V, U satisfies what we now refer to as Poisson's equation $\nabla^2 U = -4\pi\rho G$.

and a great deal of information about U can be obtained from the fact that it satisfies Laplace's equation in the region exterior to the body. However, one idea facilitated the use of this function in celestial mechanics more than any other, and that was the introduction by Legendre of a particularly useful series expansion for U. The quantity that appears in the denominator of the integrand in Eqn (9.25) is

$$|\mathbf{r}' - \mathbf{r}| = r(1 - 2\mu\alpha + \alpha^2)^{1/2},$$

where $\mu = \cos\theta$ (θ being the angle between the vectors $\mathbf{r}$ and $\mathbf{r}'$) and $\alpha = r'/r$, which we assume to be less than 1. This restriction implies that the expressions derived below are valid only in the region outside the sphere that circumscribes the body V.

Since $\alpha < 1$, we can expand $(1 - 2\mu\alpha + \alpha^2)^{-1/2}$ as

$$(1 - 2\mu\alpha + \alpha^2)^{-1/2} = \sum_{n=0}^{\infty} P_n(\mu)\alpha^n$$

for some functions P_n. Expressions for these functions can be obtained by expanding the left-hand side using the binomial theorem and equating coefficients of α^n. It turns out that P_n is a polynomial of degree n with $|P_n(\mu)| \leq 1$. The functions P_n are now known as 'Legendre polynomials', though they were referred to as 'Laplace's coefficients' for much of the nineteenth century on account of the great use Laplace made of them. If we incorporate this series into the definition of U, we obtain

$$U = G \sum_{n=0}^{\infty} \frac{U_n}{r^{n+1}},$$

where

$$U_n = \int_V \rho(\mathbf{r}')P_n(\mu)r'^n \, dV$$

of which, since $P_0(\mu) \equiv 1$, the first term is $U_0 = GM/r$, where M is the mass of the body. If the origin is taken as the centre of gravity of V, then it turns out that $U_1 = 0$ and, in fact, for a homogenous ellipsoid, it can be shown that $U_n = 0$ for all odd n.[97]

---

[97] For an ellipsoid, James MacCullagh, professor at Trinity College, Dublin, showed that $U_2 = \frac{1}{2}(A + B + C - 3I)$, where A, B and C are the moments of inertia about the coordinate axes (chosen to be the principal axes) and I is the moment of inertia about a line joining O to the point (x, y, z). This is known as MacCullagh's formula (see, for example, Battin (1987), pp. 402–3).

Laplace preferred to work with spherical polar coordinates, for which $x = r \sin\theta \cos\phi$, $y = r \sin\theta \sin\phi$, and $z = r \cos\theta$ (and similarly for the primed variables). Then, $\mu = \cos\theta \cos\theta' + \sin\theta \sin\theta' \cos(\phi - \phi')$, and the functions U_n defined above can be thought of as functions of the two variables θ and ϕ. Laplace showed that $U_n(\theta, \phi)$ satisfies

$$\frac{1}{\sin\theta} \frac{d}{d\theta} \left(\sin\theta \frac{dU_n}{d\theta} \right) + \frac{1}{\sin^2\theta} \frac{d^2U_n}{d\phi^2} + n(n+1)U_n = 0,$$

and the name 'Laplace function of the nth order' was given to any function that satisfied this differential equation. Nowadays, such functions are called 'spherical harmonics', and Laplace demonstrated how they could be used as the basis for an expansion of any function defined on the surface of a sphere. For a rotationally symmetric body, the dependence on ϕ is absent, and Laplace showed that the potential for such a body can be expanded in the form

$$U = \frac{GM}{r} \left(1 + \sum_{n=2}^{\infty} A_n \left(\frac{R}{r} \right)^n P_n(\cos\theta) \right),$$

where R is the equatorial radius. For a body with symmetry about the equatorial plane ($\theta = \pi/2$), the coefficients A_n vanish for all odd n.[98] With the aid of this general theory, Laplace could express the gravitational attraction of a spheroid which differed little from a sphere in the form of a rapidly convergent series.

Probability and statistics

One of the features of astronomy in the latter half of the eighteenth century was the use of ideas from the emerging theory of probability. This theory grew out of work on games of chance, beginning in the mid seventeenth century with the famous correspondence between the French mathematicians Blaise Pascal and Pierre de Fermat, and was developed in the eighteenth century to include ideas of statistical inference.[99] It was Laplace who broadened the discipline to include important topics such as the estimation of error and the quantification of the credibility of evidence and, in so doing, created the subject of mathematical

---

[98] From modern observations of the motion of artificial satellites it has been found that for the Earth, A_3 is not zero and, hence, that the Earth has a non-negligible pear-shape distortion. An interesting discussion of how the motion of satellites can be used to determine the coefficients A_n can be found in King-Hele (1972).

[99] Details of this correspondence can be found in, for example, Sheynin (1977a). The development of probability theory in the eighteenth century is described in detail in, for example, Daston (1988) and Hald (1998).

statistics. In his *Philosophical Essay on Probabilities* (1814)[100] Laplace emphasized the importance of the theory of probability to natural philosophy and gave a general description of his thoughts on how to determine an appropriate 'average' value from a series of astronomical observations so as to minimize the possible error.

The details of this work can be found in his 1774 *Memoir on the Probability of the Causes of Events*,[101] in which Laplace considered three observations registering the time of an astronomical event as t_1, t_2, and t_3, with $t_1 < t_2 < t_3$. First, he observed that

> it is more probable that a given observation deviates from the truth by 2 seconds than by 3 seconds, by 3 seconds than by 4 seconds, etc. The law by which this likelihood diminishes as the difference between the observation and the truth increases is unknown, however.[102]

To determine this law, Laplace considered probability as a function of error in the form of a density $\phi(x)$ (which means that the probability of the error lying between x and $x + dx$ is $\phi(x)\,dx$). He noted that $\phi(x)$ should be symmetric about $x = 0$, since it is equally likely that the error is positive or negative; $\phi(x)$ should approach zero as x increases, since the probability that the error is infinitely large is infinitely small; and $\int_{-\infty}^{\infty} \phi(x)\,dx = 1$, since the total probability of all possible events is, by definition, unity. Laplace argued that, not only must ϕ decrease as x increases, so must $d\phi/dx$, and in the absence of any reason for a different rate of decrease for these two functions, he proposed that $\phi \propto d\phi/dx$ which, given the other conditions that ϕ had to satisfy, leads to

$$\phi(x) = \tfrac{1}{2}m e^{-m|x|},$$

where m is a positive constant. This error distribution is now called the 'Laplace distribution' or the 'double exponential distribution'.

Now Laplace was faced with a problem in inverse probability. The function ϕ enabled him to compute the probability that a particular set of observations is made given knowledge of the true time of the event, but what is required is the probability that the true time of the event lies in a given range given that the set of observations is t_1, t_2, and t_3. Here he invoked his 'fundamental principle':

> If an event can be produced by a number n of different causes, the probabilities of these causes given the event are to each other as the probabilities of the event given the causes, and the probability of the existence of each of these is equal to the

---

[100] *L' essai philosophique sur les probabilités*. English translation in Laplace (1951).
[101] *La memoire sur la probabilité des causes par les événements* (see, for example, Sheynin (1977b) and Hald (1998). Stigler (1986) contains a complete translation.
[102] Quoted from Stigler (1986).

probability of the event given that cause, divided by the sum of all the probabilities of the event given each of these causes.[103]

In modern notation, with E representing the event, C_i, $i = 1, \ldots, n$, the (mutually exclusive and exhaustive) causes, and $P(A|B)$ the probability of A given B, Laplace's principle is equivalent to the equations

$$\frac{P(C_i|E)}{P(C_j|E)} = \frac{P(E|C_i)}{P(E|C_j)}, \qquad P(C_i|E) = \frac{P(E|C_i)}{\sum_{j=1}^{n} P(E|C_j)},$$

the second equation following from the first, since $\sum_{i=1}^{n} P(C_i|E) = 1$.[104] Laplace assumed that the same principle held in the case of a set of observations of a continuously varying parameter, so he could conclude that the probability of the true time of an event lying between τ and $\tau + d\tau$, given the observations t_1, t_2, and t_3, was $f(\tau) d\tau$, where

$$f(\tau) \propto \phi(\tau - t_1)\phi(\tau - t_2)\phi(\tau - t_3).$$

What criterion should be used to determine τ? Laplace considered two possibilities. In the first place, we could find the time such that it is equally probable that the event happened before or after it (which Laplace called the 'mean of probability'), or we could seek the time that minimizes the sum of the errors to be incurred multiplied by their probabilities (which he called the 'mean of error' or 'astronomical mean'). The first choice corresponds to finding τ, such that

$$\int_{-\infty}^{\tau} f(u) \, du = \int_{\tau}^{\infty} f(u) \, du,$$

and the second to saying that the minimum of the function

$$F(t) = \int_{-\infty}^{\infty} |u - t| f(u) \, du$$

is at $t = \tau$. Laplace proved that these amount to the same thing (as can be seen readily by evaluating $F'(t)$ and solving $F'(\tau) = 0$). A simple calculation shows that

$$\tau = t_2 + \frac{1}{m} \ln\left(1 + \frac{1}{3}e^{-m(t_2-t_1)} - \frac{1}{3}e^{-m(t_3-t_2)}\right),$$

and Laplace demonstrated that as $m \to 0$, $\tau \to \frac{1}{3}(t_1 + t_2 + t_3)$, i.e. to the arithmetic mean. Laplace's analysis thus suggested that the arithmetic mean, which

---

[103] Quoted from Stigler (1986).
[104] Laplace's principle is thus equivalent to a form of Bayes' theorem. Thomas Bayes' result was published posthumously in 1764, but it would appear that Laplace was ignorant of his work. For more details see Stigler (1978), Dale (1982), and Hald (1998), pp. 159–65.

was commonly used as the best estimate of the true value, is the appropriate choice only if all errors are equally probable, something which is not true. As well as not corresponding to common practice, the formula above is limited in value, since we have no way of knowing what m is. However, Laplace went on to show how to cope with the fact that m is an unknown of the problem, though here he made a mistake that complicated the solution unnecessarily, leading as it did to the need to find the root of a fifteenth-degree equation! From a practical point of view, Laplace's work was a failure since it produced no practical benefit. However, his attempts to use the calculus of probabilities to study the theory of errors was extremely significant. It would be Gauss who eventually would bring theory into line with practice.

Laplace's pioneering work on the theory of errors was not the first application of probability theory to a problem from astronomy. This appears to have been due to Daniel Bernoulli in 1734, in response to the Paris Academy's prize question on the physical cause of the inclinations of the planetary orbits to the plane of the equator. Bernoulli decided to show that the phenomenon under consideration could not simply be a coincidence:

> I will look for the two planetary orbits between which there is the greatest angle in inclination; after which I will calculate what probability there is that all the other orbits will be inclined by chance within the limits of these two. We will see that this probability is so small that it may be considered a moral impossibility.[105]

The orbits with the largest angle between them are those of the Earth and Mercury, the inclination being about $7°$. Bernoulli computed the chance of two great circles on the celestial sphere lying in a zone of width $7°$ as approximately $1/17$ and, hence, the probability that all five planetary orbits lie within this zone is $1/17^5 = 1/1\,419\,857$. Subsequently, he revised his calculations, but not his conclusion: the fact that the orbits of the planets are all in roughly the same plane is not just pure chance.

In 1776, Laplace addressed a similar problem: given n bodies projected randomly into orbits around the Sun, what is the probability that the mean inclination of the orbits to the ecliptic lies between two given values? At the time there were about sixty known comets with a mean inclination of around $46°$, about half going one way round the Sun, and half the other. Laplace concluded from his calculations (in which, for brevity, he included only twelve comets)

---

[105] Quoted from Gower (1987). Bernoulli's use of ideas from probability, and the reactions of others to it, are also discussed in Hald (1998), pp. 68–70. The concept of moral certainty and impossibility had been introduced into probability theory by Daniel's uncle, Jakob Bernoulli, in his *Ars conjectandi*, published posthumously in 1713. For an outcome to be morally certain, its probability had to be greater than 0.999, whereas if the likelihood was less than 0.001, he considered it to be morally impossible.

that it would be unreasonable to assume that whatever caused the planets to orbit in the same direction in roughly the same plane was also the cause of the phenomenon of comets.[106]

The application of statistical ideas to astronomy predates Laplace, also. In his 1748 essay on Jupiter and Saturn, Euler made an important observation: when attempting to match theory and observation it is necessary to determine an unperturbed orbit to serve as a baseline for the perturbations, and this can be done only by using observational data, which inevitably contain errors. Euler's method of minimizing the error involved so-called 'equations of condition', with more equations than there were unknowns, and it introduced the concept of multiple linear regression to astronomy.[107] Euler applied his technique to the determination of the orbital elements for Saturn, but he had little success, because his underlying planetary theory was insufficiently accurate. The first successful use of regression techniques probably was that of Mayer, who wanted to determine accurately the inclination of the equator of the Moon to the plane of its orbit. He utilized twenty-seven observations and derived twenty-seven equations of condition for three unknowns, one of which was the desired inclination. He then separated these twenty-seven equations into three groups of nine according to the value of the coefficient of the inclination, and added the equations in each group together. The resulting three equations then could be solved for the three unknowns.[108] Mayer's method was extremely simple both to understand and to use, and it became popular quickly.

Laplace extended Mayer's approach in his work on Jupiter and Saturn. After deriving a formula for the longitude of Saturn based on perturbation theory, and including the 'great inequality' for the first time, Laplace compared the resulting predictions with twenty-four observations over the period between 1591 and 1785. He obtained twenty-four equations, (1)–(24), for example, for four unknowns: the corrections required to the mean longitude, the mean annual motion, the eccentricity, and the longitude of aphelion. He then reduced these twenty-four equations to four, according to criteria designed to

---

[106] Details of the analysis can be found in Sheynin (1973a), Gillispie (1997), Chapter 5, and Hald (1998), pp. 74–7.

[107] The first mathematician to address the issue of how to determine a set of unknowns from a set of equations of greater number was Roger Cotes, the editor of the second edition of the *Principia*. Suppose observations suggest that an unknown x must satisfy the n equations of condition $a_i x = b_i, i = 1, \ldots, n$ (where for simplicity we assume that all the quantities are positive). One approach is to take the arithmetic mean of the solutions, $x = (1/n) \sum (b_i / a_i)$. However, Cotes states that the equations with the larger values of a_i are more reliable and concludes that one should take a weighted average given by $x = \sum b_i / \sum a_i$ (this is sometimes referred to as the centre of gravity of the observations). Cotes' researches on this topic were published posthumously in 1722 in his *Harmonia mensurarum* (Schmeidler (1995)).

[108] Details of Mayer's method can be found in, for example, Hald (1998), pp. 94–7.

maximize certain coefficients. First, he added them all together. Second, he subtracted the sum of equations (1)–(12) from that of (13)–(24). Third, he took the linear combination $-(1) + (3) + (4) - (7) + (10) + (11) - (14) + (17) + (18) - (20) + (23) + (24)$, and finally he constructed $(2) - (5) - (6) + (8) + (9) - (12) - (13) + (15) + (16) - (19) + (21) + (22)$. Then, he solved these four equations, applied the corrections, and checked the theory against more observations. The results were impressive, with a huge improvement over the values from Halley's 1719 tables.[109] These types of approach became standard practice until superseded by techniques based on the method of least squares in the early nineteenth century.[110]

Delambre took the opportunity afforded by Laplace's analysis to review all the observational data on Jupiter and Saturn, and published his tables of Jupiter and Saturn in 1789. They were included by Lalande in the third edition of his *Astronomy* (1792), the standard manual of practical astronomy. Laplace claimed that these were the first tables based on theory alone, in that the only observational data used were those to determine the constants of integration in the solutions of the differential equations.

The System of the World and the stability of the Solar System

The fabric of French society was torn apart by the Revolution, which began in 1789, and the effects on the scientific establishment were profound. A good deal of antiscientific rhetoric culminated in the suppression of the Academy in 1793, and Laplace left Paris in that year. However, by 1795, the new leaders recognized the value of science, and the Academy was resurrected as part of the new Institut de France. Le Bureau des Longitudes was established, charged with such tasks as calculating and publishing ephemerides and perfecting the theories of celestial mechanics – matters of great importance to the French navy

---

[109] Details of Laplace's method, including probable reasons for his selection of the linear combinations he used, can be found in Stigler (1975), Wilson (1985), p. 274, and Hald (1998), pp. 107–8. The issue of which observations to bring to bear on a calculation was also a thorny one. Since ancient times it had been generally accepted that one should select the 'best' observations (to be determined by some justifiable criteria). Bradley appears to have been the first to suggest forcefully that whole data sets should be used, rather than just a sample (Sheynin (1973b)).

[110] The use of statistical methods in astronomy up to the mid nineteenth century is described in Sheynin (1984).

through their impact on problems of navigation – and Lagrange and Laplace were both appointed to it.[111]

The Bureau also took over the administration of the new metric system that had been introduced in the years since 1791 on the recommendation of a committee on which both Lagrange and Laplace had served. This committee proposed a decimal system, with the unit of length as a 10 millionth of the quarter meridian through Dunkirk and Barcelona.[112] A plan for a universal decimal system embracing not just length, but also weights, time (including the calendar), money, navigation, and many other things, began to emerge. Laplace was a keen supporter of this grand idea. He proposed the names meter, deci-, centi-, milli-, and subdivided a right angle into $100°$ in his subsequent writings. In a lecture in 1795, Laplace put forward a justification for the decimal system and incorporated the substance of his presentation into his hugely influential book *The System of the World* (1796).[113]

Laplace's book is unique among his work on celestial mechanics in that it contains no mathematical formulae and no diagrams. It is, however, full of technical vocabulary, and addresses some of the most complex theoretical problems with which he had had to grapple; thus, it would be slightly misleading to describe the work as a popularization. Nevertheless, it was read by large numbers of the educated public and made known the achievements of the eighteenth century in advancing our understanding of the heavens. The impression readers were given was that all the theoretical challenges posed by the application of universal gravitation to the Solar System had been overcome, and that predictive astronomy was now in the highly satisfactory state in which differences between theory and observation were more than likely due to observational error than any theoretical shortcoming.

The System of the World begins with a description of the phenomena of the heavens, the laws of planetary motion, and the basic principles of mechanics. Then, Laplace described his own contributions to the theory of perturbations – special emphasis being placed on the implications for the stability of the Solar

---

[111] The Bureau des Longitudes and the Paris Observatory which, until 1854, operated virtually as a single organization, had a pervasive influence over French scientific research throughout the nineteenth century (see Davis (1986)).

[112] This required a new survey and it is often said that the reason for this choice of length unit (other ideas, such as the length of a seconds pendulum at a latitude of $45°$ had been suggested) was to get the government to pay for such a survey. However, Gillispie (1997), p. 151 argues that this is probably an unfair assessment.

[113] The French title was *L' exposition du système du monde*, and it went through six editions, the last being published posthumously in 1835. It was translated into many languages, an English translation by John Pond (Astronomer Royal 1811–35) appeared in 1809 (Laplace (1809)).

System – and the final part is devoted mainly to a brief history of astronomy. But it is the very last chapter that has received the most attention, for there Laplace embarked on a flight of fancy and provided a possible explanation for the origin of the Solar System. The fact that all the planets move in the same direction around the Sun and in roughly the same plane demonstrated, on probabilistic grounds, that a single physical process was responsible for the planetary system.

Laplace's theory – which became known as the 'nebular hypothesis' – proposed that a diffuse mass of gas and dust collapsed as it cooled and thus began to spin faster due to angular momentum conservation. Eventually, the centrifugal force at the equator exceeded the force of gravity there and a ring of matter split off, a process that repeated itself a number of times. These rings of matter condensed subsequently under gravity to form the planets and satellite systems.[114] A similar explanation was proposed independently by Immanuel Kant in 1755 in his *Universal Natural History and Theory of the Heavens*.[115] Kant's ideas were different in detail from those of Laplace, but both men accounted for the origin of the planets by an evolutionary development from a diffuse cloud – something that is accepted widely today. The idea that the Universe had evolved from a disordered nebulous mass into its final ordered state fitted in well with the eighteenth-century view of humankind having emerged from the Dark Ages and reached a state of enlightenment. Laplace put it like this:

> Whatever may have been the origin of this arrangement of the planetary system . . . it is certain that its elements are so arranged, that it must possess the greatest stability . . . . It seems that nature has disposed everything in the heavens to ensure the duration of the system . . . , to preserve the individual and insure the perpetuity of the species.[116]

People wanted to believe that the Universe was a stable entity, and Laplace's words made very satisfying reading.

---

[114] The theory went through a number of significant modifications over the course of the five editions of *The System of the World* that were published between 1796 and 1824. Laplace was prone to seek out observational findings that added support to his hypothesis, but to overlook those that presented problems (Jaki (1976)).

[115] *Allgemeine Naturgeschichte und Theorie des Himmels*. An English translation of the first two parts can be found in Kant (1969) while the third part (on possible inhabitants of other planets) is translated in Jaki (1977). Kant did not just consider the Solar System, but hypothesized a process by which the whole Universe has evolved. He identified correctly the Milky Way as a manifestation of our Solar System, being part of a disc-like galaxy, with the great circle on the celestial sphere formed by the Milky Way being for stars what the ecliptic is for planets – an idea he developed from a work by the little-known English astronomer Thomas Wright of Durham (see Paneth (1950), Hoskin (1970)) – and suggested that nebulous stars were examples of distant galaxies.

[116] LAPLACE *The System of the World*, Book V, Chapter VI. Quoted from Laplace (1809).

Laplace's 'proof' of the stability of the Solar System was read to the Academy in 1785 in the same memoir in which the resolution of the great inequality of Jupiter and Saturn was announced.[117] The fact that the elements a, e, and i of an orbit remained bounded due to the mutual interaction of two planets was by then well established, but what about the Solar System as a whole? To the first order in the planetary masses, the conservation of angular momentum (Eqn (9.15)), reduces to

$$\sum_i m_i \sqrt{a_i(1 - e_i^2)} \cos i_i = \text{constant}.$$

Neglecting fourth and higher powers in the eccentricities and inclinations, and using the fact that to this level of approximation these quantities are independent of one another, we can deduce that

$$\sum_i m_i \sqrt{a_i} \, e_i^2 = \text{constant}$$

and

$$\sum_i m_i \sqrt{a_i} \, i_i^2 = \text{constant}.$$

As these equations had to be true for all time, Laplace concluded that the mean distances, eccentricities, and inclinations underwent only periodic variations about their mean values. Various quantities in this analysis have been assumed to be small (the planetary masses in comparison to that of the Sun, and the eccentricities and inclinations of the planetary orbits), but since this was consistent with the observed phenomena, Laplace did not believe it was significant. Nor did anybody else until the latter half of the nineteenth century. There were other assumptions in Laplace's analysis. For example, it was a tacit assumption that the mean motions of the planets were incommensurable with each other, something that had to be taken as a matter of faith. Finally, the theory worked only if all the planets moved around the Sun in the same direction. This might not seem worthy of comment, since this condition manifestly is satisfied, but it emphasizes the fact that Laplace's result was to be interpreted as a statement about the actual Solar System and not about the general n-body problem. As Laplace himself put it:

> But does this remarkable property apply equally to a system of planets which move in a different sense? This is very difficult to determine. As this research has no utility in astronomy, we refrain from occupying ourselves with it.[118]

---

[117] The history of the stability question during the eighteenth century is studied in Suzuki (1996).
[118] Quoted from Suzuki (1996).

The theoretical demonstration of the stability of the Solar System was hailed as another remarkable achievement for universal gravitation. Laplace's statement of the result relied heavily on the previous research of both Lagrange and himself, and both men were considered as the architects of this triumph. The verdict of the scientific community in 1852 was summed up by Robert Grant, who wrote:

> The laws which regulate the eccentricities and inclinations of the planetary orbits, combined with the invariability of the mean distances, secure the permanence of the solar system throughout an indefinite lapse of ages, and offer to us an impressive indication of the Supreme Intelligence which presides over nature, and perpetuates her beneficent arrangements. When contemplated merely as speculative truths, they are unquestionably the most important which the transcendental analysis has disclosed to the researches of the geometer, and their complete establishment would suffice to immortalize the names of Lagrange and Laplace, even although these great geniuses possessed no other claims to the recollection of posterity.[119]

To the modern mathematician, Laplace's 'proof' is hopelessly inadequate. Laplace had nowhere considered the question of convergence for the series expansions that he truncated after a few terms. This should not be taken as a criticism of Laplace, as the concept of convergence for an infinite series was formulated only in the early nineteenth century through the work of Augustin-Louis Cauchy and Bernhard Bolzano. When it came to obtaining numerical values for particular quantities, the theoretical niceties of convergence could be neglected safely, but when the series were used to infer precise theoretical results (e.g. the stability of the Solar System for all time) the convergence of the series was absolutely essential. In fact, as Henri Poincaré would later show, the series generally do not converge, and Laplace's 'proof' is entirely without foundation.

Celestial Mechanics

Whereas *The System of the World* was written so as to make it accessible to a fairly wide audience, the same cannot be said for Laplace's work that brought together the technical aspects of astronomy – the monumental *Celestial Mechanics* – published in five thick volumes. The first four were published between 1799 and 1805, while the fifth, which contains a series of additions concerning work done after 1805, appeared in 1825.[120] *The System of the World* provided very useful background material for anybody interested in tackling

---

[119] Quoted from Grant (1966), p. 56.
[120] The full title is *Le traité de mécanique céleste*, the term 'celestial mechanics' having been introduced by Laplace.

Celestial Mechanics, and Laplace arranged for the second edition of the former work to be published in 1799 so as to accompany the launch of his *magnum opus*. The American mathematician Nathaniel Bowditch, who performed a magnificent service to astronomy with his translation of Volumes I–IV[121] wrote:

> The object of the author, in composing this work, as stated by him in his preface, was to reduce all the known phenomena of the system of the world to the law of gravity, by strict mathematical principles; and to complete the investigations of the motions of the planets, satellites, and comets, begun by Newton in his Principia. This he has accomplished, . . . [122]

Laplace's *Celestial Mechanics* took the discoveries of Newton, Clairaut, d'Alembert, Euler, Lagrange, as well as the author himself and numerous minor players, and combined them into a single comprehensive treatise. It left little for his immediate successors to add, but provided the directions for nineteenth-century research into celestial mechanics. It is worth noting that Laplace was not particularly conscientious at acknowledging the work of others, so perhaps he gave the impression that rather more of the content was his own than was, in fact, the case.

The first two volumes cover the basic principles of mechanics and gravitation, including Keplerian orbits, rigid body dynamics, and the tides,[123] together with the more straightforward aspects of perturbation theory. Volume III includes the theory necessary for accurate positional astronomy, and here Laplace applied for the first time the detailed analysis he had developed for the Jupiter–Saturn problem to all the planets; he was assisted in the numerical calculations by Alexis Bouvard. Volume IV then covers the motions of what might be termed the 'minor celestial bodies' (planetary satellites and comets).[124]

Difficult to read though Laplace's masterpiece may have been, it was recognized immediately for the outstanding quality of its science. John Playfair referred to *Celestial Mechanics* in the *Edinburgh Review* for January 1808 as 'the highest point to which man has yet ascended in the scale of intellectual attainment'. As well as the influence that the work had on mathematical astronomy in the nineteenth century, the philosophical impact was enormous. The

---

[121] A number of English translations of material from Volume I were made soon after its publication, but they were superseded by Bowditch's translation – made between 1815 and 1817 but published between 1829 and 1839. Bowditch's translation was accompanied by copious notes, diagrams, and corrections, which were indispensable for any reader wishing to follow Laplace's sometimes difficult exposition. Laplace often used remarks like 'It is easy to see that . . . ', and whenever Bowditch encountered such a phrase he knew that he had lots of hard work ahead of him! Volume IV of the translation, which was published after the translator's death, begins with a lengthy biographical memoir by Bowditch's son.
[122] From Bowditch's introduction to his translation of the *Celestial Mechanics*.
[123] Laplace's theory of the tides is described in Lamb (1932), Articles 213–21.
[124] A much more detailed description of the contents of the *Celestial Mechanics* is given in Gillispie (1997), Chapter 21.

success of eighteenth-century celestial mechanicians in reducing the Universe to differential equations suggested that the entire future was determined by the positions and velocities of all the particles in the Universe at a given time, and through his writing Laplace became the spokesperson for this mechanistic view of the Universe:

> We ought then to regard the present state of the universe as the effect of its anterior state and as the cause of the one which is to follow. Given for one instant an intelligence which could comprehend all the forces by which nature is animated and the respective situations of the beings who compose it – an intelligence sufficiently vast to submit these data to analysis – it would embrace in the same formula the movements of the greatest bodies of the universe and those of the lightest atom; for it, nothing would be uncertain and the future, as the past, would be present to its eyes. The human mind offers, in the perfection which it has been able to give to astronomy, a feeble idea of this intelligence.[125]

The idea that the future is predetermined by the past is challenging philosophically. What about free will, our apparent ability to make choices that influence the future? Many philosophers could not accept the deterministic universe of Newton and Laplace. Some, like Kant, argued that determinism and free will are not incompatible, the former being a property of the empirical world, and the latter inhabiting the world of reason, which is outside time.

Laplace was the greatest scientist of his time, and his preoccupation with research during the tumultuous political times in which he was working shows considerable single-mindedness. He survived the Terror of 1793/4 in which thousands were executed, was briefly Minister of the Interior under Napoleon, and became Grand Officer of the Légion d'honneur in 1808. When Napoleon's power began to wane, Laplace avoided declaring any public support for him and managed to remain in favour with Louis XVIII, who heaped more honours upon him. Whether this passive acceptance of the various political regimes that succeeded each other in rapid succession is worthy of criticism is hard to judge, but his scientific achievements are beyond question.

The achievements of Lagrange and Laplace were phenomenal. Newton's *Principia* and the methods of the calculus had opened up all sorts of previously unexplored areas to quantitative analysis, and throughout the eighteenth century, mechanics was a developing subject. In the early years of the nineteenth century, the situation appeared different. Apart from tweaking around the edges, rational mechanics appeared to form a coherent and complete discipline. John Toplis, the headmaster of Nottingham Grammar School in England, who translated Book I of *Celestial Mechanics* so as to draw the attention of English-speaking

---

[125] Quoted from Laplace (1951), Chapter II.

mathematicians to the power of the methods that had been developed on the Continent during the previous century, put it like this:

> It is now certain that the *Mécanique Analytique* and the *Mécanique Céleste* are the true sources from which a complete and methodical knowledge of all the properties of the equilibrium and the motion of bodies either solid or fluid ... can be obtained; it is therefore necessary that in future the elementary treatises should be composed with the view of leading to these works.[126]

Things did not improve particularly rapidly, however. In 1827, Lord Brougham, parliamentarian and supporter of radical social reform, asked Mary Somerville to write an account of Laplace's *Celestial Mechanics*. It was his opinion that 'In England there are now not twenty people who know this great work, except by name; and not a hundred who know it even by name.'[127] He believed that the combination of Somerville's writing skills and scientific knowledge could increase significantly these numbers. She accepted the challenge, and in 1831 *The Mechanism of the Heavens* was published. This work consisted both of a seventy-page *Preliminary Dissertation* that provided the background to *Celestial Mechanics* and placed Laplace's achievements in context, and Somerville's rendering of *Celestial Mechanics* itself.

Somerville treated the majority of the topics found in Laplace's great work, but some (such as comets and the tides) were omitted. Although she provided simplified demonstrations for many of Laplace's propositions (some of her own and some due to others), *The Mechanism of the Heavens* was still readable only by those with considerable mathematical expertise. Nevertheless, the book was a highly acclaimed success, and Somerville achieved something close to celebrity status. Laplace is reported to have said that there were only two women besides Mary Somerville who understood his work, a Miss Fairfax and a Mrs Greig, Fairfax being Somerville's maiden name, and Greig the surname of her first husband.[128]

Somerville identified gravity as the means by which scientists could both uncover the secrets of the past and peer into the future of the Universe. Laplace had shown, with *Celestial Mechanics*, how this could be accomplished with a detailed theoretical treatment of all the significant problems in which gravity was a factor, but Somerville understood that the process of discovery and analysis was not over, and that new problems would emerge to tax future astronomers. One source of new challenges derived from the discovery of new celestial bodies orbiting the Sun, to which we now turn.

---

[126] Quoted from Toplis (1814).   [127] Quoted from Neeley (2001), p. 75.
[128] Neeley (2001), p. 19.

10

The asteroids and the outer planets

The Titius–Bode law of planetary distances

By the eighteenth century, Newton had explained theoretically all of Kepler's laws and they had been transformed from empirical conjecture to scientific fact. Perhaps as a result, Kepler's other ideas about the possibility of a pattern in the planetary distances were treated with some seriousness. In his *Elements of Physical and Geometrical Astronomy* (1702) David Gregory wrote:

> ... supposing the distance of the earth from the Sun to be divided into ten equal parts, of these the distance of Mercury will be about four, of Venus seven, of Mars fifteen, of Jupiter fifty two, and that of Saturn ninety five.[1]

a statement repeated in 1723 by the Leibnizian philosopher Christian Freiherr von Wolff, who then went on to discuss how these distances were necessary so that the planets would not be too close to one another. Von Wolff passed on his ideas about cosmology to his young disciple Immanuel Kant, who published his theory of the Universe in 1755 (see p. 348). Kant made the observation that if one ignores Mercury and Mars because of their small masses, succeedingly distant planets from the Sun have increasing eccentricities, and he used this to argue that there might be planets beyond Saturn having eccentricities closer to that of comets. He also noted the large gap between the orbits of Mars and Jupiter and suggested that there might be another planet between them. Another person to make this observation was Johann Lambert who, in 1761, asked:

> And who knows if there are not planets missing in the large distance between Mars and Jupiter that will be discovered?[2]

---

[1] Quoted from Hoskin (1995).
[2] Quoted from Nieto (1972), a book that provides a thorough investigation into the Titius–Bode law and its subsequent modifications. The origins and early history of the law are also discussed in detail in Jaki (1972a, 1972b).

Kepler had, in fact, toyed with this idea previously but had rejected it, though only really because he came across his polyhedral hypothesis (see pp. 171–3). Newton, on the other hand, believed that the planets Jupiter and Saturn had been placed far from the inner planets, and far away from each other, because otherwise the gravitational disturbances that they would have caused might have been too great.

In 1764, the famous natural philosopher Charles Bonnet published his *Contemplation of Nature* which was translated from the original French into many languages. A German translation was made in 1766 by Johann Daniel Titius. Oddly, Titius added notes to Bonnet's work simply by including his own remarks in the main text, and between paragraphs 6 and 8 of Part I, Chapter 4, he inserted:

> Given the distance from the Sun to Saturn as 100 units, then Mercury is distant 4 such units from the Sun; Venus $4 + 3 = 7$ of the same; the Earth $4 + 6 = 10$; Mars $4 + 12 = 16$. But see, from Mars to Jupiter there comes forth a departure from this so exact expression. From Mars follows a place of $4 + 24 = 28$ such units, where at present neither a chief nor a neighbouring planet is to be seen. And shall the Builder have left this place empty? Never! Let us confidently wager that, without doubt, this place belongs to the as yet still undiscovered satellites of Mars; let us add that perhaps Jupiter also has several around itself that until now have not been seen with any glass. Above this, to us unrevealed, position arises Jupiter's domain of $4 + 48 = 52$; and Saturn's at $4 + 96 = 100$ units. What a praiseworthy relation![3]

Six years later, Titius produced a second edition of his translation in which he incorporated his old and new additions as translator's footnotes. It seems clear that the origins of Titius' statement were the observations of Gregory, together with Lambert's belief in an empty orbit between Mars and Jupiter.

The Titius law at first received little attention, perhaps because of the method by which it was published and also the lack of renown of the author. However, things changed in 1772, when Johann Elert Bode, a very important and well-respected astronomer, came across Titius' note and became convinced that the law described correctly the Creator's design. In the second edition of his influential astronomy book *Introduction to the Study of the Heavens*,[4] Bode added a footnote that, both from the reasoning used and from similarities in the original German, demonstrates clearly that he took the law from Titius, though there is no acknowledgement of Titius in Bode's work.[5] From other comments made by Bode, it is clear that he envisioned an undiscovered 'main' planet between Mars and Jupiter, whereas Titius had thought of a 'neighbouring'

---

[3] Quoted from Nieto (1972).
[4] *Anleitung zur Kenntniss des gestirnten Himmels.*
[5] Bode did acknowledge Titius in 1784, following the elevation of the status of the law that resulted from the discovery of Uranus.

planet or satellite. For many years the law was not highly thought of, though Bode mentioned it many times in his writings. However, things changed with the sensational discovery of a new planet in 1781 by William Herschel.

The Georgian planet

In an attempt to establish his career as a musician, Friedrich Wilhelm Herschel moved to England from his native Hannover in 1757, eventually setting up in Bath in 1766.[6] Before he made his remarkable discovery, astronomy was simply a hobby (albeit one to which he devoted considerable time) and he made thousands of systematic observations with his incredibly powerful home-made telescopes. Most astronomers of his day were involved in the great post-Newton project of matching observations with the theoretical predictions of universal gravitation. By contrast, Herschel's interest was the stars and the structure of the Cosmos. The existence of an amateur astronomer living in southwest England armed with some of the most powerful observing instruments in the world came as quite a surprise to the astronomical community. Unsurprisingly, finding a new planet led to instant fame and worldwide recognition.

On Tuesday, 13 March 1781, Herschel was proceeding with a systematic search of the heavens for double stars in an attempt to detect stellar parallax.[7] He recorded the sighting of a mysterious object in his journal:

> In the quartile near ζ Tauri the lowest of two is a curious either nebulous star or perhaps a comet.[8]

Herschel knew that he was not just observing an ordinary star, because it varied in size when he altered the magnification on his telescope, but it did not immediately occur to him that he might be looking at a new planet. Given that new

---

[6] Many biographies of Herschel exist; one that concentrates on his scientific achievements is Armitage (1962) and of particular note is the description of his and his sister Caroline's life and work, told through their correspondence, by Herschel's granddaughter (Lubbock (1933)). Useful background information can also be found in Hunt (1982). William Herschel, as he was known, became an English subject in 1793.

[7] Two stars that appeared close together but were in fact very different distances away from the Earth should move relative to each other as the Earth orbits the Sun. Herschel was unsuccessful in his attempts to measure parallax, though had he lived in more southerly latitudes he might have detected it for α Centauri. Parallax was first detected in the 1830s. Thomas Henderson, working at the Cape of Good Hope in 1833, detected a parallax for α Centauri but decided to wait until more observations were available before announcing his discovery. F. G. W. Struve gave some preliminary parallax calculations for Vega in 1837, but the first person to provide an accurate value for the distance to a star based on a systematic observing programme was Bessel in 1838. His figure of $0.314''$ for the parallax of 61 Cygni (corresponding to a distance of about 660 000 AU) is out by about 10 per cent (see Hirshfeld (2001)).

[8] Quoted from Alexander (1965).

comets were being discovered fairly regularly and that the number of planets had remained constant since the beginning of recorded history (excepting the change in status of the Earth) this is perhaps not surprising. The appearance of the object clearly aroused his interest, though, and 4 days later he added:

> I looked for the comet or nebulous star and found that it is a comet, for it has changed its place.[9]

Regular, consistent observations were needed to determine the orbit of the 'comet', and Herschel informed the astronomical community of his findings quickly. This would also ensure that he was given priority over the discovery.

The Astronomer Royal, Nevil Maskelyne, looked immediately for the new object. He had some difficulty, because the instruments at the Royal Observatory, being designed for accurate observations of known objects, were not as powerful as Herschel's home-made telescopes. However, he soon found what he was looking for due to the object's motion and, being more than a little curious as to the nature of the instruments that had enabled a musician from Bath to make such a discovery, he wrote to Herschel on 23 April with numerous questions and queries. It is clear from his opening remarks that Maskelyne immediately was aware of the possibility that the object Herschel had observed was a planet:

> I am to acknowledge my obligation to you for the communication of your discovery of the present comet, or planet, I don't know which to call it. It is as likely to be a regular planet moving in an orbit nearly circular round the sun as a comet moving in a very eccentric ellipse. I have not yet seen any coma or tail to it.[10]

Attempts to compute a parabolic orbit for the new body failed, the difference between predictions and observations grew rapidly, and it soon became clear that it was not a comet. A number of astronomers concluded that they were dealing with a planet, and proceeded to work out circular orbits (Lexell, Euler's colleague in St Petersburg who, by chance, was in England at the time, claimed later to have been the first). Over the next few years, efforts were concentrated on working out elliptic elements for the orbit.[11] The semi-major axis of the new planet was found to be a little over 19 times that of the Earth. If there was a planet outside the orbit of Saturn, then the Titius–Bode law would give a radius

---

[9] Quoted from Alexander (1965).

[10] Quoted from Lubbock (1933). When, 3 days later, Herschel presented a report of his discovery to the Royal Society, he entitled it simply *Account of a Comet*. This report is reproduced in Hunt (1982).

[11] A table of the various elliptic elements that were computed can be found in Alexander (1965), p. 49

of the orbit at $4 + 192 = 196$ of the units used by Titius, or in astronomical units, at 19.6 AU, only 2 or 3 per cent above the computed values.

The name for the new planet was a matter of great debate. Various names were suggested, including Cybele (the wife of Saturn), Hypercronius ('above Saturn'), Oceanus (the mythological river surrounding the Earth), even Neptune (the brother of Jupiter). Herschel, from 1782 onwards, always referred to it as the 'Georgian planet', in honour of George III[12] and this was how it was known in Britain for many years. In France, the planet was called 'Herschel' or 'Herschel's planet', but the name that gained acceptance eventually was that suggested by Bode: Uranus. In classical mythology, Saturn was the father of Jupiter, and so it was natural to chose the name of Saturn's father for the planet.

Uranus is quite a bright object in the sky; occasionally, it is visible to the naked eye, and Bode asked himself the obvious question: why had no one seen it before? He began searching through old star catalogues to see if it had been observed, but mistaken for a star. He was successful, and many old observations were found over the decades following the discovery of Uranus, the oldest being a sighting due to Flamsteed in 1690.[13] These observations helped enormously in the determination of the orbit of Uranus, as they extended greatly the amount of the orbit for which data were available (the period of the orbit of Uranus orbit is about 84 years).

Jupiter and Saturn were known to have moons, and so it was natural that Herschel should look to see if Uranus was also accompanied by a satellite system. In 1787, he discovered that this is, indeed, the case, when he discovered the two satellites now called Titania and Oberon.[14] He spent many years observing these objects and determined that their orbits were inclined at about 90° to the ecliptic. Laplace realized the significance of this for the planet itself:

---

[12] In 1782, George III granted Herschel a pension of £200 per annum (the Astronomer Royal only received £300) and as a result he could devote all his time to astronomy. The king later helped fund Herschel's telescope-making, and also provided £50 per annum so that Herschel's sister Caroline could become his assistant. She later became well known as a successful comet-hunter. Herschel was knighted in 1816 by George III's son, the Prince Regent. Herschel's astronomical career is described in Ronan (1967).

[13] A list of twenty-two pre-discovery sightings of Uranus (the last being found in 1864) is given in Alexander (1965), p. 90. A twenty-third was unearthed more recently (Rawlins (1968)). The notoriously disorganized Pierre Charles Lemonnier observed Uranus six times in 8 days in 1769 without realizing that it was not a star, and his reputation was tarnished severely as a result. However, in his defence, it is worth noting that Lemonnier's telescope was too small to resolve the disc of Uranus, and so the only way for him to recognize that what he was observing was not a star would have been to note its motion. Lemonnier was unfortunate in that while he was observing Uranus it was near its stationary point and so its motion was much reduced from its mean value (see Baum and Sheehan (1997), pp. 55 and 79).

[14] Many more have been found subsequently: two in 1851, one in 1948 and, following the visit of the spacecraft Voyager 2 in 1986, numerous smaller satellites.

... if the various satellites of a planet move in a plane greatly inclined to that of its orbit, it can be inferred that they are kept in that plane by the action of the planet's equator, and that therefore the planet rotates around an axis nearly perpendicular to the plane of the orbits of its satellites. It may therefore be affirmed that the planet Uranus, all of whose satellites move in a single plane almost perpendicular to the ecliptic, itself turns on an axis very little inclined to the ecliptic.[15]

The addition of a seventh planet to the Solar System was a hugely significant event, for many reasons. In the first place, it changed completely Herschel's financial circumstances and allowed him to dedicate his life to astronomy. His agenda was markedly different from that of his contemporaries, concerned as he was with grand questions concerning the physical nature and ultimate fate of the Universe. Herschel constructed large telescopes that could peer deep into outer space and built a speculative cosmology based on what he observed. He transformed the starry heavens from a static backdrop against which to measure planetary positions, into a vast dynamic region in which stars evolved from clouds of nebulous material. In so doing he became the pioneer of modern sidereal astronomy.[16]

The size of the known Solar System was increased greatly by the discovery of Uranus but, more importantly, its extent became less certain. Psychologically, the case for the orbit of Saturn representing the outermost limit of the influence of the Sun had been compelling prior to 1781, since no new planets had ever been discovered, even following 150 years of telescopic observation. But if there were more planets than those visible with the naked eye, why only one? The possibility that there were more planets waiting to be discovered had to be taken seriously – perhaps the Titius–Bode law would indicate where to look.

The discovery of a new planet at just over 19 AU, compared to the prediction of 19.6 AU from the Titius–Bode law, was enough to convince many of the validity of the law, despite there being no physical basis for it. If the law did govern planetary distances, then what was to be made of the absence of any planet at a distance of 2.8 AU from the Sun? Many people, including Bode, became convinced of the existence of a planet between Mars and Jupiter, and the search began.[17]

---

[15] LAPLACE *Celestial Mechanics*, Volume II, Book 5, Chapter 3. Quoted from Alexander (1965), p. 76.

[16] The effect of the discovery of Uranus on Herschel is discussed in Schaffer (1981). Schaffer (1981).

[17] It is claimed often that the philosopher Hegel logically 'proved' the impossibility of anything between Mars and Jupiter, only to look rather foolish when such an object was discovered. However, this is not true (Pinkard (2000), pp. 107–8). In August 1801, Hegel defended his habilitation thesis (a requirement in German universities before the right to lecture was granted) in which he highlighted the fact that people were searching for a planet on the basis of the Titius–Bode law and then went on to describe certain numerological speculations of Plato (from the *Timaeus* (see, for example, Knorr (1990)) in which no planet was missing. He did not, however, endorse Plato's view.

The search for a planet between Mars and Jupiter

Baron Franz von Zach, the Hungarian-born director of the Duke of Saxe-Gotha's observatory at Seeberg, began a methodical search for the 'missing' planet in 1787, but after 13 unsuccessful years he decided to enlist some help. In 1800, he formed a 'planet-hunting' club – von Zach called them jokingly the 'celestial police' – and they tried to enlist others to join.[18] However, they were all beaten to the discovery of an object orbiting between Mars and Jupiter by one of those whose help they intended to request – Giuseppe Piazzi – based in Palermo, Sicily. On 1 January 1801, he observed a faint object in the constellation Taurus that over successive nights moved with respect to the stars. Piazzi thought he had discovered a new planet, but was cautious. He wrote to Bode and Lalande claiming to have observed a comet, though one without a tail, but to his friend Barnaba Oriani from the Brera Observatory in Milan he said:

> I have announced this star as a comet, but since it is not accompanied by any nebulosity and, further, since its movement is so slow and rather uniform, it has occurred to me several times that it might be something better than a comet. But I have been careful not to advance this supposition in public.[19]

Before anybody could confirm his observations, the object approached too near the Sun and became lost from view.

Men like von Zach wanted to believe that the long search was over. Even though no one except Piazzi had observed the new object, he published an article in his monthly journal entitled *On a Long Supposed, Now Probably Discovered, New Primary Planet of our Solar System Between Mars and Jupiter.*[20] Piazzi's data were extremely limited, and the determination of an orbit proved difficult.[21] Parabolic, circular, and elliptical orbits were all tried, but nobody could reconcile theory satisfactorily with the available observations. It was predicted that the new object would emerge from the glare of the Sun around September 1801, but despite considerable effort, astronomers failed to relocate it.

By then the object had a name. Piazzi wanted to call it Ceres Ferdinandea – Ceres was the patron goddess of Sicily and King Ferdinand of Naples and Sicily was Piazzi's patron. Piazzi's desire to honour a royal patron by attaching his name to a new celestial body was as unsuccessful as Herschel's was, and Ceres became the accepted name. But where was Ceres? The answer was provided by the brilliant young mathematician, Carl Friedrich Gauss.[22]

---

[18] Von Zach's astronomical career is discussed briefly in Cunningham (1988).
[19] Quoted from Hoskin (1995).
[20] The article appeared in Volume 3 (1801) of von Zach's *Monatliche Correspondenz*.
[21] It took a while for the observational data to become widely available. The complete set was first published in the *Monatliche Correspondenz* in September 1801.
[22] Details of Gauss' astronomical work can be found in Forbes (1971b, 1978).

Gauss became interested in astronomy while he was a student in Göttingen, and sought help and advice from von Zach. His first published article – on the calculation of the date of Easter – appeared in von Zach's journal.[23] It was through his reading of this journal that Gauss became aware of the discovery of Ceres and the need to compute an orbit for it. He decided to tackle the problem but, rather than simply use established procedures, worked through September and October 1801 on a completely new method for orbit determination. Time was of the essence and by November he was able to solve in an hour a problem that had taken Euler 3 days.[24]

Gauss sent the results of his numerical computations – which suggested that Ceres was several degrees away from the positions predicted by other astronomers – to von Zach, and on 31 December 1801, von Zach relocated Ceres almost exactly where Gauss had predicted. Heinrich Wilhelm Olbers – another member of the 'celestial police' and a friend of Gauss – saw it the following evening. According to Gauss, the mean distance of the orbit of Ceres was 2.77 AU compared to the Titius–Bode figure of 2.8 AU. The law was confirmed!

Not all was as it should be, however. First, there was the problem of the size of Ceres: it was tiny. Herschel concluded from his observations that its diameter was less than 261 km (this is actually about $3\frac{1}{2}$ times too small), in stark contrast to the sizes of the other planets. Then there was the inclination of the orbit to the ecliptic, which, at over 10°, was far greater than that of any other planet. More serious, though, was Olbers' discovery a few months later of a second object, subsequently called Pallas, the mean distance of which was calculated by Gauss as 2.67 AU. The whole basis for the search for a planet between Mars and Jupiter had been the Titius–Bode law and, if this was to be believed, there was only space for one such planet. Herschel wrote in 1802 (note that Titius had been forgotten):

---

[23] *Monatliche Correspondenz*, **2** (1800), 121–30. Easter Sunday is defined as the first Sunday following the full moon that occurs on or after 21 March (the so-called paschal full moon), though it should be emphasized that the moon in this definition is in fact a notional object, the behaviour of which is designed closely to mimic that of the actual Moon and that is governed by simple mathematical rules based around the Metonic cycle (19 years $\approx$ 235 synodic months). Gauss' method is remarkably elegant: let y be the year and let $m = 23$, $n = 4$ (this is appropriate for the Gregorian calendar between 1800 and 1899, other values are needed for other centuries), then calculate $a = y$ (mod 19) (i.e. a is the remainder when y is divided by 19), $b = y$ (mod 4), $c = y$ (mod 7), $d = (19a + m)$ (mod 30), $e = (2b + 4c + 6d + n)$ (mod 7). The date of Easter Sunday is given then by $(22 + d + e)$ March. The paschal full moon is $(22 + d)$ March and the addition of e makes sure that the date refers to a Sunday. A detailed discussion of algorithms for computing the date of Easter, together with a brief history, can be found in Richards (1998), Chapter 29.

[24] The method Gauss used is described in Marsden (1995). It was substantially different from that which he eventually published in 1809.

> There is a certain regularity in the arrangement of planetary orbits which has been pointed out by a very intelligent astronomer, so long ago as 1772; but this, by the admission of the two new stars into the order of planets, would be completely overturned; whereas, if they are of a different species, it may still remain established.[25]

Herschel described the appearance of Ceres and Pallas in telescopes as 'asteroidal' (star-like) and suggested that they should be referred to as 'asteroids'. Piazzi thought the term 'planetoids' more appropriate; after all, they were not stars at all. Others thought that the word 'planet' was fine. In the end, Herschel's suggestion was taken up.

Olbers believed in the Titius–Bode law, and to save it he suggested that the small planets were remnants of a single large planet that had disintegrated.[26] If this were true, there ought to be many more of these minor planets, since tens of thousands of objects of the size of Ceres and Pallas would be needed to make up a planet the size of Mercury. Olbers' idea was helped by the discovery in 1804 of Juno (at 2.67 AU) by Karl Ludwig Harding and Vesta (at 2.36 AU) in 1807 by Olbers himself. The search for asteroids continued throughout the nineteenth century and thousands have been found.

Just as in the case of Uranus, the admission of a host of minor planets into the Solar System had many significant repercussions. Soon after the initial success with Ceres, it became abundantly clear that there were many more similar discoveries to be made and, with the prospect of worldwide fame for the discoverer, numerous astronomers began systematic observations of the stars in the zodiacal belt – this was precisely how Harding chanced upon Juno. The star charts prepared by the Berlin Academy from 1830 onwards were to play a crucial role in the discovery of Neptune.

As far as theoretical astronomy was concerned, the asteroids were directly responsible for initiating Gauss' work on orbit determination, and they also led to some challenging problems in perturbation theory. One outcome of this was the re-evaluation of the mass of Jupiter from the measured deviations in asteroidal orbits.[27] Moreover, with Olbers' disintegration hypothesis given the blessing of no less an authority than Lagrange, the 'truth' of the Titius–Bode law became such a widely held belief that it became a basis for future planetary astronomy.

---

[25] Quoted from Herschel (1802).

[26] The suggestion that a major planet had once occupied the space between Mars and Jupiter but had been shattered by the impact of a comet had been made in the mid eighteenth century (but never published) by Thomas Wright of Durham (see Hoskin (1995)). Nowadays, it is thought more likely that the asteroids are evidence of a planet that failed to form (Goodwin and Gribbin (2001)).

[27] An increase of about 2 per cent over Laplace's value (Clerke (1908)).

One possible problem with the law was noted in 1790 by Johann Friedrich Wurm who wrote it in the form

$$r_n = 4 + 3 \times 2^n, \tag{10.1}$$

where $r_n, n = 0, 1, \ldots$, correspond to the radii of the orbits of the planets (in the units of Titius), beginning with $n = 0$ corresponding to Venus. The Titius–Bode law required taking a value of $n = -\infty$ for Mercury, whereas if the law was a proper geometric progression, it should predict an orbit of Mercury at 0.55 AU by taking $n = -1$. In 1802, Gauss – apparently ignorant of Wurm's work – pointed out that there should be an infinite number of allowed orbits between Mercury and Venus. To most people, this imperfection was not considered important, however.

Wurm also noted that a better fit could be obtained for the whole planetary system if one took the more general equation

$$r_n = a + b \times 2^n, \qquad n = -\infty, 1, 2, \ldots \tag{10.2}$$

and set $a = 3.87$ and $b = 2.93$. He speculated that many more planets might exist outside the orbit of Uranus, and used this formula to calculate their distances. The huge orbits that were implied for these hypothetical planets did not worry him, but he was troubled by the enormous periods that followed from Kepler's third law. Apparently, he thought it unreasonable that a planet should have an orbital period that was greater than the 26 000-year period of the precession of the Earth! [28] More importantly, Wurm was the first to realize that, if the law had a dynamical origin, then it was probable that a generalized law should exist that encompassed the satellite systems of Jupiter and Saturn. In fact, Wurm found that laws of the form in Eqn (10.2) gave reasonably good values for the then known satellite systems of these planets if the unit of distance in each case were taken as the radius of the parent planet and values of $a = 3.0$, $b = 3.0$ were taken for Jupiter, and $a = 4.5$, $b = 1.6$ for Saturn. In the latter case, however, the fourth and sixth orbits were empty.

Another modification to the law as regards the satellite systems of Jupiter and Saturn was made by Ludwig Wilhelm Gilbert in 1802. He realized that there was no need for the law to be given in terms of powers of 2. Thus, he wrote the law in the form

$$r_n = a + bc^n, \qquad n = -\infty, 1, 2, \ldots$$

and with the units of the radius of Saturn, he applied this equation to the satellite system of Saturn with values of $a = 3.08$, $b = 0.872$, and $c = 2.08$. Similar

---

[28] Jaki (1972a).

observations were made by James Challis in 1828.[29] Challis is remembered for something entirely different, however. As we shall see, he failed spectacularly to discover Neptune.

Orbit determination

One of Gauss' notable characteristics was his perfectionism, and, while his procedure for determining Ceres' orbit, hastily derived in 1801, proved more than adequate for performing numerical calculations, he did not consider it of a standard suitable for publication. The world had to wait another 8 years for the carefully polished theory to appear in Gauss' *Theory of the Motion of the Heavenly Bodies Moving About the Sun in Conic Sections,* usually referred to as the *Theoria motus.*[30] When it did arrive, though, the astronomical community soon realized that Gauss had revolutionized the problem of orbit determination. The book contains a complete system of algorithms for computing the orbit of a body from three or four observations, and ends with Gauss' version of the method of least squares.

There are two separate problems in the accurate determination of the orbit of a celestial body. The first, and more difficult, part of the process is the creation of an approximate set of orbital elements that allow the object to be tracked and more observational data to be obtained. The extra information then can be used to improve the orbit, and we will consider this, second, problem first.

The unknowns in the orbit improvement problem are the differences between the approximate and actual elements, which are small quantities and, thus, the resulting equations can be linearized. Typically, we would then have a set of $n > 6$ linear equations for unknowns $x_1, \ldots, x_6$ of the form

$$a_{i1}x_1 + \ldots + a_{i6}x_6 = b_i, \qquad i = 1, \ldots, n.$$

A given choice of unknowns will give rise to right-hand sides $b_i + e_i$ in each of these equations, e_i being the errors. In 1805, motivated primarily by the practical desire for algebraic simplicity, Legendre proposed that one should minimize the sum of the squares of the errors; previously, it had been common

---

[29] Published in Challis (1830).

[30] English translation in Gauss (1963). From the Preface: 'Several astronomers wished me to publish the methods employed in these calculations immediately after the second discovery of Ceres; but many things ... prevented my complying at the time with these friendly solicitations. I ... have no cause to regret the delay. For, the methods first employed have undergone so many and such great changes, that scarcely any resemblance remains between the method in which the orbit of Ceres was first computed, and the form given in this work.' Gauss did send a brief manuscript containing his early ideas to Olbers in 1802 (Dunnington (1955)).

practice to try and make $\sum_i |e_i|$ as small as possible. In other words, one should minimize

$$s = \sum_{i=1}^{n}(a_{i1}x_1 + \ldots + a_{i6}x_6 - b_i)^2,$$

which can be achieved by setting all the derivatives $\partial s/\partial x_j$, $j = 1, \ldots, 6$ equal to zero. This leads to a 6×6 linear system of equations which can be solved by standard methods.

Legendre pointed out that if one has a set of observations corresponding to just one unknown, then the method of least squares suggests that the arithmetic mean of the observations is the most appropriate choice, exactly as one would like, and in contrast to the conclusions from Laplace's theory of errors (see pp. 343–4). Gauss took this property of the arithmetic mean as his starting point:

> It has been customary certainly to regard as an axiom the hypothesis that if any quantity has been determined by several direct observations, made under the same circumstances and with equal care, the arithmetical mean of the observed values affords the most probable value, if not rigorously, yet very nearly at least, so that it is always safe to adhere to it.[31]

Working from this assumption, he calculated that the error distribution should have density

$$\phi(x) = \frac{h}{\sqrt{\pi}}\,e^{-h^2x^2},$$

in which the parameter h can be considered as a measure of the precision of the data, since the probability that $x \in (-a, a)$ for a distribution with parameter h is the same as the probability that $x \in (-\frac{a}{c}, \frac{a}{c})$ for a distribution with parameter ch ($c > 0$).[32]

Gauss then went on to examine the consequences of his form for ϕ. In particular, if one has a set of observations t_i, then the probability of the unknown true value lying between τ and $\tau + d\tau$ is proportional to $\exp[-h^2 \sum_i(t_i - \tau)^2]$ and so the most probable value of τ is found by minimizing $\sum_i(t_i - \tau)^2$. Thus, Gauss provided a probabilistic justification for the method of least squares. He also showed how a set of observations, each of different accuracy, should

[31] GAUSS *Theoria motus*, Article 177 (Gauss (1963)).
[32] In modern language $h\sigma = 1/\sqrt{2}$, where σ is the standard deviation of the distribution. Many names have been used for Gauss' error distribution; nowadays we refer to it as the 'normal distribution'.

be treated by minimizing a weighted sum of the squares of the errors. Gauss' treatment of least squares occupies only a small part of the *Theoria motus*.[33] The much greater part was concerned with problems associated with the initial determination of an orbit from only a few observations.

Prior to the discovery of Ceres, there was little need for methods to determine general elliptic orbits. Apart from Uranus, for which a circular orbit proved a satisfactory first approximation, the only bodies for which orbits had been required were comets. During the period of visibility of a comet, the path it follows is, to all intents and purposes, parabolic, and this can be used to simplify the determination procedure. The first person to produce a general method for obtaining parabolic orbital parameters based on observations, was Newton, who published an approximate graphical solution to 'this exceedingly difficult problem' in the *Principia*.[34] Newton's method was adapted and used to great effect by Halley, who in 1705 published his researches into the orbits of twenty-four comets. Interest in the problem of orbit determination for comets continued throughout the eighteenth century, but the first systematic analytic procedure that could be considered practical was due to Laplace.

In 1774, Laplace became involved in a dispute with the Croatian Rudjer Bošković over the validity of the latter's method of orbit determination.[35] Laplace's procedure, read to the Academy 8 days after the discovery of Uranus, can be described as follows. If we denote the heliocentric position of the Earth by $\mathbf{R}$ and the geocentric position of the comet by ρ, then the heliocentric position of the comet is $\mathbf{r} = \mathbf{R} + \rho$. Next we write $\rho = \rho\mathbf{e}$, differentiate twice, and use the equation of motion (Eqn (8.8), p. 274) for both the Earth and the comet (neglecting the masses of the Earth and comet compared to the Sun):

$$-\frac{\mu}{r^3}(\mathbf{R} + \rho) + \frac{\mu}{R^3}\mathbf{R} = \ddot{\rho}\mathbf{e} + 2\dot{\rho}\dot{\mathbf{e}} + \rho\ddot{\mathbf{e}}.$$

In this equation, $\mathbf{R}$ can be assumed known from tables, $\mathbf{e}$ can be found from

---

[33] Gauss tried to establish priority for the discovery of the least squares principle by claiming that he had been using it in his calculations since 1795 (when he was only 18!). This greatly irritated Legendre and led to some bitterness between the two men (see Plackett (1972), Stigler (1981)). The historical development of the least squares method is described in Hald (1998), Section 6.9 and Chapter 18.

[34] NEWTON *Principia*, Book III, Propositions 40 and 41. The complexity of Newton's procedure meant that few astronomers actually used it in practice, but its utility was demonstrated by the great Russian scientist, Aleksei Nikolaevich Krylov, the author of a Russian translation of the *Principia* (1915) (see Kriloff (1925)).

[35] Bošković had proposed a method for determining cometary orbits in 1746 and he presented an improved version to the Academy in 1771. Laplace described Bošković's approach as 'faulty, illusory, and erroneous' (Gillispie (1997), p. 97).

observation, and r is given in terms of ρ by

$$r^2 = \mathbf{r} \cdot \mathbf{r} = (\mathbf{R} + \boldsymbol{\rho}) \cdot (\mathbf{R} + \boldsymbol{\rho}) = R^2 + \rho^2 + 2\rho \mathbf{R} \cdot \mathbf{e}.$$

The unknowns are ρ, $\dot{\rho}$, $\ddot{\rho}$, $\dot{\mathbf{e}}$, and $\ddot{\mathbf{e}}$. The last two of these can be approximated numerically at some given time, τ, from a set of closely spaced observations, and then we have three scalar equations for three scalar unknowns $\rho(\tau)$, $\dot{\rho}(\tau)$, and $\ddot{\rho}(\tau)$. For a general elliptic orbit, this yields an eighth-order algebraic equation for $\rho(\tau)$, which has to be solved numerically, though some reduction is possible if a parabolic orbit is assumed. We can then calculate $\mathbf{r}(\tau)$ and $\dot{\mathbf{r}}(\tau)$ from which the elements are determined easily.[36]

Laplace's method suffers from a number of weaknesses, not least of which is the need to determine the derivatives of observational data. If only three data points are available, this will probably lead to significant errors. Despite this and other problems, Laplace's procedure was understood easily and became popular quickly.

It was superseded by a method proposed by Olbers in 1797,[37] which in turn was eclipsed by the brilliance of Gauss. Both Olbers and Gauss took as their starting point a fact that had been noted in 1733 by Pierre Bouguer. Since an orbiting body moves in a plane (neglecting perturbations), its heliocentric position vector $\mathbf{r}_i$ at three different times t_i, $i = 1, 2, 3$, are linearly related, i.e.

$$\mathbf{r}_2 = c_1 \mathbf{r}_1 + c_3 \mathbf{r}_3, \tag{10.3}$$

where c_1 and c_3 are some unknown constants. Crossing this equation with $\mathbf{r}_3$ and $\mathbf{r}_1$ shows that

$$c_1 = |\mathbf{r}_2 \times \mathbf{r}_3| / |\mathbf{r}_1 \times \mathbf{r}_3|, \qquad c_3 = |\mathbf{r}_2 \times \mathbf{r}_1| / |\mathbf{r}_1 \times \mathbf{r}_3|,$$

and, since $|\mathbf{a} \times \mathbf{b}|$ is twice the area of the triangle having $\mathbf{a}$ and $\mathbf{b}$ as two of its sides, it follows that c_1 and c_3 are the ratios of the areas of the triangles formed by the Sun and the respective positions of the body. Bouguer made the approximation

$$c_1 = (t_3 - t_2)/(t_3 - t_1), \qquad c_3 = (t_2 - t_1)/(t_3 - t_1),$$

which, due to Kepler's second law, is equivalent to replacing the areas of the relevant triangles by the areas of the corresponding sectors.

---

[36] See, for example, Escobal (1965), Section 3.7.2.
[37] *Abhandlung über die leichteste und bequemste Methode die Bahn eines Cometen zu berechnen.* Brief extracts are translated in Shapley and Howarth (1929).

If sufficiently accurate values for c_1 and c_3 are known, the solution can then be obtained. Since $\mathbf{r}_i = \mathbf{R}_i + \boldsymbol{\rho}_i$, Eqn (10.3) can be thought of as three scalar equations for the three unknowns $|\boldsymbol{\rho}_i|$. Once these geocentric distances have been determined, the heliocentric position vectors $\mathbf{r}_i$ and, hence, the spatial coordinates of three points on the orbit, are known. Any two of these points then can be used to recover the orbital elements and the third point used as a check. The process then can be repeated until the required accuracy is obtained. However, Bouguer's approximate formulas for c_1 and c_3 are not sufficiently accurate in practice, since small changes in c_1 and c_3 can be shown to lead to large changes in the calculated heliocentric positions.

In the *Theoria motus*, Gauss explained very clearly the nature of the problem of determining an orbit from three latitude and longitude observations. He pointed out that such a procedure will be impractical if the orbit lies close to the ecliptic plane. This is because if the orbit were precisely in the ecliptic, then three longitude observations would not be sufficient to determine the four remaining elements (i and Ω being redundant). For orbits close to the ecliptic, it is better to use four longitude and two latitude data. Gauss treated this problem separately and we will not consider it further here.

For the three-observation problem, the observations that are used for the initial orbit determination should, Gauss said, not be too close (or the inevitable inaccuracies in the observations will affect the calculation too much) and not be separated too widely (or the resulting elements will be poor approximations). What is too close and too far apart must be judged from experience. Gauss claimed, however, that he had used his method successfully to determine the orbit of Juno from observations covering a heliocentric motion of $7° 35'$ and had computed the orbit of Ceres from observations embracing $62° 55'$ of arc.

It is possible, in principle, to reduce the six equations in six unknowns to a single equation for each of the elements, but Gauss noted that the complicated nature of the equations makes this impossible in practice. Instead, he did the next-best thing. There are many ways in which the problem can be reduced to two equations of the form $X(P, Q) = 0$ and $Y(P, Q) = 0$ in terms of two unknown quantities P and Q. These need not necessarily be two of the elements, but the elements must be calculated readily once P and Q have been found. The selection of P and Q is crucial to the success of the orbit determination process. First, one would like the functions X and Y to be as simple as possible; second, it is desirable that approximations to P and Q are calculated easily, and third, the numerical solution of the equations $X = Y = 0$ should be efficient and, in particular, as insensitive as possible to numerical error.

Gauss selected

$$P = \frac{c_3}{c_1}, \qquad Q = 2(c_1 + c_3 - 1)r_2^3,$$

and he showed that small inaccuracies in the values of P and Q will lead to small errors in the heliocentric positions. Moreover, P and Q are approximated well by

$$\frac{t_2 - t_1}{t_3 - t_2} \quad \text{and} \quad \mu(t_3 - t_2)(t_2 - t_1),$$

respectively (the latter result being far from obvious). In determining the elements from P and Q, the same eighth-order algebraic equation arises as in Laplace's method, but Gauss transformed this into what is now sometimes referred to as 'Gauss' equation',

$$\alpha \sin^4 z = \sin(z - \beta),$$

where the unknown z is the angle between the heliocentric and geocentric position vectors at the middle observation, and α and β are known functions of P and Q. This equation has two or four real roots, and Gauss explained how to select the appropriate solution.

The *Theoria motus* is the work of a mathematical genius, fully aware that he is developing tools for practical use. As an example, we will consider Gauss' solution to what is often called 'Lambert's problem', i.e. the determination of an orbit from two points $\mathbf{r}_1$, $\mathbf{r}_2$, and the time, τ, between them. It is a surprising fact that τ depends only on the semi-major axis a, the sum of the distances $r_1 + r_2 = s$, for example, and the length of the chord $|\mathbf{r}_2 - \mathbf{r}_1| = d$, say. Gauss described the history of this result, now known as Lambert's theorem, thus:

> This formula appears to have been first discovered, for the parabola, by the illustrious Euler, who nevertheless subsequently neglected it, and did not extend it to the ellipse and hyperbola: they are mistaken, therefore, who attribute the formula to the illustrious Lambert, although the merit cannot be denied this geometer, of having independently obtained the expression when buried in oblivion, and of having extended it to the remaining conic sections.[38]

In fact, the first analytic demonstration of the result was due to Lagrange in 1778, a year after Lambert's death, though Gauss failed to mention this. For an ellipse, Lagrange obtained

$$\tau \sqrt{\frac{\mu}{a^3}} = \alpha - \sin \alpha - (\beta - \sin \beta),$$

---

[38] GAUSS *Theoria motus*, Article 106 (Gauss (1963)).

where α and β are determined from $s + d = 2a(1 - \cos\alpha)$ and $s - d = 2a$ $(1 - \cos\beta)$, and from this Gauss deduced that

$$a = \frac{r_1 + r_2 - 2\sqrt{r_1 r_2}\cos f \cos g}{2\sin^2 g},$$

where f is half the angle between $\mathbf{r}_1$ and $\mathbf{r}_2$, and the unknown g is half the difference between the eccentric anomalies corresponding to the two positions.

This formula is, however, totally unsuitable for practical use since, if the positions are close together, the right-hand side is the difference of two almost equal quantities divided by a small quantity. So Gauss introduced a quantity l defined by

$$2(1 + 2l)\cos f = \sqrt{\frac{r_1}{r_2}} + \sqrt{\frac{r_2}{r_1}},$$

in terms of which, a can be represented in a suitable form. The problem manifest in the computation of a from the original formula now appears in the calculation of l, but Gauss got round this by defining a new quantity ω through

$$\tan\left(\tfrac{1}{4}\pi + \omega\right) = \left(\frac{r_2}{r_1}\right)^{1/4},$$

and then $l\cos f = \sin^2\tfrac{1}{2}f + \tan^2 2\omega$, which is not sensitive to numerical error.

Gauss showed that g could be determined from the simultaneous solution of the two equations

$$y^2 = \frac{m^2}{l + \sin^2\tfrac{1}{2}g}, \qquad y^3 - y^2 = m^2\frac{2g - \sin 2g}{\sin^3 g},$$

where $m^2 = \mu\tau/(2\cos f\sqrt{r_1 r_2})^3$, and which (of course) Gauss showed how to accomplish accurately and efficiently. The elegant practicality of the whole approach is brought home by the fact that y turns out to be nothing other than the ratio of the area of the sector swept out to that of the associated triangle – precisely the quantity one needs to determine the correct values for the unknown constants in Eqn (10.3).

The methods of Gauss, Olbers, and Laplace were all used, generalized and improved upon during the nineteenth century;[39] here, we will mention just one significant event. In 1889, Josiah Willard Gibbs, Professor of Mathematical Physics at Yale, published a paper on orbit determination which represents the first use of vector algebra in celestial mechanics. Gibbs was one of those

---

[39] See Marsden (1995) for details.

chiefly responsible for the creation of modern vector algebra, but he was not an astronomer. He wrote his paper simply so as to demonstrate to astronomers the usefulness of vector methods.[40]

Problems with Uranus

The discovery of Uranus was a success for Herschel's observational astronomy. The discovery of the asteroids was a joint effort between observation and theory. However, the discovery of Neptune in 1846 was a triumph for theory alone.[41]

Following the discovery of Uranus, it was soon realized that its orbit was not quite as predicted. By the end of the eighteenth century, observational astronomy had developed to the extent that planetary positions could be measured to within a few arc-seconds, and Uranus was expected to fit with theory with the same precision as everything else. For the determination of the orbit of Uranus, the 'old' (i.e. pre-1781) sightings were invaluable. In 1784, Father Placidus Fixlmillner, director of the observatory at Kremsmünster in Austria, computed an orbit using Flamsteed's 1690 observation, that of Mayer in 1756 and two post-1781 measurements. The resulting predictions were satisfactory, but not for long; by 1788, errors of over 30″ were being reported.

One obvious source of error was the fact that the perturbing effect of Jupiter and Saturn had not been taken into account. The required mathematical analysis, based on Laplace's theory, was undertaken by Delambre (among others) and he won a prize from the Paris Academy in 1790 for his efforts. The new orbit was published in 1791 and was a significant improvement; it was still producing satisfactory agreement with observation in 1798. But after 1800, discrepancies began to appear.

During this time, many more 'old' observations of Uranus were unearthed and, armed with these, Alexis Bouvard (employed by Laplace as a 'calculator') began work on a new orbit in 1820. He had at his disposal observations spread over a 130-year period, but found he could produce an orbit that fitted well the 'old' data, or one that agreed with the post-discovery observations – but not both. Rather than accept failure, he suggested that the pre-1781 observations were inaccurate, and decided to produce new tables based only on recent observations.

[40] The history of the development of vectorial methods is told in Crowe (1994).
[41] Much has been written about the discovery of Neptune. The most comprehensive account is Grosser (1962) which also contains an extensive bibliography. Numerous articles are cited in the following pages of this book. The discovery of Neptune has been described in many books that have appeared since 1962, including Ronan (1967), Lyttleton (1968), Hoyt (1980), Baum and Sheehan (1997), and Standage (2000).

Bouvard was, in effect, attributing errors to previous distinguished astronomers an order of magnitude greater than any that had ever been found, and he seems to have realized that his solution was not built on solid foundations, since he wrote in the introduction to his tables:

> I leave to the future the task of discovering whether the difficulty of reconciling the two systems results from the inaccuracy of the ancient observations, or whether it depends on some extraneous and unknown influence which may have acted on the planet.[42]

Notwithstanding his dubious assumptions, which were the subject of serious criticism from fellow astronomers, Bouvard's new tables fit well with new observations of Uranus, but only for a few years.

When William Herschel died in 1822, the planet he had discovered over 40 years previously was still a source of great frustration. During the 1820s, the true heliocentric longitude of Uranus gradually got further and further ahead of the predicted position, with errors approaching $20''$ in 1825. Then, strangely, things started to improve and around 1829–30, the longitude of Uranus was pretty much as given by Bouvard's tables. This temporary respite was short-lived, though, and Uranus then began to fall behind its predicted position at an alarming rate; by 1832, the errors were nearly $30''$. What was going on? Many hypotheses for the strange behaviour of Uranus were suggested. Perhaps it was due to some resistance from the medium in which Uranus moved. Encke had used this same idea to attempt to explain the anomalous motion of a comet (comet Encke), the period of 3.3 years of which appeared to diminish by about $2\frac{1}{2}$ h each cycle.[43] But why would only some bodies be affected and not others? Other suggestions included the existence of an enormous, as yet undetected, satellite. Even ignoring the fact that one would have expected such a satellite to have been visible through a telescope, perturbations caused by such a body would have had a much shorter period than those observed. Another possibility was that the tables that had been constructed contained errors. It was impossible to produce error-free tables and Bouvard's were no exception. Mistakes were found, but they did not solve the problem.

A more serious candidate for the source of the errors was Newton's law of gravity itself. Maybe it needed modifying when the distances involved were as great as those that were relevant to the study of Uranus. While this could not be ruled out, most astronomers were not keen on the idea. Every time universal

---

[42] Quoted from Grosser (1962).
[43] This affect (and a similar phenomenon for Halley's comet, the period of which increases by about 4.1 days each cycle) is believed now to be due to the ejection of gas when the objects are strongly irradiated by the Sun.

gravitation had been put to the test before, it had triumphed. It was a brave man who would bet against it. Instead, attention became focused on the possibility of there being a planet beyond Uranus that was causing perturbations to its orbit.[44]

The possibility of such a planet existing was not new, but speculation about a trans-Uranian planet increased the more Uranus 'misbehaved'. As far back as 1758, Clairaut had hypothesized that the gravitational effect of planets outside Saturn might sensibly perturb the orbit of Halley's comet, an idea taken up in the 1830s by astronomers such as Jean Valz and Friedrich Nicolai from the observatories in Marseille and Mannheim, respectively. Bouvard himself thought this the most likely cause for the failure of his tables actually to predict the position of Uranus. The change in attitudes to the existence of a planet beyond Uranus during the 1830s is illustrated graphically in the writing of Mary Somerville. In *The Mechanism of the Heavens* (1831), Uranus represented the outermost object in the Solar System. In the first edition of *On The Connexion of the Physical Sciences* (1834), the issue is avoided but, in the third edition (1836), the possibility of the perturbing influence of some unseen planet is mentioned explicitly.[45]

In 1834, Thomas J. Hussey, a clergyman and amateur astronomer from England who had discussed the problem of Uranus with Bouvard, suggested to Airy (the director of the Cambridge Observatory at the time) that such a planet might be discovered through the use of mathematics. His idea was that perturbation theory could be used to predict an approximate orbit for the planet, and then the sky searched in the vicinity of the predicted position. Hussey did not have the technical ability, but what did Airy think? We get an idea of what he thought from his reply:

> I have often thought of the irregularity of Uranus, and since the receipt of your letter have looked more carefully to it. It is a puzzling subject, but I give it as my opinion, without hesitation, that it is not yet in such a state as to give the smallest hope of making out the nature of any external action on the planet.[46]

---

[44] The fact that astronomers chose, by and large, to accept the absolute truth of universal gravitation even though it failed to predict the correct motion of Uranus has made the discovery of Neptune an important topic in the philosophy of science. The issue is examined in detail in Bamford (1996) in which the author concludes that in the 1840s there was little reason to believe that any trans-Uranian planet would have been discovered by accident, and so it was entirely rational to consider the existence of a new planet as being a more likely cause for the erratic motion of Uranus rather than any imperfection in Newton's law of gravity.

[45] *On The Connexion of the Physical Sciences* was a greatly expanded version of Somerville's *Preliminary Dissertation* (described on p. 353) which had served as an introduction to *The Mechanism of the Heavens*. In it, she described the current state of physical science. The work was hugely popular, selling over 15 000 copies through numerous editions.

[46] Quoted from Grosser (1962).

Airy believed that a more detailed examination of the perturbations due to Saturn was the answer but, even if the irregularities were caused by a new planet, he thought that Uranus would have to complete a few orbits (each of 84 years) before it would be possible theoretically to predict its nature. He repeated these same misgivings in 1837 in a letter to Bouvard's nephew Eugène, who was working on new tables of Uranus for his uncle. This negative attitude towards the existence of a trans-Uranian planet was to be very significant. Airy became Astronomer Royal in 1835 and had a considerable influence over many aspects of British science. He was a meticulous administrator and an extremely hard-working public servant, but he was not one to indulge in scientific flights of fancy.

Others were not so cautious. Friedrich Wilhelm Bessel became very enthusiastic about the possibility of discovering a new outer planet:

> I think the time will come when the solution of Uranus' mystery may be furnished by a new planet whose elements would be known according to its action on Uranus, and verified according to that which it exercises on Saturn.[47]

Unfortunately, his health prevented him from getting very far with his investigations. Johann Heinrich Mädler, director of the Dorpat Observatory in Estonia, wrote in his book *Popular Astronomy* (1841):

> we may even express the hope that analysis will at some future time realize in this her highest triumph, a discovery made with the mind's eye, in regions where sight itself was unable to penetrate.[48]

As it happened, thanks to the heroic efforts of two brilliant young men – John Couch Adams and Urbain Leverrier – Mädler got what he wanted and Airy's pessimism was shown to be misplaced.

Adams and Leverrier

Adams was a largely self-taught child prodigy, born into the family of a tenant farmer from Cornwall.[49] He was educated at St John's College, Cambridge, where he studied mathematics and tried to avoid the tempting distractions of astronomy in case they interfered with his studies. Clearly, he had some success in this, since he graduated as Senior Wrangler (the student with the highest score in the hugely demanding Mathematical Tripos) in 1843, reportedly with twice

---

[47] Quoted from Hanson (1962).
[48] Quoted from Grosser (1962).
[49] For a full-length biography of Adams, see Harrison (1994), written by his great-great-niece. Biographical details can also be found in Smart (1947).

as many marks as the second-placed student. In 1841, Adams read a report written 10 years earlier by Airy on the then current state of astronomy. Airy's description of the problem with Uranus clearly inspired him, since a week later he wrote in his diary:

> 1841, July 3. Formed a design, in the beginning of this week, of investigating, as soon as possible after taking my degree, the irregularities in the motion of Uranus which are yet unaccounted for; in order to find whether they may be attributed to the action of an undiscovered planet beyond it; and if possible then to determine approximately the elements of its orbit etc., which would probably lead to its discovery.[50]

After graduating, he put his plan into action. By the 1840s, Bouvard's tables were out by upwards of 70″ in longitude, and Airy had shown that they were in error for planetary distances, too. Thus, Adams first constructed new tables of the errant planet taking into account more terms[51] in the perturbations of Jupiter and Saturn, and a more accurate value for the mass of Jupiter. As he probably suspected, the problems with the orbit of Uranus did not disappear.

To make progress, Adams was faced with an inverse problem. Instead of having to calculate the disturbing effect of one planet on another, he had to find the orbit and mass of a planet that would produce the observed irregularities in the orbit of Uranus – a much harder task. Assuming the existence of a trans-Uranian planet, there were two distinct causes for the errors in Bouvard's tables. As well as its obvious perturbing effect, the undiscovered planet also influenced all the observations of Uranus from which its underlying elements had been determined, and so these had to be considered erroneous. Adams decided to assume that the two planets moved in the same plane, and then he needed to consider the necessary correction to the semi-major axis, δa, the eccentricity, δe, the longitude of perihelion, $\delta\bar{\omega}$, and the mean longitude at epoch, $\delta\epsilon$. On top of this, there were the actual perturbations due to the hypothetical planet, which depended on five parameters: the four elements a', e', $\bar{\omega}'$, ϵ', of the planet and its mass, m'. The discrepancies D between the observed and tabulated longitudes of Uranus could then be written as the sum of two complicated functions in the form

$$D = U(\delta a, \delta e, \delta\bar{\omega}, \delta\epsilon) + N(m', a', e', \bar{\omega}', \epsilon'), \qquad (10.4)$$

---

[50] Quoted from Harrison (1994). Adams later recalled that his ideas were in part motivated by reading Somerville's suggestion of a trans-Uranian planet in *On The Connexion of the Physical Sciences*.

[51] These extra terms were first pointed out by Peter Hansen (Grant (1966), p. 170).

involving nine unknowns. These unknowns can then theoretically be determined by performing a least squares analysis on at least nine equations of condition formed from the observed discrepancies. The calculations involved are horrendous, however. It is worth noting that all this analysis had to be performed after the effects of Jupiter and Saturn had been accounted for. Hence, Adams' preliminary work on these perturbations was in no way a wasted effort.

Adams began by making some simplifying assumptions to see if a planet outside Uranus could have the right qualitative effect on Uranus. If it could, then he would go on to make more detailed calculations. He thus assumed a circular orbit for the new planet ($e' = 0$) and, in the absence of any other rationale, took the semi-major axis of the orbit to be twice that of Uranus, $a' = 2a = 38.4$ AU, consistent with the Titius–Bode law. With help from his brother George, he carried out the necessary calculations, and concluded that the irregularities of Uranus were consistent with the exterior planet hypothesis.

Adams now needed more astronomical data. He asked Airy's successor as Plumian Professor of astronomy and director of the Cambridge Observatory, James Challis,[52] to write to the Astronomer Royal (Airy) on his behalf, which he did, and Airy obliged with all the observational data on Uranus from 1754 to 1830. Adams then embarked on a more detailed analysis in 1844. He now assumed that the eccentricity was small but nonzero, and neglected terms containing squares or higher powers of e' in Eqn (10.4) but retained his assumption that $a'/a = 2$. The quantity a' entered the equations in the form of complicated infinite series in powers of a/a', and the only way to make the problem tractable was to assign a value to this ratio.

By September 1845, Adams had a solution in the form of corrections to the orbit of Uranus and elements for the orbit of a new planet. Now he could predict where the undiscovered planet should be, and all he needed was somebody to conduct a telescopic search of the appropriate area of sky. If he had thought that this last step would be straightforward, he was sadly mistaken. Adams told Challis the results of his calculations, including the predicted geocentric longitude of the new planet for 30 September but, despite Challis' respect for Adams as a mathematician, he did not take the opportunity to search for the planet. He did, however, provide Adams with a letter of introduction so that he could visit Airy at the Royal Observatory in Greenwich.

Such a meeting never took place, but Adams did leave a summary of his results at the Royal Observatory:

---

[52] Challis (in 1825) and Airy (in 1823) were both themselves Senior Wrangler.

According to my calculations, the observed irregularities in the motion of Uranus may be accounted for by supposing the existence of an exterior planet, the mass and orbit of which are as follows:–

Mean Distance (assumed nearly in accordance with Bode's law)	38.4
Mean Sidereal Motion in 365.25 days	1° 30′.9
Mean Longitude, 1st October 1845	323° 34′
Longitude of Perihelion	315° 55′
Eccentricity	0.1610
Mass (that of the sun being unity)	0.000 165 6

There then followed a list of the remaining errors in mean longitude, and Adams continued:

The errors are small, except for Flamsteed's observation of 1690. This being an isolated observation, very distant from the rest, I thought it best not to use it in forming the equations of condition. It is not improbable however that this error might be destroyed by a small change in the assumed mean motion of the new planet.[53]

Airy replied with typical caution. There was something rather *ad hoc* about the use of Bode's law in the assumption for the mean distance, and Airy needed more evidence to be convinced. He acknowledged that the mean errors reported by Adams were 'extremely satisfactory', but he asked if the theory could also predict correctly the observed irregularities in the orbital radius of Uranus.[54] Unfortunately, Adams did not provide an answer – for a number of reasons. In the first place, he was busy redoing his calculations for a different mean distance in an attempt to reduce the error for Flamsteed's 1690 observation. Second, he did not appreciate the importance Airy attached to the question and, third, he thought the answer trivial. When he later corresponded with Airy on the matter, Adams argued that variations in the longitude θ were linked to variations in orbital distance r through the equation (see Eqn (8.6), p. 269)

$$r^2\dot\theta = \sqrt{\mu a(1 - e^2)}.$$

This represents the conservation of angular momentum for a planet moving in an elliptic orbit, and Adams believed that even for the perturbed motion of

---

[53] Adams' handwritten note is reproduced in Jones (1947). Adams preferred to work in terms of perturbations of the mean longitude of Uranus rather than those of the true longitude, which are more easily determined from observations. His justification for the method he used to convert between the variations in the two quantities was rather dubious, however (see Lyttleton (1968), p. 229).

[54] The fact that the theoretical distance between the Sun and Uranus was less than that computed from observations made between 1833 and 1836 had been pointed out by Airy in 1838 (Smart (1947)).

Uranus the variations in the right-hand side would be 'very small'; hence, any significant decrease in $\dot{\theta}$ (as had been observed) must be accompanied by an associated increase in r. In fact, Adams' argument is flawed.[55] The variation in angular momentum is, in fact, of the same order as the variation in θ, and so correcting the longitude does not necessarily produce the required changes in the orbital radius. Whatever the reasons behind Adams' failure to respond to Airy's enquiry, the result was that his predictions lay idly on the desks of two of the most accomplished astronomers in England.

Meanwhile, in France, Uranus was also occupying the minds of astronomers. Eugéne Bouvard's new tables had not made the problems go away, and François Arago, director of the Paris Observatory, asked the talented Leverrier to investigate. Like Adams, Leverrier was an exceptionally skilled theoretician, but there the similarity ended. In 1845, Adams was just a graduate student, with no publications to his name and no reputation. Leverrier was 8 years older and had an established reputation with two significant papers already published on perturbation theory,[56] as well as some highly respected work on the theory of Mercury.

Just like Adams, Leverrier began by re-examining the motion of Uranus. His analysis was the most thorough to date (he claimed to have included all perturbative inequalities the magnitude of which was greater than $\frac{1}{20}''$) but he reached the same conclusion as everybody else who had analysed the problem: there was something else affecting the orbit of Uranus. When he read Leverrier's paper, Airy was impressed, and for the first time considered that the theory of Uranus was at least placed on a satisfactory foundation. Although he now realized he had been wrong about Saturn being the solution, he does not appear to have taken Adams' calculations any more seriously.

Leverrier continued his analysis and, in June 1846, presented his second memoir on Uranus. He discussed the various possibilities for the irregularities in the motion of Uranus and, like many others before him, came to the conclusion that the most likely source for the peculiar motion of Uranus was a trans-Uranian planet orbiting close to the ecliptic plane:

> Is it possible that Uranus' inequalities may be due to a planet located in the ecliptic, at a mean distance double that of Uranus? And were this so, where is the planet now? What is its mass? What are the elements of its orbit?[57]

---

[55] This was pointed out by J. E. Littlewood (1953).

[56] One on the stability of the Solar System and one on the influence of Jupiter on comets (see, for example, Grosser (1962) pp. 61–9 for more details). Leverrier actually began his scientific career as a chemist, working with Gay-Lussac for 3 years before accepting a position in astronomy at L' École Polytechnique in 1837.

[57] Quoted from Hanson (1962).

Faced then with the same inverse problem as Adams, Leverrier had to make some assumptions about the unknown elements of this hypothetical planet. Of the nine unknowns in Eqn (10.4), the four corrections to the orbit of Uranus cause little difficulty, since they are all small quantities and squares and higher powers can be neglected. The equations are thus linear in these quantities, which then can be eliminated easily. Of the remaining unknowns, e' and $\bar{\omega}'$ can be replaced by $h' = e' \cos \bar{\omega}'$ and $k' = e' \sin \bar{\omega}'$, and then h' and k' occur linearly in the equations and similarly can be eliminated. The elements a' and ϵ' are much more troublesome and, like Adams, Leverrier took the orbit to have a mean distance consistent with Bode's law as a first approximation. This left an equation that related the longitude of the new planet, ϵ', to its mass, m'. Leverrier knew that the mass could not be so large as to cause effects on those planets inside the orbit of Uranus and, of course, it had to be positive. This placed restrictions on the possible location of the disturbing planet. Then he could select the position that gave the best results by minimizing the associated errors, and concluded (after numerous lengthy calculations) that, on 1 January 1847, the planet would have heliocentric longitude $325 \pm 10°$.[58]

Leverrier's conclusions were very similar to those of Adams, but Leverrier's were in the public domain. When Airy received a copy of Leverrier's analysis, he was the only person who knew that there were, in fact, two calculations of the position of a supposed perturbing planet in existence, and they produced very nearly the same answer. Airy wrote:

> ... there are two remarkable calculations. One is by Adams of St John's (which in manuscript reached me first). The other is by Le Verrier. Both have arrived at the same result, viz. that the present longitude of the said disturber must be somewhere near 325°.[59]

Airy also wrote to Leverrier, but he made no mention of Adams. Instead, he asked for the same information for which he had asked Adams the previous year. Did the analysis account for the irregularities in orbital distance? Airy was clearly doubtful on this point, but Leverrier replied immediately and confirmed that the orbital radius did not represent a problem. Airy was again impressed by Leverrier's confident and clear explanations, and became convinced that the theoretical calculations he had in his possession were to be taken seriously. He decided that the search for the new planet should now begin.

Airy decided that the best way to search for the planet was to map the area of sky near the predicted position, then to repeat the exercise a few weeks later, and a few weeks after that. The data from the three surveys could then be compared,

---

[58] For more details of Leverrier's procedure (see, for example, Grant (1966), pp. 179–83).
[59] Letter to William Whewell, 25 June 1846. Quoted from Standage (2000), p. 91.

and if one of the 'stars' was found to have moved, the quarry would have been found. He decided also that the most appropriate telescope to use for the search was the Northumberland telescope at the Cambridge Observatory, rather than the smaller instruments at the Royal Observatory.

Airy asked Challis to conduct a search of an area of sky measuring 30° in longitude by 10° in latitude, centred on the position predicted by Adams and Leverrier. This represented an enormous amount of work and was, in fact, an extremely inefficient method, given that a quick result was desired. Nevertheless, Challis began his search on 29 July 1846 and even decided to record the positions of stars much fainter than the new planet was predicted to be. Had Adams' predictions been well known, many amateur astronomers might have pointed their telescopes at the appropriate area of the heavens and the chances of discovery would have been much improved, but it would appear that Airy and Challis were both keen for the honour of discovering a new planet to go to Cambridge. Leverrier's prediction did appear in the British press in August.

In France, a search mounted by the Paris Observatory in early August was soon terminated. Tempting as the possibility of finding a new planet must have been, there was no guarantee of success, and it was hard to justify utilizing valuable resources on what might turn out to be a wild-goose chase. On 31 August, Leverrier's third paper on Uranus was presented to the Academy. In it, he treated the value of the longitude at epoch, ϵ', which he had calculated previously, and the semi-major axis, a', which he had assumed previously, as approximations. He thus introduced two new unknowns into the problem, (i.e. the differences between the true values, and the values previously used) and assumed that these were small quantities. Lengthy calculations produced orbital elements for the disturbing planet as follows:[60]

Semi-major axis	36.154 AU	($a/a' = 0.531$)
Eccentricity	0.10761	
Longitude of perihelion	284° 45′	
Mass	1/9300	
True heliocentric longitude, 1 January 1847	326° 32′	

Leverrier included also an estimate of the size of the new planet, suggesting that it should have a disc measuring about 3.3″ across. Provided one had a sufficiently powerful telescope, searching for a 3″ disc was much easier than trying to detect a very slow motion between successive observations.

During 1846, Adams was also busy refining his calculations. In September, he sent the results of his most recent attempt at the problem of Uranus to

---

[60] Quoted from Grosser (1962).

the Royal Observatory. His initial calculations had predicted an eccentricity that was rather large in comparison to the other planets, and by reducing the assumed semi-major axis he was able to derive an orbit with a more reasonable eccentricity as well as one that slightly better fitted the data. His revised orbital elements were:[61]

Semi-major axis	37.25 AU ($a/a' = 0.515$)
Eccentricity	0.120 62
Longitude of perihelion	299° 11′
Mass	1/6666
True heliocentric longitude, 1 January 1847	329° 57′

Adams expressed also the opinion that by decreasing the semi-major axis still further, an even better fit with observation might be achieved, and he suggested that $a/a' = 0.57$ (corresponding to about 33.7 AU) would probably deliver the best results. As it happened, Airy was away in Germany and Adams' letter was not forwarded to him.

Neptune

Neither Adams nor Leverrier was having much luck with their respective astronomical establishments. Adams could, at least, be thankful that Challis was at last searching for the new planet, but Leverrier was having no success in getting his fellow countrymen to scan the heavens in search of something that might not exist. On 18 September, Leverrier wrote asking for help to an assistant at the Berlin Observatory – Johann Gottfried Galle – who had sent him a copy of his doctoral dissertation the previous year. On 23 September, the same day he received the letter, Galle and an astronomy student, Heinrich d'Arrest, turned the Fraunhofer telescope to the place indicated by Leverrier. They decided to compare their observations, as they made them, with the Berlin Academy's new star atlas, and after a few hours found a 'star' within 1° of the predicted position that was not in the atlas. The following day (accompanied now by the Observatory's director, Encke) they observed the 'star' again. It had moved by exactly the amount predicted by the calculations! One can only imagine the excitement that these men must have felt. Once they knew they were looking at a planet, it became easier to discern a disc (as Encke put it, 'the disc can be

---

[61] Adams actually supplied the longitude values for 1 October 1846; the figure for 1 January 1847 is included here for easy comparison with Leverrier's data. The values are from Grosser (1962).

recognized only when one knows that it exists'[62]) and they managed to measure its size as between 2.2 and 3.2''. Galle wrote to Leverrier on 25 September:

> The planet whose position you have pointed out *actually exists*. The same day that I received your letter, I found a star of the eighth magnitude which was not shown on the excellent chart . . . published by the Royal Academy of Berlin. The observations made the following day determined that this was the sought-for planet.[63]

Despite the magnitude of the scientific achievement represented by Galle's discovery, there were a few people in England who perhaps were a little disappointed. The discovery of a trans-Uranian planet was a triumph for the power of the theory of gravitation, but having the discovery taken from under your nose must have been a blow. Worse still, Challis looked back over his observations and found he had seen the new planet twice during his search! Subsequent investigations revealed other 'old' sightings. Adams himself soon put any disappointment to one side and decided instead to use the now known position of the planet to calculate a more accurate orbit. The official spokesman for British astronomy – Airy – was still abroad,[64] but Sir John Herschel, a member of the Board of Visitors of the Royal Observatory who was by then aware of Adams' work, decided that the young Cambridge mathematician's contribution should be put on record.[65] On 1 October 1846, he wrote a letter to the British journal the *Athenaeum* (published on 3 October) in which, after praising Leverrier for his remarkable work, he pointed out that Adams had tackled the same problem independently and arrived at very similar conclusions, thus providing important corroboration for Leverrier's approach. This simple act was to have consequences far greater than Herschel could have imagined.

---

[62] Letter to H. C. Schumacher, the editor of the *Astronomische Nachrichten*, 26 September 1846. The letter is translated in full in Shapley and Howarth (1929).

[63] Quoted from Grosser (1962).

[64] Airy received the news of the discovery on 29 September while in Germany, several days before the news reached England. According to Grosser (1962) he did not write to anyone about it for more than 2 weeks, but Chapman (1988) cites two letters that Airy wrote on 30 September concerning the discovery, one to his assistant at Greenwich and the other to Challis.

[65] John Herschel was the only child of William Herschel. In 1812, together with fellow Cambridge undergraduates Charles Babbage and George Peacock, he formed the Analytical Society to promote the superior continental methods of analysis, or, as Babbage put it, 'to uphold the principles of pure D-ism in opposition to the Dot-age of the University' (Cannon 1961). (This was the beginning of a long battle (see Becher (1980)).) John Herschel, Senior Wrangler in 1813, had a distinguished scientific career and was knighted in 1831. His *Treatise on Astronomy*, first published in 1833, in which Herschel described the reasoning underpinning the application of Newtonian principles to the heavens without using mathematical terminology, was a huge success. It appeared in seven editions, was then expanded into the *Outlines of Astronomy* in 1849 which went through twelve more editions, the last appearing in 1873. It was translated into many languages and John Herschel became one of the world's most well-known astronomers.

When Herschel's article began to circulate in France there was uproar. Who was this Adams working away secretly in England? Why was he trying to steal Leverrier's glory? Leverrier himself was furious, and wrote to Airy asking why he had not mentioned Adams' work prior to the discovery. There were vitriolic attacks in the press, with Airy, Challis and Herschel accused of plotting to undermine Leverrier's claim as the discoverer. The name for the new planet now became a banner behind which to rally. After initially suggesting Neptune, Leverrier decided that he wanted the planet named after himself, and Arago agreed. Many astronomers were distinctly unhappy about naming celestial bodies after their discoverers and various alternatives were suggested. This was, of course, interpreted as yet another attempt to belittle Leverrier's achievement.

Back in England, things were not much better for Airy and Challis. When the truth became apparent, British astronomers were quick to criticize the two men for their secretive behaviour. At a meeting of the Royal Society in November, Airy, Challis, and Adams all gave accounts of their respective roles in the affair. Airy tried to distance himself from the controversy, claiming he had acted as dictated by his official capacity. Challis only succeeded in looking foolish. Adams, on the other hand, gained respect. It was clear that he was a superb theoretician who, in the opinion of many, had been robbed of the discovery only by Challis' incompetence. A slogan began to circulate around Cambridge:

> When Airy was told, he wouldn't believe it;
> When Challis saw, he couldn't perceive it.[66]

Herschel was horrified at all this. He was obviously disappointed that the discovery had not been made at Cambridge, but he did not think he had behaved improperly. In November 1846 he wrote:

> I mourn over the loss to England and to Cambridge of a discovery which ought to be theirs every inch of it, but I have said enough about it to get heartily abused in France and I don't want to get hated in England for saying more.[67]

---

[66] Quoted from Standage (2000), p. 129.
[67] Letter to William Whewell, 6 November 1846. Quoted from Smith (1989). This article describes the controversy surrounding the discovery of Neptune and examines in detail the actions of Airy, Challis and Herschel (on this topic, see also, for example, Grant (1966), pp. 193–200, Pannekoek (1953), Grosser (1962), Chapman (1988), and *The Neptune Scandal*, Chapter VI of Ronan (1967)). Airy's actions have come in for serious criticism, but there is a recognition that his public duties as Astronomer Royal put him in a difficult position. History has been less kind to Challis. When Airy found out about earlier observations of Neptune he remarked 'Let no one after this blame Challis' (Jones (1947)). However, Jones (1947) describes Challis' attempts to justify his failure to discover Neptune as 'pitiable'. Mary Roseveare, a niece of Adams, later recalled that 'as children we put most blame in the end on Professor Challis' (Harrison (1994), p. 71).

Moreover, he had hoped that the discovery would be used to inspire the public, and the continual bickering was distinctly unhelpful. The war of words continued for some time, but eventually bridges were built, civility returned, and a consensus emerged (outside France at least) for the name 'Neptune'. When tempers had cooled and the evidence was considered carefully, it was appreciated that Adams deserved some credit for the discovery. Adams and Leverrier first met in June 1847, and became friends. In 1876, Leverrier was awarded the Gold Medal from the Royal Astronomical Society (for the second time) and the address was made by the then president, none other than Adams himself.

No sooner had the controversy over the discovery died down, than doubt was cast on Adams' and Leverrier's calculations following the publication of elements for the orbit of Neptune. Since the two men had predicted the longitude of Neptune to within $2\frac{1}{2}$ and $1°$, respectively, it was expected that the orbital elements on which they had based their predictions similarly would be close to the actual elements of Neptune. Preliminary calculations by Adams based on observations by Challis suggested that this was not the case, and this was confirmed in 1847 by Sears Cook Walker, an astronomer from the US Naval Observatory in Washington D.C. After some preliminary calculations of elements for Neptune, Walker decided to examine old catalogues to see if the planet previously had been recorded as a star, and found what he wanted in J.-J. de Lalande's star catalogue. It appeared likely that Lalande had observed Neptune on 10 May 1795, and in February 1847 Walker worked out elements for the orbit of Neptune utilizing this observation. He communicated his findings to Leverrier, who presented them to the Academy of Sciences in March.[68] Sensationally, they were quite different from those predicted by Adams and Leverrier:[69]

---

[68] The 10 May 1795 observation of Neptune was also unearthed by Adolf Cornelius Peterson of the Altona Observatory in Germany at around the same time, and Leverrier reported both findings to the Academy. The status of this observation as a prediscovery sighting of Neptune was uncertain, however, since it was marked in the catalogue with a colon, Lalande's way of indicating a doubtful position. Investigations of the original manuscripts revealed that Neptune actually had been seen on both 8 and 10 May (by Lalande's nephew Michel according to Baum and Sheehan (1997)). There was a small discrepancy between the two observations, consistent with the planet's motion, but Lalande simply had assumed that the first observation was inaccurate and had published the second as doubtful! Other prediscovery sightings came to light subsequently. Remarkably, a recent study has shown that it is more than likely that Galileo observed Neptune in December 1612 and January 1613, during which period Neptune appeared very close to Jupiter (Kowal and Drake (1980)).

[69] The quoted values are from Hubbell and Smith (1992). The mass of Neptune was not known with any accuracy until extensive observations had been made of Neptune's satellite Triton, which was discovered by the amateur astronomer William Lassell not long after Neptune itself was discovered. On 10 October 1846, Lassell reported seeing a tiny star within a few diameters of the planet. He suspected that this might be a satellite, and finally managed to confirm its status in the following July.

Semi-major axis	30.250 42 AU
Eccentricity	0.008 840 7
Longitude of perihelion	0° 12′ 25″ .51
True heliocentric longitude, 1 January 1847	328° 7′ 56″ .64

Quite different except in one important respect. The heliocentric longitude around the time of discovery was almost exactly the same, and this was, of course, the crucial figure when searching for the planet.

Some argued, on the basis of Walker's elements, that the coincidence of the theoretical predictions and the actual position of Neptune were purely fortuitous, the leading proponent of this 'happy accident' thesis being Benjamin Peirce, Professor of Astronomy and Mathematics at Harvard University.[70] Unsurprisingly, such ideas were attacked vigorously by Leverrier and Sir John Herschel. In fact, the value assumed by Adams and Leverrier for the semi-major axis was largely irrelevant. Provided what was chosen was not too far away from reality, the method they used compensated by predicting an erroneous mass and eccentricity, so that the gravitational pull of the hypothetical planet has the same effect on Uranus. It was fortunate that at the time Adams and Leverrier were working, Neptune was as close as it gets to Uranus, but then this was why the motion of Uranus was so erratic – the reason for tackling the problem in the first place. The predicted orbits of Adams and Leverrier, together with the actual orbits of Uranus and Neptune, are illustrated schematically in Figure 10.1. Neptune's orbit is in reality an almost perfect circle, whereas the two predicted orbits have a much greater eccentricity. But since Adams and Leverrier both began with a semi-major axis that was significantly too long, the large eccentricity has the effect of making the actual distances from the Sun around the time of discovery quite close to the true ones. Adams' and Leverrier's disturbing planets are both still too far away, but their masses were larger, so that the gravitational effect, which is all that was important in the calculations, was roughly the same as that of Neptune itself.[71]

With the knowledge of the existence of Neptune, astronomers could look back at the problems they had faced in producing tables for Uranus. Figure 10.2 shows schematically the orbit of Uranus in a frame of reference in which Neptune is fixed. The orbital period of Uranus is about 30 685 days, and that

---

[70] Peirce's arguments are discussed in, for example, Hoyt (1980), Hubbell and Smith (1992).
[71] Ernest William Brown, Professor of Mathematics at Yale University, conjectured that much of Adams' and Leverrier's complicated analysis was redundant and that, given knowledge of the time of conjunction, little further effort would be needed to find the planet (Brown (1931)). This idea was taken up by J. E. Littlewood and R. A. Lyttleton, who asked the question 'What is the simplest theoretical approach and minimum amount of calculation that could have led to the discovery of the planet?' Details of their analyses can be found in Littlewood (1953b) and Lyttleton (1968), Chapter 7.

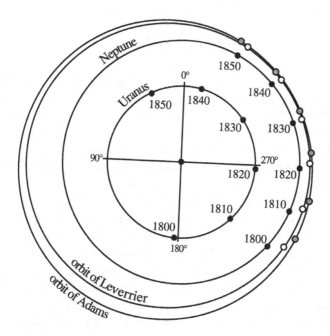

Fig. 10.1. The predicted orbits of Adams and Leverrier.

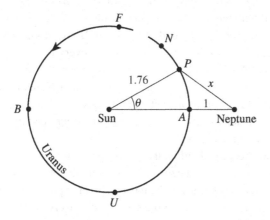

Fig. 10.2. The orbit of Uranus relative to Neptune.

of Neptune about 60 190 days; hence, the period of Uranus relative to Neptune is $(1/30685 - 1/60190)^{-1}$ days, or approximately 171.4 years. Uranus was at is closest to Neptune in 1822, and so the angle θ marked in the figure, giving the point on the orbit relative to the point of closest approach, can be represented by

$$\theta = \frac{\text{year} - 1822}{171.4} \times 360°.$$

The positions on the orbit indicated are F (Flamsteed's observation, 1690, $\theta = -277°$), B (point of maximum separation, 1736, $\theta = -180°$), U (discovery of Uranus, 1781, $\theta = -86°$), A (closest approach, 1822, $\theta = 0°$), and N (discovery of Neptune, 1846, $\theta = 50°$).

The ratio of the radii of the orbits of the two planets is about 1.57 : 1 and so in units chosen so that the minimum distance between them is unity, the radius of the orbit of Uranus is 1.76 (since $2.76/1.76 \approx 1.57$). Of significance for the motion of Uranus is the gravitational attraction of Neptune, which is inversely proportional to the square of the distance between the two planets. Thus, if the maximum pull is scaled to 1, the pull when Uranus is at P is $1/x^2$, where $x^2 = 1.76^2 + 2.76^2 - 2 \times 1.76 \times 2.76 \cos \theta$. It follows that at F and U the effect of Neptune is about one-tenth of its maximum value, with a minimum value at B roughly half of this. When Neptune was discovered, its gravitational effect on Uranus was just over 20 per cent of its maximum value.[72]

In the absence of any knowledge of the existence of Neptune then, the pre-discovery observations were the best on which to base an elliptic orbit for Uranus, as largely they were unaffected by the undiscovered planet. After correcting for the effects of Jupiter and Saturn, such an ellipse would, of course, still not fit well with the motion of Uranus while it was close to Neptune, but it would represent accurately the underlying orbit. Bouvard's decision to discard the 'old' observations in favour of the post-discovery ones was thus a rather unfortunate one! While he had the choice as to which data to use, he had no control over the relative positions of Uranus and Neptune when he was preparing his tables. As it happened, his timing was the worst possible, since the maximum interaction between the two planets occurred almost immediately after publication of the tables.

Perhaps the theoretical difficulty of the problem solved by Adams and Leverrier sometimes is overstated, and there is no doubt that, from a technical point of view, their work was not the most brilliant application of mathematics to a problem in astronomy; nevertheless, the discovery of Neptune ranks as one of the greatest achievements in the history of astronomy. Its psychological impact was immense, and it led to a new wave of optimism about what might

---

[72] The fact that the gravitational force is a maximum when the two planets are closest does not, of course, imply that the perturbations in the elements take their maximum values at this point. It is, however, reasonable to conclude that data collected during a period when the perturbing force is large is unlikely to lead to an accurate determination of the underlying elliptic orbit.

be achieved in the future. To the public, the idea that a new planet could be discovered by somebody sitting at a desk and writing endless unintelligible mathematical formulae, rather than peering at the heavens through a powerful telescope, gave the discovery miracle status. The astronomical community understood the basic premise on which the calculations of Adams and Leverrier were based, but were no less impressed. In his address to the Royal Astronomical Society in November 1846, Airy expressed his admiration in the following terms:

> ... the motions of Uranus, examined by philosophers who were fully impressed with the universality of the law of gravitation, have long exhibited the effects of some disturbing body: mathematicians have at length ventured on the task of ascertaining where such a body could be; they have pointed out that the supposition of a disturbing body moving in a certain orbit, precisely indicated by them, would entirely explain the observed disturbances of Uranus: they have expressed their conviction, with a firmness which I must characterise as wonderful, that the disturbing planet would be found exactly in a certain spot, and presenting exactly a certain appearance; and in that spot, and with that appearance, the planet has been found. Nothing in the whole history of astronomy can be compared with this.[73]

Pluto

Leverrier did not think that Neptune was the end of the story, but rather the beginning of a new era of 'invisible astronomy'. He believed that the inverse technique he had developed one day would be used to discover new planets yet further from the Sun. As he put it:

> This success must allow us to hope that after 30 or 40 years of observations of the new planet, it will be possible to use it, in turn, to discover the next planet in order of distance from the sun. And then the next; soon the planets will, unfortunately, become invisible, due to their immense distance from the sun, but their orbits will, in the following centuries, be worked out with great accuracy, through the use of theory.[74]

Once the mass of Neptune and the elements of its orbit had been determined accurately from observations in the late nineteenth century, astronomers could check to see whether it did account for the irregularities in the motion of Uranus. Results seemed to suggest that there was still a discrepancy, albeit a tiny one, and some began to speculate that this was due to a planet exterior to Neptune.

---

[73] Monthly Notices of the Royal Astronomical Society, **VII** (1846).
[74] Quoted from Standage (2000), pp. 156–7.

As Adams and Leverrier had found, determining the elements of such a perturbing planet was impossible unless some assumption was made concerning the semi-major axis of the orbit. Both men had based their initial calculations on the Titius–Bode law, but given the apparent failure of this law for Neptune, this was now hardly an attractive option. A completely different approach was taken by the chief assistant of the US Nautical Almanac Office, David Peck Todd, who in 1877 made the first serious attempt to predict theoretically the orbit of a trans-Neptunian planet. He examined the discrepancies in the predicted positions of Uranus and guessed the period of the new planet from the times between successive peaks in these residuals using a method similar to that described by John Herschel in his *Outlines of Astronomy* (1849). The orbital radius then followed from Kepler's third law. He concluded that there was a planet at 52 AU, the longitude of which in August 1877 was $170 \pm 10°$. He searched in vain for thirty nights.

Another stab at guessing the distance from the Sun was made by the Scottish astronomer, George Forbes, based on an argument put forward by Camille Flammarion, a onetime assistant of Leverrier, in the 1879 edition of his *Popular Astronomy*. Flammarion argued that a clustering of cometary aphelia at a certain distance would be suggestive of a planet orbiting at that distance, and in 1880 Forbes used existing data to propose two trans-Neptunian planets, at approximately 100 and 300 AU. He received some dubious encouragement for his predictions from Pliny Earle Chase, Professor of Philosophy at Haverford College near Philadelphia, who noted that the radii of Forbes' two planets fit well with the formula $\pi^n/32$ (with $n = 7$ and 8, respectively), which he had noted produces numbers that agree fairly well with some of the other planets. An unsuccessful search for the closer of the two was mounted during the period 1888–92 by Isaac Roberts, one of the pioneers of telescopic photography.

During the late nineteenth and early twentieth centuries there were many other attempts to predict a new planet beyond Neptune, some more sensible than others.[75] In the light of future events, two of these were significant: those of William Henry Pickering and Percival Lowell. Pickering announced the first of his many predictions concerning trans-Neptunian planets late in 1908. Using, like Todd before him, Herschel's graphical method for analysing the residuals of Uranus he deduced the existence of a planet at 51.9 AU with a mass twice that of the Earth. A photographic search of the relevant portion of the sky found

---

[75] See Grosser (1964) for a detailed account.

no sign of Pickering's 'Planet O'. Despite this lack of success, and criticism of his methods, Pickering maintained his belief in the existence of a planet beyond Neptune that was responsible for perturbations in Uranus, and his work prompted Lowell, who had been working on the same problem on and off since 1905, to renew his activity.[76]

In the early 1890s, Lowell decided to give up his business career so that he could devote his time to astronomy. In particular, he was fascinated by Giovanni Schiaparelli's 1877 report of 'canals' on Mars and wanted to continue observations of these faint lines on the Martian surface now that the Italian astronomer's eyesight meant he could no longer pursue his study. To this end, Lowell built an observatory near the town of Flagstaff in Arizona at an altitude of over 2000 m, where the air was particularly clear and the conditions for astronomical observations, excellent. Lowell became famous in the 1890s following his advocacy of the possibility of intelligent canal-building life on Mars.[77]

In contrast to all other trans-Neptunian planet hunters, Lowell decided that the only way to attack the problem was through solving the appropriate mathematical equations and he made his first calculations in 1905 using what was essentially Leverrier's 1846 technique. The results were unsatisfactory in that he got very different answers depending on whose tables of residuals for Uranus he used. Pickering's Planet O prediction in 1908 persuaded Lowell to revisit his 'Planet X', and he decided to see whether he could use the simpler graphical method of Pickering to good effect. He concluded that this was not the answer, writing at the end of his copy of Pickering's paper: 'This planet is very properly designated O [and] is nothing at all.'[78]

The only way to progress was to go back to Leverrier's theory and proceed with more rigour, including terms in the analysis up to the squares in the eccentricities. During the 5 years from 1910, Lowell gradually developed and refined his procedures and, finally, achieved what he believed to be a satisfactory result in 1915. He concluded that:[79]

---

[76] For full details of the discovery of a trans-Neptunian planet, see Hoyt (1980). Further information on Pickering's efforts can be found in Hoyt (1976) and Sadler (1990). Pickering proposed at least seven new planets that he designated O, P, Q, R, S, T, U, with distances between 5.79 and 6250 AU, and masses between 0.045 and 20 000 times that of the Earth.

[77] Schiaparelli used the word *canali*, meaning channels. However, the English-language press translated this as canals, with the implied involvement of some intelligent lifeform. In fact, the channels are illusory.

[78] Quoted from Hoyt (1980), p. 104.

[79] Extracts from Lowell's *Memoir on a Trans-Neptunian Planet*, published in Volume 1 of the *Memoirs of the Lowell Observatory* (1915). Quoted from Hoyt (1980).

- By the most rigorous method, that of least squares throughout, taking the perturbative action through the first powers of the eccentricities, the outstanding squares of the residuals from 1750 to 1903 have been reduced 71% by the admission of an outside perturbing body.
- The second part of the investigation, in which solutions were made for the second powers of the eccentricities as well, gave comfortable results.
- The investigation disclosed two possible solutions in each case, one with ϵ around 0° and one with ϵ around 180°; and that this duality of possible places would necessarily always be the case.
- On the whole, the best solutions for the two gave:

ϵ'	22°.1	205°.0
a'	43.0	44.7
m' (mass of sun $=$ 50 000)	1.00	1.14
e'	0.202	0.195
$\bar{\omega}'$	203°.8	19°.6
heliocentric longitude 0 July 1914	84°.0	262°.6

- It indicates for the unknown a mass between Neptune's and the earth's; a visibility of the 12–13 magnitude according to albedo; and a disc of more that 1″ in diameter.
- From the analogy of other members of the solar family, in which eccentricity and inclination are usually correlated, the inclination of its orbit to the plane of the ecliptic should be about 10°. This renders it more difficult to find.

Lowell co-ordinated three unsuccessful searches for a planet beyond Neptune, the third during the last 3 years of his life.[80] His will specified that the search should be continued at the Flagstaff Observatory until the planet was found.

Pickering continued with his predictions. Analysis of the motion of Neptune, which had begun to deviate from its tabular longitude, led him in 1919 to refine his elements for Planet O, which he now placed at 55.1 AU. Further searches were made, but the results were once again negative. By March 1929, Pickering had become disheartened:

> ... it has been suggested in some newspaper articles that all the large observatories would be hunting for planet O when it comes into opposition this February, I wish to state here that I myself have the gravest doubts on that point. Indeed if anyone is hunting for it I shall be much gratified.[81]

As it happened, the fourth systematic search for Lowell's Planet X began in April 1929 under the direction of Vesto Melvin Slipher, and the man charged with comparing the 14 × 17 in. photographic plates (which covered an area of sky about 12 by 14°) to detect moving bodies was a young assistant at the

---

[80] The first was carried out during 1905–7 and the second in 1911.
[81] Quoted from Grosser (1964).

observatory, Clyde Tombaugh.[82] A special blink comparator was used (in which the viewer sees two images alternately in the eyepiece – stars remain fixed but other objects appear to jump about), but even then the task was daunting, with some of the plates showing over half a million stars! On 18 February 1930, Tombaugh found two faint images of an unknown planet on photographic plates produced in January, and subsequent photographs provided confirmation that the object was orbiting beyond Neptune. The announcement of the discovery was withheld until 13 March, the anniversary of both the discovery of Uranus and the birth of Lowell, by which time those working on the project were convinced that Planet X had at last been found.

A telegram was dispatched to the Harvard College Observatory on 12 March which the next day was used to inform the rest of the world:

> Systematic search begun years ago supplementing Lowell's investigation for Trans-Neptunian planet has revealed object which since seven weeks has in rate of motion and path consistently conformed to Trans-Neptunian body at approximate distance he assigned. Fifteenth magnitude. Position March twelve days three hours GMT was seven seconds of time West from Delta Geminorum, agreeing with Lowell's predicted longitude.[83]

Astronomers everywhere turned their attention to this new member of the solar family. Numerous suggestions were made for a name, but astronomers settled finally on Pluto, the brother of Jupiter and Neptune. Preliminary orbits allowed other observatories to locate Pluto on previous photographic images; it appeared on plates produced at the Königstuhl Observatory and at Harvard in 1914, at the Mount Wilson Observatory in 1919 and 1925, at the Yerkes Observatory in 1921 and 1927, and at Belgium's Royal Observatory in Ucce in 1927. It also transpired that Pluto had been photographed from Flagstaff twice during the first month of Tombaugh's search. These prediscovery sightings then allowed a much more accurate set of elements to be produced; these are compared with the predictions of Lowell and Pickering in Table 10.1.[84]

Just like Adams' and Leverrier's predictions nearly 100 years previously, the longitudes of Planets X and O were quite close to that of the newly discovered planet. Most of the other elements were also in the right ball park; indeed, the agreement was rather better than Adams and Leverrier had achieved, though the mean distance of Planet O was substantially in error. Following the discovery, opinions were divided as to the nature of the new planet. Some thought that Pluto was Lowell's Planet X, though Pickering defended his own prediction. However, the majority were of the opinion that the discovery of Pluto was

[82] For Tombaugh's own account of the search, see Tombaugh and Moore (1980).
[83] Quoted from Hoyt (1980), p. 196.
[84] Adapted from Hoyt (1980), p. 221.

Table 10.1. *A comparison of the elements of Pluto with predictions of Lowell and Pickering.*

Element	X (Lowell)	O (Pickering)	Pluto (1930)
a	43.0	55.1	39.5
e	0.202	0.31	0.248
i	$10°\pm$	$15°\pm$	$17°.1$
Ω		$100°\pm$	$109°.4$
$\overline{\omega}$	$204°.9$	$280°.1$	$223°.4$
τ	1991.2	2129.1	1989.8
Longitude 1930.0	$102°.7$	$102°.6$	$108°.5$
m (mass of Earth $= 1$)	6.6	2.0	< 0.7
Magnitude	12–13	15	15

pure chance, since there was no way that the object that had been discovered possibly could have been predicted from perturbations in the orbits of Uranus and Neptune.[85] A vigorous defence of both Lowell and Pickering against this new 'happy accident' hypothesis was mounted by Victor Kourganoff in his doctoral thesis in 1940,[86] and many were convinced. However, the final word came in 1978 with the discovery of a moon orbiting Pluto – Charon – by James W. Christy.[87] This allowed the first direct and accurate calculation of the mass of Pluto, a quantity that had been reduced gradually over the years since the planet's discovery. It transpires that the Pluto–Charon system has a mass of about 0.002 times that of the Earth, over 3000 times smaller than the mass predicted by Lowell. This tiny object is not responsible for the reported perturbations in Uranus, and the discovery of Pluto must go down as being due entirely to the systematic nature of the search rather than to any predictions based on celestial mechanics. Lowell does deserve some credit for the discovery of Pluto, though, since he created the research programme that eventually led to its discovery. His analytical methods of prediction were sound, but the data

---

[85] In the early days after the discovery of Pluto, the chief proponent of this idea was Yale's E. W. Brown (Brown (1930)). Lowell was defended by Crommelin (1931): 'Is Lowell's forecast then, wholly illusory? I find it hard to admit this in the face of so many accordances ... '. Hoyt (1980) describes the ensuing controversy in detail. It is worth noting that Tombaugh himself did not believe that Lowell's prediction had anything to do with his discovery: 'It has been stated by some writers of the Pluto story that Percival Lowell's calculations greatly aided my finding Pluto. Quite to the contrary, when I found out in the latter part of 1929 how Lowell had drastically changed his predicted position of Planet X from Libra to Gemini, this indicated to me that considerable uncertainty was involved, and I could not take the prediction seriously' (Littmann (1988), p. 83).

[86] This did not become known widely until it was translated into English in 1951 (Reaves (1951)).

[87] A short account of the discovery by Christy himself can be found in Littmann (1988), pp. 175–7 (see also Stern and Mitton (1998), Chapter 2).

on which he based them were not. By contrast, Pickering's so-called graphical methods were essentially unsound.

This begs the question of whether there is another planet out there waiting to be discovered, which is the cause of the small calculated irregularities in the motion of Uranus and Neptune. This is an open question that has caused much debate, but on balance it would appear that there is no need for such a planet to exist to account for the motions of the known members of the Solar System.[88]

---

[88] See Littmann (1988), Chapter 13.

11

New methods

Patterns and resonances

The discovery of Neptune was a body blow for the Titius–Bode law and it had long since been discarded when Pluto was discovered at a distance of 39.5 AU instead of the law's prediction of 77.2 AU. A comparison of actual planetary distances and those given by the original law of Titius are shown in Table 11.1. The discrediting of the Titius–Bode law did not, however, destroy the psychological belief in the existence of a certain regularity in the structure of the Solar System that was deeply held by many people, and attempts to find patterns relating planetary distances continued well into the twentieth century.[1] For example, the American astronomer Daniel Kirkwood suggested that if r_n is the distance of the nth planet from the Sun (or the nth satellite from its respective host), then

$$\frac{x_{n+1} - x_n}{x_n - x_{n-1}} = \text{constant}, \qquad x_n \equiv \left(\frac{r_n^2 + r_{n+1}^2}{2}\right)^{1/2}.$$

This happens to work particularly well in the case of the satellite system of Saturn.

---

[1] Many of these are described in Nieto (1972). None were in particularly good agreement with the facts until Mary Adela Blagg suggested that planetary or satellite distances were of the form $r_n = A(1.7275)^n[B + f(\alpha + n\beta)]$ in which A, B, α and β are constants to be determined for each satellite system, and f is an empirically determined periodic function that is the same for all systems (Blagg (1913)). A similar, and apparently independent, analysis was carried out by Richardson (1945). With an appropriate choice of constants, the fit to the planetary and Jovian systems is excellent, but it is hard to believe that there is anything significant in this approach. The fit that can be achieved between planetary spacings and a simple numerical relationship containing two (or more) parameters often appears remarkable, but a statistical analysis shows that pretty much any set of spacings can be accommodated in such a model (Murray and Dermott (1999), pp. 5–9). Recent discoveries of planetary systems orbiting other stars have revived interest in the Titius–Bode law (Goodwin and Gribbin (2001), p. 59).

Table 11.1. *A comparison of the Titius–Bode law with observation.*
The planets in parentheses were not known in 1766. Also listed are
the values predicted by Kepler's regular polyhedron construction.

Planet	n in (10.1)	Distance	Titius–Bode	Kepler
Mercury	$-\infty$	0.39	0.4	0.56
Venus	0	0.72	0.7	0.79
Earth	1	1.00	1.0	1.00
Mars	2	1.52	1.6	1.26
(Ceres)	3	2.77	2.8	
Jupiter	4	5.20	5.2	3.77
Saturn	5	9.55	10.0	6.54
(Uranus)	6	19.2	19.6	
(Neptune)	7	30.1	38.8	
(Pluto)	8	39.5	77.2	

Kirkwood rose to prominence in the astronomical world in 1849 when he published what became known as 'Kirkwood's analogy', another simple numerical relation between properties of planets. From reading about Laplace's nebular hypothesis on the formation of the Solar System, Kirkwood developed the concept of the diameter of the sphere of attraction of a planet. This was the width of the original nebulous ring from which the planet had formed, and was defined as the distance between the two points on either side of a planet at which the gravitational attraction of the next planet was equal and opposite to that of the planet itself. With this quantity as D, and n as the number of axial rotations of a planet per orbit (i.e. the ratio of its sidereal year to its sidereal day), Kirkwood claimed that n^2/D^3 was constant for all planets.

The evidence in favour was fairly meagre. Since the calculation of D requires knowledge of planets on either side of the one under consideration, nothing could be said about Mercury or Neptune. The axial rotation rate of Uranus was not tied down with any degree of accuracy, and Mars and Jupiter had the asteroid belt as one of their neighbours. This only left Venus, Earth, and Saturn for which the calculation could be made to support the hypothesis. Things could be made to work for Mars and Jupiter if the existence of a planet in the gap between them was assumed, and the orbital radius came out to be about 2.9 AU, which then fitted well with the earlier suggestion made by Olbers to rescue the Titius–Bode law.

Many in the United States, including leading scientists, became convinced that Kirkwood had discovered a new law of nature comparable to Kepler's third law. Of particular note was the relationship with the nebular hypothesis, supporters of which had been declining as problems with it emerged. Kirkwood's analogy was seen as a manifestation of Laplace's theory, and through this association the two provided mutual support. It was hoped that in time somebody would establish a firm link between them and so provide a physical basis for the analogy, just as Newton had for Kepler's third law. On the other side of the Atlantic, the reaction was rather different, however. In England in particular, where the nebular hypothesis had very little support, the analogy was afforded a similar status to that of the Titius–Bode law.[2]

Although his analogy came to nothing, Kirkwood's search for simple numerical relationships amonge planetary phenomena did not go unrewarded. He examined the distribution of the orbital radii of asteroids that had been discovered and found that they were not distributed uniformly around their mean, with clustering around some values. In 1866, when 88 asteroids were known, he suggested a reason for this uneven density, and reiterated it with more confidence in 1876 when he had data for 169 asteroids: the intervals in which the numbers of such bodies were very small (now known as 'Kirkwood gaps') are centred on those places at which the orbital period would be in simple ratio with that of Jupiter. In such a case, conjunctions with Jupiter would occur continually at the same positions in the asteroid's orbit and so, Kirkwood argued, its orbit would become more and more eccentric until it was brought into contact with other bodies and displaced into a different orbit. The evidence Kirkwood used to support his argument is shown in Table 11.2.[3]

Clearly, resonances with Jupiter are significant, but Kirkwood's explanation can be only part of the story since other resonances are marked by a clumping of asteroid orbits. This includes the 1 : 1 resonance, where the asteroids orbit with the same period as Jupiter at one of the Lagrange points for the Sun–Jupiter system (see p. 326), and the 3 : 2 resonance, in which there is a clustering of asteroids known as 'the Hildas'.

Resonances abound in the Solar System – we have discussed in Chapter 9 the approximate 5 : 2 ratio between the mean motions of Jupiter and Saturn (p. 332) – and they have both dynamical and theoretical significance. Perhaps

---

[2] More on the history of Kirkwood's analogy can be found in Numbers (1973).

[3] Adapted from that given in Kirkwood's *The Asteroids between Mars and Jupiter* published in the *Annual Report of the Smithsonian Institution* (1876) and reprinted in Shapley and Howarth (1929). Similar ideas were used by Kirkwood to explain gaps in Saturn's rings. Here it is commensurability with the moons of Saturn that is important. Kirkwood showed in 1867 that a body in the Cassini division – the largest of the gaps in the ring system of Saturn, discovered by G. D. Cassini in 1676 – would have a period in simple ratio with four of the eight moons of Saturn. Recently, the structure of the ring system of Saturn was used to predict the existence of an eighteenth satellite (Showalter (1991)).

Table 11.2. *The distribution of the 169 asteroids the*
elements of which were known in 1876.
The final column shows simple ratios between the period
of Jupiter to that of an asteroid (given in parentheses). In
the 2.75–2.85 AU range, eighteen of the nineteen asteroids
are in the first half of the interval, thus demonstrating the
lack of asteroids in a 5 : 2 resonance with Jupiter.

Distance range (AU)	No. of asteroids	Period at midpoint (yrs)	Resonance
< 2.25	2		
2.25–2.35	7	3.49	7 : 2 (3.39)
2.35–2.45	32	3.72	
2.45–2.55	3	3.95	3 : 1 (3.95)
2.55–2.65	26	4.19	
2.65–2.75	28	4.44	
2.75–2.85	19	4.69	5 : 2 (4.74)
2.85–2.95	6	4.94	
2.95–3.05	10	5.20	
3.05–3.15	19	5.46	
3.15–3.25	9	5.72	
3.25–3.35	1	5.99	2 : 1 (5.93)
3.35–3.45	3	6.27	
3.45–3.55	2	6.55	
> 3.55	2		

the most striking is that between three of the moons of Jupiter – Io, Europa, and
Ganymede – the orbital periods of which are approximately in the ratio $1 : 2 : 4$.
Denoting their mean motions by n_I, n_E, and n_G, respectively, we have[4]

$$\frac{n_I}{n_E} = 2.007\,294\,411, \qquad \frac{n_G}{n_E} = \tfrac{1}{2}(1 - 0.007\,294\,410).$$

Hence, to a remarkable degree of accuracy, and certainly within the limits of
observational error, $n_I - 3n_E + 2n_G = 0$, a relation first noted by Laplace. The
physical cause of resonances such as this is still a subject of debate, but some
of its consequences are clear. One effect is that triple conjunctions of the three
satellites are impossible, since whenever two of them are in conjunction the
third is always at least 60° away.[5]

Another example, though one that has been understood fully only recently,
is the $3 : 2$ resonance between Neptune and Pluto, discovered in 1965.[6] The

[4] Data from Murray and Dermott (1999).
[5] See Murray and Dermott (1999), Section 8.17.
[6] Cohen and Hubbard (1965) discovered the resonance, not by observation but via a numerical
integration of the system over a period of 120 000 years. Their calculation was extended to
cover 4.5 million years by Williams and Benson (1971) and the results confirmed.

orbit of Pluto is highly eccentric ($e = 0.25$) and, for a significant proportion of its 248-year orbit, is actually closer to the Sun than Neptune, but the two planets are never close together because the orbital resonance maximizes their separation at conjunction.[7]

When it comes to analysing resonant orbital motion as illustrated by the examples above, the mathematical techniques available at the beginning of the nineteenth century run into immediate problems. The thorn in the side of perturbation theory is the presence in the expansions for the elements of small divisors, and these are caused precisely by the presence of resonances between orbital periods. In order to make theoretical progress it was necessary to develop new techniques, and the pioneers were the mathematicians Carl Gustav Jacobi and William Rowan Hamilton, from Germany and Ireland, respectively, who took the analytical mechanics begun by Euler and Lagrange to a new level.

Hamilton and Jacobi

Hamilton developed his general methods for dynamics from a study of optical systems. At the beginning of the nineteenth century, optics mostly was treated geometrically. Hamilton developed an analytic theory much as Euler and Lagrange had done for mechanics. He generalized his results subsequently to cover the theory of dynamics and published two fundamental papers in 1834/5. Hamilton's respect for Lagrange is clear from the introductory remarks to the first of these:

> Lagrange has perhaps done more than any other analyst . . . by showing that the most varied consequences respecting the motions of systems of bodies may be derived from one radical formula; the beauty of the method so suiting the dignity of the results, as to make of his great work a kind of scientific poem.

Hamilton's dynamics revolved around what he called the 'principal function S' – now usually referred to as the 'action' – that is the integral of the Lagrangian function $L = T - V$ (i.e. $S = \int L \, dt$), and his fundamental dynamical principle was that the actual motion between two given points at two given times is such that this action is stationary. Hamilton's principle of stationary action can be shown to be equivalent to Lagrange's equations

$$\frac{d}{dt} \frac{\partial L}{\partial \dot{q}_\alpha} - \frac{\partial L}{\partial q_\alpha} = 0, \qquad \alpha = 1, \ldots, N, \qquad (11.1)$$

---

[7] The minimum separation is about 2.6 billion km (Stern and Mitton (1998), p. 141).

where, in general, $L = L(q_\alpha, \dot{q}_\alpha, t)$, with t appearing explicitly and L no longer necessarily equal to $T(q_\alpha, \dot{q}_\alpha) - V(q_\alpha)$. Following Poisson, Hamilton introduced new independent variables p_α, where

$$p_\alpha = \frac{\partial L}{\partial \dot{q}_\alpha} \quad \text{and} \quad \dot{p}_\alpha = \frac{\partial L}{\partial q_\alpha}, \tag{11.2}$$

the second equation in Eqn (11.2) following from the first because of Eqn (11.1). The variable, p_α is the called the 'momentum conjugate' to q_α. In principle, we can solve for the N quantities $\dot{q}_\alpha$ in terms of the p_αs, the q_αs, and t, and then construct a new function $\mathcal{H}(p_\alpha, q_\alpha, t)$, which is now referred to as 'the Hamiltonian', defined by

$$\mathcal{H}(p_\alpha, q_\alpha, t) = \sum_{\alpha=1}^{N} p_\alpha \dot{q}_\alpha - L.$$

Differentiation (making use of Eqn (11.2)) shows that $d\mathcal{H}/dt = \partial \mathcal{H}/\partial t = -\partial L/\partial t$ and, hence, that if t does not appear explicitly in the Lagrangian, it does not appear in the Hamiltonian either, and $\mathcal{H}$ is a constant. In such a situation, called a 'conservative system', the Hamiltonian is shown readily to be the total energy in the system, i.e. $\mathcal{H} = T + V$.

It also follows that

$$\dot{q}_\alpha = \frac{\partial \mathcal{H}}{\partial p_\alpha}, \qquad \dot{p}_\alpha = -\frac{\partial \mathcal{H}}{\partial q_\alpha}, \qquad \alpha = 1, \dots N, \tag{11.3}$$

and these are Hamilton's canonical equations of motion for a system with N degrees of freedom. Whereas Lagrange's formulation results in N second-order differential equations, the Hamiltonian formulation delivers $2N$ first-order equations. For the n-body problem, we can write $\mathbf{q}_i = \mathbf{r}_i$ and $\mathbf{p}_i = m_i \dot{\mathbf{q}}_i$, and then the equations of motion (Eqn (9.13)) are simply $\dot{\mathbf{p}}_i = -\nabla_i V$. These can be written in canonical form with the Hamiltonian

$$\mathcal{H}(\mathbf{p}_i, \mathbf{q}_i) = \sum_{i=1}^{n} \frac{\mathbf{p}_i \cdot \mathbf{p}_i}{2m_i} + V(\mathbf{q}_i).$$

Stimulated by Hamilton's work, Jacobi considered how to choose the coordinates for a particular problem so that the integration of Hamilton's canonical equations is as simple as possible. This leads to the concept of a canonical transformation,[8] a transformation from variables p_α, q_α to P_α, Q_α, that preserves the Hamiltonian nature of the equations. In other words, after applying the

---

[8] Sometimes referred to as a 'contact transformation'. For a thorough treatment, see, for example, Whittaker (1937), pp. 292 ff. or Boccaletti and Pucacco (1996), pp. 76 ff.

transformation we obtain

$$\dot{Q}_\alpha = \frac{\partial \mathcal{H}^*}{\partial P_\alpha}, \qquad \dot{P}_\alpha = -\frac{\partial \mathcal{H}^*}{\partial Q_\alpha},$$

for some function $\mathcal{H}^*(P_\alpha, Q_\alpha, t)$. A necessary and sufficient condition for a transformation to be canonical is that

$$\sum_\alpha p_\alpha \, dq_\alpha - \sum_\alpha P_\alpha \, dQ_\alpha = dW,$$

in which the time t is assumed constant and W is an arbitrary function of t and $2N$ of the $4N$ scalar variables $p_\alpha, q_\alpha, P_\alpha$, and Q_α. The new and old Hamiltonians are related via

$$\mathcal{H}^* = \mathcal{H} + \frac{\partial W}{\partial t}. \tag{11.4}$$

Another approach to such transformations is via generating functions. Given a function $W_1(q_\alpha, Q_\alpha, t)$ we can construct $p_\alpha = \partial W_1/\partial q_\alpha$, $P_\alpha = -\partial W_1/\partial Q_\alpha$ and then this defines a canonical transformation. Other possibilities are

$$
\begin{aligned}
p_\alpha &= \partial W_2/\partial q_\alpha, & Q_\alpha &= \partial W_2/\partial P_\alpha & &\text{with} \quad W_2(q_\alpha, P_\alpha, t), \\
q_\alpha &= -\partial W_3/\partial p_\alpha, & P_\alpha &= -\partial W_3/\partial Q_\alpha & &\text{with} \quad W_3(p_\alpha, Q_\alpha, t), \\
q_\alpha &= -\partial W_4/\partial p_\alpha, & Q_\alpha &= \partial W_4/\partial P_\alpha & &\text{with} \quad W_4(p_\alpha, P_\alpha, t).
\end{aligned}
$$

In each case, the new Hamiltonian satisfies Eqn (11.4).[9] As an example with important applications in celestial mechanics, consider the transformation defined by

$$P_\alpha = \sqrt{2p_\alpha} \cos q_\alpha, \qquad Q_\alpha = \sqrt{2p_\alpha} \sin q_\alpha. \tag{11.5}$$

It is easy to show that $p_\alpha \, dq_\alpha - P_\alpha \, dQ_\alpha = d(-\frac{1}{2} p_\alpha \sin 2q_\alpha)$ and, hence, that this is a canonical transformation. It is generated by $W_2(q_\alpha, P_\alpha) = \frac{1}{2} \sum_\alpha P_\alpha^2 \tan q_\alpha$.

If a canonical transformation can be found such that the resulting Hamiltonian does not depend on a particular pair of the new conjugate variables P_α and Q_α, then the order of the system of equations can be reduced by 2, since these variables – known as 'ignorable coordinates' – will be constants of the motion. In the extreme case, suppose we can find a function W such that, after applying the transformation, the new Hamiltonian is zero. Then we will have solved the problem completely, since all the coordinates P_α and Q_α will be constant. If we let the generating function W_1 which yields this solution be $S(q_\alpha, Q_\alpha, t)$ so

---

[9] If W_1, W_2, W_3, and W_4 generate the same transformation then $W_2 = W_1 + \sum_\alpha P_\alpha Q_\alpha$, $W_3 = W_1 - \sum_\alpha p_\alpha q_\alpha$, and $W_4 = W_1 + \sum_\alpha P_\alpha Q_\alpha - \sum_\alpha p_\alpha q_\alpha$.

that $p_\alpha = \partial S/\partial q_\alpha$, we have, from Eqn (11.4),

$$\mathcal{H}\left(\frac{\partial S}{\partial q_\alpha}, q_\alpha, t\right) + \frac{\partial S}{\partial t} = 0,$$

a nonlinear first-order partial differential equation for the function S, known as the 'Hamilton–Jacobi equation'.

This function S turns out to be nothing other than Hamilton's principal function: the action. Like the Lagrangian, L, it characterizes a dynamical system. But, whereas L serves only to generate the equations of motion, S actually solves them. Unfortunately, and unsurprisingly, the determination of S is no easier than solving the equations of motion, but the existence of S is at the heart of all dynamical problems. Hamilton and Jacobi revolutionized thinking about mechanics during the nineteenth century, and their methods were absorbed gradually into the work of celestial mechanicians. Hamilton–Jacobi theory formed the basis for Poincaré's landmark research at the end of the century.

Perturbation theory in the nineteenth century

Perturbation theory in celestial mechanics is essentially the attempt to solve

$$\ddot{\mathbf{r}} + \frac{\mu}{r^3}\mathbf{r} = \nabla R, \tag{11.6}$$

where R is the disturbing function, considered small, and expanded in terms of small parameters. If $R \equiv 0$, we have the standard two-body problem with (setting $G = 1$ for simplicity) μ, the sum of the masses of the bodies, and $\mathbf{r}$, the position of one of the bodies with respect to the other. In the simplest case, the perturbation comes from one other body, the position of which is $\mathbf{r}'$ and the mass of which is m, for example, and then, from Eqn (9.3),

$$R = m\left(\frac{1}{|\mathbf{r} - \mathbf{r}'|} - \frac{\mathbf{r} \cdot \mathbf{r}'}{r'^3}\right).$$

In the problems of planetary motion, R is small because m is small relative to μ, whereas in the lunar theory, R is small because the two terms in parentheses – representing the effect of the Sun on the Moon and on the Earth, respectively – are very similar due to the smallness of the quantity r/r'.

There are two distinct approaches to perturbation theory: absolute perturbations, and variation of orbital parameters. In the former, one seeks equations for the small variations in the coordinates about some approximate orbit, whereas in the latter, one seeks equations that determine how the elements of the orbit vary with time. Variation of orbital parameters is particularly useful for discovering long-period inequalities such as that which exists between Jupiter and Saturn, and is appropriate more generally in planetary theory. In the early nineteenth century, it was the planetary theory that received the most attention from theoreticians, but in the latter part of the century there was a shift toward the lunar theory. This stemmed partly from the difficulty astronomers found in reconciling theory and observation for the Moon, and partly it reflected a change in the nature of mathematical astronomy. In 1800, celestial mechanics was pursued for one purpose only, i.e. to provide the theory necessary to produce accurate tables with which to predict future positions of heavenly bodies. By 1900, another competing goal had emerged as mathematicians began to analyse the equations of celestial mechanics in their own right to see what general conclusions could be drawn about the dynamics of the Solar System. The mathematical tools needed to do this came to the fore in attempts to understand the motion of the Moon.

In the early nineteenth century, many people, including Johann Karl Burckhardt, Philippe Comte de Pontécoulant, and Leverrier, extended Laplace's general approach by expanding the disturbing function to higher and higher powers in the eccentricity and inclination. The algebra becomes horrendous, and some turned to direct integration of the equations of motion ('mechanical quadrature'). Significant theoretical progress was made by Peter Hansen, who succeeded Encke as the director of the observatory at Seeberg when the latter moved to Berlin in 1825. The standard procedure for analysing planetary perturbations resulted in long-period inequalities that had to be applied to the mean longitude and short period variations in the true longitude. The two effects thus had to be calculated separately. Hansen's theory improved matters by treating both effects as variations in mean longitude. The method was based on an unperturbed orbit that was an ellipse of fixed dimensions located in the moving plane of the actual orbit and with a perigee the longitude of which was a linear function of time. The theory is difficult, but the series that result are convergent more rapidly than in previous methods and, as a result, it can be applied to orbits of greater eccentricity and inclination.[10] Hansen applied his technique to the Jupiter–Saturn problem and to the lunar theory.

Following his success with Neptune, Leverrier embarked on an ambitious programme to produce extremely accurate tables for all the planets. Between 1858 and 1861, he published theories for the four inner planets, Mercury, Venus, Earth, and Mars; all represented significant improvements over existing tables. The case of Venus, however, illustrates clearly the problems inherent in the use of perturbation theory. Laplace previously had calculated perturbations up to

---

[10] Details can be found in Brouwer and Clemence (1961), Chapter XIV.

e^3, i^3, but Airy discovered the fifth-order 239-year perturbation due to the near resonance in the mean motions of the Earth and Venus ($13n_E - 8n_V \approx 1/239$ revolutions per year). The amplitude of this inequality in the longitude of Venus is about $3''$, which compares with $1''.4$ for the maximum amplitude of the e^3, i^3 terms, and $0''.1$ for the fourth-order terms.

Jupiter, Saturn, Uranus, and Neptune present a more difficult challenge but, nevertheless, Leverrier succeeded in producing accurate theories. In the case of Neptune, the work was completed by A. J.-B. Gaillot a month after Leverrier's death. Leverrier's planetary theories are a magnificent testimony to the legacy of Laplace. They represent, with some minor modifications, the motions of the planets over hundreds of years with errors of at most a few arc-seconds. And yet they are somewhat unsatisfactory. The masses of many of the planets were not known with any real precision, and the last step in Leverrier's construction of each planetary theory was to tweak these masses slightly so as to minimize the discrepancies between theory and observation. Unfortunately, this meant that the same planet sometimes was accorded a different mass in each of the separate theories.

The first to develop a consistent set of planetary theories was the American astronomer, Simon Newcomb.[11] Newcomb made significant technical improvements to the underlying Laplacian theory, which reduced the labours involved in computing perturbations to high order in e and i. Tables based on his theory for the four inner planets appeared in 1898. In the same year, Hill produced tables for the outer planets based on Hansen's method. The planetary theories of Newcomb and Hill are marginally superior to those of Leverrier, and were not improved upon until the advent of the electronic computer.[12]

There was still a significant problem in the theory for Mercury, which has so far been glossed over. In order to get theory and observation to agree, Leverrier found that he had to alter significantly the theoretical value for the rate at which the perihelion of Mercury moved round the Sun. Newcomb was confronted by the same problem, but could not explain the discrepancy. At the time, this may have appeared a rather small problem but, as it turned out, its resolution had profound consequences. This is the subject of Chapter 12.

Newcomb was involved in the first significant astronomical discovery made by American astronomers, who in the second half of the nineteenth century

---

[11] Newcomb's early career is described in Norberg (1978).

[12] Unlike Newcomb and Hill, Leverrier was not influenced by Hansen – some copies of the latter's work presented to Leverrier remained with their pages uncut!, (see Williams (1945), who examines carefully the differences between Leverrier's and Newcomb's tables). Newcomb described Hansen as the 'greatest living master of celestial mechanics since Laplace' (Roseveare (1982), p. 53).

were beginning to compete with their European counterparts. The architect of the discovery, though, was an amateur astronomer who had spent time working as a 'calculator' for first Benjamin Apthorp Gould and then Benjamin Peirce at the US Coast Survey. Seth Chandler established, by reducing thousands of extant observations, that the latitude of a point on the surface of the Earth (defined in terms of the declination of the celestial pole) undergoes a small but measurable variation.

It had been known since Euler's work on the dynamics of rigid bodies that, if the Earth rotated about an axis that was inclined slightly to its polar axis, then it would wobble, and the period of such an oscillation had been computed to be approximately 10 months. People searched in vain for just such an effect to explain small but persistent errors in their observations. Chandler was no theoretician and did not set out to find a variation with a 10-month period. His phenomenal abilities at reducing data led him in 1891 to the discovery that there was a variation in latitude, but that its period was not the theoretical value of 306 days – it was 427 days. Newcomb made Chandler's discovery believable when he pointed out that the Earth was not perfectly rigid as the Eulerian theory supposed, and that the fluidity of the ocean and the elasticity of the solid earth would both serve to lengthen the period of the 'Chandler wobble'. Chandler went on to discover that the variation in latitude actually was rather more complicated, with the superposition of two oscillations, one with a period of 427 days and the other an annual fluctuation.[13]

Lunar theory

The fact that the position of the Moon can be measured extremely accurately against the backdrop of the fixed stars means that the many irregularities in its motion are relatively easy to detect. Before the advent of the telescope, for example, Ptolemy discovered the evection, and Tycho Brahe, the variation. In the nineteenth century, the ease with which observers could uncover new inequalities posed a considerable challenge for theoreticians. To add to the difficulties, the series by which these irregularities were determined typically converged slowly, and the selection of coordinate system was of crucial importance.

---

[13] For a popular account of Chandler's discovery and an insight into American astronomy in the nineteenth century, see Carter and Carter (2002). The magnitude of the wobble varies but never amounts to more than $0''.7$ (corresponding to about 15 m at the pole). Modern numerical simulations suggest that the Chandler wobble is excited by a combination of atmospheric and oceanic processes, with pressure fluctuations on the ocean floor being the dominant mechanism (Gross (2000)).

In the lunar theory, the motion of the disturbing mass (the Sun) is assumed known. This is because the motion of the Sun around the Earth is given to a very high degree of approximation by an ellipse with the centre of gravity of the Earth–Moon system at one focus. The effects of the planets on the Moon are very small and can be treated separately. In fact, as Laplace showed, the effects of the planets on the Earth (and, hence, on the centre of gravity of the Earth–Moon system) are more significant, and even when these effects are too small to be observed in the motion of the Earth, their indirect action on the Moon may be appreciable, as in the case of the secular acceleration of the Moon (see p. 334).

Laplace's lunar theory, published in 1802, built on Clairaut's successful analysis of the apogee of the Moon, and included many small perturbative effects such as that due to the equatorial bulge of the Earth, as well as a theoretical treatment of the secular acceleration of the Moon. The small parameters in the theory are $m = n'/n$, e, e', and i, where a dash is used for solar parameters, and then the equation of centre is a first-order quantity, whereas variation, evection, and the annual equation are second-order quantities. Laplace included fifteen inequalities at third order and 'the most important of the fourth order'. This theory reduced errors by a factor of 3, to about $30''$.

Lunar theory had advanced significantly since the mid eighteenth century, but the magnitudes of many of the small inequalities had to be determined empirically, only the arguments came from theory. Laplace was convinced that his theory could be extended so that these amplitudes were determined analytically, and so in 1818 he suggested that this topic might be the subject of a prize question from the Paris Academy. The question was:

> To formulate, by the theory of universal gravitation alone, and borrowing from the observations only the arbitrary elements, tables of lunar movement as accurate as our best ones at present.[14]

Two entries were received, one from Marie-Charles-Théodore de Damoiseau, based in Paris, and one from the Italian astronomers Giovanni Plana and Francesco Carlini from Turin and Milan, respectively.

Damoiseau, director of the observatory at L' École Militaire in Paris, used Laplace's method but carried the analysis much further. The tables produced from his theory, published in 1824, were accurate to about $4''$, but did not represent much theoretical progress. Plana and Carlini, who also built on Laplace's approach, but included some ingenious modifications, began their collaboration

---

[14] Quoted from Tagliaferri and Tucci (1999). The prize was to be awarded in 1820.

on the lunar theory in 1811. In contrast to most of their contemporaries, they developed the theory algebraically throughout, without substituting any known numerical values. Plana and Carlini had doubts about whether to enter the prize competition, thinking that Damoiseau (who was known to be working on lunar theory) would be favoured due to his nationality, but in the end did submit an entry. Laplace subsequently was critical of their approach, and Plana in particular engaged in a bitter dispute for many years. When it came to writing up the theory for publication, Plana attempted to convert the prize-winning paper into a complete theory of the Moon which emphasized and corrected what he believed were the shortcomings in Laplace's approach. Eventually, Carlini got so exasperated with Plana that he pulled out of the project, and the monumental *Theory of the Moon's Motion,* running to over 2 500 pages, was published eventually in 1832 under the authorship of Plana alone.[15]

Poisson implemented an iterative method for the lunar theory in 1833 based on the method of variation of orbital parameters. He assumed, as a first approximation, that a, e, and i were constant, and that the mean longitude of perigee and of the ascending node were linear functions of time. Substituting these into the full equations, he obtained more accurate expressions for the elements that then could be used to repeat the process. The calculations quickly become unmanageable and Poisson's theory did not get very far. Others who made significant contributions to the development of lunar theory in the mid nineteenth century included John Lubbock and Pontécoulant.[16]

Hansen's lunar theory was published in 1838,[17] with associated tables in 1857. He discovered new inequalities due to the direct action of the planets, the largest of which is due to Venus, and arises from the fact that $16n_E + n_M - 18n_V$ is a small quantity. Hansen computed an amplitude of 27.4″ and a period of 273 years, though subsequent analysis showed that this was about double the correct value.[18] As far as Robert Grant was concerned, Hansen's theory in 1852 accounted completely for all outstanding errors in the tables and the lunar theory was now

---

[15] *La théorie du mouvement de la lune* (see Tagliaferri and Tucci (1999)).

[16] A survey of eighteenth- and nineteenth-century lunar theories is given in Brown (1960), Chapter 12.

[17] *Fundamenta nova investigationis orbitae verae quam luna perlustrat.* Revised versions appeared in 1861 and 1864 (Morando (1995)). One novelty Hansen considered in 1856 was that the centre of gravity of the Moon was slightly further from the Earth than the geometrical centre of the Moon. As well as having an effect on the Moon's inequalities, this implied also an increased gravitational force on the far side, and some people speculated that the Moon might have an atmosphere on that side, and that perhaps even life existed (see Roseveare (1982), p. 54).

[18] See Brown (1960), Article 312.

divested of all serious embarrassment; and in its present state it undoubtedly
constitutes one of the noblest monuments of intellectual research which the annals
of science offer to our contemplation. From the age of Hipparchus down to the
present day, the complicated movements of the moon have formed the subject of
anxious enquiry. One by one have her numerous inequalities been detected, and
their laws ascertained, until the astronomer is finally enabled to predict her place
with all the accuracy called for by the most refined appliances of modern
observation. Perhaps no other part of astronomy exhibits so many unequivocal
triumphs of the theory of gravitation as the researches connected with the moon's
motion.[19]

Of all the entirely algebraic theories of the Moon, the most complete was that
of Charles Delaunay, who worked within a Hamiltonian framework to generate,
for the first time, solutions in the form of pure trigonometric series, devoid of
secular terms. The work took 20 years to finish and eventually was published
in two thick volumes of the *Les mémoires* of the Paris Academy in 1860 and
1867, in total amounting to over 1800 pages. The choice of variables in a lunar
theory was critical, and Delaunay used a canonical set, now called 'Delaunay
variables'. If we consider the Lagrange brackets formed from the six elements
(see p. 329)

$$\sigma = \epsilon - \bar{\omega}, \qquad \omega = \bar{\omega} - \Omega, \qquad \Omega,$$

$$L = \sqrt{\mu a}, \qquad G = \sqrt{\mu a(1 - e^2)}, \qquad H = G \cos i,$$

then it can be shown that

$$(\sigma, L) = (\omega, G) = (\Omega, H) = 1, \qquad (L, \sigma) = (G, \omega) = (H, \Omega) = -1,$$

and all the other Lagrange brackets are zero. With this choice of elements,
Eqns (9.23) reduce to the Hamiltonian form of Eqn (11.3) with $\mathbf{q} = (\sigma, \omega, \Omega)$,
$\mathbf{p} = (L, G, H)$, and the Hamiltonian given by minus the disturbing function, R.

There is a serious defect with this set of equations (just the same problem
applies to Lagrange's planetary Eqns (9.24)). The way one would like to proceed
is to expand the disturbing function in a series of periodic terms, the coefficients
of which are functions of a, e, and i, and the arguments of which are functions
of ϵ, $\bar{\omega}$, and Ω. But ϵ always appears in the form $nt + \epsilon$, which is a function of a
via Kepler's third law $n^2 a^3 = \mu$. Hence, a appears explicitly in the coefficients
and implicitly in the arguments. This leads to the time t appearing as a factor
in the coefficients of the final series for ϵ, which is undesirable. To get round
this[20] we can define the mean anomaly by $l = \sigma + \int n \, dt$ and then (with $g = \omega$

[19] Quoted from Grant (1966), p. 121.
[20] See, for example, Brouwer and Clemence (1961), p. 290.

and $h = \Omega$) we obtain a Hamiltonian system with $\mathbf{q} = (l, g, h)$, $\mathbf{p} = (L, G, H)$, and

$$\mathcal{H} = -\frac{\mu^2}{2L^2} - R.$$

The set l, g, h, L, G, H are the variables that Delaunay used.[21]

Delaunay's method can be summarized as follows. The disturbing function is expanded in terms of the small parameters m, e, e', i, and a/a', so that $\mathcal{H}$ is written as the sum of a non-periodic part $\mathcal{H}_0$ and a series of periodic terms. The equations of motion are then solved for the simpler Hamiltonian

$$\mathcal{H} = \mathcal{H}_0 + A \cos \theta,$$

where $\cos \theta$ is one single periodic term from the expansion, chosen for its importance. This solution is accomplished by means of a canonical transformation that removes the $\cos \theta$ term. This transformation is then applied to the complete Hamiltonian with the effect that the $\cos \theta$ term disappears and, although the argument θ reappears in the transformed series for R, its coefficient will be of higher order.[22]

By repeating this process, the dependence of the Hamiltonian on the angular variables l, g, and h can be removed systematically – up to some order of approximation – and then the problem is solved. However, each step of the process requires a hugely laborious algebraic calculation, and Delaunay's theory – which was taken to eighth order in m, e, and i, sixth order in e', and fourth order in a/a' – involved a series expansion of 320 terms and required fifty-seven transformations![23]

Delaunay's lunar theory, like that of Hansen's, reduced errors to about 1 arc-second but, in terms of accuracy, it was not the unmitigated success that Grant had described. Newcomb transformed Hansen's theory into a form that allowed term-by-term comparison with that of Delaunay, and he found that the two approaches essentially were equivalent. He demonstrated that, while the work of Hansen and Delaunay had improved the accuracy of the short-period inequalities considerably, the same could not be said for those of long period, which were not represented much better than they had been by Laplace.

---

[21] They were probably first found by Tisserand (Goroff (1993), Section 2.7). If $R = 0$, so that we have the two-body problem, it follows immediately from the fact that the Hamiltonian is a function of L alone, that g, h, L, G, and H are all constant, and $\dot{l} = \partial\mathcal{H}/\partial L = (\mu/a^3)^{1/2} = n$.

[22] For more details see, for example, Brouwer and Clemence (1961), pp. 541–59.

[23] Delaunay's work has a significant place in the history of computer algebra. In 1970, a group at Boeing Scientific Research Laboratories in Seattle tested their computer algebra software on Delaunay's procedure. Remarkably, they found only three errors, two of which were consequences of the first (see Pavelle, Rothstein, and Fitch (1981)).

Hansen's tables might be accurate for the current epoch, but the agreement with ancient observations was poor.[24]

In 1877, Newcomb became director of the Nautical Almanac Office, and asked Hill for assistance in producing new tables for the Moon and the planets. Hill, a keen admirer of Euler, reintroduced an idea that Euler had first tried in the 1770s by describing the motion of the Moon with respect to a rotating set of axes (in this case the axes rotated with the mean angular speed of the Sun). With certain simplifying assumptions, Hill obtained a differential equation dependent solely on the ratio m of the mean motion of the Sun to that of the Moon, a quantity the numerical value of which was well established. He found a particular solution to this equation in the form of an oval, symmetric with respect to the rotating axes and with the longer axis perpendicular to the direction of the Sun, the first periodic solution to the three-body problem since Lagrange's discovery of special exact solutions in 1772. With the numerical value of m inserted, this solution, known as 'Hill's intermediate orbit' or 'variational curve', represents a good approximate orbit for the Moon which then could be used as the basis for a perturbation analysis in which successive terms decay rapidly.

By means of some ingenious transformations, Hill managed to reduce the resulting perturbation problem to that of solving what is now generally known as 'Hill's equation': $\ddot{x} + \phi(t)x = 0$, where ϕ is a periodic function. Hill expanded ϕ as a Fourier series, the coefficients of which were related to an infinite determinant, a concept for which no formal theory existed.[25] The results depended on the convergence of this infinite determinant – something Hill was not able to establish – but which was later proved by Poincaré. Poincaré was influenced greatly by Hill's approach and, in particular, he employed the idea of using non-trivial periodic solutions as the starting point for a perturbation theory to great effect.

Hill's theory was turned by E. W. Brown into a practical method for computing lunar positions. Brown completed his theory in 1908, and the tables derived from it were published in 1919. These served as the basis for British and American lunar ephemerides until 1959. The Hill–Brown theory represents an order of magnitude increase in accuracy over that of Delaunay and the pinnacle of success for quantitative perturbation techniques. In the hands of some of the world's finest mathematical astronomers, perturbation theory had become a hugely powerful tool with which to study Solar System dynamics, but

---

[24] Extracts from Newcomb's *Researches on the Motion of the Moon* (1878) are reprinted in Shapley and Howarth (1929).

[25] The same determinant appears in work done by Adams on the lunar theory in 1868, though this was not published (Barrow-Green (1997), p. 27).

as a theory by which to explain phenomena, was somewhat unsatisfactory. The theory was incredibly complex and provided little insight into the underlying physics; accurate numbers could be churned out, but the qualitative nature of the solutions was lost amid pages and pages of calculations.

The circular restricted three-body problem

In Hill's radical lunar theory, the intermediate orbit was essentially a solution to what is now known as the 'circular restricted three-body problem'. In this problem, two of the bodies – the Sun and Earth for the lunar theory – called primaries and with masses m_1 and m_2, are assumed to describe circular orbits around their common centre of gravity, and then the aim is to determine the motion of the third body of negligible mass (the Moon) within the resulting gravitational field. This problem had been studied first by Euler, and significant progress was achieved by Jacobi when he found a new invariant quantity.

In a coordinate frame rotating with angular velocity ω, the equation of motion for the small mass, setting $G = 1$ for convenience is,

$$\ddot{\mathbf{r}}_3 = -2\boldsymbol{\omega} \times \dot{\mathbf{r}}_3 + \boldsymbol{\omega} \times (\boldsymbol{\omega} \times \mathbf{r}_3) + \frac{m_1}{r_{13}^3}\mathbf{r}_{13} + \frac{m_2}{r_{23}^3}\mathbf{r}_{23}. \qquad (11.7)$$

The second term on the right-hand side is referred to as the 'centrifugal force' and is due simply to the rotation of the coordinate system. By contrast, the first term is due to the motion within the rotating frame of reference and is called the 'Coriolis force'.

If we write $\mathbf{r}_3 = x\mathbf{i} + y\mathbf{j} + z\mathbf{k}$, take $\boldsymbol{\omega} = \omega\mathbf{k}$ so that the axis of rotation is perpendicular to the x–y plane with rotation rate ω, and define

$$U = \tfrac{1}{2}\omega^2(x^2 + y^2) + \frac{m_1}{r_{13}} + \frac{m_2}{r_{23}},$$

then Eqn (11.7) reduces to

$$\ddot{\mathbf{r}}_3 - 2\omega\dot{y}\,\mathbf{i} + 2\omega\dot{x}\,\mathbf{j} = \nabla U. \qquad (11.8)$$

This is a system of order 6, but Jacobi showed that the assumption of circular orbits for the primaries (in which case their coordinates do not change with time) allows us to reduce the order by one.

We choose the origin at the centre of gravity of primaries, which are a distance a apart, with the x axis passing through the primaries (and rotating with them) so that

$$\mathbf{r}_1 = -\frac{m_2}{m_1 + m_2}a\mathbf{i}, \qquad \mathbf{r}_2 = \frac{m_1}{m_1 + m_2}a\mathbf{i},$$

and ω is the angular rotation rate of the primaries determined from the two-body problem. If we dot Eqn (11.8) with $\dot{\mathbf{r}}_3$, we get $\dot{\mathbf{r}}_3 \cdot \ddot{\mathbf{r}}_3 = \dot{\mathbf{r}}_3 \cdot \nabla U$ which can be integrated since U is a function of x, y, z only to give

$$\dot{x}^2 + \dot{y}^2 + \dot{z}^2 = 2U - C, \tag{11.9}$$

where C is a constant of integration. This is Jacobi's integral, sometimes referred to as the 'integral of relative energy'.[26] In terms of non-rotating coordinates (ξ, η, ζ) defined by

$$x = \xi \cos \omega t + \eta \sin \omega t, \quad y = -\xi \sin \omega t + \eta \cos \omega t, \quad z = \zeta, \tag{11.10}$$

Jacobi's integral takes the form[27]

$$\dot{\xi}^2 + \dot{\eta}^2 + \dot{\zeta}^2 + 2(\dot{\xi}\eta - \dot{\eta}\xi) = \frac{2m_1}{r_{13}} + \frac{2m_2}{r_{23}} - C.$$

The equations of the circular restricted three-body problem can be written in canonical form by writing $\mathbf{p} = (\dot{\xi}, \dot{\eta}, \dot{\zeta})$, $\mathbf{q} = (\xi, \eta, \zeta)$, and defining the Hamiltonian to be

$$\mathcal{H}(\mathbf{p}, \mathbf{q}, t) = \tfrac{1}{2}\mathbf{p} \cdot \mathbf{p} - \frac{m_1}{r_{13}(\mathbf{q}, t)} - \frac{m_2}{r_{23}(\mathbf{q}, t)}.$$

Note that $\mathcal{H}$ depends on t explicitly since r_{13} and r_{23}, when written in terms of ξ, η, ζ, do. Hence, $\mathcal{H}$ is not the total energy of the system and is not a constant of the motion. However, if we apply the canonical transformation generated (see p. 403) by

$$W_3 = -p_1(Q_1 \cos \omega t - Q_2 \sin \omega t) - p_2(Q_1 \sin \omega t + Q_2 \cos \omega t) - p_3 Q_3,$$

where $\mathbf{P} = (\dot{x} - \omega y, \dot{y} + \omega x, \dot{z})$, $\mathbf{Q} = (x, y, z)$, which is equivalent to the change of variables given in Eqn (11.10), we obtain a new Hamiltonian $\mathcal{H}^* = \mathcal{H} + \partial W_3 / \partial t$ that does not depend explicitly on t and is therefore constant. In fact,

$$\mathcal{H}^* = \tfrac{1}{2}(\dot{x}^2 + \dot{y}^2 + \dot{z}^2) - U,$$

and so we recover Jacobi's integral.

---

[26] This name is somewhat misleading as the total energy of the two primaries is constant, but the energy associated with the small mass is not.

[27] First given by Jacobi in *Sur le mouvement d'un point et sur un cas particulier du problème des trois corps* (1836).

Tisserand's criterion, and Hill curves

An example of the application of Jacobi's integral is in Tisserand's criterion for the identification of comets. In their passage round the Sun, comets are often perturbed significantly by the actions of the planets, particularly Jupiter. If we make the approximation that the orbit of Jupiter is a circle, then C must remain constant for the comet during its encounter with the planet. With certain other simplifying assumptions,[28] none of which introduce any significant numerical errors provided data for the comet are obtained when it is far from Jupiter, it can be shown that the instantaneous elements of the comet's orbit must satisfy

$$\frac{1}{a} + 2\sqrt{a(1 - e^2)} \cos i = \text{constant},$$

in which the unit of length is the Sun–Jupiter distance. The quantity on the left-hand side thus represents a defining characteristic of a cometary orbit and can be used to check whether two comets observed at different times are one and the same object.

Jacobi's integral also can be used to restrict the region of space in which the small mass can move. It is clear from Eqn (11.9) that if $2U = C$, the speed of the small mass (with respect to the two primaries) will be zero and thus, for a given C, the surface $2U = C$ forms a boundary to the motion. This idea was used by Hill in his ground-breaking work on the lunar theory to show that (neglecting the eccentricity of the Earth) the Moon could never escape from its orbit around the Earth. The nature of some of the surfaces that Hill described is illustrated in Figure 11.1, where we have further simplified the problem by assuming that all three masses move in the same plane. In this case, the Hill surfaces reduce to curves.

If we fix units so that $m_1 + m_2 = a = \omega^2 = 1$, then the Hill curves are given by

$$x^2 + y^2 + \frac{2(1 - m_2)}{\sqrt{(x + m_2)^2 + y^2}} + \frac{2m_2}{\sqrt{(x - 1 + m_2)^2 + y^2}} = C,$$

and the Figure 11.1 shows a selection of such curves when $m_2 = 0.1$. The points A and B are the two primaries and the curves are obviously symmetric with respect to the line AB. The value of C is a minimum at the point labelled L_4 and increases without bound as one approaches either A, B, or infinity. It follows, therefore, that if, for a particular system, C is sufficiently large and the mass m_3 is close to either A or B, then it will remain in the vicinity of that primary. The points L_1–L_4 are stationary points of the function $U(x, y)$

---

[28] See Murray and Dermott (1999), Section 3.4, for more details.

New methods

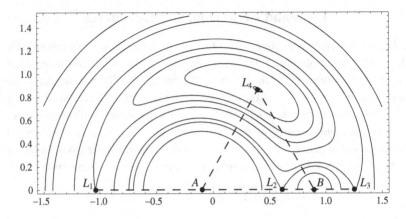

Fig. 11.1. Hill curves.

and, hence, at these points $\partial U/\partial x = \partial U/\partial y = 0$. It follows from Eqn (11.8) that we have $\ddot{x} = \ddot{y} = \ddot{z} = 0$ at these points. The points L_1–L_4 are therefore equilibrium points; they are, in fact, precisely the Lagrange points shown in Figure 9.5 (p. 326).

Hill's work contains numerous flashes of originality and provided inspiration for many celestial mechanicians toward the end of the nineteenth century. As well as leading to an accurate lunar theory, Hill's research showed how information about the motion of celestial bodies could be obtained by examining the qualitative, rather than the quantitative, behaviour of solutions to differential equations. In the hands of Henri Poincaré, this idea would revolutionize the study of dynamics.

Poincaré

After an initial flirtation with engineering, Poincaré embarked on a distinguished mathematical career with the receipt of his doctorate in 1879.[29] He began teaching at the University of Paris 2 years later. Poincaré was endowed with a phenomenal memory and had the capacity to think through difficult problems in his head; only then would he put pen to paper. As a result, his written work was often hastily prepared, requiring corrections and explanations. As soon as

[29] Brief biographical details are given in Goroff (1993), an essay that provides much useful background to Poincaré's work.

Poincaré satisfied himself that he had overcome the fundamental difficulties in a problem, he would move on and leave others to fill-in the details.

Between 1881 and 1886, Poincaré published a four-part memoir in which he created the qualitative theory of differential equations.[30] His reasons for approaching such equations this way were stated clearly:

> Moreover, this qualitative study has in itself an interest of the first order. Several very important questions of analysis and mechanics reduce to it. Take for example, the three body problem: one can ask if one of the bodies will remain within a certain region of the sky or even if it will move away indefinitely; if the distance between two bodies will infinitely increase or diminish, or even if it will remain within certain limits? Could one not ask a thousand questions of this type which would be resolved when one can construct qualitatively the trajectories of the three bodies? And if one considers a greater number of bodies, what is the question of the invariability of the elements of the planets, if not a real question of qualitative geometry, since to show that the major axis has no secular variations shows that it constantly oscillates between certain limits.[31]

Poincaré illustrated the difference between qualitative and quantitative phenomena with reference to the solution of algebraic equations. One can establish the number of real roots without solving the equation – this is qualitative information – but the values of these solutions require a quantitative study.

Hamilton's equations show that in a dynamical system, position and momentum should be considered equally important, in contrast to the more usual equations of motion in which position is treated as the fundamental quantity. This symmetry leads naturally to the concept of phase space, in which the state of a system is given in terms of the positions and momenta of all the constituent particles. Thus, for the n-body problem, the phase space has $6n$ dimensions. A single point in this $6n$-dimensional space defines the state of motion of the entire system.

To illustrate this concept, Figure 11.2 shows solutions to the two-body problem in both real and phase space. On the left are plotted a circular orbit ($e = 0$), elliptic orbits with eccentricities 0.25, 0.5, and 0.75, and a parabolic orbit ($e = 1$). In each case, the arrow indicates the direction of motion, the origin corresponds to a focus, and the parameter $\ell = 1$, so that the curves have polar equation $r = 1/(1 + e \cos \theta)$. If we utilize the conservation of angular momentum condition $r^2 \dot{\theta} = $ constant, the two-body problem can be reduced to a

---

[30] *Le mémoire sur les courbes définies par une équation différentielle.* Some previous work on the qualitative theory of linear second-order differential equations, due to Charles Sturm, had appeared in 1836 (Barrow-Green (1997), p. 30).
[31] Quoted from Barrow-Green (1997), p. 30.

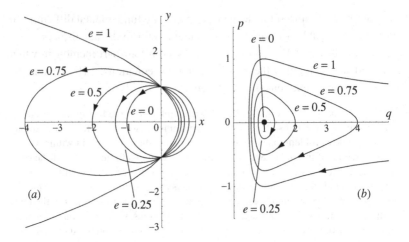

Fig. 11.2. Solutions to the two-body problem in (*a*) real and (*b*) phase space.

system with just one degree of freedom. If we write $q = r$ and $p = \dot{r}$, then the problem has the Hamiltonian formulation

$$\mathcal{H}(p, q) = \tfrac{1}{2}\left(p^2 + 1/q^2\right) - 1/q, \quad \dot{q} = \frac{\partial \mathcal{H}}{\partial p}, \quad \dot{p} = -\frac{\partial \mathcal{H}}{\partial q},$$

where we have set $\mu = 1$ for convenience. The Hamiltonian is a constant of the motion (the total energy) that can be shown to equal $\tfrac{1}{2}(e^2 - 1)$, and the curves $p^2 + 1/q^2 - 2/q = e^2 - 1$ for the same five values of e are shown on the right of the figure. Each curve represents a solution to the two-body problem with a different total energy, and the arrows indicate the direction the curves are traversed as time increases.

When $e = 0$ the solution in phase space is simply the point $q = 1, p = 0$. The circular orbit is thus an example of a *fixed point*. The elliptic orbits correspond to closed curves that surround this fixed point, and the parabolic orbit serves to divide phase space into two regions: one containing the closed periodic orbits, the other occupied by curves representing hyperbolic orbits (not shown). Crucially, the curves in phase space do not intersect or branch in two. If they did, at (p_0, q_0), for example, there would be two possible solutions to the differential equation, both starting from the same state $p = p_0, q = q_0$. For a wide class of differential equations, including all those of interest here, it is known that this cannot be the case; given the initial data, the solutions are unique. Knowledge of the location of any fixed points or curves that bound regions of phase space can thus be used to determine qualitatively the nature of all the solutions to a particular differential equation.

In the vicinity of a fixed point, the phase portrait need not take the form of closed curves as it does in Figure 11.2(*b*) (such a fixed point is called a centre). Poincaré realized that, in two dimensions, fixed points of differential equations could, in general, be points to which all nearby solutions tend as time tends to infinity (or alternatively as $t \to -\infty$), called 'nodes', or they could be saddle points, where the nearby trajectories are hyperbolic in nature and resemble map contours in the vicinity of a col.

Unfortunately, the situation usually is much more complex than that shown in Figure 11.2, since only in the most straightforward examples is the phase space two-dimensional. For the planar restricted three-body problem there are two degrees of freedom, but the two position and two momentum variables are related by Jacobi's integral, so the solutions can be represented by curves in a three-dimensional phase space.

This was the problem Poincaré studied in his remarkable memoir *On the Three-Body Problem and the Equations of Dynamics,*[32] published in November 1890, which transformed the subject of dynamics and introduced into mathematics the phenomenon now called 'chaos'. By drawing diagrams and thinking geometrically rather than performing lengthy calculations, Poincaré revealed more about the nature of the solutions to the problems of celestial mechanics than any of his predecessors, and his conclusions shattered the Laplacian paradigm in which the current state of the Universe could be used to predict the future.

King Oscar's sixtieth birthday competition

Like many great pieces of work before it, Poincaré's memoir was prompted by the establishment of a prize competition. Gösta Mittag-Leffler, mathematics professor at the newly formed University in Stockholm, proposed a contest as part of the sixtieth birthday celebrations for Oscar II, king of Sweden and Norway. Oscar had distinguished himself as a mathematics student at the University of Uppsala, and since then had provided financial support for a number of activities including, in 1882, the founding of the journal *Acta mathematica*, of which Mittag-Leffler was the editor-in-chief. The planning for the contest began in 1884, though Oscar would not be 60 until 1889. A prize commission was formed consisting of Mittag-Leffler and two of his former teachers (both among Europe's leading mathematicians), Charles Hermite in Paris and Karl Weierstrass in Berlin.

---

[32] *Sur le probléme des trois corps et les équations de la dynamique* (1890).

For the contest, mathematicians were invited to address one of four issues at the frontier of mathematical research, though they could submit also an entry on a self-selected topic. Entries had to be submitted – anonymously, accompanied by a sealed envelope containing the author's name – to Mittag-Leffler before 1 June 1888, and the winning paper would be published in *Acta mathematica*. The first question was:

> A system being given of a number whatever of particles attracting one another mutually according to Newton's laws, it is proposed, on the assumption that there never takes place an impact of two particles to expand the coordinates of each particle in a series proceeding according to some known functions of time and converging uniformly for any space and time.[33]

One of Weierstrass' reasons for asking this question was that he believed a method of solution had been developed previously by Peter Lejeune Dirichlet at the University of Göttingen shortly before his death.[34]

All the members of the commission knew that Poincaré was going to enter the competition; the only uncertainty was which question he would choose to answer. It turned out to be the first. In fact, Poincaré only considered the restricted three-body problem, but Mittag-Leffler, Weierstrass, and Hermite agreed unanimously that his long paper (158 printed pages) should win the prize. Although entries were supposed to have been anonymous, Poincaré had sent his with a signed covering letter so the commission knew exactly whose work they were dealing with. All three recognized the quality of the memoir, but understanding the details was not helped by Poincaré's style. Hermite commented:

> But it must be acknowledged, in this work as in almost all his researches, Poincaré shows the way and gives the signs, but leaves much to be done to fill the gaps and complete his work.[35]

Mittag-Leffler took it upon himself to correspond with Poincaré to try to clarify some of the difficulties and as a result Poincaré produced some additional notes, amounting to a further ninety-three printed pages!

---

[33] The complete announcement of the competition was published widely, including in *Nature* on 30 July 1885; it is quoted in full in Barrow-Green (1997), Appendix 2. Twelve entries were received prior to the closing date. Five entrants addressed question 1, one tackled question 3 concerning certain first-order nonlinear differential equations, and the remainder chose their own topics.

[34] Dirichlet had told his student Leopold Kronecker about his novel method for solving differential equations. Kronecker was, in 1885, a professor in Berlin and a bitter rival of Weierstrass and reacted angrily to having not been included in the prize commission. He sent a long list of complaints to Mittag-Leffler and later complained to the Berlin Academy of Sciences that Weierstrass' remarks concerning Dirichlet were inaccurate.

[35] Hermite to Mittag-Leffler in 1888. Quoted from Barrow-Green (1997), p. 65.

The result was announced following the king's birthday on 21 January 1889, and Mittag-Leffler hoped to have the winning memoir published in October. In fact, Volume 13 of *Acta mathematica* containing Poincaré's work did not appear until the following year, and what was eventually published was significantly different from the submitted entry. The problem was that Poincaré discovered that he had made a serious error that affected much of the contents of his article. He wrote a revised memoir, correcting his error and those that arose from it and at the same time incorporated the extra notes he had prepared for the first version into the main text. In the process, he produced an article with key results that had not appeared anywhere in his prize-winning submission. That is what appeared in November 1890, though the story of its production was known only to a select few, since Mittag-Leffler took it upon himself to suppress all information concerning the whole business, including destroying all printed copies of the original version he could lay his hands on.[36] The nature of Poincaré's error and the key conclusions of his memoir will be described in due course (pp. 426–7); suffice it to say that Weierstrass got rather more than he had bargained for.

New Methods of Celestial Mechanics

Poincaré's corrected memoir formed the basis of a much grander work – *New Methods of Celestial Mechanics* – published in three volumes between 1892 and 1899.[37] *New Methods* introduced many original and powerful techniques of relevance to Solar System dynamics and of fundamental importance in mathematics generally. Poincaré was presented with the Gold Medal of the Royal Astronomical Society in 1900, and George Howard Darwin predicted in the presidential address that 'for half a century to come it will be the mine from which humbler investigators will excavate their materials'.[38] Fifty years was a serious underestimate.

Its preface shows clearly Poincaré's attitude to the problems of celestial mechanics:

---

[36] The story behind the publication of Poincaré's prize-winning memoir became known widely only in 1985 when Richard McGehee from the University of Minnesota was studying in the archives of the Mittag-Leffler Institute in Stockholm. He came across one of the recalled copies of Volume 13 of *Acta mathematica* marked in Swedish: 'This whole edition was destroyed. M.L.' (Diacu and Holmes (1996), pp. 48–50). Incidentally, Poincaré agreed to pay for the first printing even though this amounted to considerably more than he received in prize money!

[37] *Les méthodes nouvelles de la mécanique céleste*. English translation in Poincaré (1993).
Poincaré's corrected memoir is studied in detail in Barrow-Green (1997).

[38] Quoted from Goroff (1993), Section 5.5.

The three-body problem is of such importance in astronomy, and is at the same time so difficult, that all efforts of geometers have long been directed toward it. ... At the beginning of this century the achievements of Lagrange and Laplace and, more recently, Leverrier's calculations, have added such a degree of perfection to [perturbation theory] that until now [it has] been sufficient for practical use. ... It is certain, nevertheless, that [it] will not always be adequate ....

The final goal of celestial mechanics is to resolve the great problem of determining if Newton's law alone explains all astronomical phenomena. The only means of deciding is to make the most precise observations, and then compare them to calculated results. This calculation can only be approximate, and it ... is therefore useless to ask more precision from calculation than from observations, but neither should we ask less. Furthermore, the approximation with which we can content ourselves today will be insufficient in several centuries.

However, and here is the crunch,

We must not believe that in order to obtain precise ephemerides over a great number of years it will suffice to calculate a greater number of terms in the developments implicit in the old methods.

This is because, as Poincaré had demonstrated in his prize-winning memoir, the series that arise in perturbation theory do not in general converge. Convergence, or the lack of it, was of great importance, since only through analysing such issues could one put a theoretical bound on the error incurred by using just a few terms. This information would then help to determine whether any discrepancy between theory and observation was due to imperfections in the theory.

In Poincaré's formulation of the restricted three-body problem, the masses of the primaries are denoted by μ and $1 - \mu$. The case $\mu = 0$ corresponds to a two-body problem and so, for small μ, the problem can be thought of as a perturbation of a completely integrable problem. It is thus appropriate to expand the Hamiltonian in powers of μ:

$$\mathcal{H} = \mathcal{H}_0 + \mu \mathcal{H}_1 + \mu^2 \mathcal{H}_2 + \cdots . \tag{11.11}$$

Since the problem is integrable completely when $\mu = 0$, we can choose variables p_α and q_α so that $\mathcal{H}_0 = \mathcal{H}_0(p_\alpha)$ (in which case $\dot{p}_\alpha = 0 \Rightarrow p_\alpha = \text{const}$, and $\dot{q}_\alpha = \partial \mathcal{H}_0 / \partial p_\alpha \Rightarrow \dot{q}_\alpha = \text{const}$). One such choice is the Delaunay variables $\mathbf{q} = (l, g, h)$, $\mathbf{p} = (L, G, H)$, where $L = \sqrt{a}$, $G = L\sqrt{1 - e^2}$, $H = G \cos i$, l is the mean anomaly, h is the longitude of ascending node, $g + h$ is the longitude of perihelion, and then each of the terms $\mathcal{H}_i$ is periodic in the variables l, g, h. The study of Hamiltonian systems in which the Hamiltonian is of the form of Eqn (11.11), with $\mathcal{H}_0 = \mathcal{H}_0(p_\alpha)$ and $\mathcal{H}_i$ periodic in the variables q_α, is called the 'general problem of dynamics' by Poincaré. Nowadays, the variables p_α are called 'action variables', and q_α are 'angle variables'.

When expanding the perturbing function, Poincaré found that it was often convenient to work in terms of a different set of variables obtained from the Delaunay variables through two canonical transformations. First, we replace the angle variables l, g, h by $l + g + h, -g - h, -h$, and then the action variables L, G, H have to be replaced by $L, L - G, G - H$ so as to preserve the Hamiltonian nature of the system. Then we can apply the transformation defined by Eqn (11.5) to the second and third conjugate pairs, to obtain what are now referred to as 'Poincaré variables':

$$l + g + h \qquad\qquad L$$
$$-\sqrt{2(L - G)}\sin(g + h) \quad \sqrt{2(L - G)}\cos(g + h)$$
$$-\sqrt{2(G - H)}\sin h \quad\quad \sqrt{2(G - H)}\cos h$$

This new set of canonical elements turns out to be particularly advantageous when e and i are small.

Poincaré devoted considerable effort toward finding periodic solutions to the restricted three-body problem. The first such solutions to be found were the straight line and equilateral triangle solutions of Euler and Lagrange, respectively. Hill's intermediate orbit for the lunar theory was another example, which Poincaré generalized in a paper in 1884 when he showed that a whole family of periodic solutions existed for the case when two of the masses were very small. These solutions could be classified into three different categories and, in *New Methods*, Poincaré described their relation to two-body solutions. First-type solutions, in which the eccentricity is small and the inclination is zero, are continuations of circular solutions to the two-body problem to the case $\mu > 0$, elliptic two-body orbits generate periodic solutions of the second type, and third-type solutions come from elliptic two-body solutions when the small mass μ is not in the orbital plane. Hill's periodic solution is an example of the first type. In all three cases, the mutual distances between the bodies are periodic functions of time, so that at the end of a period the bodies are in the same position relative to each other. However, the system will have turned through a certain angle. In order for the coordinates of the bodies to be periodic functions of time, we would need to refer them to a set of uniformly rotating axes.

Periodic orbits are special because we can know everything about them by studying only a finite amount of time. But is knowledge of them useful? In general, the probability of a given initial state leading to a periodic orbit is zero, but Poincaré realized that they might usefully be used as intermediate orbits in the style of Hill, and held out the hope that one can always find a periodic solution that approximates any given solution to a given accuracy over as long a time as desired. He described periodic solutions as 'the only breach by which we may attempt to enter an area heretofore deemed inaccessible'.[39]

There are conditions under which periodic solutions do not persist when μ is increased from zero, and this can also be used to advantage. Suppose we have three bodies of mass $1 - \mu$, μ, and zero, and suppose in the first instance that $\mu = 0$, with the two massless bodies orbiting the third in circles with mean motions n and n' ($n' > n$). This system is periodic with respect to axes rotating with angular speed n, with period $2\pi/(n' - n)$. Poincaré showed that periodic first-type solutions will exist when μ is small but non-zero, provided $n'/(n' - n)$ is not equal to an integer. In other words, we cannot have the case

$$\frac{n}{n'} = \frac{j-1}{j}$$

for an integer j. However, what happens if $n'/(n' - n)$ is not integral, but only nearly so? In this case, there will be a first-type solution, but it will exhibit a large irregularity. In the case, of the Saturnian moons Hyperion and Titan, their mean motions are roughly in the ratio $3:4$ ($n_H/n_T = 0.749\,43\ldots$) and we would expect this to lead to a considerable inequality in their motion. Whatever the actual motion of the two satellites, the form of this inequality will be established most easily by studying the appropriate periodic solution.[40]

Poincaré continued to be occupied with periodic orbits for the rest of his life. In 1912, he published a paper containing a theorem impling the existence of an infinite number of periodic solutions in the restricted three-body problem for any value of the mass parameter μ. He was unable to prove the result, however, and had resigned himself to publishing the work in an unfinished state, hoping that others would be able to complete what he had started. His hopes were realized, sooner than he might have expected, when, shortly after Poincaré's death, George David Birkhoff provided a brilliant proof.[41] Rather earlier than this (in 1897) and stimulated by Poincaré's results, George Darwin took the first steps in what is now a flourishing area of research: the numerical construction of periodic orbits. Building on Hill's idea for partitioning space according to the value of the Jacobi constant, Darwin made a systematic and painstaking search

[39] POINCARÉ *Les méthodes nouvelles*, Section 36. Poincaré's use of probabilistic concepts to argue that certain classes of solution were infinitely unlikely (have zero measure as we would now say) was itself a radical innovation (Aubin and Dalmedico (2002)).

[40] Poincaré described this idea, used by Tisserand in 1886, and other implications of periodic solutions for the satellite systems of the outer planets in Sections 49–50 of *Les méthodes nouvelles*.

[41] Birkhoff's proof is outlined in Barrow-Green (1997), Section 7.4.2.

for periodic solutions to the restricted three-body problem and discussed their stability.[42]

In the third volume of *New Methods*, Poincaré described what has become a basic tool in the qualitative study of differential equations: the Poincaré return map. Suppose we have a periodic orbit within phase space which, for simplicity, we will assume to be three-dimensional (as it is in the planar-restricted three-body problem). Take a surface S (called a 'Poincaré section'), such that the periodic orbit crosses the surface at M, for example. Viewed on S, the periodic orbit takes the form of a fixed point. Now, consider a curve in phase space that passes through a point M_0 on S. Provided M_0 is sufficiently close to M, this trajectory will cross S again, e.g. at M_1. Poincaré calls M_1 the 'consequent of M_0', and we refer to the function mapping M_0 to M_1 as 'Poincaré's return map'. The point M_1 will have its own consequent, M_2, and so on, and the collection $M_0, M_1, M_2, \ldots$, is called the 'orbit of the point M_0'. With this device, the study of the region of phase space near periodic orbits becomes the study of the region of the Poincaré section near a fixed point.

The new problem is set in a space of one dimension less than the original one, but the nature of the possible behaviour near fixed points is now much more complicated. For one particular class of fixed points (which Poincaré referred to as unstable), there is a family of points that get closer and closer to the fixed point under the action of the return map (referred to by Poincaré as the 'second family of asymptotic solutions' and now called the 'stable manifold', S, for example) and there is a different family of points (Poincaré's first family of asymptotic solutions; the unstable manifold, U, for example) that approach the fixed point under the action of the inverse of the return map, i.e. as time runs backwards. The curves S and U are invariant in the sense that, for any point on S, its consequent is also on S, and similarly for U.

If S and U intersect at some point T, then the orbit of T must lie on both S and U and so must approach the unstable fixed point both as time increases and as it decreases. Poincaré called these 'doubly asymptotic solutions' and T a 'homoclinic point'. If T is on both S and U, then so is its consequent, and so on. Thus, if S and U intersect once, they must intersect infinitely many times. The simplest possibility (and the only one for completely integrable systems) is

[42] Darwin's paper was entitled *Periodic Orbits* (see Barrow-Green (1997), Section 8.4.1 for more details). Darwin's results were discussed by Poincaré in the third volume of *Les méthodes nouvelles*, Sections 381–4. Until the advent of computers, this type of work was incredibly time-consuming. An attempt to trace a continuous family of periodic orbits from those encircling both primaries to those just orbiting one primary was made by Moulton (1920). In the Introduction he wrote: 'The amount of labor [Chapter XVI] cost can scarcely be overestimated'.

that $\mathcal{S} = \mathcal{U}$, in which case the stable and unstable manifolds represent a smooth connection between the unstable fixed point and itself.

The consequences of $\mathcal{S}$ and $\mathcal{U}$ actually crossing each other leads to some very odd behaviour indeed, and Poincaré believed initially that this was impossible. In his original prize entry he had shown that if the Hamiltonian for the restricted three-body problem written in the form of Eqn (11.11) was suitably truncated, then the resulting system was completely integrable and its Poincaré map possessed unstable fixed points, the stable and unstable manifolds of which had to coincide. Then, he claimed a stability result on the assumption that $\mathcal{S}$ and $\mathcal{U}$ must still coincide when the truncated terms in the Hamiltonian were restored. The entries to the prize competition were read initially by Lars Edvard Phragmén, an editor of *Acta mathematica*, and he corresponded with Poincaré (via Mittag-Leffler) over parts of the manuscript he considered unclear. Sometime during this process, perhaps prompted by Phragmén, Poincaré realized he had made a serious error.[43]

Poincaré realized suddenly that so-called transverse homoclinic points could exist, with devastating consequences for the dynamics of the three-body problem. The existence of a single such point implies the existence of infinitely many such points all intertwined in an extremely complicated fashion, known as a 'homoclinic tangle':

> When we try to represent the figure formed by these two curves and their infinitely many intersections, each corresponding to a doubly asymptotic solution, these intersections form a type of trellis, tissue, or grid with infinitely fine mesh. Neither of the two curves must ever cut across itself again, but it must bend back upon itself in a very complex manner in order to cut across all of the meshes in the grid an infinite number of times.
>
> The complexity of this figure is striking, and I shall not even try to draw it. Nothing is more suitable for providing us with an idea of the complex nature of the three-body problem . . . .[44]

In correcting his error, Poincaré changed completely the nature of his original memoir. Instead of demonstrating stability, he was now opening up the possibility of all sorts of weird and wonderful behaviour. The nature of the trajectories in a homoclinic tangle showed that solutions corresponding to initial conditions that were very close together could separate extremely quickly; here was the first glimpse of what we now term 'chaos'. He had understood that, in systems modelled by differential equations (e.g. the equations of celestial

[43] Phragmén's queries and Poincaré's response are described in Andersson (1994).
[44] POINCARÉ *Les méthodes nouvelles*, Section 397. More on Poincaré's discovery of homoclinic tangles can be found in Holmes (1990) and Diacu and Holmes (1996), the latter being less demanding mathematically.

mechanics), the qualitative behaviour of the solutions often was highly complex, with extreme sensitivity to initial data. The deterministic nature of the equations did not imply that accurate predictions were possible. In his 1903 essay *Science and Method*, he incorporated this fundamental property of dynamical systems into his definition of chance:

> A very slight cause, which escapes us, determines a considerable effect which we cannot help seeing, and then we say this effect is due to chance. If we could know exactly the laws of nature and the situation of the universe at the initial instant, we should be able to predict exactly the situation of this same universe at a succeeding instant. But even when the natural laws should have no further secret for us, we could know the initial situation only *approximately*. If that permits us to foresee the subsequent situation *with the same degree of approximation*, this is all we require, we say the phenomenon has been predicted, that it is ruled by laws. But this is not always the case; it may happen that slight differences in the initial conditions produce very great differences in the final phenomena; a slight error in the former would make an enormous error in the latter. Prediction becomes impossible and we have the fortuitous phenomenon.[45]

Bold and brilliant as it may have been, Poincaré's work was only the first step into a whole new mathematical world, but after his death the abstract study of dynamical systems did not, as one might expect, flourish. Perhaps this was because the ideas that Poincaré had unearthed ran counter to the generally accepted view of the exact sciences, and until the consequences of his ideas were understood more fully, they easily were ignored. It was not until the work of Birkhoff – whose book *Dynamical Systems* was published in 1927 – that Poincaré's results were improved upon significantly, and then the next major advances had to wait until the 1950s, spurred by the translation of Birkhoff's book into Russian. Since then, however, Poincaré's creation (with the addition of the catchy label 'chaos theory' and the advent of the computer) has become one of the most popular branches of mathematical research.

The stability of the Solar System

At the end of the nineteenth century, greater mathematical awareness meant that the stability of the Solar System was, if anything, less clear than it had been at the end of the eighteenth, following the initial 'proof' of stability by Laplace (see p. 349). One area of concern was the question of the magnitude of oscillations caused by perturbations. For example, Laplace had shown that the inclinations

---

[45] POINCARÉ *Science and Method*. Translation from Poincaré (1929), p. 397.

of planetary orbits were periodic, but the amplitudes still could be considerable. The first successful determination of amplitudes was by Lagrange. He grouped the planets into two groups: the outer planets beginning with Jupiter, and the inner planets Mercury to Mars. For each of the inner group he computed the effects of the outer group. The values of the masses used by Lagrange for many of the planets were inaccurate (particularly for Venus where his value was 50 per cent too large). Leverrier improved matters by including all the planets at once, and using better mass values, but his results were not very different from those Lagrange had found. Leverrier showed that oscillations in i could be large for small planets near about 2 AU (i.e. near the asteroid belt) and also in a region between Mars and Venus.

The issue of real concern, however, was to do with the convergence of the series used to represent the motion. Laplace's original result was improved on by Lagrange, who showed that, to all powers of e and i, but still only first-order in the masses, the mean distances were subject only to periodic inequalities. Analysis to second order was carried out by Poisson (1809) who showed that, although pure powers of t did not occur, it was possible to get terms such as $t \sin(\alpha t + \beta)$. This suggests that deviations can become large but, since $\sin(\alpha t + \beta)$ passes periodically through zero, the solution continually will return to its value with this term absent. This idea leads to an alternative concept of stability from that understood by Lagrange and Laplace: if a system returns arbitrarily close to its initial state infinitely often, Poincaré described it as being 'Poisson-stable'. If nothing else, Poisson's work showed that the stability of the Solar System according to Lagrange and Laplace was on shaky foundations.

It had been appreciated during the nineteenth century that the presence of coefficients containing powers of t could just be because the series was being expanded with respect to inappropriate angular frequencies. Thus, perhaps one could remove all the troublesome terms by using more suitable angular variables. This, in essence, was the underlying motivation behind the theories of Hansen, Delaunay, Newcomb, and Hill. The complexity that resulted from this type of approach made it almost impossible to tackle the key question of whether the resulting series actually converge. Poisson's result, and those of Lagrange and Laplace before him, assumed that the mean motions of the perturbed and perturbing bodies were incommensurable, and this could be justified by appealing to that fact that the probability of there being a rational ratio between mean motions is zero. However, rational numbers can approximate irrational ones with arbitrary accuracy, and so there will always be the problem of small divisors in the perturbation expansions.

Significant progress was made by the Finnish astronomer Hugo Gyldén, director of the observatory at Stockholm, and the Swedish astronomers Anders Lindstedt and Karl Bohlin.[46] Gyldén developed a technique for overcoming the problem of small divisors and establishing convergence that ultimately was unsuccessful but which stimulated the work of Lindstedt and Poincaré. Gyldén's style made his work difficult to appreciate (Poincaré described it as 'a bit repelling'[47]) and Lindstedt tried to clarify Gyldén's arguments. Bohlin improved Delaunay's method by removing the need to perform numerous changes of variable. The resulting procedure, which involved expanding the Hamiltonian in powers of $\sqrt{\mu}$, was very similar to one devised by Poincaré, and was Poincaré's preferred method for dealing with cases in which the mean motions led to problems with small divisors.

The second volume of *New Methods* was devoted to the analysis of the types of series that result from perturbation theory, and Poincaré utilized the power of the Hamilton–Jacobi theory to completely solve the problem of removing the secular terms from the expansions. The key lies in casting the problem in action-angle variables. Both Gyldén and Lindstedt had studied differential equations of the form

$$\ddot{x} + n^2 x = f(x, t),$$

where $f(x, t)$ is a power series in x with small coefficients that are periodic functions of t. Lindstedt attempted to show that one could always construct the solution as a purely trigonometric series, and illustrated his idea with the simple example of $\ddot{x} + n^2 x = a + bx$. Rather than look for a solution with x expanded in terms of the frequency n, clearly it makes more sense to define a new frequency $n_1 = \sqrt{n^2 - b}$ and solve $\ddot{x} + n_1^2 x = a$. Lindstedt devised an iterative procedure for the general case, but the increased complexity at each stage meant he could not prove convergence.[48]

By recasting the problem into the scheme of his 'general problem of dynamics', Poincaré demonstrated that Lindstedt's goal was, indeed, possible but that, contrary to the opinions and hopes of astronomers, no reliable qualitative conclusions could be drawn from the existence of these trigonometric series, since, in general, they did not converge. This was not to say that the series were not useful – the first few terms in such series might well provide good numerical approximations. As Poincaré was at pains to point out, series that did

---

[46] Inspired by Mittag-Leffler, Sweden became something of a centre for research into the mathematical problems of celestial mechanics (see, for example, Gårding (1998)).

[47] In a letter to Mittag-Leffler in 1889 (Goroff (1993)).

[48] See Gårding (1998), p. 103 for more details.

not converge in the rigorous mathematical sense had been used by astronomers to great effect for well over 100 years:[49]

> Between mathematicians and astronomers some misunderstanding exists with respect to the meaning of the term "convergence". Mathematicians ... stipulate that a series is convergent if the sum of the terms tends to a predetermined limit even if the first terms decrease very slowly. Conversely, astronomers are in the habit of saying that a series converges whenever the first twenty terms, for example, decrease rapidly even if the following terms might increase indefinitely.
>
> ... Both rules are legitimate; the first for theoretical research and the second for numerical applications. Both must prevail, but in two entirely separate domains of which the boundaries must be accurately defined.
>
> Astronomers do not always know these boundaries accurately but rarely exceed them; the approximation with which they are satisfied usually keeps them far on this side of the boundary. In addition, their instinct guides them and, if they are wrong, a check on the actual observation promptly reveals their error....[50]

The lack of convergence of the series solutions to the problems of celestial mechanics did not diminish the usefulness of perturbation theory as a tool for predicting the future positions of the planets, but it did show that there was a limit to the accuracy of such an approach, and that it could not be used to say anything about the long-term stability of the Solar System.

Poincaré's results showed that, in general, perturbation series diverge. But that does not rule out the possibility that, for the special parameter values associated with the Solar System, the relevant series might, in fact, be convergent. Poincaré was unable to settle this question, but came to the conclusion that it was highly improbable that special choices of the parameters would lead to convergence. Further progress on this point had to wait until a whole new mathematical framework, known as 'KAM theory', had been created. In the early 1960s, it was shown that if we were to select the initial state for a Solar System at random, there would be a positive probability that the motion will be confined within definite limits for all time, assuming that the masses of the planets are sufficiently small.[51] However, the parameter values for the actual Solar System do not satisfy the necessary conditions of the theorem.

---

[49] Poincaré illustrated his remarks with reference to the series $\sum_n 1000^n/n!$ and $\sum_n n!/1000^n$. The first converges even though the first 1000 terms increase, whereas the second diverges even though the first 1000 terms decrease.

[50] POINCARÉ *Les méthodes nouvelles*, Section 118.

[51] This result is due to Arnold: *Small Denominators and Problems of Stability of Motion in Classical and Celestial Mechanics* (1963). The KAM theory is named after its three co-creators. The initial steps were taken by Andrei Nikolaevich Kolmogorov in 1954, and the theory was improved and perfected over the next 8 years by Vladimir Igorevich Arnold and Jürgen Moser. A general description of the theory can be found in Diacu and Holmes (1996), Chapter 5.

Not all Poincaré's results concerning stability were negative in nature. His famous recurrence theorem, based on the concept of integral invariants, states that in any Hamiltonian system, the system will return arbitrarily close to its initial configuration infinitely often, provided the motion is restricted to a finite region of phase space. There is a set of initial states for which the theorem fails, but the probability of the initial setup being in this set is zero. Hill had demonstrated previously (see Figure 11.1) that, for the circular restricted three-body problem, there were finite regions of space from which the massless body cannot escape. Poincaré showed that, in the case of a body constrained to move in the vicinity of one of the primaries A or B, the corresponding region of phase space was also finite, and, hence, the motion is Poisson-stable. In this case, the motion was also bounded for all time, but Poincaré still did not claim complete stability, since for this he required a third condition, i.e. that none of the bodies collide with one another (something he had simply assumed).[52]

In 1898, Poincaré wrote a popular article in which he summarized the stability question as follows:

> People who are interested in advances in celestial mechanics, but who can only follow them from afar must experience some surprise in seeing how many times the stability of the solar system has been proven.
>
> Lagrange established it first, Poisson demonstrated it again, more demonstrators have come since, more will yet come. Were the old proofs insufficient, or is it the new ones that are superfluous?
>
> These peoples' surprise will undoubtably redouble if one tells them that one day, perhaps, a mathematician may show, by rigorous argument, that the solar system is unstable.
>
> This really could happen; there would be nothing contradictory about it, and the old demonstrations would still retain their value. It is just that they are actually only successive approximations; thus they do not pretend to rigorously bound the elements of the orbits between narrow limits from which they can never escape.[53]

For Laplace, the stability of the Solar System was a fundamental question about the physical Universe. To Poincaré, it was a question of rigorous mathematics. The expressions that Laplace and his successors used to investigate stability consisted of the first few terms of series which, in general, diverge and, as such, they can tell us nothing about the behaviour of the Solar System for all time. On the other hand, Poincaré's theoretical conclusions reveal a great deal

---

[52] Another definition of stability that was developed around this time was that of Alexander Liapunov. Liapunov stability implies that solutions that pass through neighbouring points in phase space remain close together for all time and is thus more restrictive than the notion of stability used by Poincaré.

[53] Poincaré's article *On the Stability of the Solar System* appeared in English translation in *Nature* (1858). Quoted from Goroff (1993), Section 4.6.

about the structure of solutions to differential equations but say little about
the actual Solar System. Poincaré sought to prove rigorous results valid for
interacting point masses in the limit as $t \to \infty$, notwithstanding the fact that
the planets and their moons have finite size, are affected by tidal forces, and
are not perfectly spherical. He was aware that the tidal forces operating in the
Solar System act like frictional forces and generate heat, thus removing energy
from the system. This is arguably more important from a practical point of view
than any theoretical conclusion drawn from an idealized problem. To Poincaré,
however, this did not make the analysis of the n-body problem in the limit as
$t \to \infty$ any less interesting.

From the time of Eudoxus to the end of the nineteenth century, there have
been scientists who sensibly can be given the label 'mathematical astronomer'.
They practised mathematics through applications to astronomy and sought to
understand the heavens with the aid of mathematics. Poincaré brought the distinction between the two disciplines into sharp focus and, from his day to this,
the nature of a piece of research distinguishes clearly the mathematician from
the astronomer.

The n-body problem and its solution

One of the best known (and most misunderstood) of Poincaré's results concerns
the non-existence of any new integrals for the restricted three-body problem.
The basic equations for the n-body problem (a system of order $6n$) and the
ten integrals that can be obtained have already been described (see pp. 324–5).
In 1887, Heinrich Bruns showed that when rectangular coordinates are used for
the dependent variables, no other algebraic integrals exist for the three-body
problem. Poincaré generalized Bruns' result to exclude the possibility of any
new transcendental integral existing, in the case when two of the masses are
small compared to that of the third and the dependent variables are the Delaunay
elements.[54]

The theorems of Bruns and Poincaré, and the various subsequent improvements, demonstrate that the three-body problem is not completely integrable,
but this does not imply (as is sometimes believed) that it cannot be solved in a

---

[54] Bruns' paper contained an error that Poincaré later corrected (Barrow-Green (1997), p. 164).
Many generalizations of Bruns' and Poincaré's results have appeared, and the existence of
integrals for the n-body problem is still a topic of mathematical research (see, for example,
Julliard-Tosel (2000)). At the beginning of the twentieth century, F. R. Moulton wrote: 'The
practical importance of the theorems of Bruns and Poincaré have often been overrated by those
who have forgotten the conditions under which they have been proved to be true' (Moulton
(1970), Article 147).

different way. In particular, it does not follow that Weierstrass' question asking for a solution in terms of a convergent power series cannot be answered. In fact, it was shown by Paul Painlevé in 1895 that such a solution *was* possible. Painlevé was investigating singularities in differential equations, and established that the only possible singularities in the three-body problem were due to collisions between the masses. Given some initial conditions that do not lead to a collision in the future, a convergent series solution must exist.[55]

The existence of a solution is one thing, finding it is another. In the final volume of his *Celestial Mechanics* (1896) Tisserand wrote:

> The rigorous solution of the three-body problem is no further advanced today than during the time of Lagrange, and one could say that it is manifestly impossible.[56]

This was overly pessimistic. A complete theoretical solution to the three-body problem was developed during the first decade of the twentieth century by Karl Sundmann from the Helsinki observatory, and published in 1912. He obtained a series solution in powers of $t^{1/3}$ that converged provided the total angular momentum of the system was nonzero. Sundmann's method did not work in the case of a triple collision, but he showed that this could only happen if the angular momentum was zero. However, when only two of the masses collide – a binary collision – he was able analytically to extend his solution beyond the singularity, effectively allowing the bodies to bounce off one another. This process is called 'regularization'.[57] An improved and simplified theory was developed by the Italian mathematician Tullio Levi-Civita in 1920.

Given the nature of the problem Sundmann was solving – one that had defeated the greatest mathematicians and astronomers since its formulation by Newton – his method was remarkably simple, depending only on well-established results in the theory of differential equations.[58] It may appear something of a paradox that Sundmann's solution is not particularly well known. After all, he achieved what Poincaré had not: he actually answered one of the

---

[55] Painlevé could not extend his result to the case of more than three bodies, and conjectured that singularities not due to collisions (he called them 'pseudocollisions') do exist in the n-body problem when $n > 3$. In 1908, the Swedish mathematician and astronomer Edvard Hugo von Zeipel showed that a pseudocollision corresponds to a situation in which one of the masses disappears off to infinity in finite time (McGehee (1986)). The first example of a noncollision singularity was constructed as part of his Ph.D. thesis by Zhihong (Jeff) Xia in 1987 (Xia (1992)) for the case $n = 5$, and his solution can be generalized to any $n > 4$. The ideas underlying Xia's construction are described in Saari and Xia (1995). The existence of pseudocollisions when $n = 4$ is still an open question (Diacu and Holmes (1996)).

[56] Quoted from Barrow-Green (1997), p. 190.

[57] For a brief sketch of Sundmann's ideas, see, for example, Saari (1990), Hall and Josić (2000).

[58] Sundmann's method did not apply to the case of more than three bodies, and it took almost 70 years before a solution to the general case was found by a Chinese research student at the University of Cincinnati, Quidong Wang (Wang (1991)).

questions set for Oscar II's prize. But the reason is simple; the series in the solution are so slowly convergent that they provide no qualitative information about the behaviour of the system and, from a practical point of view, the solution is totally useless. It has been estimated that, in order to achieve the accuracy required to compare with observations, one would need to take $10^{8\,000\,000}$ terms in Sundmann's series![59]

Chaos in the Solar System

Poincaré saw that the equations of celestial mechanics allow for some very strange types of solutions. But do any celestial bodies actually exhibit such behaviour? The answer to this question is: Yes, though our understanding of the extent to which Solar System dynamics is chaotic is by no means complete.[60] What is clear is that the phenomenon of chaos is tied closely to the dynamics of resonances.

The most commonly cited example of chaos in the Solar System is the motion of Hyperion, a small and irregularly shaped (about $380 \times 290 \times 230$ km) satellite of Saturn. Little was known about Hyperion until the Voyager 2 spacecraft passed close by it in 1981, but it turns out that its spin rate and orientation change markedly during the course of a few orbital periods (each about 21 days). Given sufficient time, one would expect any aspherical satellite to present the same face to its parent planet due to the action of tidal forces, just as in the case of our Moon (see p. 319). Most major satellites in the Solar System display such a 1 : 1 spin–orbit resonance, though there are exceptions.[61] In the case of Pluto and Charon, where the masses of the two bodies are much closer than in other planet–satellite systems, both keep the same face toward the other. They are said to be 'totally tidally despun'.

The process by which synchronous rotation is reached is very slow and, in the case of Hyperion – where the moon is far from its host planet – the tidal forces are very weak and the appropriate timescale is the age of the Solar System. A phase plane analysis of the spin–orbit resonance problem shows that the synchronous state is an island of stability surrounded by a chaotic region, the size of which depends on numerous factors, including the eccentricity of

---

[59] Goroff (1993), Section 2.5. The philosophical question of whether Sundmann's convergent power series should even be classified as a solution is discussed in Diacu (1996).

[60] The brief details given here are based largely on Wisdom (1987).

[61] Most notable is Mercury, which exhibits a 3 : 2 spin–orbit resonance (Petengill and Dyce (1965)). It was in order to explain this unusual phenomenon that Goldreich and Peale (1966) developed a dynamical theory of spin–orbit coupling, subsequently used to explain the motion of Hyperion.

the orbit and the shape of the satellite. Thus, it appears that all irregularly shaped satellites must pass through a period of chaotic motion before reaching a resonant state. In the case of Hyperion, the approximate 4 : 3 orbital resonance with the much larger moon Titan results in a very large chaotic region and, while the rotation of Hyperion may one day stabilize, it is highly unlikely that it will ever achieve a 1 : 1 spin–orbit resonance.[62]

Another example of chaos at work in the Solar System is the creation of the Kirkwood gaps in the asteroid belt. We have seen already that these are associated with orbital resonances between asteroids and Jupiter, but that some such resonances appear to be stable, while others are unstable. The first dynamic analysis of asteroid motion near to resonance actually was undertaken by Poincaré in 1902 for the case of the 2 : 1 resonance with Jupiter, using an averaging procedure to isolate the long-period variations. The averaging method – which was developed by Gauss and Lagrange – is not justified rigorously since it is based on perturbation series which, as we have seen, are not necessarily valid for large values of the time. Nevertheless, Poincaré's method predicts qualitatively the behaviour of an asteroid near resonance over a few centuries, but if one is interested in hundreds of thousands of years another approach is required.

With the advent of the computer, numerical techniques have been developed (ultimately based on a non-rigorous averaging procedure) that enable qualitative information about such timescales to be obtained. Computations for the 3 : 1 resonance show a chaotic region of phase space characterized by large excursions in eccentricity. An asteroid can spend 100 000 years orbiting with an eccentricity of less than 0.1 and then 'suddenly' jump to an orbit with $e > 0.3$, subsequently returning to its more normal state. The significance of this is that when the eccentricity is so large, the orbit of the asteroid crosses that of Mars and a close encounter could see the orbit of the asteroid changed considerably. It seems likely that this is the mechanism by which the 3 : 1 Kirkwood gap was formed.[63]

The dynamics of the 2 : 1 resonance (where there is a gap) and the 3 : 2 resonance (where we find the Hilda asteroids) are more difficult to analyse than the 3 : 1 case. Indeed, the qualitative nature of the two problems, as predicted

---

[62] Following the discovery of the satellite's irregular shape, the tumbling motion of Hyperion was predicted by Wisdom, Peale, and Mignard (1984) and confirmed through the observations of James Klavetter in 1987 (Peterson (1993), Chapter 9). See Murray and Dermott (1999), Chapter 5, for a mathematical treatment of the spin–orbit coupling problem.

[63] These computations were based on a two-dimensional analysis. When three-dimensional effects are taken into consideration, larger excursions in eccentricity are found, with some orbits becoming Earth-crossing, thus providing an even stronger mechanism for clearing the gap (Wisdom (1987)).

by Poincaré's method, are identical. The method that worked for the 3 : 1 case fails here because it is necessary to include much higher-order terms in the perturbation series. It appears that the best way to tackle the problem is by direct numerical integration of the equations of motion. This was done in the 1980s using the Digital Orrery, a purpose-built computer for solving problems of celestial mechanics designed by Gerald Jay Sussman, and sure enough, one finds a large chaotic region of phase space near the 2 : 1 resonance, but the 3 : 2 resonance is dominated by stable behaviour.[64]

From the beginning of the eighteenth, to the end of the twentieth, century predictive astronomy was built around progressively more exact analyses of the equations of motion resulting from Newton's theory of universal gravitation. Poincaré achieved a deeper understanding of the nature of these equations, and was the first to appreciate that this incremental process had to come to an end, as there was a theoretical limit to the predictive power of Newtonian theory. Research into celestial mechanics in the twentieth century had a radically different focus to that which had gone before.

There was another major surprise in store for astronomers; indeed, for the whole of the scientific community. In creating general relativity, Einstein showed that Newtonian mechanics, the bedrock of mathematical astronomy, was an approximation valid only for speeds much less that that of light. Lagrange had described Newton as the luckiest (as well as the greatest) of all scientists because he had created the science of the Universe and this could only be done once, but he was wrong![65] Apart from making people think more deeply about space and time, relativity theory had little effect on planetary astronomy and did not really become an important predictive tool until it entered into the field of cosmology in the latter half of the twentieth century. There was, however, one notable exception, and to understand this we must return to Leverrier and his work on the planet Mercury.

---

[64] Many of the mathematical techniques used to investigate chaotic dynamics in the Solar System are described in Murray and Dermott (1999), Chapter 9.
[65] Infeld (1973).

12

Mercury and relativity

The anomalous advance of the perihelion of Mercury

In the early stages of Leverrier's career in astronomy, Arago, the director of the Paris Observatory, suggested to him that he work on the theory of Mercury's orbit. Here was a problem with which the ambitious Leverrier felt he could make his name, and he immediately took up the challenge. Many astronomers previously had tackled the problem with only limited success, and tables of Mercury were notoriously inaccurate.[1]

Difficulties with Mercury stemmed from two different sources. On the one hand there was not much good observational data to work with because of the proximity of the planet to the Sun. This was, however, mitigated against by the presence of a small number of extremely accurate observations made during transits. The other problem lay in the fact that the eccentricity and inclination of the orbit of Mercury are relatively large ($e \approx 0.21$, $i \approx 7°$) and so expansions in powers of e and i in the theory are not as rapidly convergent as they are for the other planets.

Problems with tables of the motion of Mercury were highlighted by their poor performance at predicting transits. Tables in use in 1707 were a day out in predicting the Mercury transit in that year; by the time of the transit in 1753, improvements had reduced the error in prediction to a few hours, and by 1786 there was just under 1 h between the predicted and actual start of a transit. Leverrier's first sorties with Mercury led to a new set of tables that predicted the 8 May 1845 transit to within 16 s. This was a considerable improvement but, given the accuracy of the other planetary theories at the time, it was still considered unacceptable. Further refinements were necessary,

---

[1] A detailed technical account of the resolution of problems in the theory of Mercury is given in Roseveare (1982). A more popular account can be found in Baum and Sheehan (1997).

therefore, but it was at this point that Leverrier was tempted away from Mercury, again by Arago, to bring his analytical skills to bear on the errant motion of Uranus.

Following the discovery of Neptune, Leverrier had honours heaped upon him. A Chair in Celestial Mechanics was created for him in Paris and in 1854 he became the director of the Paris Observatory, which he ran in much the same way as Airy ran the Royal Observatory. His authoritarian style led to friction with his staff and he was removed from his position in 1870, only to be reinstated 2 years later on the death of his successor, Charles Delaunay. His success with Neptune had significant consequences for Leverrier's attitude to any discrepancies between theory and observation when it came to Mercury – tinkering with universal gravitation simply could not be countenanced.

> If the tables [of Mercury's positions] do not strictly agree with the group of observations, we will certainly not be tempted into charging the law of universal gravitation with inadequacy. These days, this principle has acquired such a degree of certainty that we would not allow it to be altered; if it meets an event which cannot be explained completely, it is not the principle itself which takes the blame but rather some inaccuracy in the working or some material cause whose existence has escaped us.[2]

As a first step toward finding out what was causing the errors in the predicted position of Mercury, Leverrier re-examined the solar theory. This needed to be error-free, since the transit observations that provided the bulk of the data against which to check the Mercury theory obviously were given with respect to the position of the Sun. By making a small change to the accepted value of the solar parallax, and by increasing the accepted values for the masses of the Earth and Mars by 10 per cent, Leverrier found that he could produce better agreement between observation and theory; in particular, he was able now to give a satisfactory quantitative prediction for the regression of the nodes of Venus and the advance of the perihelion of Mars.

Now, when he turned his attention back to Mercury, Leverrier was sure that he had accounted for all possible sources of error, and was confident that the transit data with which he was working gave the positions of Mercury to within $1''$ of arc. However, he found that while theory was within this tolerance at November transits (when Mercury is near its ascending node) this was not the case during May transits (with the planet close to its descending node). He found that there was a discrepancy that got progressively worse from one May transit to the next, amounting to about $13''$ over a 92-year period.

---

[2] LEVERRIER *Les nouvelles recherches sur les mouvements des planètes* (1849). Quoted from Roseveare (1982).

Table 12.1. *Contributions to the advance
of the perihelion of Mercury due to the
other planets (measured in arc-seconds
per century).*

Planet	
Venus	280.6
Earth	83.6
Mars	2.6
Jupiter	152.6
Saturn	7.2
Uranus	0.1
Total	526.7

Mercury's orbit is, of course, perturbed by the other planets. These perturbations can be divided into two types (see Eqn (9.18), p. 330): a secular term
proportional to the time t, and a sum of periodic terms. Moreover, to first-order
in the planetary masses, we can consider the orbit of the perturbing planet as
a Keplerian ellipse with fixed elements. If, in Eqn (9.24), we retain only those
parts of the disturbing function R that give rise to secular perturbations, then we
find that the variations in the elements $\bar{\omega}$ and e of Mercury's orbit (the longitude
of perihelion and the eccentricity, respectively) are given (to lowest order in the
eccentricities) by:[3]

$$\frac{d\bar{\omega}}{dt} = nm' \left(A + B\frac{e'}{e} \cos(\bar{\omega} - \bar{\omega}') \right),$$

$$\frac{de}{dt} = nm'Be' \sin(\bar{\omega} - \bar{\omega}'),$$

$$(12.1)$$

in which e' and $\bar{\omega}'$ are the corresponding elements of the perturbing body, n is
the mean motion of Mercury, m' the disturbing mass, and A and B are constants
depending on the ratio of the semi-major axes of Mercury and the perturbing
planet.

From these equations, Leverrier could calculate the effect of all the other
members of the Solar System on the perihelion and eccentricity of Mercury
separately. For the perihelion advance he found a total contribution of just
under 530'' per century (about $1\frac{1}{4}''$ per orbit) made up as shown in Table 12.1.[4]

---

[3] See Eqns (6.170) and (6.171) of Murray and Dermott (1999).
[4] The numbers are quoted from Roseveare (1982). It is possible to reproduce these numbers fairly
easily and accurately by treating each outer planet as a uniform ring of matter with the
appropriate mass density (see Price and Rush (1979)).

By far the largest contribution is due to Venus, the nearest neighbour to Mercury, but it is noteworthy that Jupiter contributes over a quarter of the total despite being so far away. Similarly, Leverrier could compute the effect of the other planets on the eccentricity of Mercury.

Now, the true longitude λ is related to the mean longitude l via the expansion

$$\lambda = l + 2e\sin(l - \bar{\omega}) + \tfrac{5}{4}e^2\sin(2l - 2\bar{\omega}) + O(e^3),$$

and by differentiating this with respect to time and substituting values for $d\lambda/dt$ and dl/dt calculated from observations, Leverrier could relate the values of $d\bar{\omega}/dt$ and de/dt. The results did not quite match up with the theoretical values he had computed, and Leverrier was forced to look for a theoretical explanation – the discrepancy could not be plausibly accounted for by inaccuracies in the observations. One possibility was to change the accepted value of the mass of Venus, but while this could be used to reconcile the motion of Mercury, it merely shifted errors to the theory of the Earth. He wrote:

> ... it is impossible to determine Venus' mass so as to account both for Mercury's transits and the observed obliquity of the ecliptic. If one determines this value so as to resolve the discordance in Mercury, it only reappears in the theory of the earth—and conversely. Moreover, adjustments in this value force anomalies into other conditional equations basic to the theories of Mercury and earth ... One can shape these theories to fit the observations, provided that in a century Mercury's perihelion turns not merely $527''$ as a result of the combined actions of the other planets, as [Newtonian theory] requires, but rather $527'' + 38''$. There is, then, with the perihelion of Mercury, a progressive displacement reaching $38''$ per century, and this is not explained.[5]

More detailed calculations only sought to confirm Leverrier's conclusion. Everything fitted together nicely provided he included the additional $38''$ in the perihelion advance. Leverrier's arguments were extremely convincing and most astronomers believed that the perihelion of Mercury did advance by $565''$ rather than $527''$ per century, but what was the cause? Attention became focused on the idea of an additional planet inside the orbit of Mercury.

Such an idea was not new. Ever since the discovery of Uranus and the asteroids, the search for new planets had been a popular activity, and the possibility that an unknown planet close to the Sun might be detected as a dark spot on the solar disc, encouraged Heinrich Schwabe – a pharmacist and amateur

---

[5] From a letter written by Leverrier to Hervé Faye, the secretary of the Paris Academy, published in the *C. R. Acad. Sci.* in 1859. Quoted from Hanson (1962). Also published in the *Mon. Not. R. Astr. Soc.* **20** (1859) under the title: 'Suspected Existence of a Zone of Asteroids Revolving Between Mercury and the Sun'.

astronomer from the German town of Dessau – to begin systematically to observe spots on the sun in 1826. Schwabe's efforts did not repay him with the sought-after new planet, but after recording sunspots for 17 years, he began to suspect that there was a regular variation in their number and went on to achieve fame as the discoverer of the 11-year sunspot cycle. In the United States, Kirkwood and Walker both saw the opportunity to extend the former's analogy (see p. 398) with the inclusion of an interior neighbour for Mercury. From Kirkwood's formula, Walker (in 1850) calculated that such a planet should have a mean distance of 0.2 AU.

There were, however, some serious objections to the existence of an intra-Mercurial planet. First, such a planet had never been seen crossing the solar disc, despite the regular observations of men such as Schwabe. More serious, perhaps, was the fact that any planet very close to the Sun should shine brilliantly during a solar eclipse, yet throughout recorded history there had been no such sighting. The objection to a single planet – comparable to Mercury in size and circling within its orbit – was powerful, and Leverrier soon discarded such a notion. But something was causing the anomalous advance in the perihelion of Mercury, and he decided that the most likely answer lay in the existence a number of small bodies orbiting between the planet and the Sun. Such a group of intra-Mercurial asteroids could be distributed in such a way as to account for the necessary motion of perihelion without introducing significant (and unobserved) periodic inequalities into the motion of Mercury.

This still begged the question of why no such objects had been seen traversing the face of the Sun but, with universal gravitation a matter of faith, something had to give. Leverrier encouraged astronomers to become more assiduous in their studies of the surface of the Sun and to observe carefully every tiny spot:

> We cannot establish their existence other than by observing their motion across the sun's disc; this discussion should make astronomers the more zealous to study each day the sun's surface. It is important that every regular spot appearing on the sun's disc, however tiny, be carefully followed for a few months to determine its nature through familiarity with its motion.[6]

Vulcan

Almost immediately, successful observations of intra-Mercurial bodies were reported. Over succeeding years, the number of such 'sightings' was so great that, in 1882, Tisserand felt the need to write over 40 pages on the subject.[7]

[6] Leverrier to Faye in 1859. Quoted from Hanson (1962).
[7] La notice sur les planètes intra-mercurielles, published in *Annu. Bur. Longit.* (1882).

The vast majority of these reports were easily dismissed, but one in particular caught Leverrier's eye. On 22 December 1859, Edmond Lescarbault, village physician at Orgères and amateur astronomer, wrote to the great man reporting an observation he had made on 26 March of that year. He summarized the details of his observations, noting that the dot he had seen was perfectly circular and about one-quarter the size of Mercury (the transit of which he had observed on 8 May 1845). He then made the prediction that someday

> someone will again observe the transit of a perfectly round, tiny, black dot, traversing a plane-line inclined at an angle of between about $5\frac{1}{3}°$ to $7\frac{1}{3}°$; the orbit described by this plane-line cutting that of the earth at about 183° from north to south; and unless there is an enormous eccentricity in the black dot's orbit, it should be visible on the sun's disc for $4\frac{1}{2}$ hours. This black dot will very probably be the planet whose path I observed on 26 March 1859, and it will be possible to calculate all the elements of its orbit. I have reason to believe that its distance is less than that of Mercury, and that this body is the planet, or one of the planets, whose existence in the heavens you, Monsieur Directeur, had, some months ago, predicted in the neighbourhood of the sun, by the marvelous power of your calculations; which enabled you also to recognize the conditions of the existence of Neptune, and fix its place at the confines of our planetary system, and to trace its path across the depths of space.[8]

The description of the observation was sufficiently detailed for Leverrier to conclude that he ought to check out this unknown observer for himself, and decided to pay him an unscheduled visit. The high and mighty Leverrier cross-examined the shy and obsequious Lescarbault on every aspect of his observing practice, and on the precise details of the particular observation in question.[9] He may well have journeyed to Orgères to satisfy himself that he did not need to trouble himself further with Dr Lescarbault, but Leverrier returned to Paris convinced that a planet had been found orbiting between Mercury and the Sun. At a public meeting of the Academy of Sciences on 2 January 1860, Leverrier informed his audience of Lescarbault's letter and their subsequent meeting, and suggested that the details of his observations ought to be admitted to science.

This was sufficient to send shock waves through the scientific community and beyond. In France, Leverrier's reputation was further enhanced, and Lescarbault was catapulted to unwanted fame. On Leverrier's advice, Napoleon III awarded Lescarbault la Légion d'houneur, though the recipient did not travel to receive it in person. The romantic picture of an unpretentious amateur being responsible

---

[8]　Lescarbault to Leverrier. Published in the *C. R. Acad. Sci.* **50** (1859). Quoted from Baum and Sheehan (1997).
[9]　This bizarre meeting subsequently was described at length by the Abbé François Moigno, with the characters referred to as the lion and the lamb (see Baum and Sheehan (1997), pp. 150–4).

for an epoch-making discovery – just as Herschel had been 80 years previously – was appealing to people the world over, and Lescarbault became a rather unlikely hero. The planet he had discovered soon acquired a name – it was 'Vulcan'.[10]

From Lescarbault's numbers, Leverrier calculated an approximate orbit for Vulcan. He placed the planet at 0.147 AU from the Sun in a circular orbit inclined to the ecliptic by just over 12° and with a synodic period of about 20 days. There were at least three important problems with this. First, Vulcan should have been visible clearly through a telescope when at elongation (which Leverrier calculated to be 8° from the Sun), and during a solar eclipse it would have shone like a bright star. Second, it should pass in front of the Sun at least twice, and possibly four times, each year. Finally, its mass (which Leverrier put at one-seventeenth that of Mercury), was insufficient to cause the necessary perihelion advance.[11] In order to maintain a belief in the existence of this planet, therefore, one had to assume that it was unusually dim and that there were many more such bodies as yet undetected.

Far-fetched as it was, this is precisely what Leverrier did assume. His psychological attachment to the existence of asteroids within the orbit of Mercury was so great that he had no choice. Based on the unquestionable truth of universal gravitation, he had 'proved' theoretically that such matter must be present, much as he had 'proved' the existence of Neptune. If this meant that some established ideas about celestial phenomena had to change, then so be it. A proof is a proof, after all.

Leverrier's commitment to intra-Mercurial planets, and to Vulcan in particular, encouraged others. Reports of sightings started to flood in, including many claims that Vulcan had been seen prior to Lescarbault's observation. But not everyone was prepared to accept one man's theoretical calculations as a reason for turning a blind eye to all manner of unexplained facts that came with the assumption of the existence of Vulcan. Foremost among the critics was another French astronomer, Emmanuel Liais, who was working in Brazil at the time of Lescarbault's observation. Liais was not one to be bullied by Leverrier's

---

[10] In 1846, another French scientist, Jacques Babinet, had hypothesized the existence of gaseous masses close to the Sun as a way of explaining the large prominences that had been observed around the Sun during the solar eclipse of 1842. He had used the name 'Vulcan' for this intra-Mercurial matter. Interestingly, Babinet (in 1848) had sided with those like Peirce who believed that the Neptune discovered by Galle was not the planet predicted by theory, and had suggested that the differences between the actual orbit of Neptune and those that had been predicted by Adams and Leverrier were due to the presence of another planet outside the orbit of Neptune. He chose Hypérion for its name.

[11] Leverrier could estimate the required mass of a disturbing planet as a function of its orbital radius by setting $e' = 0$ in Eqn (12.1).

reputation, and was one of those who, after the realization that the actual orbit of Neptune was rather different from those of Adams' and Leverrier's hypothetical bodies, had claimed that Galle's successful discovery had been just good fortune. Liais had good grounds for being sceptical about Vulcan, for he had been observing the Sun at precisely the same time that Lescarbault reported having seen his circular black spot, and he had seen nothing.[12]

Gradually, Vulcan mania died down. New 'sightings' appeared sufficiently often to rekindle the fire briefly, but no corroborated evidence for the existence of an intra-Mercurial planet ever materialized.[13] Leverrier's confidence in the theoretical necessity for planets or asteroids between the Sun and Mercury never diminished, however. In 1874, he wrote:

> There is, without doubt, in the neighbourhood of Mercury, and between that planet and the sun, matter hitherto unknown. Does it consist of one, or of several small planets, or of asteroids, or even cosmic dust? Theory alone cannot decide this point.[14]

But even the great Leverrier could not conjure up what did not exist. In retrospect, we know that the road to success in the theory of Mercury was to question precisely that which Leverrier simply could not: Newton's theory of gravity.

Following Leverrier's death in 1877, Tisserand became the leading French practitioner of celestial mechanics. In his 1882 review of intra-Mercurial planets[15] he concluded:

- The anomalous advance in Mercury's perihelion is not caused by a single interior planet.
- If planets do exist inside Mercury's orbit, of comparable size to Lescarbault's, than there cannot be many of them, certainly not enough to produce the required perturbations in Mercury's motion.
- The most likely explanation for the perihelion advance lies in Leverrier's original suggestion of a ring of tiny asteroids between Mercury and the sun.

Vulcan had always been a red herring as far as the perihelion of Mercury was concerned. Even if it had been discovered, the missing 38″ would have been

---

[12] Liais's criticisms appeared in Sur la nouvelle planète announcé par M. Lescarbault, published in *Astr. Nachr.* **52** (1860).

[13] Detailed descriptions of the continuing searches for Vulcan throughout the late nineteenth century can be found in, for example, Fontenrose (1973) and Baum and Sheehan (1997).

[14] LEVERRIER La théorie nouvelle du mouvement de la planète neptune: remarques sur l'ensemble des théories des huit planètes principales . . . , published in *C. R. Acad. Sci.* **79** (1874). Quoted from Baum and Sheehan (1997).

[15] See note. 7 on p. 441.

reduced only slightly, but such was the lure of a new planet that huge amounts of energy, and considerable resources, were channelled into the search for this elusive object.

Alternative hypotheses

We have seen already (on p. 406) that Newcomb's theory for Mercury confirmed Leverrier's discovery of the anomalous advance of the perihelion. Newcomb was concerned particularly with determining accurately the masses of the planets and, in 1882, he computed the mass of Venus from a number of different effects. The results were all in agreement, except that calculated from the theoretical value of the perihelion shift of Mercury:

> We must, therefore, conclude that the discordance between the observed and theoretical motions of the perihelion of Mercury, first pointed out by Leverrier, really exists, and is indeed larger than he supposed.[16]

Newcomb calculated 42.95″ per century.[17]

While Newcomb could not disprove that intra-Mercurial matter was the cause, he was sceptical. Another possibility he considered was that the Sun is oblate. It had been known since the mid eighteenth century that an oblate central body leads to a rotation of the line of apsides of an orbit of a satellite, and Charles Walmesley had shown that the oblateness of Jupiter was consistent with the motion of its moons. However, measurements of the diameter of the Sun did not reveal any significant oblateness and, moreover, there was no dynamical reason why there should be any, since the rotation rate is too small.

Unlike many of his contemporaries, Newcomb was prepared to countenance a different solution – modification of the inverse square law. Back in the early eighteenth century, Euler and Clairaut had, of course, tried just such an approach to resolve the problem with the motion of the lunar apogee. They had suggested adding to the force of gravity a term proportional to the inverse cube or inverse fourth power of the distance, and in the late nineteenth century a number of dynamical theories of gravity were proposed that led to similar conclusions.

One class of such theories – so-called pulsation theories – explained gravitational forces as being due to oscillatory motions in the ether (considered as an incompressible fluid) with extremely short period. The main proponent of

---

[16] NEWCOMB *Discussion and Results of Observations on Transits of Mercury from 1677 to 1881* (1882). Quoted from Roseveare (1982).
[17] Roseveare (1982), p. 41.

these hydrodynamical models was the Norwegian Carl Bjerknes, who tried to develop a general fluid theory that would explain electromagnetic and gravitational interactions. One follower was James Challis (of Neptune infamy), who wrote:

> Again, the mathematical theory seems to indicate that, if the approximation were carried further, the law of gravity would be found to be expressed by such a function as $m/r^2 + \mu/r^4$, μ being excessively small. This second term, if at all sensible, would be most likely to be detected in the action of the sun on Mercury. . . . The proper course in this case is no doubt that which M. Leverrier has recommended, *viz.* to endeavour to ascertain whether there are small bodies circulating between the sun and Mercury, to which the motion of Mercury's perihelion may be ascribed. But should this explanation fail, it might, I think, be reasonably questioned whether the law of gravity is so absolutely that of the inverse square as has been generally assumed.[18]

Other theories treated the ether as an elastic medium, and it was also suggested that gravity was caused by unidirectional, rather than pulsatory, flows of ether, but perhaps the most famous of all the mechanical theories of gravitation discussed around this time was that originally due to Georges-Louis Le Sage. Le Sage had proposed that gravity is caused by the action of tiny solid particles[19] moving through the ether in all directions. The force of gravity is due to an imbalance of these particles impacting on a body from different sides, and this occurs when one body is shadowed by another. Thus, an apple falls to the ground not because it is attracted by the Earth, but because the Earth shields one side from the gravific particles. The theory was debated by such distinguished figures as William Thomson, James Clerk Maxwell, and Poincaré, and a mathematical treatment was given by Darwin in 1905.[20]

A common feature in the quantitative analysis of all these approaches was that the force law for gravitation differed from a pure inverse square law by the addition of very small terms proportional to higher negative powers of the distance between particles. Newcomb calculated the magnitude that such terms would have to have in order to account for the advance in the perihelion of Mercury, and concluded that, for an inverse cube term, the coefficient would have to be 16×10^{-8} times that for the inverse square term.[21] Small as this may seem, it leads to forces that are significant over very small distances and contradicts the experimental findings of Henry Cavendish. In 1798, he used

---

[18] CHALLIS *The Force of Gravity* (1859). Quoted from Roseveare (1982).
[19] Le Sage called them 'ultra-mundane corpuscles' (Whittaker (1951), p. 31).
[20] *The Analogy Between Lesage's Theory of Gravitation and the Repulsion of Light.*
[21] See Roseveare (1982), p. 113

experimental apparatus designed by John Michell[22] to measure the mean density of the Earth. This involved the attraction of lead spheres over distances measured in inches, and if there were extra terms in the expression for the gravitational force of the type hypothesized by Clairaut, they would dominate at such distances and lead to far greater attraction than was measured by Cavendish. Newcomb thus ruled out modifying the inverse square law in this way.

Another possibility emerged from the emerging field of electrodynamics.[23] In order to 'save the phenomena' that were being discovered in relation to electricity and magnetism, a number of physicists assumed that moving charges exert forces that, in addition to the electrostatic part appearing in Coulomb's law,[24] are dependent on the velocities of the charges involved. Theories based on this type of assumption became quite successful, and their gravitational equivalents became the subject of scrutiny. One such law, based on the electrodynamic law of Wilhelm Weber, expressed the magnitude of gravitational attraction between bodies a distance, r, apart as

$$F = \frac{Gm_1m_2}{r^2}(1 - h^{-2}\dot{r} + 2rh^{-2}\ddot{r}).$$

In the electrodynamic context, h had been measured and determined to be almost exactly $\sqrt{2}$ times the speed of light.

If this gravitation law is applied to the perihelion problem, it produces an advance of just over $6''$ per century for Mercury. However, if one treats h as a parameter that takes a different value in the theory of gravity from that which it has in the electrodynamic theory, the whole of the anomalous advance of the perihelion of Mercury can be reproduced by taking $h = 174\,000\,\mathrm{m\,s^{-1}}$. The use

---

[22] Michell (the 'father of seismology') began his scientific career as a geologist and, following the Lisbon earthquake of 1755, published a paper on the causes of such phenomena. It was in the context of geology that he designed his experiment to determine the density of the Earth. Later, he became interested in astronomy and here also he made significant contributions. He used a statistical analysis to show that the number of double stars observed was more than could be expected if they were just chance alignments of stars. He concluded that most such combinations were, in fact, pairs orbiting each other. Michell also hypothesized the existence of massive stars from which light could not escape but the presence of which could be inferred from the effect of their gravity on other stars – black holes as we would now call them.

[23] A detailed history of nineteenth-century electrodynamical theories can be found in Whittaker (1951).

[24] The first to claim that the law of electrical attraction had the same form as that for gravity appears to have been Joseph Priestley in 1767, based on the absence of electrical force inside a charged spherical shell, though the inverse square nature of the attraction had been suspected earlier by Daniel Bernoulli (Whittaker (1951), p. 53). The first person to measure the force directly was Charles Coulomb.

of this law in celestial mechanics was investigated by Tisserand in 1872,[25] but
Newcomb in 1882 was not prepared to draw any firm conclusions, particularly
since a number of objections to Weber's electrodynamic theory had been voiced
in the physics community.

Another way of modifying the inverse square law in order to bring Mercury
into line was suggested by Asaph Hall in 1894.[26] He proposed that the force
of gravitation was proportional to $r^{-(2+\delta)}$, where δ was a very small quantity.
In *The Elements of the Four Inner Planets and the Fundamental Constants of
Astronomy* (1895), Newcomb calculated that the perihelion advance of Mercury
could be accounted for by taking $\delta = 1.574 \times 10^{-7}$, and he thought that this
approach was more satisfactory than solutions based on velocity-dependent
force laws.[27]

Newcomb's planetary theories suggested that there were three other secular
variations, the observed value of which disagreed from theory by more than one
could reasonably expect. These were, in decreasing order of error, the motion
of the node of Venus, the advance of the perihelion of Mars, and the variation
in the eccentricity of Mercury. Newcomb's *ad hoc* value for δ improved the
agreement with the perihelial motion of Mars, but made the situation worse
for the nodal line of Venus. This latter phenomenon has a curious status in this
phase of the history of astronomy. It was significant in the ensuing debates
concerning resolutions to the Mercury problem, but it was demonstrated subsequently that the prediction of Newcomb's theory was in error (a possibility that
Newcomb himself had entertained), probably due to systematic errors in old
observations.[28]

By making small changes to the planetary masses, Newcomb managed to
bring all the troublesome secular inequalities into line, but this then led –
essentially through Kepler's third law (Eqn (8.9)) – to an unacceptable value for

[25] There was some confusion over whether h or $h/\sqrt{2}$ was the speed of propagation of gravity
(see Roseveare (1982), p. 123). Whittaker (1951), p. 208, reports Tisserand as finding an
advance of $14''$, but this is the value obtained by taking h equal to the speed of light.
[26] Hall is most famous for his discovery of the two satellites of Mars in 1877 (see Gingerich
(1970), (1992), Chapter 23).
[27] The determination of the appropriate value of δ was based on Newton's precession theorem
(see p. 269). If the magnitude of the central force varies like $r^{-(2+\delta)}$, with $\delta \ll 1$, then the
advance per revolution is $180\delta + O(\delta^2)^\circ$. Since Mercury makes 415.2 revolutions per
century, we need $\delta = \Delta\theta/(180 \times 415.2 \times 3600)$, where $\Delta\theta$ is the perihelion advance
measured in arc-seconds per century. Newton's result is valid only in the limit as the
eccentricity of the orbit tends to zero. In fact, for Mercury, $e \approx 0.21$, but it turns out that the
error this generates in the amount of precession for a given δ is only about 1 per cent (Valluri,
Wilson, and Harper (1997)).
[28] Duncombe (1958). The fact that Einstein's general theory of relativity did not account for this
aspect of the motion of Venus was used by some as evidence against it (see, for example, Poor
(1921)), but there is no evidence that it concerned Einstein.

the solar parallax. There was sufficient uncertainty over the solar parallax data for Newcomb not immediately to discard this approach, but Hall's hypothesis finally became untenable when the Hill–Brown lunar theory improved significantly the theoretical values for lunar perigee and nodal motion, neither of the new values being compatible with Newcomb's proposed theory.

Newcomb reverted to inverse square gravity and the conclusion that the advance of the perihelion of Mercury had to be caused in some way by the attraction of matter. Along with most of the astronomical establishment, he became a supporter of the zodiacal light hypothesis proposed by Hugo von Seeliger in 1906.[29] The zodiacal light is a faint glow sometimes visible before dawn or after sunset, rising from the point at which the Sun is about to rise or has set, and forming an approximate triangular shape pointing along the zodiac. The first systematic observations of this phenomenon were made by G. D. Cassini beginning in 1683. Cassini suggested that the light was the result of the scattering of sunlight by lots of small particles, though a number of other hypotheses were suggested over the next 200 years or so. That the light was derived from the Sun was confirmed in 1874 by Arthur Wright, who found that the light had the same spectrum as sunlight and was polarized in the plane of the ecliptic, and it is now generally agreed that the zodiacal light is caused by dust particles in or around the ecliptic plane.

Thus, there was observational evidence for the presence of extra mass in the Solar System between the Sun and the Earth; the problem was how to quantify this. If it was assumed that this matter was responsible for the whole of the anomalous advance of Mercury, one could determine an appropriate mass distribution, but this then had to be consistent with the low luminosity of the observed phenomenon. In 1906, Seeliger proposed a solution that involved two ellipsoids of matter: one close to the Sun (and, hence, probably invisible) which was largely responsible for the perihelion advance of Mercury, and one extending to the Earth and which was the cause of the zodiacal light. Added to this was an arbitrary rotation of the Solar System with respect to the fixed stars, the need for which was related ultimately to the need for the theory to fit with Newcomb's (erroneous) numbers for the nodes of Venus.

Seeliger's solution worked. It was backed up to some extent by observational evidence, but included a number of *ad hoc* assumptions that could not be tested independently. In the absence of anything better, it was accepted by many astronomers, but when Einstein's sensational explanation materialized, arguments based on zodiacal light matter faded from view fairly quickly.

---

[29] Details of Seeliger's work are given in Roseveare (1982), Chapter 4.

Einstein

Unlike most of the people we have had cause to mention, Einstein was not driven by the technical problems of celestial mechanics. His general theory of relativity – a radically new theory of gravitation – did much more than solve the problem of the perihelion of Mercury: it changed profoundly our understanding of space and time, and led to a completely new way of viewing the Universe. However, the moment at which Einstein realized that general relativity did explain quantitatively the anomalous behaviour of Mercury was hugely significant. Abraham Pais has described the discovery as 'by far the strongest emotional experience in Einstein's scientific life, perhaps in all his life', and Einstein wrote 'for three days I was beside myself with joyous excitement'.[30] The agreement between his fledgling theory and this barely observable astronomical phenomenon was confirmation that the years of development had not been in vain; he had to be right.

Albert Einstein was born in Ulm, southern Germany, less than 100 km from Kepler's birthplace.[31] He was a very bright student at school, showing a particular aptitude for mathematics, but the learning environment was not one from which he took any pleasure. He was much happier reading for himself and discussing science and philosophy with friends. His parents moved to Italy in 1894 and, against their wishes, he dropped out of school and followed 6 months later. He told his parents that he intended to take the entrance exam for the Federal Institute of Technology (ETH) in Zürich, which he did in October 1895. He failed. Einstein then went to study at a Swiss secondary school where he gained a high school diploma that entitled him to enroll at the ETH. This he did in 1896, the same year in which he gave up his German citizenship (he was granted Swiss citizenship in 1901).

Examinations were not Einstein's forte, but his progress was eased by his friend and fellow student Marcel Grossman, whose clear and meticulous lecture notes were a real help! This was by no means the last time Einstein's association with Grossman would prove beneficial. In August 1900, he graduated, but without a job. Then in June 1902, after a couple of short teaching posts and with the benefit of the influence of Grossman's father, he became a technical expert third class at the Swiss patent office in Bern. Here he found time both to do his job – which was undemanding but quite interesting – and pursue his interests in physics. He began writing scientific papers on topics such as

---

[30] See Pais (1982), p. 253.

[31] Many biographies of Einstein have been written. The brief details given here are gleaned from the hugely impressive technical work of Pais (1982), and the popular biography of White and Gribbin (1993).

intermolecular forces, thermodynamics, and statistical physics. The period 1902 to 1904 were marked by a number of events of a personal nature – the death of his father, marriage to Mileva Marič, and the birth of a son – but throughout this period he was thinking hard about ideas of fundamental scientific importance. In 1905, his *annus mirabilis*, he stunned the world of physics with papers on the quantization of light, the nature of Brownian motion, and special relativity.

Special relativity

Newtonian mechanics is based firmly on a principle of relativity. The laws of motion remain unchanged if we change from coordinates x, y, z to x', y', z' via the transformation

$$x' = x - vt, \quad y' = y, \quad z' = z, \tag{12.2}$$

in which v is a constant representing the speed of one frame of reference with respect to the other. The set of Eqns (12.2) is an example of a Galilean transformation.[32] Newton's laws are thus true in the assumed absolute space and in any frame moving with constant velocity with respect to it (so-called inertial frames). There is thus no way of telling which inertial frame one is in, and absolute space appears to have no physical significance.

This was not the case, however, for the great synthesis of electromagnetism and optics achieved by Maxwell in the 1850s and 1860s. Based on the equations for electromagnetism that bear his name, Maxwell investigated the properties of electromagnetic waves in a non-conducting medium and found that the speed of such disturbances was given by $c = 1/\sqrt{K\mu}$, where K and μ are, in modern terminology, the electric permittivity and magnetic permeability of the medium, respectively. The quantity c can be measured experimentally and it was found to be approximately the same as the speed of light, a quantity that had been measured by totally independent techniques. As Maxwell put it, 'the properties of the electromagnetic medium are identical with those of the luminiferous medium'.[33]

Maxwell, like the vast majority of his contemporaries, was committed to the concept of an ether. He was in no doubt that the space between the heavenly bodies was filled with some substance that was in a state of absolute rest and

---

[32] The adjective 'Galilean' was introduced in 1909 (see Pais (1982), p. 140). In general, such a transformation takes the form $x' = Rx + vt + b$, $t' = t + \tau$, in which R is a 3×3 rotation matrix and R, v, b and τ are independent of t, and $x = (x, y, z)$. Since a rotation can be described by three angles, a general Galilean transformation is determined by ten parameters. In Eqn (12.2) all but one of these parameters is zero.

[33] Maxwell (1998), Part IV, Chapter XX.

served as the medium through which light propagates.[34] His equations predicted that electromagnetic waves should propagate through the ether with a
definite speed, c, and, hence, his theory was not invariant under Galilean transformations. In principle, electromagnetic phenomena therefore could be used to
distinguish between inertial frames. In particular, the Earth moves through the
ether at a known rate, so one should be able to measure the difference between
the speed of light relative to the Earth in directions parallel and perpendicular
to its motion.

Maxwell was of the opinion that the effect would be too small to measure,
but this pessimism was not shared by Albert Michelson, an expert on the experimental determination of the speed of light. Michelson designed a device – the
Michelson interferometer – with which to measure the 'ether wind', and in an
article published in 1881 he reported that he had failed to find any effect. Now
this could, of course, be due simply to the apparatus being incapable of resolving the phenomenon, and many remained unconvinced. As a result, Michelson
decided to repeat the experiment with a new version of his device, this time
in collaboration with Edward Morley. The experiment was carried out in the
summer of 1887 and the results were published in 1888. Again, no effect was
detected. Crucially, now the result was accepted by the scientific establishment
and it was the theory that became the subject of intense scrutiny.

The Irish physicist George FitzGerald suggested in *The Ether and the Earth's
Atmosphere* (1889) that motion through the ether might affect molecular forces
and lead to a contraction in the direction of motion. In 1892, and quite independently, the Dutch physicist Hendrik Lorentz recognized that the stationary
ether could be saved if the length of an object moving through the ether with
speed, v, is reduced by a factor (up to second order in v/c)

$$1 - v^2/2c^2. \tag{12.3}$$

Unfortunately, the FitzGerald–Lorentz contraction is not testable directly, as
any measuring device will suffer the same effect, and all indirect attempts to
observe the phenomenon ended in defeat.[35]

Lorentz made the next major advance when he realized that the contraction
hypothesis could be made a consequence of a more general principle. He discovered that in order for electromagnetic phenomena to be the same in frames
moving uniformly with respect to each other, a different time variable had to be
used in each one. In a stationary frame, t was the general time, but in any frame

---

[34] There were a number of different ether theories proposed during the nineteenth century,
associated with names such as Thomas Young, Augustin Fresnel, and G. G. Stokes, but the
concept of a stationary ether with the Earth moving through it was the easiest to reconcile with
the phenomenon of aberration.

[35] Some of the experiments that were performed are described in Whittaker (1958).

moving with speed v, a local time, t', had to be used, related to the general time by $t' = t - vx/c^2$. In 1899, Lorentz wrote down the transformations that were later (in 1904) given his name by Poincaré:[36]

$$x' = \gamma(x - vt), \quad y' = y, \quad z' = z, \quad t' = \gamma(t - vx/c^2),$$
$$\gamma = (1 - v^2/c^2)^{-1/2}.$$

It follows that $x = \gamma(x' + vt')$ and, hence, that if a body has length $l = x_2 - x_1$ in an inertial frame (at all times), then the length in the new frame (measured at time t') is $x_2' - x_1' = \gamma^{-1}(x_2 - x_1)$. Now, $\gamma^{-1} = (1 - v^2/c^2)^{1/2} = 1 - v^2/2c^2$ to second order in v/c, which agrees with Eqn (12.3). Exactly the same set of equations, together with a derivation of the FitzGerald–Lorentz contraction, was arrived at independently by Joseph Larmor and published in 1900.[37]

All the above equations are familiar to anyone conversant with special relativity, but there is still a huge conceptual leap to be made before we arrive at Einstein's formulation of the theory. Lorentz, FitzGerald, and others were trying to construct a system in which electromagnetic phenomena could be made consistent with a stationary ether, and the transformations they introduced were intended to represent some dynamical process.

More remarkable than the work described above were the contributions of Poincaré. In *La mesure du temps* (1898), he discussed the problems inherent in the way time is measured, and came to the conclusion that the concept of simultaneity could not be defined objectively.[38] In an address to the International Congress of Arts and Science in 1904, he went further and discussed the synchronization of clocks via light signals by two observers moving relative to one another, describing how the Lorentz transformations lead to the concept of time dilation. Poincaré was still embedded firmly in nineteenth-century dynamics, but he could see the possibility that something new might be around the corner. He was remarkably prescient about the form such a new dynamics might take:

> Perhaps, too, we shall have to construct an entirely new mechanics that we only succeed in catching a glimpse of, where, inertia increasing with velocity, the velocity of light would become an impassable limit. The ordinary mechanics, more simple, would remain a first approximation, since it would be true for velocities not too great, so that the old dynamics would still be found under the new. . . . I hasten to say in conclusion that we are not yet there, and as yet nothing proves that the [old] principles will not come forth from out of the fray, victorious and intact.[39]

---

[36] These transformations were derived first in 1887 by Woldemar Voigt, who noted that under them the wave equation retains its form.

[37] Pais (1982), p. 126.

[38] These ideas were discussed also in his essays *La science et l'hypothèse* (1902) and *La valuer de la science* (1905). English translations in Poincaré (1952) and Poincaré (1958), respectively.

[39] Quoted from Poincaré (1958).

Poincaré continued to work on the relativity principle and, in 1905, practically simultaneously with Einstein, created much the same mathematical framework. What Poincaré, like Lorentz before him, did not do was abandon the ether and base the whole of physics on new kinematical principles.[40] FitzGerald and Lorentz had appreciated that a new postulate was required to reconcile the classical principle of relativity with observed electromagnetic phenomena, but their contraction hypothesis was of a dynamic origin. Einstein (ignorant of Lorentz's work since 1895) took a different way out.[41] Rather than give the ether new mechanical properties, or revise the foundations of electrodynamics, he changed the structure of the underlying concepts of space and time by postulating that

(i) The laws of physics are of the same form in all inertial frames.
(ii) In any given inertial frame, the speed of light is the same whether the source is at rest or in uniform motion.

Put simply, special relativity is the application of these two principles, which appeared in June 1905 in *On the Electrodynamics of Moving Bodies*.[42]

The Lorentz transformations are derived easily from Einstein's postulates. Consider two frames x, y, z, t and x', y', z', t' with the latter moving with speed v in the x direction relative to the first, and assume that their origins coincide at time $t = t' = 0$. The first postulate implies that the relations between the coordinates are linear, and in particular that $x' = \gamma(x - vt)$, $x = \gamma(x' + vt')$, for some constant γ. Now consider a ray of light emitted from the common origin at $t = t' = 0$. Since the speed of light is c in both frames, we must have

$$c = x/t = x'/t'.$$

Combining these equations yields $\gamma = (1 - v^2/c^2)^{-1/2}$.[43]

---

[40] There is a considerable literature discussing Poincaré's role in the development of relativity theory (see Darrigol (1995) and the references cited therein). Given that, during the years up to 1905, Poincaré reduced gradually the significance of the ether in electrodynamical theory, it is surprising, perhaps, that he did not follow Einstein and eliminate it completely. Darrigol suggests that 'The most plausible answer is that he wished to maintain the notions of true space and time. These notions were so adequate in describing the course of current mechanical phenomena that it was foolish to give them up for the sake of the fourth or fifth decimal.'

[41] A thorough analysis of the philosophical positions with regard to relativity taken by Poincaré, Lorentz, and Einstein can be found in Hirosige (1976). Hirosige emphasizes the fundamental contribution to the emergence of Einstein's theory made by Ernst Mach with his criticism of the dogmatic mechanistic worldview that characterized nineteenth-century physics.

[42] *Zur Elektrodynamik bewegter Körper*, published in *Ann. Phys.* **17**. English translations can be found in, for example, Lorentz, *et al.* (1923), Stachel (1998), Miller (1998). In the latter, from which quotes reproduced here have been taken, Einstein's paper is analysed in great detail.

[43] Like the Galilean transformation, a general Lorentz transformation involves ten parameters: four for the arbitrary choice of origin in space and time, three for an arbitrary rotation of the space axes, and the remaining three as the components of the velocity of one frame with respect to the other. In the limit as $c \to \infty$, a Lorentz transformation reduces to a Galilean transformation.

Before Einstein, people tried to remove the apparent contradiction between electromagnetic theory and invariance under Galilean transformations. Einstein, on the other hand, abandoned absolute time and created a theory in which invariance under the Lorentz transformation was fundamental, and applied to all dynamical phenomena. The consequences are profound, and for many people the physics of special relativity was extremely difficult to come to terms with, even when the mathematics was understood. The idea that the temporal order of two events is observer-dependent is certainly counter-intuitive. Length contraction and time dilation are also disturbing, but perhaps the most famous consequence is the association between energy and inertial mass in the form $E = mc^2$, which Einstein derived later in 1905.[44]

The relation between the mass and velocity of charged particles had been the subject of investigation since the 1880s, when Joseph John Thomson demonstrated that an electric charge had the effect of increasing the inertia of a particle. This was due to the moving charge creating a magnetic field which, in turn, acted back on the particle. Thomson is best known for his discovery of the electron in 1897 and his hypothesis that these tiny electrically charged particles were the building blocks of atoms. Around the turn of the century there were two main candidates for a theory of the electron: those of Max Abraham and Lorentz. Both predicted a variation in mass with velocity, mass being the constant of proportionality in the equation $\mathbf{F} = m\mathbf{a}$. Abraham introduced the concepts of longitudinal mass m_{L} and transverse mass m_{T}, these being the value of m when the force was in the direction of motion and perpendicular to that direction, respectively. In his original 1905 paper, Einstein obtained the expressions

$$m_{\mathrm{L}} = m_0 \gamma^3, \qquad m_{\mathrm{T}} = m_0 \gamma^2,$$

where m_0 is the 'mass of the electron, as long as its motion is slow'. He then proposed that these results must apply to all material bodies, since the electron was the building block of matter.

The problem with Einstein's approach is that $\mathbf{F} = m\mathbf{a}$ is a poor choice for the definition of force when m is a varying quantity, a much better one being Newton's $\mathbf{F} = \mathrm{d}(m v)/\mathrm{d}t$. It was Max Planck[45] who, in 1906, demonstrated that with this definition one could simply write $m = m_0 \gamma$ and define the

---

[44] In Ist die Trägheit eines Körpers von seinem Energieghalt abhängig?, published in *Ann. Phys.* **17** (English translation in Lorentz, *et al.* (1923)). Einstein showed that if a body gives off energy E in the form of radiation, then its mass diminishes by E/c^2, though his derivation has caused its fair share of controversy (see Fadner (1988)).

[45] According to Pais (1982), p. 150, Planck's was the first paper on the new relativity theory published by somebody other than Einstein. Planck was one of only a handful of physicists who realized immediately that the consequences of Einstein's ideas, whether they turned out to be correct or not, needed to be investigated (see Goldberg (1976)).

momentum via

$$\mathbf{p} = m_0 \gamma \mathbf{v}.$$

Planck's work showed that, as in other aspects of the relativity theory, Einstein's and Lorentz's predictions for the mass–velocity relation were equivalent mathematically, but Lorentz had founded his model on dynamical principles, whereas Einstein's theory was entirely kinematic.

In his first paper on special relativity, Einstein calculated the kinetic energy of the electron to be $K = m_0 c^2 (\gamma - 1)$, which becomes unphysical when $v \geq c$, and thus deduced that 'velocities greater than that of light . . . have no possibility of existence'. When combined with the mass–energy relation from the second relativity paper we get

$$E = \gamma m_0 c^2 = m_0 c^2 + \tfrac{1}{2} m_0 v^2 + \ldots$$

for the total energy of a body. When $v \ll c$, the energy due to the motion (i.e. the kinetic energy) in a given frame is thus consistent with the Newtonian value.

Astronomers were well aware that a velocity-dependent mass would have an effect on the precessional motion of planets. George Darwin, referring to a meeting held in Göttingen in 1902, recounted that 'the greater part of one day's discussion was devoted to the astronomical results which would follow from the new theory of electrons'.[46] In 1906, Poincaré developed a Lorentz-invariant form of the inverse square law and the astronomical consequences were studied by de Sitter in 1911. Poincaré's analysis had revealed a whole family of possible gravitation theories indexed by an integer n, the simplest corresponding to $n = 1$. For this theory, de Sitter calculated an advance in the perihelion of Mercury of $7''.15$ per century, but arbitrary multiples of this could be realized simply by taking larger values of n. Moreover, any advance of this magnitude could be incorporated easily into Seeliger's zodiacal light hypothesis by reducing the mass of the inner matter ellipsoid a little.

Special relativity has not changed much since its inception, except in one important regard: the creation of the concept of spacetime by Hermann Minkowski, Einstein's mathematics teacher during his ETH days. Given what has been said previously, it will perhaps come as no surprise that one of the key ingredients in Minkowski's four-dimensional picture of the Universe was provided by Poincaré, who showed in 1906 that a Lorentz transformation was equivalent mathematically to a rotation (through an imaginary angle) about the origin in a four-dimensional space with coordinates x, y, z, and $\mathrm{i}ct$, where $\mathrm{i} = \sqrt{-1}$.

---

[46] Quoted from Roseveare (1982), p. 159.

Thus, for example, we can write the Lorentz transformation $x' = \gamma(x - vt)$, $t' = \gamma(t - vx/c^2)$ as

$$\begin{pmatrix} x' \\ ict' \end{pmatrix} = \begin{pmatrix} \cos i\psi & \sin i\psi \\ -\sin i\psi & \cos i\psi \end{pmatrix} \begin{pmatrix} x \\ ict \end{pmatrix},$$

where $\psi = \cosh^{-1} \gamma$ is a real quantity.[47] Poincaré was searching for those quantities that remained invariant under the Lorentz transformation, and his geometrical interpretation revealed this. In particular, it follows immediately that the 'distance' $x^2 + y^2 + z^2 - c^2t^2$ between any point and the origin remains fixed. However, Poincaré did not believe that it would be worth while to translate relativity theory into this geometrical framework.

Minkowski took a radically different view. He saw the four-dimensional picture as representing an observer-independent reality, which he called 'The Absolute World':

> The views of space and time which I wish to lay before you have sprung from the soil of experimental physics, and therein lies their strength. They are radical. Henceforth space by itself, and time by itself, are doomed to fade away into mere shadows, and only a kind of union of the two will preserve an independent reality.[48]

In three-dimensional Euclidean space, the line element (the distance between neighbouring points in space) is $\mathrm{d}s$, where

$$\mathrm{d}s^2 = \mathrm{d}x^2 + \mathrm{d}y^2 + \mathrm{d}z^2,$$

whereas, in Minkowski's spacetime, the line element is given by

$$\mathrm{d}s^2 = \mathrm{d}x^2 + \mathrm{d}y^2 + \mathrm{d}z^2 - c^2\mathrm{d}t^2.$$

The fact that spacetime can be thought of as a four-dimensional Euclidean space with ict as one of the coordinates led Minkowski to the 'mystic formula', 3×10^5 km $= \sqrt{-1}$ s, which I leave the reader to ponder at his or her leisure!

The question of who should receive the credit for the creation of special theory has been the subject of some controversy. In 1953, Sir Edmund Whittaker published a review of the subject under the title *The Relativity Principle of Poincaré and Lorentz*, and many authors since then have claimed that Einstein's contribution was simply to bring together various ideas that were in the public

---

[47] Note that the trigonometric functions are related to the hyperbolic functions via $\cos i\psi = \cosh \psi$ and $\sin i\psi = i \sinh \psi$, and, hence, if $\cosh \psi = \gamma$, $\sinh \psi = \gamma v/c$.

[48] From *Space and Time*, an address delivered in Cologne in 1908. Translation from Lorentz, *et al.* (1923) (see also Galison (1979)).

domain.[49] A number of detailed refutations of this point of view have appeared in print subsequently,[50] and it seems clear that, while the contributions of Poincaré and Lorentz were hugely significant, it was Einstein who took the fundamental leap needed to create the new theory. Max Born, a contemporary of Einstein and protégé of Minkowski, put it like this:

> Relativity actually ought not to be connected with a single name or with a single date. It was in the air about 1900 and several great mathematicians and physicists – Larmor, Fitzgerald, Lorentz, Poincaré, to mention a few – were in possession of many of its contents. In 1905 Albert Einstein based the theory on very general principles of a philosophical character, and a few years later Hermann Minkowski gave it final logical and mathematical expression. The reason Einstein's name alone is usually connected with relativity is that his work of 1905 was only an initial step to a still more fundamental 'general relativity', which included a new theory of gravitation and opened new vistas in our understanding of the structure of the universe.[51]

General relativity

Special relativity was not the end of the road for Einstein. He knew that it represented the first step on a path toward a new theory of gravitation, something which was necessary since the Newtonian theory implied that effects could be transmitted instantaneously over finite distances, whereas relativity limited the speed of transmission to c. In a letter in 1907 he wrote that he was

> busy on a relativistic theory of the gravitational law with which I hope to account for the still unexplained secular change of the perihelion motion of Mercury.[52]

and in a review article from the same year[53] we find Einstein's first suggestion of the equivalence between a gravitational field and an accelerating frame of reference. This so-called principle of equivalence always remained, in one form or other, as the bedrock on which Einstein's theory of gravity was built.

Already in 1907, Einstein realized that a relativistic theory of gravity would predict a measurable redshift in the spectral lines of the Sun, and that light would

---

[49] Whittaker's review appeared as a chapter in Whittaker (1953). Anti-Einstein sentiment continues to surface at regular intervals (see, for example, Bjerknes (2002), a book serving as a useful source of quotations if nothing else).

[50] See, for example, Pais (1982), Chapter 8, Holton (1988), Chapter 6.

[51] Quoted from the Introduction to Born (1962). This book serves as an excellent introduction to both the special and general theories of relativity.

[52] Quoted from Roseveare (1982), p. 154.

[53] Über das Relativitätsprinzip und die aus demselben gezogenen Folgerungen, published in *Jahrb. Rad. Elektr.* **4**. English version in Schwarz (1977).

be bent by massive bodies, but over the next few years he was concerned most with the quantum theory, and gravitation was put on hold. In 1909, Einstein gained his first academic appointment as an associate professor of theoretical physics at the University of Zürich, and he was on the move again in 1911 to take up a full professorship in Prague. By then, his status was such that many institutions tried to tempt him, and he returned to Zürich (this time to the ETH) 16 months later. But he was on his travels again in 1914, this time to Berlin, where he was based until he left Germany in 1932 – never to return. It was in Prague that Einstein's serious attack on a new theory of gravitation began.

One stimulus for this new activity was Einstein's realization that the bending of light might be capable of experimental verification by considering rays of light passing close to the Sun during a solar eclipse. Einstein's quantitative approach to the bending of light given in *On the Influence of Gravitation on the Propagation of Light* (1911)[54] proceeds as follows. Since inertial mass depends on the energy through the relation $E = mc^2$, the equivalence of inertial and gravitational mass implies that the same must be true for the latter.[55] This leads to what Einstein described as a 'consequence which is of fundamental importance', i.e. that the speed of light is a function of the gravitation potential Φ and is given, to the first approximation, by

$$c = c_0\left(1 + \Phi/c_0^2\right), \tag{12.4}$$

where c_0 is the speed of light from special relativity. Thus, in order to extend special relativity to accommodate the equivalence of inertial and gravitational mass, Einstein had to modify one of the fundamental postulates of the original theory. Reconciling special relativity with gravity was not going to be easy.

Since the speed of light varies in a gravitational field, it follows from Huygens' principle[56] that it will be bent, and Einstein calculated the magnitude of the deflection for a light ray coming from infinity and passing at a distance Δ from a point mass M, as $2GM/c^2\Delta$.[57] For the Sun, Einstein computed the value $0''.83$, and remarked that 'it would be a most desirable thing if astronomers would take up the question here raised'.

---

[54] Über den Einfluss der Schwerkraft auf die Ausbreitung des Lichtes, published in *Ann. Phys.* **35**. English translation in Lorentz, *et al.* (1923).

[55] In an experiment in 1889, Eötvös determined that the difference in the ratio of inertial to gravitational mass for wood and platinum was less than 10^{-9} (see Weinberg (1972), pp. 11–13), but Einstein did not learn of this result until 1912 (Pais (1982), p. 216).

[56] In 1678, Huygens published *Le traité de la lumiere* containing his wave theory of light in which the wavefront at any instant is the envelope of spherical wavelets emanating from every point on the wavefront at the previous instant (with the understanding that the wavelets have the same speed as the overall wave). This principle enabled Huygens to establish the laws of reflection and refraction which were subsequently derived analytically by Gustav Kirchhoff.

[57] Unless specified otherwise, c will always refer to the special relativistic value.

Einstein (just like the rest of the physics community) was entirely unaware that almost exactly the same prediction had been made over 100 years before him by combining Newtonian optics and mechanics! If we envisage light as being made up of material corpuscles, then these will be attracted by gravity just like any other matter. It is straightforward to compute the deflection of a particle in a hyperbolic orbit passing close to a point mass, M, given that its speed at infinity is c, and this appears to have been done first in 1784 by Cavendish, who obtained the value $2\sin^{-1}(\epsilon/(1+\epsilon))$, where $\epsilon = GM/c^2\Delta$. This agrees with Einstein's formula, since ϵ is small. A very similar calculation was carried out in 1801 by Johann von Soldner (the only difference being that in von Soldner's calculation the speed of light was taken as c at the point of closest approach), with the same conclusion. Cavendish never published his result, but von Soldner's calculation appeared in a major astronomical journal – and was largely ignored. In 1921, Philip Lenard tried to discredit Einstein by resurrecting von Soldner's work, publishing part of the original paper in the *Annalen der Physik.*[58]

In his 1911 paper, Einstein also quantified the gravitational redshift, showing that, again to the first approximation, the spectral lines when measured on the Earth should be shifted toward the red by a frequency of Φ/c^2, Φ being the difference in potential between the surface of the Sun and the Earth. Einstein calculated this figure as 2×10^{-6}, and noted that such an effect had been observed but attributed to physical phenomena on the surface of the Sun. The measurement of the gravitational red shift is, in fact, extremely difficult, as it is masked by convection currents in the solar atmosphere, which lead to Doppler shifts of the same order of magnitude.

At this point, Einstein had gone about as far as possible toward a new theory of gravity within the kinematical framework of special relativity. He realized that the Lorentz transformations were not by themselves sufficient to crack the problem, but quite what new idea was required he did not know.

Einstein was not the only person working on gravitation at this time. In 1912, Max Abraham, a staunch opponent of relativity,[59] published a theory in which the direction of the force of gravity is given by ∇c, with the speed of light given as a function of the gravitational potential by $c = c_0(1 + 2\Phi/c_0^2)^{1/2}$, which agrees with Einstein's (Eqn (12.4)) to first order in $1/c_0^2$. Since $E = mc^2$

---

[58] Cavendish's calculation was only found among his papers early in the twentieth century (see Will (1988)). The story surrounding von Soldner's work, including a translation of his essay, is told in Jaki (1978).

[59] Einstein and Abraham became embroiled in a very public (and at times unpleasant) debate over the merits of relativity and their respective theories of gravity, and Abraham took great pleasure in the fact that Einstein was now advocating a variable c, in direct conflict with special relativity (see Pais (1982), pp. 231–2).

and the total energy of a body in a gravitational field is a function of the potential Φ, it follows that either m or c (or both) must depend on Φ. Both Einstein and Abraham had $c = c(\Phi)$, but theories in which $m = m(\Phi)$ and c retained its constant special relativity value were put forward by Gunnar Nordström and Gustav Mie. As far as Einstein was concerned, only the second of two theories put forward by Nordström needed to be taken seriously.

In all these attempts to extend relativity to include gravitation, the field was determined from a single scalar function (as it is in Newtonian mechanics) and Einstein developed his own scalar theory of a static gravitational field during the same period. Although the theory had problems, one positive outcome was the unpleasant realization that the equations of gravitation had to be nonlinear, since the gravitational field possesses energy that acts as its own source. This then led Einstein to the reluctant conclusion that the principle of equivalence could only hold locally. In other words, it is not possible, in general, to find a global coordinate transformation that removes the effect of gravity completely, but that this is possible at each point in spacetime.[60] Though he did not work it out at the time, Einstein's scalar theory of gravity leads to two-thirds of the general relativistic advance for the perihelion of Mercury.[61]

It was in 1912 that Einstein was struck by the similarities between the problems he was facing and the theory of surfaces due to Gauss. With the help of Grossman (Einstein was now back in Zürich) he learned about the 'absolute differential calculus' developed through the work of Riemann, Christoffel, Ricci, and Levi-Civita, and began to investigate whether gravitation could be described through a curved space defined via the metric

$$ds^2 = g_{\mu\nu}\, dx^\mu\, dx^\nu, \qquad (12.5)$$

where $g_{\mu\nu}$ (which depends on the coordinates x^μ) is a symmetric tensor of rank 2.[62] Here we have replaced x, y, z, and ct, by x^1, x^2, x^3, and x^4, respectively,

---

[60] For a thorough discussion of Einstein's interpretation of the principle of equivalence, see Norton (1985).

[61] See Whitrow and Morduch (1965), an article that tabulates the predictions of numerous relativistic theories of gravitation for redshift, light deflection, and perihelion advance. A similar comparison is given in Harvey (1965). Nordström's second theory (which is equivalent to the theory proposed by Littlewood (1953a)) actually leads to a regression with one-sixth the general relativistic value (Pirani (1955)). It is clear that the redshift is the least restrictive of the three tests, as the same value is predicted by almost all theories. Much more on Nordström's theory of gravity and its influence on Einstein can be found in Norton (1992, 1993).

[62] The theory of tensors is described in most textbooks on general relativity (see, for example, Weinberg (1972), pp. 93–100). Tensors are collections of quantities (components) that transform in a certain way and are the natural objects with which to express the mathematical invariance of physical laws. A subscript indicates what is known as a 'covariant transformation rule', whereas a superscript is used for 'contravariant transformations'. This distinction is superfluous for three-dimensional Cartesian tensors, as is illustrated nicely in the treatment given by Foster and Nightingale (1995).

and used a notational convenience that Einstein himself introduced in 1916, and which is now referred to as the 'Einstein summation convention'. If the index in an expression occurs twice – once as a subscript and once as a superscript (μ and ν in the above equation) – then a summation is implied. Hence, Eqn (12.5) is shorthand for $ds^2 = \sum_{\mu=1}^4 \sum_{\nu=1}^4 g_{\mu\nu}\, dx^\mu\, dx^\nu$. The difficulties Einstein faced when developing general relativity were not helped by the fact that initially his notation was much more cumbersome.

This would be a radical departure from any previous type of theory, in which the underlying structure of space and time was given to begin with. In his new approach, the geometry of spacetime was one of the things gravity had to explain. Also, it made everything much more complicated, since the single variable, c, had been replaced by the ten unknowns $g_{\mu\nu}$.[63] Clearly, Einstein found the work incredibly difficult:

> At present I occupy myself exclusively with the problem of gravitation and now believe that I shall master all difficulties with the help of a friendly mathematician here. But one thing is certain, in all my life I have labored not nearly as hard, and I have become imbued with great respect for mathematics, the subtler part of which I had in my simple-mindedness regarded as pure luxury until now. Compared with this problem, the original relativity is child's play.[64]

The collaboration with Grossman led to the publication in 1913 of an *Outline of a Generalized Theory of Relativity and of a Theory of Gravitation*, often referred to as the '*Entwurf* theory'.[65] This was Einstein's first attempt at a tensor theory, and came remarkably close to the final theory that would emerge just over 2 years later.

It was Minkowski who had first formulated the equations of special relativity in modern tensor form,[66] but Einstein initially was unimpressed by this formal simplification to the theory (he described it as 'superfluous learnedness').[67] The

---

[63] It seems likely that one of the inspirations behind Einstein's new approach was Born's work on the definition of a rigid body in special relativity (see Stachel (1980), Pais (1982), pp. 214–16, and Maltese and Orlando (1995)). The latter paper quotes Einstein, from a lecture given in 1921, as saying 'In a system of reference rotating relatively to an inertial system, the laws of disposition of rigid bodies do not correspond to the rules of Euclidean geometry on account of the Lorentz contraction; thus if we admit non-inertial systems on an equal footing, we must abandon Euclidean geometry.'

[64] From a letter to Sommerfeld in October 1912. Quoted from Pais (1982), p. 216.

[65] Entwurf einer verallgemeinerten Relativitätshtheorie und einer Theorie der Gravitation, published in *Z. Math. und Phys.* **62**. The first part of the paper was physics, written by Einstein; the second part was mathematics, written by Grossman. Some details of the content are given in Norton (1984).

[66] Die Grundgleichenungen für die elektromagnetischen Vorgänge in bewegten Körpen, published in *Göttinger Nachr.* (1908). An Appendix to this paper contained a treatment of gravitation that influenced strongly the later theory of Nordström (see Pyenson (1977)).

[67] Pais (1982), p. 152

Minkowski metric can be written

$$ds^2 = \eta_{\mu\nu}\,dx^\mu\,dx^\nu, \tag{12.6}$$

where $\eta_{\mu\nu} = 0$ if $\mu \neq \nu$ and $\eta_{\mu\mu} = 1, 1, 1,$ and -1, for $\mu = 1, 2, 3,$ and 4, respectively. The principle of equivalence as understood by Einstein in 1913 is equivalent to the condition that at any point in spacetime there is a coordinate transformation that reduces the metric tensor $g_{\mu\nu}$ to the Minkowski tensor $\eta_{\mu\nu}$. Locally, spacetime has the structure of special relativity.

In the *Entwurf* theory, the motion of a particle in free fall within a gravitational field given by Eqn (12.5) is determined from the variational principle $\delta \int ds = 0$. In other words, in the absence of any forces, objects move between two points along the shortest path just as they do in special relativity, but now length is being measured by the line element given in Eqn (12.5) and so the resulting trajectories are not straight lines. Gravity is thus no longer a force but, instead, is considered as a deformation of Minkowski's spacetime. A helpful analogy is to think of gravity as a curved surface that can be approximated locally by the tangent plane to the surface. To complete the theory, equations that relate the metric $g_{\mu\nu}$ to the distribution of matter are needed – these are the so-called 'field equations of gravitation' – and this is where Einstein and Grossman came up short.

The core of general relativity as we know it today is that all frames of reference should be treated equally, none are preferred.

> The general laws of nature are to be expressed by equations which hold good for all systems of coordinates, that is, are co-variant with respect to any substitutions whatever (generally co-variant). . . . this requirement . . . takes away from space and time the last remnant of physical objectivity.[68]

However, the field equations in the *Entwurf* paper generally are not co-variant, but covariant only with regard to linear transformations, and Einstein thought the theory marred by an 'ugly dark spot'.[69] Although he defended it in public, Einstein was never fully convinced by the *Entwurf* theory, and it was not received well by other physicists. To make matters worse, he went on to 'prove' that suitable, generally covariant, field equations were not attainable since they violated causality, though he later realized that his arguments were flawed.[70]

---

[68] EINSTEIN *Die Grundlage der allgemeinen Relativitätstheorie* (1916). English translation in Lorentz, *et al.* (1923).
[69] From a letter to Lorentz in August 1913. Quoted from Norton (1984).
[70] A mathematical treatment of Einstein's arguments against general covariance is given in Earman and Glymour (1978b); (see also Howard and Norton (1993)).

Einstein and Grossman did not comment on the effect of their theory on the perihelion of Mercury, though Einstein did work it out subsequently in collaboration with an old friend from his student days Michele Besso.[71] Their value of 18″ per century was not published, but was confirmed by Johannes Droste in 1915. In fact, throughout the development of general relativity, Einstein rarely mentioned perihelion shifts, but paid much more attention to light deflection and redshift calculations. This was probably because, since the effects of other matter may well be important, he was not confident that a new theory of gravitation eventually would yield all of the missing 43″. The same was true of the other competitors in the race for a new theory of gravity. However, immediately after Einstein arrived at the generally covariant field equations for gravity in 1915, he wrote to Sommerfeld and gave three reasons for turning his back on the *Entwurf* theory, one of which was the fact that it did not give the correct advance for the perihelion of Mercury.[72]

When Einstein moved to Berlin in 1914, there were two theories of gravitation deserving attention: his and Nordström's. The latter was much simpler but did not satisfy Einstein's principle of equivalence. Moreover, since c is constant in the Nordström approach, it predicts no deflection for light passing close to the Sun and, hence, there was the possibility of deciding between the alternatives through observation.

On astronomical matters Einstein usually turned for help to his friend, Erwin Freundlich from the Berlin Observatory.[73] On this occasion, Freundlich decided to travel to the Crimea to observe the solar eclipse due to occur in August 1914. Unfortunately, the First World War curtailed his efforts, as he was captured by the Russians and had his equipment confiscated. Following an exchange of prisoners, he was soon back in Berlin, and there, Freundlich directed his efforts toward demolishing the zodiacal light hypothesis. If this explanation for the anomalous motion of Mercury was accepted, then predictions of perihelial shifts could not be used to distinguish between theories of gravity. Thus, the fact that Nordström's theory gave a regression of 7″ per century just meant that the interplanetary dust that caused the zodiacal light had to account for 50″ rather than 43″. His calculations showed that the density of matter required by Seeliger was simply not possible given the low luminosity of the zodiacal light itself, but they were refuted by Seeliger. Freundlich was not part of the scientific establishment, and his criticism of one of the most powerful men in the German astronomical community had a serious repercussion: he lost his job.

The year 1914 was a turbulent one in Einstein's personal life, with the separation from his wife, Mileva, and the War (which Einstein opposed vehemently)

---

[71] See Earman and Janssen (1993), Mehra (1998).   [72] See Earman and Glymour (1978a).
[73] Einstein's association with Freundlich is described in Pyenson (1976).

inevitably led to considerable upheaval in Berlin. But Einstein remained incredibly productive during the war years, and it was during this period that the general theory of relativity was completed. In 1917, Einstein fell quite seriously ill with various ailments brought on by overwork. He was nursed back to health by his cousin Elsa Löwenthal, whom he subsequently married.

One thing that was missing in the Einstein–Grossman theory, but which Einstein derived in 1914, was the relativistic equation of free fall (the geodesic equation), though this was well established in the theory of Riemannian geometry.[74] The principle of equivalence implies that in such a situation one can find local coordinates ξ^α, such that

$$\frac{d^2\xi^\alpha}{ds^2} = 0,$$

with ds^2 given by its special relativistic form (Eqn (12.6)). Now consider another coordinate system x^μ. We have, using Einstein's summation convention,

$$\frac{d^2\xi^\alpha}{ds^2} = \frac{\partial}{\partial x^\nu}\left(\frac{\partial\xi^\alpha}{\partial x^\mu}\frac{dx^\mu}{ds}\right)\frac{dx^\nu}{ds}.$$

If we evaluate the derivative and multiply by $\partial x^\lambda/\partial\xi^\alpha$, we obtain the geodesic equations

$$\frac{d^2x^\lambda}{ds^2} + \Gamma^\lambda_{\mu\nu}\frac{dx^\mu}{ds}\frac{dx^\nu}{ds} = 0, \tag{12.7}$$

where

$$\Gamma^\lambda_{\mu\nu} = \frac{\partial x^\lambda}{\partial\xi^\alpha}\frac{\partial^2\xi^\alpha}{\partial x^\mu\partial x^\nu} \tag{12.8}$$

is the Christoffel symbol of the second kind, also called the 'affine connection', or simply the 'connection coefficient'. Applying the same transformation to the line element ds – which must be invariant under a coordinate transformation, since it is a property of the underlying spacetime – we find that $ds^2 = g_{\mu\nu}\,dx^\mu\,dx^\nu$, where the metric tensor takes the form

$$g_{\mu\nu} = \frac{\partial\xi^\alpha}{\partial x^\mu}\frac{\partial\xi^\beta}{\partial x^\nu}\eta_{\alpha\beta}.$$

Within the framework of relativity, there are no forces acting on an object in free fall, and so the effect of gravity is determined by the affine connection. By itself, this is not particularly useful, since Eqn (12.8) requires knowledge of the local inertial coordinates at each point. However, it can be shown by direct

---

[74] See, for example, Kline (1972), Chapter 37.

calculation that

$$\Gamma^{\lambda}_{\mu\nu} = \tfrac{1}{2} g^{\lambda\sigma} \left(\frac{\partial g_{\nu\sigma}}{\partial x^{\mu}} + \frac{\partial g_{\mu\sigma}}{\partial x^{\nu}} - \frac{\partial g_{\mu\nu}}{\partial x^{\sigma}} \right)$$

and, hence, the metric tensor can be considered as the gravitational potential.[75]

The final twists and turns in the creation of general relativity are evident from a series of four dramatic communications made by Einstein to the Prussian Academy during November 1915.[76] The first was on 4 November and the second, a week later. In these, Einstein began the return toward general covariance – which had been his goal right from the start – but he was still held back by some of his previous misconceptions. It was Mercury that provided the key. A calculation of the perihelial motion based on the equations with which he was now working yielded 43″ per century for the advance of Mercury. Overjoyed, he wrote that his theory

> explains . . . quantitatively . . . the secular rotation of the orbit of Mercury, discovered by Leverrier, . . . without the need of any special hypothesis.[77]

In his communication of this result on 18 November, he also noted that his new theory gave a deflection of light which was twice his earlier prediction. There were still some problems, but the quantitative success with Mercury meant there was no doubt he was on the right track. In the course of his analysis, Einstein realized crucially that the form of the equations in the case of a weak, static field could be more general than he had supposed hitherto. With the stumbling block removed, Einstein soon grasped the true nature of the solution he was looking for and, on 25 November, communicated the final form for the field equations, concluding that ' . . . the general theory of relativity is closed as a logical structure'.[78] The change to the final set of equations did not affect the calculations of perihelion shift and light deflection of the week before.

---

[75] See Weinberg (1972), p. 75. The relation between $\Gamma^{\lambda}_{\mu\nu}$ and $g_{\mu\nu}$ is given in Grossman's part of the *Entwurf* paper. Note that $g^{\mu\nu}$ is the inverse of $g_{\mu\nu}$, i.e. $g^{\sigma\nu} g_{\mu\sigma} = \delta^{\nu}_{\mu}$, where δ^{ν}_{μ} is the Kronecker delta.

[76] See Pais (1982), Section 14c, Norton (1984), Mehra (1998).

[77] Quoted from Pais (1982), p. 253.

[78] Quoted from Pais (1982), p. 256. Remarkably, the same field equations were derived by Hilbert almost simultaneously, though he was in pursuit of a much grander objective – an axiomatic theory of the world. Einstein had spent a week in Göttingen earlier in 1915 lecturing on general relativity and had got to know Hilbert. During November, Einstein and Hilbert were in constant correspondence (postcards sent between Göttingen and Berlin would be delivered the following day) and there has been much discussion on Hilbert's influence on Einstein's derivation of his field equations. It would, however, appear unlikely that Hilbert influenced Einstein at all during this period, other than through his encouragement. For further details, see Earman and Glymour (1978a), Pais (1982), Section 14d.

In Newtonian theory, the gravitational potential Φ satisfies Poisson's equation $\nabla^2\Phi = -4\pi\rho G$ (see p. 339), where ρ is the mass density of the source of the gravitational field. In general relativity, gravity is determined from the tensor $g_{\mu\nu}$ and so, if we want any chance of recovering the Newtonian formula in an appropriate limit, the role of the mass density must be played also by a second-order tensor. This is the energy–momentum stress tensor $T_{\mu\nu}$ first introduced by Minkowski, which serves to define the distribution of energy and momentum throughout space-time. Where the energy–momentum tensor is zero, general relativity must reduce to special relativity and, hence, whatever is to appear in place of $\nabla^2\Phi$ must in such cases be zero. This provides a vital clue.

The most general tensor that can be constructed from the metric tensor and its first and second derivatives, linear in the second derivatives, is the Riemann–Christoffel curvature tensor $R^{\lambda}{}_{\mu\nu\kappa}$, defined by

$$R^{\lambda}{}_{\mu\kappa\nu} = \frac{\partial\Gamma^{\lambda}_{\mu\kappa}}{\partial x^{\nu}} - \frac{\partial\Gamma^{\lambda}_{\mu\nu}}{\partial x^{\kappa}} + \Gamma^{\eta}_{\mu\kappa}\Gamma^{\lambda}_{\nu\eta} - \Gamma^{\eta}_{\mu\nu}\Gamma^{\lambda}_{\kappa\eta}.$$

To obtain a tensor of rank two, we can form the contraction

$$R_{\mu\nu} = R^{\lambda}{}_{\mu\lambda\nu},$$

which is known as the 'Ricci tensor', and we can also make use of

$$R = g^{\mu\nu}R_{\mu\nu},$$

which is the curvature scalar.

The crucial property of the curvature tensor for our discussion here is that if we are in the situation of special relativity, where the metric tensor is $\eta_{\mu\nu}$, then $R^{\lambda}{}_{\mu\kappa\nu} = R_{\mu\nu} = R = 0$. On 11 November, Einstein proposed that the field equations of general relativity are

$$R_{\mu\nu} = -\kappa T_{\mu\nu},$$

where κ is a constant, though he knew that this could be true only if an additional hypothesis about the electromagnetic nature of matter was made. This then reduces to

$$R_{\mu\nu} = 0 \tag{12.9}$$

in empty space. Einstein had been here before, but he and Grossman had been led off the scent earlier with their erroneous belief that Eqn (12.9) did not have the correct form in the Newtonian limit.

When Einstein managed to show that Eqn (12.9) led to exactly the right advance for the perihelion of Mercury, he knew he was very close, and it did not take him long to appreciate that no additional assumptions were required if he wrote $R_{\mu\nu} = -\kappa(T_{\mu\nu} - \frac{1}{2}g_{\mu\nu}T)$, where $T = T_\mu^\mu$, or equivalently,

$$R_{\mu\nu} - \tfrac{1}{2}g_{\mu\nu}R = -\kappa T_{\mu\nu}. \qquad (12.10)$$

These are the field equations of general relativity from which the metric tensor is to be determined. An analysis of the Newtonian limit reveals that the coupling constant κ must have the value $8\pi G/c^4$. Note that in empty space we still have Eqn (12.9).[79]

Let us turn now to the motion of Mercury.[80] This is essentially a one-body problem, as we can neglect the effects of the mass of Mercury on the gravitational field of the Sun. Rather than following Einstein and using the method of successive approximations, we will begin our calculation from the exact solution to Einstein's field equations found in 1916 by Karl Schwarzschild.[81] If we assume that the field of the Sun is static and isotropic, then we obtain from Eqn (12.9), with coordinates (r, θ, ϕ) which, far from the origin, are the usual spherical polar coordinates,

$$ds^2 = \frac{dr^2}{1 - \alpha/r} + r^2(d\theta^2 + \sin^2\theta \, d\phi^2) - \left(1 - \frac{\alpha}{r}\right)c^2 dt^2. \quad (12.11)$$

Here, α is a constant of integration that depends on the mass M of the Sun. It is possible to determine α by comparing the form of the geodesic Eqns (12.7) in the case of a weak, static field with the equivalent Newtonian equation $\ddot{\mathbf{r}} = \nabla(GM/r)$. This shows that the coefficient of $c^2 dt^2$ in the metric must approach $1 - 2GM/c^2r$ as $r \to \infty$, and, hence, $\alpha = 2GM/c^2$.

A lengthy calculation shows that with this value for α, the geodesics for the metric (Eqn (12.11)) in the case $\theta = \frac{1}{2}\pi$ are the solutions to the equation

$$\frac{d^2u}{d\phi^2} + u = \frac{GM}{h^2} + 3GM\frac{u^2}{c^2},$$

---

[79] In 1917, Einstein introduced an extra term $-\lambda g_{\mu\nu}$ into the left-hand side of Eqn (12.10) because his original equations led to the conclusion that the Universe either was contracting or expanding, contrary to the then accepted paradigm of a static universe. The quantity λ has come to be known as the 'cosmological constant'. When Hubble later showed that the Universe is, in fact, expanding, Einstein abandoned this approach (see Pais (1982), pp. 285–8).

[80] The derivation of the perihelion advance given here follows closely that in Foster and Nightingale (1995), who credit the method to Møller (1972), Section 12.2.

[81] Schwarzschild was serving in the German army on the Russian front at the time, and his paper to the Prussian Academy was read by Einstein in January 1916. Schwarzschild died 4 months later. The same solution was found independently by Droste, though his derivation did not appear until 1917. Einstein's procedure, which was based on the technique that he and Besso had devised previously to calculate the perihelion advance from the *Entwurf* theory, is described in detail in Earman and Janssen (1993).

in which $u = 1/r$ and $h = r^2\dot{\phi}$. This is the same as the Newtonian Eqn (9.1), except for the presence of the final relativistic term and has as a first integral

$$\left(\frac{du}{d\phi}\right)^2 + u^2 = \frac{2GM}{h^2}u + \frac{2GM}{c^2}u^3 + E, \qquad (12.12)$$

where E is a constant of integration. At aphelion and perihelion, $du/d\phi = 0$ and then

$$\alpha u^3 - u^2 + \beta u + E = 0,$$

where $\alpha u = 2GM/c^2 r \ll 1$ and $\beta = 2GM/h^2$. This is a cubic equation, two of the roots of which (e.g. u_1 and u_2) are the reciprocal distances at aphelion and perihelion, respectively. Given that the sum of the roots must be $1/\alpha$, the third root u_3 must be large. It follows from Eqn (12.12) that

$$\frac{du}{d\phi} = [\alpha(u - u_1)(u_2 - u)(u_3 - u)]^{1/2}.$$

If we replace u_3 by $1/\alpha - u_1 - u_2$, we find that, to first order in α,

$$\frac{d\phi}{du} = \left[1 + \tfrac{1}{2}\alpha(u_1 + u_2 + u)\right][(u - u_1)(u_2 - u)]^{-1/2}$$

$$= \left[1 + \tfrac{3}{2}\alpha v_+ + \tfrac{1}{2}\alpha(u - v_+)\right]\left[v_-^2 - (u - v_+)^2\right]^{-1/2},$$

where $v_\pm = \tfrac{1}{2}(u_2 \pm u_1)$. The angle between aphelion and the next perihelion is

$$\int_{u_1}^{u_2} \frac{d\phi}{du}\, du = \pi\left(1 + \tfrac{3}{2}\alpha v_+\right).$$

If we substitute the Newtonian values $r_1 = a(1 + e)$ and $r_2 = a(1 - e)$, we find that the advance in the perihelion per revolution (which is given by doubling this and subtracting 2π) is

$$\frac{6\pi GM}{c^2 a(1 - e^2)}.$$

For the Sun, $GM/c^2 \approx 1475$ m, and for Mercury $a(1 - e^2) \approx 55.46 \times 10^9$ m so that the advance comes out to be $0.1034''$ per revolution, or a little under $43''$ per century. To see just how well general relativity performs with regard to perihelion shifts, Table 12.2 shows the values predicted for the motions of

Table 12.2. *Comparison of general relativistic predictions for perihelion motion with observation and computation (seconds of arc per century).*

Planet	$\Delta\theta$ (gen. rel.)	$\Delta\theta$ (empirical)	$\Delta\theta$ (calculated)
Mercury	42.98	43.11 ± 0.45	42.980 ± 0.001
Venus	8.63	8.4 ± 4.8	8.618 ± 0.041
Earth	3.84	5.0 ± 1.2	3.846 ± 0.012

the perihelia of the three inner planets compared with modern empirical and computational data.[82] The level of agreement is impressive.

If general relativity is accepted as a successor to the universal gravitation of Newton, then the anomalous advance in the perihelion of Mercury is explained, and our understanding of the motions of the bodies in the Solar System essentially is complete. But in 1915 it was going to take more than one experimental success for people willingly to accept a theory that reinterprets the whole fabric of space and time. Moreover, as we have seen, there are a number of other possible influences on the motion of Mercury that are hard to quantify. The advance in technology in the second half of the twentieth century has led to numerous new tests for general relativity and the theory has passed each one (with many competing theories failing), and so the grounds for accepting Einstein's theory are now quite strong.[83] But this was not the case when it was published.[84]

[82] With the exception of the general relativistic value for Mercury, which is from Nobili and Will (1986), the numbers in the first two columns are from Whitrow and Morduch (1965). The value of 43.03″, which is often quoted as the prediction of general relativity for the advance of Mercury, was derived by Clemence (1947) based on unconventional values for certain astronomical constants. The large error in the observational data for Venus is due to the fact that the eccentricity of Venus is very small and, hence, the determination of the perihelion is more difficult. The computations reported in the final column are from Taylor and Wheeler (2000) and were made by Myles Standish at the NASA Jet Propulsion Laboratory by computing the motions of the four inner planets over a 400-year period with and without relativistic effects. There is also a general relativistic effect due to the rotation of the Sun, but this turns out to be negligible (Whitrow and Morduch (1965)).

[83] See Will (1979) for a detailed discussion. In the latter part of the twentieth century, general relativity has become an important tool in the construction of accurate ephemerides (see Brumberg (1991)) and plays an essential role in the operation of the Global Positioning System (Taylor and Wheeler (2000)).

[84] There was considerable opposition within Germany, where Einstein's nationality (Swiss) and religion (Jewish) both worked against him. Ernst Gehrcke was particularly active in trying to discredit Einstein, and was responsible for the reprinting of a paper published originally in 1902 by Paul Gerber. Gerber had developed (though his derivation was somewhat unclear) a velocity-dependent force law that gave exactly the same formula for the perihelion advance of Mercury as general relativity. A somewhat unsatisfactory debate ensued involving, among others, Seeliger and Einstein (see Roseveare (1982), Section 6.6). Note that the special relativistic advance of 7″ should be added to Gerber's value, since his theory treats gravitation alone, whereas general relativity already contains the special theory.

In his first full-scale exposition of general relativity, Einstein wrote:

These equations, which proceed by the method of pure mathematics, from the requirement of the general theory of relativity, give us ... to a first approximation Newton's law of attraction, and to a second approximation the explanation of the motion of the perihelion of the planet Mercury discovered by Leverrier.... These facts must, in my opinion, be taken as convincing proof of the correctness of the theory.[85]

Others disagreed. For example, Max von Laue objected on the grounds that the calculations assumed that the Sun and Mercury were point masses, and, unlike in the Newtonian theory, there was no reason to believe that one could treat extended bodies this way. Einstein recognized that more experimental confirmation should be sought, and at the end of the paper we find his calculations for the gravitational redshift, the bending of light (which is now double his original 1911 value), and the motion of Mercury.

For reasons described earlier (see p. 460), red-shift observations were insufficiently precise to be of much use.[86] On the other hand, observations of light deflection made in 1919 made Einstein an international superstar. Given the negative reception that special relativity had in England[87] and the fact that the First World War had made German journals virtually inaccessible to British scientists, it is perhaps surprising that it was a British team that provided the first new empirical justification for Einstein's theory.[88] There was at least one significant champion of Einstein's in England – the secretary of the Royal Astronomical Society, Arthur Stanley Eddington – and he persuaded the Astronomer Royal, Frank Watson Dyson, to support two expeditions to measure the deflection of light during the solar eclipse of May 1919. One team set out to Sobral in Brazil, and another (led by Eddington) travelled to Principe, off the west coast of Africa. The results from these expeditions were far from conclusive. The Sobral group reported a deflection of $1''.98 \pm 0''.16$ and those in Principe found $1''.61 \pm 0''.40$, compared with the general relativistic value of $1''.75$, but Dyson announced proudly that Einstein's prediction had been confirmed.[89]

---

[85] EINSTEIN *Die Grundlage der allgemeinen Relativitätstheorie* (1916). English translation in Lorentz, *et al.* (1923).

[86] A history of the gravitational red shift problem from 1896, through the predictions of general relativity, and including a survey of subsequent attempts to quantify it, can be found in Forbes (1961) (see also Weinberg (1972), p. 81).

[87] See Goldberg (1970).

[88] These issues are explored in Earman and Glymour (1980) and Sponsel (2002).

[89] Details of the British expeditions and the many other attempts to measure light deflection during solar eclipses (both before and after 1919) can be found in von Klüber (1960). There is a technical sense in which the bending of light is a less sensitive test of general relativity than perihelion shifts (see Duff (1974)).

On 7 November 1919, the London *Times* carried an article[90] headed 'Revolution in Science/New Theory of the Universe/Newtonian Ideas Overthrown' and *The New York Times* carried some equally dramatic headlines 2 days later.

General relativity is a highly technical mathematical theory, yet it is grounded firmly on experiment. The guiding principles behind the theory – that all frames of reference are to be treated equally, special relativity should be recovered in the absence of gravitation, and Newtonian mechanics is appropriate when gravitational fields are weak – all have a sound empirical basis. Moreover, as more experimental tests have been performed, confidence in general relativity has grown. But the nonlinear nature of the theory makes it seriously challenging, and there is still plenty more to discover lying hidden within the deceptively simple looking Eqn (12.10).

A final thought

Some would say that, by reducing celestial motions to equations, science has robbed the heavens of their beauty and wonder.

> When I heard the learn'd astronomer,
> When the proofs, the figures, were ranged in columns before me,
> When I was shown the charts and the diagrams, to add, divide, and
>   measure them,
> When I, sitting, heard the astronomer, where he lectured with much
>   applause in the lecture-room,
> How soon, unaccountable, I became tired and sick,
> Till rising and gliding out, I wander'd off by myself,
> In the mystical moist night-air, and from time to time,
> Look'd up in perfect silence at the stars.
>   'When I heard the Learn'd Astronomer'
>   *by Walt Whitman (1819–92)*

However, I will leave the last word to Richard Feynman, a man who contributed perhaps more than anyone else to enhancing the public understanding of science in the twentieth century.

> To those who do not know mathematics it is difficult to get across a real feeling as
> to the beauty, the deepest beauty, of nature. . . . It is too bad that it has to be
> mathematics, and that mathematics is hard for some people. It is reputed – I do not
> know if it is true – that when one of the kings was trying to learn geometry from

---

[90] Reproduced in Pais (1982), p. 307.

Euclid he complained that it was difficult. And Euclid said, "There is no royal road to geometry". And there is no royal road. Physicists cannot make a conversion to any other language. If you want to learn about nature, to appreciate nature, it is necessary to understand the language that she speaks in. She offers her information only in one form; we are not so unhumble as to demand that she change before we pay any attention.[91]

---

[91] Feynman (1967).

References

Aaboe, A. (1958). On Babylonian planetary theories. *Centaurus* **5**, 209–77.

(1964). On period relations in Babylonian astronomy. *Centaurus* **10**, 213–31.

(1972). Remarks on the theoretical treatment of eclipses in antiquity. *Journal for the History of Astronomy* **3**, 105–18.

(1974). Scientific astronomy in antiquity. *Philosophical Transactions of the Royal Society of London*, **A 276**, 21–42.

(1980). Observation and theory in Babylonian astronomy. *Centaurus* **24**, 14–35.

Abbud, F. (1962). The planetary theory of Ibn al-Shāṭir: reduction of the geometric models to numerical tables. *Isis* **53**, 492–9.

Abetti, G. (1954). *The History of Astronomy*. Sidgwick and Jackson.

Abhyankar, K. D. (2000). Babylonian source of Āryabhaṭa's planetary constants. *Indian Journal of History of Science* **35**(3), 185–8.

Abraham, G. (1982). Algebraic formulae for the Moon, Saturn and Jupiter in the Pancasiddhantika. *Archive for History of Exact Sciences* **26**, 287–97.

Africa, T. W. (1961). Copenicus' relation to Aristarchus and Pythagoras. *Isis* **52**, 403–9.

Aiton, E. J. (1954). Galileo's theory of the tides. *Annals of Science* **10**, 44–57.

(1965). Galileo and the theory of the tides. *Isis* **56**, 56–63.

(1969). Kepler's second law of planetary motion. *Isis* **60**, 75–90.

(1972). *The Vortex Theory of Planetary Motions*. Macdonald.

(1973). Infinitessimals and the area law. In F. Krafft, K. Meyer, and B. Sticker (eds), *Internationales Kepler-Symposium, Weil der Stadt 1971*, pp. 285–305. Gerstenberg.

(1978). Kepler's path to the construction and rejection of his first oval orbit for Mars. *Annals of Science* **35**, 173–90.

(1981). Celestial spheres and circles. *History of Science* **19**, 75–114.

(1987). Peurbach's *Theoricae novae planetarum*: A translation with commentary. *Osiris*, 2nd Series **3**, 5–44.

(1989). The Cartesian vortex theory. In R. Taton and C. Wilson (eds), *Planetary astronomy from the Renaissance to the rise of astrophysics. Part A: Tycho Brahe to Newton*, Volume 2 of *The General History of Astronomy*. Cambridge University Press.

Alexander, A. F. O. (1962). *The Planet Saturn. A History of Observation, Theory and Discovery*. Faber and Faber.

(1965). *The Planet Uranus. A History of Observation, Theory and Discovery.* Faber and Faber.

Andersson, K. G. (1994). Poincaré's disocvery of homoclinic points. *Archive for History of Exact Sciences* **48**, 133–47.

Applebaum, W. (1975). Jeremiah Horrocks. In *Dictionary of Scientific Biography.* Charles Scribner's Sons.

Ariew, R. (1987). The phases of Venus before 1610. *Studies in History and Philosophy of Science* **18**, 81–92.

Aristotle (1999). *Metaphysics.* Green Lion Press. Trans. Joe Sachs.

Armitage, A. (1947). The deviation of falling bodies. *Annals of Science* **5**, 342–51.

(1950). 'Borrell's hypothesis' and the rise of celestial mechanics. *Annals of Science* **6**, 268–82.

(1962). *William Herschel.* Thomas Nelson and Sons Ltd.

(1966). *John Kepler.* Faber and Faber.

Ashbrook, J. (1984). *The Astronomical Scrapbook. Skywatchers, Pioneers, and Seekers in Astronomy.* Cambridge University Press and Sky Publishing Corporation.

Aubin, D. and A. D. Dalmedico (2002). Writing the history of dynamical systems and chaos: *Longue durée* and revolution, disciplines and cultures. *Historia Mathematica* **29**, 273–339.

Babb, Jr., S. E. (1977). Accuracy of planetary theories, particularly for Mars. *Isis* **68**, 426–34.

Balachandra Rao, S. (2000). *Indian Astronomy. An Introduction.* Universities Press.

Baldwin, M. R. (1985). Magnetism and the anti-Copernican polemic. *Journal for the History of Astronomy* **16**, 155–74.

Bamford, G. (1996). Popper and his commentators on the discovery of Neptune: A close shave for the law of gravitation? *Studies in History and Philosophy of Science* **27**(2), 207–32.

Banville, J. (1976). *Doctor Copernicus.* Martin Secker and Warburg Ltd.

(1981). *Kepler.* Martin Secker and Warburg Ltd.

Barker, P. and B. R. Goldstein (1988). The role of comets in the Copernican revolution. *Studies in History and Philosophy of Science* **19**, 299–319.

Barrow-Green, J. (1997). *Poincaré and the Three Body Problem,* Volume 11 of *History of Mathematics.* American Mathematical Society.

Battin, R. H. (1987). *An Introduction to the Mathematics and Methods of Astrodynamics.* AIAA.

Baum, R. and W. Sheehan (1997). *In Search of Planet Vulcan. The Ghost in Newton's Clockwork Universe.* Plenum Trade.

Baumgardt, C. (1952). *Johannes Kepler: Life and Letters.* Victor Gollancz.

Becher, H. W. (1980). William Whewell and Cambridge mathematics. *Historical Studies in the Physical Sciences* **11**(1), 1–48.

Beller, E. (1988). Ancient Jewish mathematical astronomy. *Archive for History of Exact Sciences* **38**, 51–66.

Benjamin, F. S. and G. J. Toomer (1971). *Campanus of Novarra and Medieval Planetary Theory.* University of Wisconsin Press.

Bennett, J. A. (1989). Magnetical philosophy and astronomy from Wilkins to Hooke. In R. Taton and C. Wilson (eds), *Planetary astronomy from the Renaissance to the rise of astrophysics. Part A: Tycho Brahe to Newton*, Volume 2 of *The General History of Astronomy*. Cambridge University Press.

Berggren, J. L. and R. S. D. Thomas (1996). *Euclid's* Phaenomena: *A Translation and Study of a Hellenistic Treatise in Spherical Astronomy*. Garland Publishing.

Berry, A. (1961). *A Short History of Astronomy*. Dover Publications.

Bjerknes, C. J. (2002). *Albert Einstein the Incorrigible Plagiarist*. XTX Inc., Downers Groves.

Blagg, M. A. (1913). On a suggested substitue for Bode's law. *Monthly Notices of the Royal Astronomical Society* **73**, 414–22.

Blair, A. (1990). Tycho Brahe's critique of Copernicus and the Copernican system. *Journal for the History of Ideas* **51**, 355–77.

Boas Hall, M. (1994). *The Scientific Renaissance 1450–1630*. Dover Publications. Originally pub. 1962.

Boccaletti, D. and G. Pucacco (1996). *Theory of Orbits. Volume 1: Integrable Systems and Non-perturbative Methods*. Springer-Verlag.

Bochner, S. (1966). *The Role of Mathematics in the Rise of Science*. Princeton University Press.

Bork, A. (1987). Newton and comets. *American Journal of Physics* **55**, 1089–95.

Born, M. (1962). *Einstein's Theory of Relativity* (rev. edn). Dover Publications.

Boscovich, R. J. (1966). *A Theory of Natural Philosophy*. MIT Press. Trans. J. M. Child.

Boyer, C. B. (1941). Early estimates of the velocity of light. *Isis* **33**, 24–40.

(1959). *The History of the Calculus and its Conceptual Development*. Dover Publications. First pub. 1949 by Hafner Publishing Company.

(1989). *A History of Mathematics* (2nd edn). Rev. Uta C. Merzbach. John Wiley and Sons.

Brackenridge, J. B. (1982). Kepler, elliptical orbits, and celestial circularity: A study in the persistence of metaphysical commitment. Parts I and II. *Annals of Science* **39**, 117–43, 265–95.

(1995). *The Key to Newton's Dynamics. The Kepler Problem and the* Principia. University of California Press.

Britton, J. P. (1969). Ptolemy's determination of the obliquity of the ecliptic. *Centaurus* **14**(1), 29–41.

(1992). Models and precision: The quality of Ptolemy's observations and parameters. In *Sources and Studies in the History and Philosophy of Classical Science*, Vol. 1. Garland.

Broughton, P. (1985). The first predicted return of Comet Halley. *Journal for the History of Astronomy* **16**, 123–33.

Brouwer, D. and G. M. Clemence (1961). *Methods of Celestial Mechanics*. Academic Press.

Brown, E. W. (1930). On the prediction of transneptunian planets from the perturbations of Uranus. *Proceedings of the National Academy of Sciences* **16**, 364–71.

(1931). On a criterion for the prediction of an unknown planet. *Monthly Notices of the Royal Astronomical Society* **92**, 80.

(1960). *An Introductory Treatise on the Lunar Theory*. Dover Publications. Originally pub. Cambridge University Press 1896.

Brumberg, V. A. (1991). *Essential Relativistic Celestial Mechanics*. Adam Hilger.

Burstyn, H. L. (1962). Galileo's attempt to prove that the earth moves. *Isis* **53**, 161–85.

Calinger, R. (1996). Leonhard Euler: The first St. Petersburg years (1727–1741). *Historia mathematica* **23**, 121–66.

Cannon, W. F. (1961). John Herschel and the idea of science. *Journal for the History of Ideas* **22**, 215–39.

Caparrini, S. (2002). The discovery of the vector representation of moments and angular velocity. *Archive for History of Exact Sciences* **56**, 151–81.

Carter, B. and M. S. Carter (2002). *Latitude. How American Astronomers Solved the Mystery of the Variation.* Naval Institute Press.

Caspar, M. (1993). *Kepler.* Dover Publications. English translation by C. Doris Hellman. First pub. Abelard-Schuman, 1959. Bibliographical references added by Owen Gingerich.

Challis, J. (1830). On the extension of Bode's empirical law of the distances of the planets from the sun, to the distances of the satellites from their respective primaries. *Transactions of the Cambridge Philosophical Society* **3**, 171–83.

Chandler, P. (1975). Clairaut's critique of Newtonian attraction: some insights into his philosophy of science. *Annals of Science* **32**, 369–78.

Chandrasekar, S. (1995). *Newton's* Principia *for the Common Reader.* Clarendon Press.

Chapin, S. L. (1995). The shape of the earth. In R. Taton and C. Wilson (eds), *Planetary Astronomy from the Renaissance to the Rise of Astrophysics. Part B: The Eighteenth and Nineteenth Centuries.* In *The General History of Astronomy.* Volume 2. Cambridge University Press.

Chapman, A. (1988). Public research and private duty: George Biddell Airy and the search for Neptune. *Journal for the History of Astronomy* **19**, 121–39.

— (1990). Jeremiah Horrocks, the transit of Venus, and the 'New Astronomy' in early seventeenth-century England. *Quarterly Journal of the Royal Astronomical Society* **31**, 333–57.

Christianson, J. (1967). Tycho Brahe at the University of Copenhagen, 1559–1562. *Isis* **58**, 198–203.

Christianson, J. R. (1979). Tycho Brahe's German treatise on the comet of 1577: A study in science and politics. *Isis* **70**, 110–40.

— (2000). *On Tycho's Island. Tycho Brahe and his Assistants 1570–1601.* Cambridge University Press.

Clarke, D. M. (1977). The impact rules of Descartes' physics. *Isis* **68**, 55–66.

Clemence, G. M. (1947). The relativity effect in planetary motions. *Reviews of Modern Physics* **19**, 361–4.

Clerke, A. M. (1908). *A Popular History of Astronomy During the Nineteenth Century* (4th edn). Adam and Charles Black.

Cohen, C. J. and E. C. Hubbard (1965). Libration of the close approaches of Pluto and Neptune. *The Astronomical Journal* **70**, 10–13.

Cohen, I. B. (1975). *Isaac Newton's* Theory of the Moon's Motion *(1702) with a Bibliographical and Historical Introduction.* Dawson and Sons Ltd.

— (1985). *Revolution in Science.* Harvard University Press.

Colwell, P. (1993). *Solving Kepler's Equation over Three Centuries.* Willmann-Bell.

Coolidge, J. L. (1949). *The Mathematics of Great Amateurs.* Clarendon Press.

Copernicus, N. (1992). *On the Revolutions.* The Johns Hopkins University Press. Paperback edn. Trans. and commentary by Edward Rosen.

Crombie, A. C. (1959). *Augustine to Galileo. Volumes I and II.* Heinemann.

Crommelin, A. C. D. (1931). The discovery of Pluto. *Monthly Notices of the Royal Astronomical Society* **91**, 380.

Crowe, M. J. (1994). *A History of Vector Analysis. The Evolution of the Idea of a Vectorial System*. Dover Publications. Corr. edn. Originally pub. University of Notre Dame Press 1967.

Cunningham, C. J. (1988). The Baron and his Celestial Police. *Sky and Telescope* **75**, 271–72.

Cushing, J. T. (1998). *Philospophical Concepts in Physics. The Historical Relationship Between Philosophy and Scientific Theories*. Cambridge University Press.

Dale, A. I. (1982). Bayes or Laplace? An examination of the origin and early application of Bayes' theorem. *Archive for History of Exact Sciences* **27**, 23–47.

Dampier, W. C. (1965). *A History of Science and its Relation with Philosophy and Religion*. Cambridge University Press.

Darrigol, O. (1995). Henri Poincaré's criticism of *fin de siècle* electrodynamics. *Studies in History and Philosophy of Modern Physics* **26**, 1–44.

Darwin, G. H. (1962). *The Tides and Kindred Phenomena in the Solar System*. W. H. Freeman and Company. Originally pub. 1898.

Daston, L. (1988). *Classical Probability in the Enlightenment*. Princeton University Press.

Davis, A. E. L. (2003). The mathematics of the area law: Kepler's successful proof in *Epitome astronomiae Copernicanae* (1621). *Archive for History of Exact Sciences* **57**, 355–93.

Davis, J. L. (1986). The influence of astronomy on the character of physics in mid-nineteenth century France. *Historical Studies in the Physical and Biological Sciences* **16**, 59–82.

Débarbat, S. and C. Wilson (1989). The Galilean satellites of Jupiter from Galileo to Cassini, Römer and Bradley. In R. Taton and C. Wilson (eds), *Planetary Astronomy from the Renaissance to the Rise of Astrophysics. Part A: Tycho Brahe to Newton*. In *The General History of Astronomy*. Volume 2. Cambridge University Press.

Densmore, D. (1996). *Newton's* Principia: *The Central Argument*. Trans. and illustrations by William H. Donahue. Green Lion Press.

De Santillana, G. (1961). *The Crime of Galileo*. Mercury Books.

De Solla Price, D. J. (1962). Contra-Copernicus: A critical re-estimation of the mathematical planetary theory of Ptolemy, Copernicus, and Kepler. In M. Clagett (ed.), *Critical problems in the history of science*, pp. 197–218. The University of Wisconsin Press.

Diacu, F. (1996). The solution of the *n*-body problem. *The Mathematical Intelligencer* **18**(3), 66–70.

Diacu, F. and P. Holmes (1996). *Celestial Encounters. The Origins of Chaos and Stability*. Princeton University Press.

Dias, P. M. C. (1999). Euler's 'harmony' between the principles of 'rest' and 'least action'. the conceptual making of analytical dynamics. *Archive for History of Exact Sciences* **54**, 67–86.

Di Bono, M. (1995). Copernicus, Amico, Fracastoro and Ṭūsī's device: Observations on the use and transmission of a model. *Journal for the History of Astronomy* **26**, 133–54.

Dicks, D. R. (1970). *Early Greek Astronomy to Aristotle*. Thames and Hudson.

Dijksterhuis, E. J. (1961). *The Mechanization of the World Picture.* Oxford University Press.

Diller, A. (1949). The ancient measurements of the earth. *Isis* **40**, 6–9.

Dobrzycki, J. (1973a). Nicolaus Copernicus, his life and work. In B. Bieńkowska (ed.), *The scientific world of Copernicus.* D. Reidel Publishing Company.

—— (ed.) (1973b). *The Reception of Copernicus' Heliocentric Theory.* D. Reidel Publishing Company. Proceedings of a symposium organized by the Nicolas Copernicus Committee of the International Union of the History and Philosophy of Science.

Dobrzycki, J. and L. Szczucki (1989). On the transmission of Copernicus' *Commentariolus* in the sixteenth century. *Journal for the History of Astronomy* **20**, 25–28.

Dobson, G. J. (1998). Newton's problems with rigid body dynamics in the light of his treatment of the precession of the equinoxes. *Archive for History of Exact Sciences* **53**, 125–45.

—— (2001). On lemmas 1 and 2 to proposition 39 of book 3 of Newton's *Principia*. *Archive for History of Exact Sciences* **55**, 345–63.

Dohrn-van Rossum, G. (1996). *History of the Hour. Clocks and Modern Temporal Orders.* University of Chicago Press. Trans. Thomas Dunlap.

Donahue, W. H. (1988). Kepler's fabricated figures: Covering up the mess in the *New Astronomy. Journal for the History of Astronomy* **19**, 217–37.

Drake, S. (1957). *Discoveries and Opinions of Galileo.* Doubleday Anchor Books.

—— (1978). Ptolemy, Galileo, and scientific method. *Studies in History and Philosophy of Science* **9**, 99–115.

—— (1987). Galileo's steps to full Copernicanism, and back. *Studies in History and Philosophy of Science* **18**, 93–105.

—— (1988). The tower argument and the *Dialogue. Annals of Science* **45**, 295–302.

Dreyer, J. L. E. (1953). *A History of Astronomy from Thales to Kepler* (2nd edn). Dover Publications. Originally pub. under the title *History of the Planetary Systems from Thales to Kepler* 1906.

Duff, M. J. (1974). On the significance of perihelion shift calculations. *General Relativity and Gravitation* **5**(4), 441–52.

Dugas, R. (1988). *A History of Mechanics.* Dover Publications. Originally pub. 1955.

Duhem, P. (1969). *To Save the Phenomena: an Essay on the Idea of Physical Theory from Plato to Galileo.* Chicago University Press. Originally pub. 1908.

Duncombe, R. L. (1958). Motion of Venus 1750–1949. *Astronomical Papers Prepared for the Use of the American Ephemerides and Nautical Almanac* **16**(1), 1–258.

Dunham, W. (1999). *Euler. The Master of Us All.* The Mathematical Association of America.

Dunnington, G. W. (1955). *Carl Friedrich Gauss: Titan of Science.* Hafner Publishing Co.

Dutka, J. (1993). 'Eratosthenes' measurement of the earth reconsidered'. *Archive for History of Exact Sciences* **46**, 55–66.

Earman, J. and C. Glymour (1978a). Einstein and Hilbert: Two months in the history of general relativity. *Archive for History of Exact Sciences* **19**, 291–308.

—— (1978b). Lost in the tensors: Einstein's struggles with covariance principles 1912–1916. *Studies in History and Philosophy of Science* **9**, 251–78.

—— (1980). Relativity and eclipses: The British eclipse expeditions of 1919 and their predecessors. *Historical Studies in the Physical Sciences* **11**, 49–85.

Earman, J. and M. Janssen (1993). Einstein's explanation of the motion of Mercury's perihelion. In J. Earman, M. Janssen, and J. D. Norton (eds), *The Attraction of Gravitation: New Studies in the History of General relativity*, pp. 129–72. Birkhäuser.

Eastwood, B. S. (1997). Astronomy in Christian Latin Europe c.500–c.1150. *Journal for the History of Astronomy* **28**, 235–58.

Escobal, P. R. (1965). *Methods of Orbit Determination*. John Wiley and Sons, New York.

Euclid (1956). *The Elements*. Dover Publications. Introduction and commentary T. L. Heath.

Evans, J. (1984). On the function and the probable origin of Ptolemy's equant. *American Journal of Physics* **52**, 1080–9.

—— (1987). On the origin of the Ptolemaic star catalogue: Parts 1 and 2. *Journal for the History of Astronomy* **18**, 155–72, 233–78.

—— (1998). *The History and Practice of Ancient Astronomy*. Oxford University Press.

Fadner, W. L. (1988). Did Einstein really discover $E = mc^2$? *American Journal of Physics* **56**, 114–22.

Fauvel, J. and J. Gray (eds) (1987). *The History of Mathematics: A Reader*. Macmillan Press.

Ferguson, K. (2002). *Tycho and Kepler. The Unlikely Partnership that Forever Changed our Understanding of the Heavens*. Walker and Company.

Fermor, J. (1997). Timing the sun in Egypt and Mesopotamia. *Vistas in Astronomy* **41**, 157–67.

Fernie, J. D. (1996). The extraordinary and short-lived career of Jeremiah Horrocks. *American Scientist* **84**(2), 114–17.

Ferrers, N. M. (Ed.) (1970). *Mathematical Papers of George Green*. Chelsea Publishing Company. Originally pub. 1871.

Feynman, R. P. (1967). *The Character of Physical Law*. MIT Press.

Field, J. V. (1988). *Kepler's Geometrical Cosmology*. The Athlone Press.

Finocchiaro, M. A. (1989). *The Galileo Affair. A Documentary History*. University of California Press.

Fischer, I. (1975). Another look at Eratosthenes' and Posidonius' determination of the earth's circumference. *Quarterly Journal of the Royal Astronomical Society* **16**, 152–67.

Fontenrose, R. (1973). In search of Vulcan. *Journal for the History of Astronomy* **4**, 145–58.

Forbes, E. G. (1961). A history of the solar red shift problem. *Annals of Science* **17**, 129–64.

—— (1970). Tobias Mayer's contributions to the development of lunar theory. *Journal for the History of Astronomy* **1**, 144–54.

—— (1971a). *The Euler–Mayer Correspondence (1751–1755). A New Perspective on Eighteenth-Century Advances in the Lunar Theory*. Macmillan.

—— (1971b). Gauss and the discovery of Ceres. *Journal for the History of Astronomy* **2**, 195–99.

—— (1978). The astronomical work of Carl Friedrich Gauss (1777–1855). *Historia mathematica* **5**, 167–81.

Forbes, E. G. and C. Wilson (1995). The solar tables of Lacaille and the lunar tables of Mayer. In R. Taton and C. Wilson (eds), *Planetary Astronomy from the Renaissance* to the Rise of Astrophysics. Part B: The Eighteenth and Nineteenth Centuries, Volume 2 of *The General History of Astronomy*. Cambridge University Press.

Foster, J. and J. D. Nightingale (1995). *A Short Course in General Relativity* (2nd edn). Springer-Verlag.

Fowler, D. H. (1987). *The Mathematics of Platos's Academy*. Clarendon Press.

Franseen, M. (1993). Did King Alfonso of Castile really want to advise God against the Ptolemaic system? The legend in history. *Studies in History and Philosophy of Science* **24**, 313–25.

Fraser, C. G. (1983). J. L. Lagrange's early contributions to the principles and methods of mechanics. *Archive for History of Exact Sciences* **28**, 197–241. Reprinted in Craig G. Fraser (1997). *Calculus and Analytical Mechanics in the Age of Enlightenment*. Variorum Reprints.

(1985a). D'Alembert's principle: the original formulation and application in Jean d'Alembert's *Traité de Dynamique*. Parts 1 and 2. *Centaurus* **28**, 31–61, 145–59. Reprinted in Craig G. Fraser (1997). *Calculus and Analytical Mechanics in the Age of Enlightenment*. Variorum Reprints.

(1985b). J. L. Lagrange's changing approach to the foundations of the calculus of variations. *Archive for History of Exact Sciences* **32**, 151–91. Reprinted in Craig G. Fraser (1997). *Calculus and Analytical Mechanics in the Age of Enlightenment*. Variorum Reprints.

(1990). Lagrange's analytical mathematics, its Cartesian origins and reception in Comte's positive philosophy. *Studies in History and Philosophy of Science* **21**, 243–56. Reprinted in Craig G. Fraser (1997). *Calculus and Analytical Mechanics in the Age of Enlightenment*. Variorum Reprints.

Gabbey, A. (1980). Huygens and mechanics. In H. J. M. Bos, M. J. S. Rudwick, H. A. M. Snelders, and R. P. W. Visser (eds), *Studies on Christiaan Huygens. Invited Papers from the Symposium on the Life and Work of Christiaan Huygens, Amsterdam, 22–25 August 1979*, pp. 166–99. Swets and Zeitlinger.

Galilei, G. (1953). *Dialogue on the Great World Systems*. Trans. Thomas Salusbury. Rev. Giorgio de Santillana. University of Chicago Press.

(1962). *Dialogue Concerning the Two Chief World Systems—Ptolemaic and Copernican*. Trans. Stillman Drake. University of California Press.

Galison, P. L. (1979). Minkowski's space-time: From visual thinking to the absolute world. *Historical Studies in the Physical Sciences* **10**, 85–121.

Gårding, L. (1998). *Mathematics and Mathematicians. Mathematics in Sweden Before 1950*. In *History of Mathematics*. Vol. 13. American Mathematical Society.

Garisto, R. (1991). An error in Isaac Newton's determination of planetary properties. *American Journal of Physics* **59**, 42–8.

Gauss, K. F. (1963). *Theory of the Motion of the Heavenly Bodies Moving about the Sun in Conic Sections*. Trans. Charles Henry Davis 1857. Dover Publications.

Gaythorpe, S. B. (1957). Jeremiah Horrocks and his 'New Theory of the Moon'. *Journal of the British Astronomical Association* **67**, 134–44.

Gilbert, W. (1958). *De Magnete*. Trans. P. Fleury Mottelay. Dover Publications.

Gillispie, C. C. (1997). *Pierre-Simon Laplace, 1749–1827. A Life in Exact Science*. Princeton University Press.

Gingerich, O. (1970). The satellites of Mars: Prediction and discovery. *Journal for the History of Astronomy* **1**, 109–15.

(1972). Johannes Kepler and the New Astronomy. *Quarterly Journal of the Royal Astronomical Society* **13**, 346–73.

(1973a). Copernicus and Tycho. *Scientific American* **229**(6), 87–101.

(1973b). Kepler's treatment of redundant observations or, the computer versus Kepler revisited. In F. Krafft, K. Meyer, and B. Sticker (eds), *Internationales Kepler-Symposium, Weil der Stadt 1971*, pp. 307–14. Gerstenberg.

(1989). Johannes Kepler. In R. Taton and C. Wilson (eds), *Planetary Astronomy from the Renaissance to the Rise of Astrophysics. Part A: Tycho Brahe to Newton*. In *The General History of Astronomy*. Volume 2. Cambridge University Press.

(1992). *The Great Copernicus Chase and Other Adventures in Astronomical History*. Cambridge University Press.

(2002). The trouble with Ptolemy. *Isis* **93**, 70–4.

Gingerich, O. and R. S. Westman (1981). A reattribution of the Tychonic annotations in copies of Copernicus's 'De Revolutionibus'. *Journal for the History of Astronomy* **12**, 53–4.

(1988). The Wittich connection: Priority and conflict in late sixteenth-century cosmology. *Transactions of the American Philosophical Society* **78**(7), 1–148.

Goldberg, S. (1970). In defense of the ether: The British response to Einstein's special theory of relativity. *Historical Studies in the Physical Sciences* **2**, 89–125.

(1976). Max Planck's philosophy of nature and his elaboration of the special theory of relativity. *Historical Studies in the Physical Sciences* **7**, 125–60.

Goldreich, P. and S. J. Peale (1966). Spin-orbit coupling in the solar system. *The Astronomical Journal* **71**, 425–38.

Goldstein, B. R. (1964). On the theory of trepidation. *Centaurus* **10**, 232–47.

(1967). The Arabic version of Ptolemy's *Planetary Hypotheses*. *Transactions of the American Philosophical Society* **57**, 3–55.

(1969). Some medieval reports of Venus and Mercury transits. *Centaurus* **14**(1), 49–59. Reprinted in B. R. Goldstein (1985). *Theory and Observation in Ancient and Medieval Astronomy*. Variorum Reprints.

(1971). *Al-Biṭrūjī: On the Principles of Astronomy*. Yale University Press.

(1972). Levi ben Gerson's lunar model. *Centaurus* **16**, 257–84.

(1974). The astronomical tables of Levi ben Gerson. *Transactions of the Connecticut Academy of Arts and Sciences* **45**, 1–285.

(1975). Levi ben Gerson's analysis of precession. *Journal for the History of Astronomy* **6**, 31–41. Reprinted in B. R. Goldstein (1985). *Theory and Observation in Ancient and Medieval Astronomy*. Variorum Reprints.

(1980). The status of models in ancient and medieval astronomy. *Centaurus* **24**, 132–47.

(1983). The obliquity of the ecliptic in ancient Greek astronomy. *Archives Internationales d'Histoire des Sciences* **33**, 3–14. Reprinted in B. R. Goldstein (1985). *Theory and Observation in Ancient and Medieval Astronomy*. Variorum Reprints.

(1984). Eratosthenes on the 'measurement' of the earth. *Historia mathematica* **11**, 411–16.

(1985). *The Astronomy of Levi ben Gerson (1288–1344)*. Springer-Verlag.

(1997). Saving the phenomena: The background to Ptolemy's planetary theory. *Journal for the History of Astronomy* **28**, 1–12.

Goldstein, B. R. and A. C. Bowen (1983). A new view of early Greek astronomy. *Isis* **74**, 330–40. Reprinted in B. R. Goldstein (1985). *Theory and Observation in Ancient and Medieval Astronomy.* Variorum Reprints.

Goldstein, B. R. and N. Swerdlow (1970). Planetary distances and sizes in an anonymous arabic treatise preserved in Bodleian Ms. Marsh 621. *Centaurus* **15**(2), 135–70.

Goldstein, S. J. (1982). Problems raised by Ptolemy's lunar theory. *Journal for the History of Astronomy* **13**, 195–205.

Goldstine, H. H. (1980). *A History of the Calculus of Variations from the 17th Through the 19th Century.* Springer-Verlag.

Goodwin, S. and J. Gribbin (2001). *XTL. Extraterrestrial Life and How to Find It.* Cassell and Co.

Goroff, D. L. (1993). Henri Poincaré and the birth of chaos theory: An introduction to the English translation of *Les méthodes nouvelles de la mécanique céleste.* In *Poincaré* (1993) below. American Institute of Physics.

Gower, B. (1987). Planets and probability: Daniel Bernouilli on the inclinations of the planetary orbits. *Studies in History and Philosophy of Science* **18**, 441–54.

Grafton, A. (1992). Kepler as a reader. *Journal for the History of Ideas* **53**, 561–72.

Grant, E. (1971). *Nicole Oresme and the Kinematics of Circular Motion.* University of Wisconsin Press.

(1987). Celestial orbs in the Latin Middle Ages. *Isis* **78**, 153–73.

(1994). *Planets, Stars, & Orbs. The Medieval Cosmos, 1200–1687.* Cambridge University Press.

Grant, R. (1966). *History of Physical Astronomy from the Earliest Ages to the Middle of the Nineteenth Century.* Originally pub. 1852. Johnson Reprint Collection.

Grattan-Guiness, I. (1995). Why did George Green write his essay of 1828 on electricity and magnetism. *The American Mathematical Monthly* **102**, 387–96.

Greenberg, J. L. (1995). *The Problem of the Earth's Shape from Newton to Clairaut.* Cambridge University Press.

Grimsley, R. (1963). *Jean d'Alembert (1717–83).* Clarendon Press.

Gross, R. S. (2000). The excitation of the Chandler wobble. *Geophysical Research Letters* **27**(15), 2329–32.

Grosser, M. (1962). *The Discovery of Neptune.* Harvard University Press.

(1964). The search for a planet beyond Neptune. *Isis* **55**, 163–83.

Guicciardini, N. (1996). An episode in the history of dynamics: Jakob Hermann's proof (1716–1717) of Proposition 1, Book 1, of Newton's *Principia. Historia mathematica* **23**, 167–81.

(1998). Did Newton use his calculus in the *Principia? Centaurus* **40**, 303–44.

(1999). *Reading the Principia: The Debate on Newton's Mathematical Methods for Natural Philosophy from 1687 to 1736.* Cambridge University Press.

Hairer, E. and G. Wanner (1996). *Analysis by its History.* Springer-Verlag.

Hairetdinova, N. G. (1970). On the oriental sources of the Regiomontanus' trigonometrical treatise. *Archives Internationales d'Histoire des Sciences* **23**, 61–6.

Hald, A. (1998). *A History of Mathematical Statistics from 1750 to 1930.* Wiley-Interscience.

Hall, A. R. (1999). *Isaac Newton. Eighteenth-Century Perspectives.* Oxford University Press.

Hall, A. R. and M. Boas Hall (eds) (1969). *The Correspondence of Henry Oldenburg.* Volume VI. University of Wisconsin Press.

Hall, R. W. and K. Josić (2000). Planetary motion and the duality of force laws. *SIAM Review* **42**, 115–24.

Hankins, T. L. (1967). The reception of Newton's second law of motion in the eighteenth century. *Archives Internationales d'Histoire des Sciences* **20**, 43–65.

(1970). *Jean d'Alembert. Science and the Enlightenment.* Clarendon Press.

(1985). *Science and the Enlightenmemnt.* Cambridge University Press.

Hanson, N. R. (1960). The mathematical power of epicyclical astronomy. *Isis* **51**, 150–8.

(1962). Leverrier: the zenith and nadir of Newtonian mechanics. *Isis* **53**(3), 359–78.

Harrison, H. M. (1994). *Voyager in Space and Time. The Life of John Couch Adams.* The Book Guild Ltd.

Hartner, W. (1969). Eclipse periods and Thales' prediction of a solar eclipse. Historic truth and modern myth. *Centaurus* **14**(1), 60–71.

Harvey, A. L. (1965). Brief review of Lorentz-covariant scalar theories of gravitation. *American Journal of Physics* **33**, 449–60.

Heath, T. L. (1913). *Aristarchus of Samos, the Ancient Copernicus.* Clarendon Press.

(1932). *Greek Astronomy.* J. M. Dent and Sons.

Hellman, C. D. (1944). *The Comet of 1577: Its Place in the History of Astronomy. Columbia University Studies in the Social Sciences.* Volume 510. AMS Press.

Herivel, J. W. (1960). Newton's discovery of the law of centrifugal force. *Isis* **51**, 546–53.

(1965a). *The Background to Newton's* Principia, *A Study of Newton's Dynamical Researches in the Years 1664–1684.* Clarendon Press.

(1965b). Newton's first solution to the problem of Kepler motion. *The British Journal for the History of Science* **2**, 350–4.

Herschel, W. (1802). Observations on the two lately discovered celestial bodies. *Philosophical Transactions of the Royal Society of London* **92**, 213–32.

Hine, W. L. (1973). Mersenne and Copernicanism. *Isis* **64**, 18–32.

Hirosige, T. (1976). The ether problem, the mechanistic worldview, and the origins of the theory of relativity. *Historical Studies in the Physical Sciences* **7**, 3–82.

Hirshfeld, A. W. (2001). *Parallax. The Race to Measure the Cosmos.* W. H. Freeman and Co.

Hockey, T. (1999). *Galileo's Planet. Observing Jupiter Before Photography.* Institute of Physics.

Hogendjik, J. P. (1988). Three Islamic lunar crescent visibility tables. *Journal for the History of Astronomy* **19**, 29–44.

Holmes, P. (1990). Poincaré, celestial mechanics, dynamical-systems theory and 'chaos'. *Physics Reports* **3**, 137–63.

Holton, G. (1988). *Thematic Origins of Scientific Thought* (rev. edn). Harvard University Press.

Hoskin, M. A. (1962). Mining all within: Clarke's notes to Rohault's *Traité de Physique. The Thomist* **24**, 353–63.

(1970). The cosmology of Thomas Wright of Durham. *Journal for the History of Astronomy* **1**, 44–52.

(1995). The discovery of Uranus, the Titius–Bode law, and the asteroids. In R. Taton and C. Wilson (eds), *Planetary Astronomy from the Renaissance to the rise of*

Astrophysics. Part B: The Eighteenth and Nineteenth Centuries. The General History of Astronomy. Volume 2. Cambridge University Press.

Howard, D. and J. D. Norton (1993). Out of the labyrinth? Einstein, Hertz, and the Göttingen answer to the hole argument. In J. Earman, M. Janssen, and J. D. Norton (eds), *The Attraction of Gravitation: New Studies in the History of General relativity,* pp. 30–62. Birkhäuser.

Howell, K. J. (1998). The role of biblical interpretation in the cosmology of Tycho Brahe. *Studies in History and Philosophy of Science* **29**, 515–37.

Hoyle, F. (1974). The work of Nicolaus Copernicus. *Proceedings of the Royal Society of London,* **A 336**, 105–14.

Hoyt, W. G. (1976). W. H. Pickering's planetary predictions and the discovery of Pluto. *Isis* **67**, 551–64.

(1980). *Planets X and Pluto.* University of Arizona Press.

Hubbell, J. G. and R. W. Smith (1992). Neptune in America: Negotiating a discovery. *Journal for the History of Astronomy* **23**, 261–91.

Hughes, B. (1967). *Regiomontanus on Triangles.* University of Wisconsin Press.

Hunt, G. (ed.) (1982). *Uranus and the Outer Planets.* Cambridge University Press.

Hutchison, K. (1990). Sunspots, Galileo, and the orbit of the earth. *Isis* **81**, 68–74.

Iltis, C. (1977). Madame du Châtelet's metaphysics and mechanics. *Studies in History and Philosophy of Science* **8**, 29–48.

Infeld, L. (1973). From Copernicus to Einstein. In B. Bieńkowska (ed.), *The scientific world of Copernicus.* D. Reidel Publishing Company.

Itard, J. (1975). Lagrange. In *Dictionary of Scientific Biography.* Charles Scribner's Sons.

Jacobsen, T. S. (1999). *Planetary Systems from the Ancient Greeks to Kepler.* University of Washington Press.

Jaki, S. L. (1972a). The early history of the Titius–Bode law. *American Journal of Physics* **40**, 1014–23.

(1972b). The original formulation of the Titius–Bode law. *Journal for the History of Astronomy* **3**, 136–8.

(1976). The five forms of Laplace's cosmogony. *American Journal of Physics* **44**, 4–11.

(1977). An English translation of the third part of Kant's *Universal Natural History and Theory of the Heavens.* In W. Yourgrau and A. D. Breck (eds), *Cosmology, History, and Theology,* pp. 387–403. Plenum Press.

(1978). Johann Georg von Soldner and the gravitational bending of light, with an English translation of his essay on it published in 1801. *Foundations of Physics* **8**, 927–50.

Jammer, M. (1957). *Concepts of Force.* Harvard University Press.

Jardine, N. (1982). The significance of the Copernican orbs. *Journal for the History of Astronomy* **13**, 168–94.

(1984). *The Birth of History and Philosophy of Science. Kepler's* A Defence of Tycho against Ursus *with Essays on its Provenance and Significance.* Cambridge University Press.

Johnson, F. R. (1937). *Astronomical Thought in Renaissance England.* The Johns Hopkins Press.

Johnson, J. H. (1931). The discovery of the first four satellites of Jupiter. *Journal of the British Astronomical Association* **41**, 164–71.

Jones, A. (1991a). The adaption of Babylonian methods in Greek numerical astronomy. *Isis* **82**, 440–53.

(1991b). Hipparchus's computations of solar longitudes. *Journal for the History of Astronomy* **22**, 101–25.

Jones, H. S. (1947). *John Couch Adams and the Discovery of Neptune.* Cambridge University Press.

Julliard-Tosel, E. (2000). Bruns' theorem: The proof and some generalizations. *Celestial Mechanics and Dynamical Astronomy* **76**, 241–81.

Kant, I. (1969). *Universal Natural History and Theory of the Heavens.* First pub. 1755. Trans. W. Hastie 1900. University of Michigan Press.

Katz, V. J. (1998). *A History Of Mathematics. An Introduction* (2nd edn). Addison-Wesley.

Kellner, M. (1991). On the status of the astronomy and physics in Maimonides' *Mishneh Torah* and *Guide of the Perplexed*: A chapter in the history of science. *The British Journal for the History of Science* **24**, 453–63.

Kennedy, E. S. (1956a). Parallax theory in Islamic astronomy. *Isis* **47**, 33–53.

(1956b). A survey of Islamic astronomical tables. *Transactions of the American Philosophical Society* **46**(2), 123–77.

(1984). Applied mathematics in the tenth century: Abu'l-Wafā' calculates the distance Baghdad–Mecca. *Historia mathematica* **11**, 193–206. Reprinted in *Astronomy and Astrology in the Medieval Islamic World* by E. S. Kennedy, Variorum Reprints. 1998.

(1988). Two medieval approaches to the equation of time. *Centaurus* **31**, 1–8. Reprinted in *Astronomy and Astrology in the Medieval Islamic World* by E. S. Kennedy, Variorum Reprints. 1998.

(1990). Spherical astronomy in Kāshī's Khāqānī Zīj. *Zeitschrift für Geschichte der Arabisch-Islamischen Wissenschaften* **2**, 1–46. Reprinted in *Astronomy and Astrology in the Medieval Islamic World* by E. S. Kennedy, Variorum Reprints. 1998.

Kennedy, E. S. and V. Roberts (1959). The planetary theory of Ibn al-Shāṭir. *Isis* **50**, 227–35.

Kenny, A. (1998). *A Brief History of Western Philosophy.* Blackwell Publishers.

Kepler, J. (1981). *Mysterium Cosmographicum: The Secret of the Universe.* Abaris Books. Trans. A. M. Duncan, introduction and commentary by E. J. Aiton.

(1992). *New Astronomy.* Trans. William H. Donahue. Cambridge University Press.

(1995). *Epitome of Copernican Astronomy & Harmonies of the World.* Trans. Charles Glenn Wallis. Prometheus Books.

(1997). *The Harmony of the World.* Trans. E. J. Aiton, A. M. Duncan and J. V. Field. American Philosophical Society.

Kerszberg, P. (1986). The cosmological question in Newton's science. *Osiris,* 2nd Series **2**, 69–106.

Kesten, H. (1945). *Copernicus and his World.* Secker and Warburg.

Kiang, T. (1971). The past orbit of Halley's Comet. *Memoirs of the Royal Astronomical Society* **76**, 27–66.

Kidd, D. (ed.) (1997). *Aratus: Phaenomena.* Cambridge University Press.

King, D. A. (1973). Ibn Yūnus' Very Useful Tables for reckoning time by the sun. *Archive for History of Exact Sciences* **10**, 342–94. Reprinted in *Islamic Mathematical Astronomy* by David A. King, 2nd edn, 1993. Variorum Reprints.

(1974). A double-argument table for the lunar equation attributed to Ibn Yūnus. *Centaurus* **18**, 129–46. Reprinted in David A. King. (1993). *Islamic Mathematical Astronomy*. 2nd edn. Variorum Reprints.

(1975a). Ibn al-Shāṭir. In *Dictionary of Scientific Biography*. Charles Scribner's Sons.

(1975b). Ibn Yūnus. In *Dictionary of Scientific Biography*. Charles Scribner's Sons.

(1988). Ibn Yūnus on lunar crescent visibility. *Journal for the History of Astronomy* **19**, 155–68.

King-Hele, D. G. (1972). Heavenly harmony and earthly harmonics. *Quarterly Journal of the Royal Astronomical Society* **13**, 374–95.

Kline, M. (1972). *Mathematical Thought from Ancient to Modern Times*. Oxford University Press.

Knorr, W. R. (1990). Plato and Eudoxus on the planetary motions. *Journal for the History of Astronomy* **21**, 313–29.

Kollerstrom, N. (1991). Newton's two 'Moon-tests'. *The British Journal for the History of Science* **24**, 369–72.

(1995). Newton's lunar mass error. *Journal of the British Astronomical Association* **95**, 151–3.

(1999). The path of Halley's comet, and Newton's late apprehension of the law of gravity. *Annals of Science* **56**, 331–56.

(2000). *Newton's Forgotten Lunar Theory*. Green Lion Press.

Kowal, C. T. and S. Drake (1980). Galileo's observations of Neptune. *Nature* **287**, 311–3.

Koyré, A. (1955). A documentary history of the problem of fall from Kepler to Newton. *Transactions of the American Philosophical Society* **45**, 329–95.

(1973). *The Astronomical Revolution. Copernicus–Kepler–Borelli*. Hermann.

Kozhamthadam, J. (1994). *The Discovery of Kepler's Laws. The Interaction of Science, Philosophy, and Religion*. University of Notre Dame Press.

Kren, C. (1968). Homocentric astronomy in the Latin West: The *De reprobatione ecentricorum et epiciclorum* of Henry of Hesse. *Isis* **59**, 269–81.

(1969). A medieval objection to 'Ptolemy'. *The British Journal for the History of Science* **4**(4), 378–93.

(1971). The rolling device of Naṣīr al-Dīn al-Ṭūsī in the *De spera* of Nicole Oresme? *Isis* **62**, 490–8.

Kriloff, A. N. (1925). On Sir Isaac Newton's method of determining the parabolic orbit of a comet. *Monthly Notices of the Royal Astronomical Society* **85**, 640–56.

Kuhn, T. S. (1957). *The Copernican Revolution. Planetary Astronomy in the Development of Western Thought*. Harvard University Press.

Kunitzsch, P. (1974). New light on al-Battānī's Zīj. *Centaurus* **18**, 270–4.

Lamb, H. (1932). *Hydrodynamics* (6th edn). Cambridge University Press.

Lanczos, C. (1986). *The Variational Principles of Mechanics* (4th edn). Originally pub. 1970. Dover Publications.

Landels, J. G. (1983). Eudoxos' planetary theory—the earliest mathematical model? *Endeavour*, New Series **7**, 183–8.

Laplace, P. S. (1809). *The System of the World*. Trans. J. Pond.

(1951). *A Philosphical Essay on Probablities*. Originally pub. 1814. Dover Publications.

Lasserre, F. (1964). *The Birth of Mathematics in the Age of Plato*. Hutchinson.

Lattis, J. M. (1994). *Between Copernicus and Galileo. Christoph Clavius and the Collapse of Ptolemaic Cosmology*. University of Chicago Press.

Littlewood, D. E. (1953). Conformal transformations and kinematical relativity. *Proceedings of the Cambridge Philosophical Society* **49**, 90–6.

Littlewood, J. E. (1948). Newton and the attraction of a sphere. *Mathematical Gazette* **XXXII**(300). Reprinted in *A Mathematician's Miscellany* (1953). Methuen and Co.

Littlewood, J. E. (1953). The discovery of Neptune and the Adams–Airy affair. In *A Mathematician's Miscellany*. Methuen and Co.

Littmann, M. (1988). *Planets Beyond. Discovering the Outer Solar System*. John Wiley and Sons.

Lohne, J. (1960). Hooke *versus* Newton. *Centaurus* **7**, 6–52.

Lohne, J. A. (1975). Thomas Harriot. In *Dictionary of Scientific Biography*. Charles Scribner's Sons.

Lorch, R. P. (1975). The astronomy of Jābir ibn Aflaḥ. *Centaurus* **19**, 85–107.

Lorentz, H. A., A. Einstein, H. Minkowski, and H. Weyl (1923). *The Principle of Relativity. A Collection of Original Memoirs on the Special and General Theory of Relativity*. Trans. W. Perrett and G. B. Jeffery. Dover Publications.

Lubbock, C. A. (ed.) (1933). *The Herschel Chronicle. The Life-Story of William Herschel and his Sister Caroline Herschel*. Cambridge University Press.

Lyttleton, R. A. (1968). *Mysteries of the Solar System*. Clarendon Press.

Machamer, P. (ed.) (1998). *The Cambridge Companion to Galileo*. Cambridge University Press.

Maeyama, Y. (1998). Determination of the sun's orbit. Hipparchus, Ptolemy, al-Battānī, Copernicus, Tycho Brahe. *Archive for History of Exact Sciences* **53**, 1–49.

Maltese, G. and L. Orlando (1995). The definition of rigidity in the special theory of relativity and the genesis of the general theory of relativity. *Studies in History and Philosophy of Modern Physics* **26**, 263–306.

Maor, E. (2000). *June 8, 2004. Venus in Transit*. Princeton University Press.

Margolis, H. (1991). Tycho's system and Galileo's *Dialogue. Studies in History and Philosophy of Science* **22**, 259–75.

(2002). *It Started with Copernicus. How Turning the World Inside Out Led to the Scientific Revolution*. McGraw-Hill.

Marsden, B. G. (1995). Eighteenth- and nineteenth-century developments in the theory and practice of orbit determination. In R. Taton and C. Wilson (eds), *Planetary Astronomy from the Renaissance to the Rise of Astrophysics. Part B: The Eighteenth and Nineteenth Centuries*. In *The General History of Astronomy*. Vol. 2. Cambridge University Press.

Martens, R. (2000). *Kepler's Philosophy and the New Astronomy*. Princeton University Press.

Martin, D. R. (1984). Status of the Copernican theory before Kepler, Galileo and Newton. *American Journal of Physics* **52**, 982–6.

Maxwell, J. C. (1998). *A Treatise on Electricity and Magnetism* (3rd edn). Oxford Classic Texts in the Physical Sciences. Originally pub. 1891. Clarendon Press.

McCluskey, S. C. (1998). *Astronomies and Cultures in Early Medieval Europe.* Cambridge University Press.

McGehee, R. (1986). Von Zeipel's theorem on singularities in celestial mechanics. *Expositiones mathematicae* 4, 335–45.

McGuire, J. E. (1978). Existence, actuality and necessity: Newton on space and time. *Annals of Science* 35, 463–508.

McGuire, J. E. and M. Tamny (1985). Newton's astronomical apprenticeship: Notes of 1664/5. *Isis* 76, 349–65.

McMullin, E. (1987). Bruno and Copernicus. *Isis* 78, 55–74.

Mehra, J. (1998). One month in the history of the discovery of general relativity theory. *Foundations of Physics Letters* 11, 41–60.

Meli, D. B. (1991). Public claims, private worries: Newton's *Principia* and Leibniz's theory of planetary motion. *Studies in History and Philosophy of Science* 22, 415–49.

(1993). *Equivalence and Priority: Newton Versus Leibniz.* Clarendon Press.

(1998). Shadows and deception: from Borelli's *Theoricae* to the *Saggi* of the Cimento. *The British Journal for the History of Science* 31, 383–402.

Methuen, C. (1996). Maestlin's teaching of Copernicus. The evidence of his university textbook and disputations. *Isis* 87, 230–47.

Miller, A. I. (1998). *Albert Einstein's Special Theory of Relativity. Emergence (1905) and Early Interpretation (1905–1911).* Originally pub. 1981. Springer-Verlag.

Moesgaard, K. P. (1973). Copernican influence on Tycho Brahe. In J. Dobrzycki (ed.), *The Reception of Copernicus' Heliocentric Theory*, pp. 31–55. D. Reidel Publishing Company.

(1975). Tychonian observations, perfect numbers, and the date of creation: Longomontanus's solar and precessional theories. *Journal for the History of Astronomy* 6, 84–99.

(1977). Cosmology in the wake of Tycho Brahe's astronomy. In W. Yourgrau and A. D. Breck (eds), *Cosmology, History, and Theology*, pp. 293–305. Plenum Press.

Møller, C. (1972). *The Theory of Relativity* (2nd edn). Oxford University Press.

Montgomery, S. L. (1999). *The Moon and the Western Imagination.* The University of Arizona Press.

Morando, B. (1995). The golden age of celestial mechanics. In R. Taton and C. Wilson (eds), *Planetary Astronomy from the Renaissance to the Rise of Astrophysics. Part B: The Eighteenth and Nineteenth Centuries.* In *The General History of Astronomy.* Volume 2. Cambridge University Press.

Moulton, F. R. (1920). *Periodic Orbits.* Carnegie Institute.

(1970). *An Introduction to Celestial Mechanics* (2nd edn). Originally pub. 1914. Dover Publications.

Murray, C. D. and S. F. Dermott (1999). *Solar System Dynamics.* Cambridge University Press.

Naylor, R. (2003). Galileo, Copernicanism and the origins of the new science of motion. *The British Journal for the History of Science* 36(2), 151–81.

Neeley, K. A. (2001). *Mary Somerville. Science, Illumination, and the Female Mind.* Cambridge University Press.

Neugebauer, O. (1948). The astronomical origin of the theory of conic sections. *Proceedings of the American Philosophical Society* 92(3), 136–8. Reprinted in *Astronomy and History, Selected Essays* by O. Neugebauer (1983). Springer-Verlag.

(1949). The astronomy of Maimonides and its sources. *Hebrew Union College Annual* **22**, 322–63. Reprinted in O. Neugebauer (1983). *Astronomy and History, Selected Essays.* Springer-Verlag.

(1950). The alleged Babylonian discovery of the precession of the equinoxes. *Journal of the American Oriental Society* **70**, 1–8. Reprinted in O. Neugebauer. (1983). *Astronomy and History, Selected Essays.* Springer-Verlag.

(1953). On the 'hippopede' of Eudoxus. *Scripta mathematcia* **19**(4), 225–9. Reprinted in O. Neugebauer (1983). *Astronomy and History, Selected Essays.* Springer-Verlag.

(1955). Apollonius' planetary theory. *Communications on Pure and Applied Mathematics* **8**, 641–8. Reprinted in O. Neugebauer (1983). *Astronomy and History, Selected Essays.* Springer-Verlag.

(1959). The equivalence of eccentric and epicyclic motion according to Apollonius. *Scripta mathematcia* **24**, 5–21. Reprinted in O. Neugebauer (1983). *Astronomy and History, Selected Essays.* Springer-Verlag.

(1962). Thâbit ben Qurra 'On the solar year' and 'On the motion of the eighth sphere'. *Proceedings of the American Philosophical Society* **106**, 264–99.

(1968). On the planetary theory of Copernicus. *Vistas in Astronomy* **10**, 89–103. Reprinted in O. Neugebauer (1983). *Astronomy and History, Selected Essays.* Springer-Verlag.

(1969). *The Exact Sciences in Antiquity* (2nd edn). Dover Publications.

(1972). On the allegedly heliocentric theory of Venus by Heraclides Ponticus. *American Journal of Philology* **93**(4), 600–1. Reprinted in O. Neugebauer (1983). *Astronomy and History, Selected Essays.* Springer-Verlag.

(1975). *A History of Ancient Mathematical Astronomy.* 3 Vols. Springer-Verlag.

Nevalainen, J. (1996). The accuracy of the ecliptic longitude in Ptolemy's Mercury model. *Journal for the History of Astronomy* **27**, 147–60.

Newton, I. (1934). *Mathematical Principles of Natural Philosophy* (3rd edn). Trans. Andrew Motte in 1729, rev. Florian Cajori. University of California Press.

(1999). *The Principia.* Trans. I. Bernard Cohen and Anne Whitman. University of California Press.

Newton, R. R. (1977). *The Crime of Claudius Ptolemy.* Baltimore.

Nicolson, M. H. (1946). *Newton Demands the Muse: Newton's 'Opticks' and the Eighteenth-Century Poets.* Princeton University Press.

Nieto, M. M. (1972). *The Titius–Bode Law of Planetary Distances: Its History and Theory.* Pergamon.

Nobili, A. M. and C. M. Will (1986). The real value of Mercury's perihelion advance. *Nature* **320**, 39–41.

Norberg, A. L. (1978). Simon Newcomb's early astronomical career. *Isis* **69**, 209–25.

North, J. D. (1967). Medieval star catalogues and the movement of the eighth sphere. *Archives Internationales d'Histoire des Sciences* **20**, 71–83.

(1974). Thomas Harriot and the first telescopic observations of sunspots. In J. W. Shirley (ed.), *Thomas Harriot. Renaissance scientist.* Clarendon Press.

(1984). How Einstein found his field equations: 1912–1915. *Historical Studies in the Physical Sciences* **14**, 253–316.

(1985). What was Einstein's principle of equivalence? *Studies in History and Philosophy of Science* **16**, 203–46.

(1992). Einstein, Nordström and the early demise of scalar Lorentz covariant theories of gravitation. *Archive for History of Exact Sciences* **45**, 17–94.

(1993). Einstein and Nordström: Some lesser-known thought experiments in gravitation. In J. Earman, M. Janssen, and J. D. Norton (eds), *The Attraction of Gravitation: New Studies in the History of General Relativity*, pp. 3–29. Birkhäuser.

(1994). *The Fontana History of Astronomy and Cosmology*. Fontana Press.

Numbers, R. L. (1973). The American Kepler: Daniel Kirkwood and his analogy. *Journal for the History of Astronomy* **4**, 13–21.

O'Leary, D. L. (1948). *How Greek Science Passed to the Arabs*. Routledge and Kegan Paul Ltd.

Olmsted, J. W. (1942). The scientific expedition of Jean Richer to Cayenne (1672–1673). *Isis* **34**, 117–28.

O'Neil, W. M. (1969). *Fact and Theory. An Aspect of the Philosophy of Science*. Sydney University Press.

Orr, M. A. (1956). *Dante and the Early Astronomers* (2nd edn). Allan Wingate.

Pais, A. (1982). *'Subtle is the Lord...'. The Science and Life of Albert Einstein*. Oxford University Press.

Palmieri, P. (1998). Re-examining Galileo's theory of the tides. *Archive for History of Exact Sciences* **53**, 223–375.

Paneth, F. A. (1950). Thomas Wright of Durham. *Endeavour* **9**, 117–25.

Pannekoek, A. (1953). The discovery of Neptune. *Centaurus* **3**, 126–37.

(1961). *A History of Astronomy*. George Allen and Unwin Ltd.

Patterson, L. D. (1949). Hooke's gravitation theory and its influence on Newton. I: Hooke's gravitation theory. *Isis* **40**, 327–41.

(1950). Hooke's gravitation theory and its influence on Newton. II: The insufficiency of the traditional estimate. *Isis* **41**, 32–45.

Pavelle, R., M. Rothstein, and J. Fitch (1981). Computer algebra. *Scientific American* **245**(6), 102–13.

Payne-Gaposchkin, C. (1961). *Introduction to Astronomy*. Methuen. Originally pub. 1954.

Pecker, J.-C. (2001). *Understanding the Heavens. Thirty Centuries of Astronomical Ideas from Ancient Thinking to Modern Cosmology*. Springer-Verlag.

Pedersen, O. (1974). *A Survey of the Almagest*. Odense University Press.

(1981). The origins of the 'Theorica Planetarum'. *Journal for the History of Astronomy* **12**, 113–23.

(1985). In quest of Sacrobosco. *Journal for the History of Astronomy* **16**, 175–221.

(1993). *Early Physics and Astronomy. A Historical Introduction* (2nd edn). Cambridge University Press.

Petengill, G. H. and R. B. Dyce (1965). A radar determination of the rotation of the planet Mercury. *Nature* **206**, 1240.

Peterson, I. (1993). *Newton's Clock. Chaos in the Solar System*. W. H. Freeman.

Peterson, V. M. (1969). The three lunar models of Ptolemy. *Centaurus* **14**(1), 142–71.

Peterson, V. M. and O. Schmidt (1967). The determination of the longitude of the apogee of the orbit of the sun according to Hipparchus and Ptolemy. *Centaurus* **12**, 73–96.

Pingree, D. (1971). On the Greek origin of the Indian planetary model employing a double epicycle. *Journal for the History of Astronomy* **2**, 80–5.

(1972). Precession and trepidation in Indian astronomy before A.D. 1200. *Journal for the History of Astronomy* **3**, 27–35.

(1975). History of mathematical astronomy in India. In *Dictionary of Scientific Biography*, Volume XV, pp. 533–633. Charles Scribner's Sons.

Pinkard, T. (2000). *Hegel. A Biography*. Cambridge University Press.

Pirani, F. A. E. (1955). On the perihelion motion according to Littlewood's equations. *Proceedings of the Cambridge Philosophical Society* **51**, 535–7.

Plackett, R. L. (1972). Studies in the history of probability and statistics. XXIX. The discovery of the method of least squares. *Biometrika* **59**, 239–51.

Playfair, J. (1822). *The Works of John Playfair*. Archibald and Co.

Poincaré, H. (1929). *The Foundations of Science*. Trans. George Bruce Halsted. The Science Press.

(1952). *Science and Hypothesis*. Dover Publications.

(1958). *The Value of Science*. Dover Publications.

(1993). *New Methods of Celestial Mechanics*. Rev. of 1967 NASA trans. of *Méthodes nouvelles de la Mécanique Céleste* (1892–99). American Institute of Physics.

Poor, C. L. (1921). The motions of the planets and the relativity theory. *Science* **54**, 30–4.

Poulle, E. (1988). The Alfonsine Tables and Alfonso X of Castile. *Journal for the History of Astronomy* **19**, 97–113.

Pourciau, B. (1991). On Newton's proof that inverse-square orbits must be conics. *Annals of Science* **48**, 159–72.

(1997). Reading the master: Newton and the birth of celestial mechanics. *The American Mathematical Monthly* **104**, 1–19.

(2001). Newton and the notion of limit. *Historia mathematica* **28**, 18–30.

Price, M. P. and W. F. Rush (1979). Nonrelativistic contribution to Mercury's perihelion precession. *American Journal of Physics* **47**, 531–4.

Pyenson, L. (1976). Einstein's early scientific collaboration. *Historical Studies in the Physical Sciences* **7**, 83–123.

(1977). Hermann Minkowski and Einstein's special theory of relativity. *Archive for History of Exact Sciences* **17**, 71–95.

Ragep, F. J. (1993). *Naṣīr al-Dīn al-Ṭūsī's* Memoir on Astronomy *(al-Tadhkira fī'ilm al hay'a)*. Volumes 1 and 2. Springer-Verlag.

Rawlins, D. (1968). A long lost observation of Uranus: Flamsteed, 1714. *Publications of the Astronomical Society of the Pacific* **80**, 217–19.

(1982). Eratosthenes' geodesy unraveled: was there a high-accuracy Hellenistic astronomy? *Isis* **73**, 259–65.

(1987). Ancient heliocentrists, Ptolemy, and the equant. *American Journal of Physics* **55**, 235–9.

Reaves, G. (1951). Kourganoff's contributions to the history of the discovery of Pluto. *Publications of the Astronomical Society of the Pacific* **63**, 49–60.

Richards, E. G. (1998). *Mapping Time. The Calendar and its History*. Oxford University Press.

Richardson, D. E. (1945). Distances of planets from the sun and of satellites from their primaries in the satellite systems of Jupiter, Saturn and Uranus. *Popular Astronomy* **53**, 14–26.

Riddell, R. C. (1978). The latitudes of Venus and Mercury in the Almagest. *Archive for History of Exact Sciences* **19**, 95–111.

(1979). Eudoxan mathematics and Eudoxan spheres. *Archive for History of Exact Sciences* **20**, 1–19.

Roberts, V. (1957). The solar and lunar theory of Ibn al-Shāṭir. *Isis* **48**, 428–32.

(1966). The planetary theory of Ibn al-Shāṭir: Latitudes of the planets. *Isis* **57**, 208–19.

Rochberg, F. (2002). A consideration of Babylonian astronomy within the historiography of science. *Studies in History and Philosophy of Science* **33**, 661–84.

Ronan, C. A. (1967). *Their Majesties' Astronomers*. The Bodley Head.

Rosen, E. (1947). *The Naming of the Telescope*. Henry Schuman.

(1959). *Three Copernican Treatises* (2nd edn). Dover Publications.

(1965). *Kepler's Conversation with Galileo's Sidereal Messenger*. Johnson Reprint Corporation.

(1983). Was Copernicus a Neoplatonist? *Journal for the History of Ideas* **44**, 667–9.

(1985a). The dissolution of the celestial spheres. *Journal for the History of Ideas* **46**, 13–31.

(1985b). Kepler's early writings. *Journal for the History of Ideas* **46**, 449–54.

Roseveare, N. (1982). *Mercury's Perihelion from Le Verrier to Einstein*. Clarendon Press.

Rosińska, G. (1974). Naṣīr al-Dīn al-Ṭūsī and Ibn al-Shāṭir in Cracow? *Isis* **65**, 239–43.

Rouse Ball, W. W. (1908). *A Short Account of the History of Mathematics* (4th edn). Macmillan and Co.

Roy, A. E. (1978). *Orbital Motion*. Adam Hilger Ltd.

Ruffner, J. A. (1971). The curved and the straight: Cometary theory from Kepler to Hevelius. *Journal for the History of Astronomy* **2**, 178–94.

(2000). Newton's propositions on comets: Steps in transition, 1681–84. *Archive for History of Exact Sciences* **54**, 259–77.

Russell, J. L. (1964). Kepler's laws of planetary motion: 1609–1666. *The British Journal for the History of Science* **2**, 1–24.

(1973). The Copernican system in Great Britain. In J. Dobrzycki (ed.), *The Reception of Copernicus' Heliocentric Theory*, pp. 189–239. D. Reidel Publishing Company.

(1989). Catholic astronomers and the Copernican system after the condemnation of Galileo. *Annals of Science* **46**, 365–86.

Saari, D. G. (1990). A visit to the Newtonian *n*-body problem via elementary complex variables. *The American Mathematical Monthly* **97**, 105–19.

Saari, D. G. and Z. Xia (1995). Off to infinity in finite time. *Notices of the American Mathematical Society* **42**(5), 538–46.

Sadler, P. M. (1990). William Pickering's search for a planet beyond Neptune. *Journal for the History of Astronomy* **21**, 59–64.

Said, H. M. and A. Z. Khan (1981). *Al-Bīrūnī, his Times, Life and Works*. Hamdard Academy.

Saliba, G. (1987a). The role of Maragha in the development of Islamic astronomy: A scientific revolution before the Renaissance. *Revue de Synthèse* **108**, 361–73. Reprinted in George Saliba (1994). *A History of Arabic Astronomy. Planetary Theories During the Golden Age of Islam*. New York University Press.

(1987b). Theory and observation in Islamic astronomy: The work of Ibn al-Shāṭir of Damascus. *Journal for the History of Astronomy* **18**, 35–43. Reprinted in George Saliba (1994). *A history of Arabic Astronomy. Planetary Theories During the Golden Age of Islam*. New York University Press.

(1991). The astronomical tradition of Maragha: A historical survey and prospects for future reserach. *Arabic Sciences and Philosophy* **1**, 67–99. Reprinted in George Saliba. (1994). *A History of Arabic Astronomy. Planetary Theories During the Golden Age of Islam.* New York University Press.

(1994). *A History of Arabic Astronomy. Planetary Theories During the Golden Age of Islam.* New York University Press.

Saliba, G. and E. S. Kennedy (1991). The spherical case of the Ṭūsī couple. *Arabic Science and Philosophy* **1**, 285–91. Reprinted in E. S. Kennedy. (1998). *Astronomy and Astrology in the Medieval Islamic World.* Variorum Reprints.

Schaffer, S. (1981). Uranus and the establishment of Herschel's astronomy. *Journal for the History of Astronomy* **12**, 11–26.

Schmeidler, F. (1995). Astronomy and the theory of errors: from the method of averages to the method of least squares. In R. Taton and C. Wilson (eds), *Planetary Astronomy from the Renaissance to the Rise of Astrophysics. Part B: The Eighteenth and Nineteenth Centuries. The General History of Astronomy.* Volume 2. Cambridge University Press.

Schofield, C. J. (1981). *Tychonic and Semi-Tychonic World Systems.* Arno Press.

Schwarz, H. M. (1977). Einstein's comprehensive 1907 essay on relativity. *American Journal of Physics* **45**, 512–17, 811–17, 899–902.

Shapiro, A. E. (1975). Archimedes' measurement of the sun's apparent diameter. *Journal for the History of Astronomy* **6**, 75–83.

Shapley, H. and H. E. Howarth (1929). *A Source Book in Astronomy.* McGraw-Hill.

Shea, W. R. (1970). Galileo, Scheiner and the interpretation of sunspots. *Isis* **61**, 498–519.

(1998). Galileo's Copernicanism. The science and the rhetoric. In P. Machamer (ed.), *The Cambridge Companion to Galileo*, pp. 211–43. Cambridge University Press.

Shevchenko, M. (1990). An analysis of errors in the star catalogues of Ptolemy and Ulugh Beg. *Journal for the History of Astronomy* **21**, 187–201.

Sheynin, O. B. (1973a). Finite random sums. *Archive for History of Exact Sciences* **9**, 275–305.

(1973b). Mathematical treatment of astronomical observations (a historical essay). *Archive for History of Exact Sciences* **11**, 97–126.

(1977a). Early history of probability. *Archive for History of Exact Sciences* **17**, 201–59.

(1977b). Laplace's theory of errors. *Archive for History of Exact Sciences* **17**, 1–61.

(1984). On the history of the statistical method in astronomy. *Archive for History of Exact Sciences* **29**, 151–99.

Showalter, M. R. (1991). Visual detection of 1981S13, Saturn's eighteenth satellite, and its role in the Encke gap. *Nature* **351**, 709–13.

Singer, D. W. (1950). *Giordano Bruno: His Life and Thought. With Annotated Translation of His Work, On the Infinite Universe and Worlds.* Henry Schuman.

Smart, W. M. (1947). John Couch Adams and the discovery of Neptune. *Occasional Notes of the Royal Astronomical Society* **2**(11), 33–88.

(1960). *Text-book on Spherical Astronomy* (6th edn). Cambridge University Press.

Smith, A. M. (1985). Galileo's proof for the earth's motion from the movement of sunspots. *Isis* **76**, 543–51.

Smith, R. W. (1989). The Cambridge network in action: The discovery of Neptune. *Isis* **80**, 395–422.

Snelders, H. A. M. (1980). Christiaan Huygens and the concept of matter. In H. J. M. Bos, M. J. S. Rudwick, H. A. M. Snelders, and R. P. W. Visser (eds), *Studies on Christiaan Huygens. Invited Papers from the Symposium on the Life and Work of Christiaan Huygens, Amsterdam, 22–25 August 1979*, pp. 104–25. Swets and Zeitlinger.

Sobel, D. (1996). *Longitude: The True Story of a Lone Genius who Solved the Greatest Scientific Problem of his Time*. Fourth Estate.

Speiser, D. (1996). The Kepler problem from Newton to Johann Bernoulli. *Archive for History of Exact Sciences* **50**, 103–16.

Sponsel, A. (2002). Constructing a 'revolution in science': the campaign to promote a favourable reception for the 1919 solar eclipse experiments. *The British Journal for the History of Science* **35**, 439–67.

Stachel, J. (1980). Einstein and the rigidly rotating disk. In A. Held (ed.), *General Relativity and Gravitation. One Hundred Years After the Birth of Albert Einstein*. Volume 1. Plenum Press.

(ed.) (1998). *Einstein's Miraculous Year. Five Papers that Changed the Face of Physics*. Princeton University Press.

Standage, T. (2000). *The Neptune File: Planet Detectives and the Discovery of Worlds Unseen*. Allen Lane, The Penguin Press.

Steele, J. M. (2000). Eclipse prediction in Mesopotamia. *Archive for History of Exact Sciences* **54**, 421–54.

(2003). Planetary latitudes in Babylonian mathematical astronomy. *Journal for the History of Astronomy* **34**, 269–90.

Stephenson, B. (1987). *Kepler's Physical Astronomy*. Springer-Verlag.

Stern, A. and J. Mitton (1998). *Pluto and Charon. Ice Worlds on the Ragged Edge of the Solar System*. John Wiley and Sons.

Stigler, S. M. (1975). Studies in the history of probability and statistics. XXXIV. Napoleonic statistics: The work of Laplace. *Biometrika* **62**, 503–17.

(1978). Laplace's early work: chronology and citations. *Isis* **69**, 234–54.

(1981). Gauss and the invention of least squares. *The Annals of Statistics* **9**, 465–74.

(1986). Laplace's 1774 memoir on inverse probability. *Statistical Science* **1**, 359–78.

Suzuki, J. A. (1996). *A History of the Stability Problem in Celestial Mechanics from Newton to Laplace*. Ph.D. thesis, Graduate School of Arts and Sciences, Boston University.

Swerdlow, N. M. (1969). Hipparchus on the distance of the sun. *Centaurus* **14**(1), 287–305.

(1972). Aristotelian planetary theory in the Renaissance: Giovanni Batista Amico's homocentric spheres. *Journal for the History of Astronomy* **3**, 36–48.

(1973). Al-Battānī's determination of the solar distance. *Centaurus* **17**, 97–105.

(1975). The planetary theory of François Viète 1. The fundamental planetary models. *Journal for the History of Astronomy* **6**, 185–208.

(1979). Hipparchus' determination of the length of the tropical year and the rate of precession. *Archive for History of Exact Sciences* **21**, 291–309.

(1981a). On establishing the text of 'De Revolutionibus'. *Journal for the History of Astronomy* **12**, 35–46.

(1981b). Translating Copernicus. *Isis* **72**, 73–82.

(1989). Ptolemy's theory for the inferior planets. *Journal for the History of Astronomy* **20**, 29–60.

(1998a). *The Babylonian Theory of the Planets*. Princeton University Press.

(1998b). Galileo's discoveries with the telescope and their evidence for the Copernican theory. In P. Machamer (ed.), *The Cambridge Companion to Galileo*, pp. 244–70. Cambridge University Press.

(1999). Acronychal risings in Babylonian planetary theory. *Archive for History of Exact Sciences* **54**, 49–65.

Swerdlow, N. M. and O. Neugebauer (1984). *Mathematical Astronomy in Copernicus's De Revolutionibus*. Springer-Verlag.

Szczesniak, B. (1949). Note on Kepler's *Tabulae Rudolphinae* in the library of Pei-t'ang in Pekin. *Isis* **40**, 344–7.

Tagliaferri, G. and P. Tucci (1999). Carlini and Plana on the theory of the moon and their dispute with Laplace. *Annals of Science* **56**, 221–69.

Taisbak, C. M. (1974). Posidonius vindicated at all costs? Modern scholarship versus the Stoic earth measurer. *Centaurus* **18**, 253–69.

Taylor, E. F. and J. A. Wheeler (2000). *Exploring Black Holes. Introduction to General Relativity*. Addison Wesly Longman.

Terrall, M. (1992). Representing the earth's shape. The polemics surrounding Maupertius' expedition to Lapland. *Isis* **83**, 218–37.

(2003). *The Man Who Flattened the Earth: Maupertius and the Sciences in the Enlightenment*. University of Chicago Press.

Teske, A. (1972). Voltaire's 'Elements of Newtonian physics': their significance for yesterday and today. In *The History of Physics and the Philosophy of Science. Selected Essays*. Zaklad Narodowy Imienia Ossolińskich Wydawnictwo Polskiej Akademii Nauk, Wroclaw.

Thoren, V. E. (1967a). An early instance of deductive astronomy: Tycho Brahe's lunar theory. *Isis* **58**, 19–36.

(1967b). Tycho Brahe's discovery of the variation. *Centaurus* **12**(3), 151–66.

(1971). Anaxagoras, Eudoxus, and the regression of the lunar nodes. *Journal for the History of Astronomy* **2**, 23–8.

(1988). Prosthaphaeresis revisted. *Historia mathematica* **15**, 32–9.

(1990). *The Lord of Uraniborg. A Biography of Tycho Brahe*. Cambridge University Press.

Thorndike, L. (1949). *The* Sphere *of Sacrobosco and its Commentators*. University of Chicago Press.

Thurston, H. (1994). *Early Astronomy*. Springer.

(2002). Greek mathematical astronomy reconsidered. *Isis* **93**, 58–69.

Todhunter, I. (1962). *A History of the Mathematical Theories of Attraction and the Figure of the Earth*. First pub. by Macmillan and Company (1873). Dover Publications.

Tombaugh, C. W. and P. Moore (1980). *Out of the Darkness. The Planet Pluto*. Stackpole Books.

Toomer, G. J. (1967). The size of the lunar epicycle according to Hipparchus. *Centaurus* **12**(3), 145–50.

(1968). A survey of the Toledan tables. *Osiris* **15**, 5–174.

(1969). The solar theory of al-Zarqal: a history of errors. *Centaurus* **14**, 306–36.

(1974). The chord table of Hipparchus and the early history of Greek trigonometry. *Centaurus* **18**, 6–28.

(1984). *Ptolemy's* Almagest. Duckworth.

Toplis, J. (1814). *A Treatise on Analytical Mechanics; Being the First Book of the Mecanique Celeste of M. le Comte Laplace.* Nottingham.

Toulmin, S. and J. Goodfield (1965). *The Fabric of the Heavens. The Development of Astronomy and Dynamics.* Originally pub. (1961). Harper and Row.

Turnbull, H. W. (ed.) (1961). *The Correspondence of Isaac Newton.* Cambridge University Press.

Valluri, S. R., C. Wilson, and W. Harper (1997). Newton's apsidal precession theorem and eccentric orbits. *Journal for the History of Astronomy* **28**, 13–27.

Van Brummelen, G. (1994). Lunar and planetary interpolation tables in Ptolemy's *Almagest. Journal for the History of Astronomy* **25**, 297–311.

Van der Waerden, B. L. (1974). The earliest form of the epicycle theory. *Journal for the History of Astronomy* **5**, 175–85.

(1982). The motion of Venus, Mercury and the sun in early Greek astronomy. *Archive for History of Exact Sciences* **26**, 99–113.

Van Helden, A. (1974a). 'Annulo Cingitur': The solution of the problem of Saturn. *Journal for the History of Astronomy* **5**, 155–74.

(1974b). Saturn and his anses. *Journal for the History of Astronomy* **5**, 105–21.

(1985). *Measuring the Universe. Cosmic Dimensions from Aristarchus to Halley.* University of Chicago Press.

(1989). The telescope and cosmic dimensions. In R. Taton and C. Wilson (eds), *Planetary Astronomy from the Renaissance to the Rise of Astrophysics. Part A: Tycho Brahe to Newton. The General History of Astronomy.* Volume 2. Cambridge University Press.

Veselovsky, I. N. (1973). Copernicus and Naṣīr al-Dīn al-Ṭūsī. *Journal for the History of Astronomy* **4**, 128–30.

Voelkel, J. R. (1999). *Johannes Kepler and the New Astronomy.* Oxford University Press.

Voltaire (1967). *The Elements of Sir Isaac Newton's Philosophy.* Trans. John Hanna (1738). Frank Cass and Co, Ltd.

Von Klüber, H. (1960). The determination of Einstein's light-deflection in the gravitational field of the sun. *Vistas in Astronomy* **3**, 47–77.

Waff, C. B. (1976). Isaac Newton, the motion of the lunar apogee, and the establishment of the inverse square law. *Vistas in Astronomy* **20**, 99–103.

(1986). Comet Halley's first expected return: English public apprehensions, 1755–58. *Journal for the History of Astronomy* **17**, 1–37.

(1995a). Clairaut and the motion of the lunar apse: the inverse-square law undergoes a test. In R. Taton and C. Wilson (eds), *Planetary Astronomy from the Renaissance to the Rise of Astrophysics. Part B: The Eighteenth and Nineteenth Centuries. The General History of Astronomy.* Volume 2. Cambridge University Press.

(1995b). Predicting the mid-eighteenth-century return of Halley's Comet. In R. Taton and C. Wilson (eds), *Planetary Astronomy from the Renaissance to the Rise of Astrophysics. Part B: The Eighteenth and Nineteenth Centuries. The General History of Astronomy.* Volume 2. Cambridge University Press.

Wall, B. E. (1975). The anatomy of a precursor. *Studies in History and Philosophy of Science* **6**, 201–28.

Wang, Q.-D. (1991). The global solution of the N-body problem. *Celestial Mechanics and Dynamical Astronomy* **50**, 73–88.

Waterhouse, W. C. (1972). The discovery of the regular solids. *Archive for History of Exact Sciences* **9**, 212–21.

Watson, G. N. (1944). *A Treatise on the Theory of Bessel Functions* (2nd edn). Cambridge University Press.

Weinberg, S. (1972). *Gravitation and Cosmology: Principles and Applications of the General Theory of Relativity*. John Wiley and Sons.

Weinstock, R. (1982). Dismantling a centuries-old myth: Newton's *Principia* and inverse-square orbits. *American Journal of Physics* **50**, 610–17.

— (1984). Newton's *Principia* and the external gravitational field of a spherically symmetric mass distribution. *American Journal of Physics* **52**, 883–90.

Wesley, W. (1978). The accuracy of Tycho Brahe's instruments. *Journal for the History of Astronomy* **9**, 42–53.

Westfall, R. S. (1967). Hooke and the law of universal gravitation. *The British Journal for the History of Science* **3**, 245–61.

— (1971). *Force in Newton's physics. The Science of Dynamics in the Seventeenth Century*. Macdonald.

— (1973). Newton and the fudge factor. *Science* **179**, 751–8.

— (1980). *Never at Rest. A Biography of Isaac Newton*. Cambridge University Press.

— (1993). *The Life of Isaac Newton*. Cambridge University Press.

Westman, R. S. (1973). The comet and the cosmos: Kepler, Mästlin and the Copernican hypothesis. In J. Dobrzycki (ed.), *The Reception of Copernicus' Heliocentric Theory*, pp. 7–30. D. Reidel Publishing Company.

— (1975). The Melanchthon circle, Rheticus, and the Wittenberg interpretation of the Copernican theory. *Isis* **66**, 165–93.

— (1980). The astronomer's role in the sixteenth century: A preliminary study. *History of Science* **18**, 105–47.

White, M. and J. Gribbin (1993). *Einstein. A Life in Science*. Simon and Schuster.

Whiteside, D. T. (1964). Newton's early thoughts on planetary motion: A fresh look. *The British Journal for the History of Science* **2**, 117–37.

— (1970a). Before the *Principia*: The maturing of Newton's thoughts on dynamical astronomy, 1664–1684. *Journal for the History of Astronomy* **1**, 5–19.

— (1970b). The mathematical principles underlying Newton's *Principia Mathematica*. *Journal for the History of Astronomy* **1**, 116–38.

— (1974). Keplerian planetary eggs, laid and unlaid, 1600–1605. *Journal for the History of Astronomy* **5**, 1–21.

— (1976). Newton's lunar theory: from high hope to disenchantment. *Vistas in Astronomy* **19**, 317–28.

Whitrow, G. J. and G. E. Morduch (1965). Relativistic theories of gravitation. A comparative analysis with particular reference to astronomical tests. *Vistas in Astronomy* **6**, 1–67.

Whittaker, E. T. (1937). *A Treatise on the Analytical Dynamics of Particles and Rigid Bodies* (4th edn). Cambridge University Press.

— (1951). *A History of the Theories of Aether and Electricity. The Classical Theories* (rev. edn). Thomas Nelson and Sons.

(1953). *A History of the Theories of Aether and Electricity. The Modern Theories 1900–1926.* Thomas Nelson and Sons.

(1958). *From Euclid to Eddington. A Study of Conceptions of the External World.* Originally pub. 1949. Dover Publications.

Will, C. M. (1979). The confrontation between gravitation theory and experiment. In S. W. Hawking and W. Israel (eds), *General Relativity. An Einstein Centenary Survey,* pp. 24–89. Cambridge University Press.

(1988). Henry Cavendish, Johann von Soldner, and the deflection of light. *American Journal of Physics* **56**, 413–15.

Williams, J. G. and G. S. Benson (1971). Resonances in the Neptune–Pluto system. *The Astronomical Journal* **76**, 167–77.

Williams, K. P. (1945). A comparison of the solar theories of Newcomb and LeVerrier with conversion tables for the nineteenth century. *Publications of the Kirkwood Observatory of Indiana University* **8**.

Wilson, C. A. (1968). Kepler's derivation of the elliptical path. *Isis* **59**, 5–25. Reprinted in *Astronomy from Kepler to Newton* by C. A. Wilson (1989). Variorum Reprints.

(1970). From Kepler's laws, so-called, to universal gravitation: Empirical factors. *Archive for History of Exact Sciences* **6**, 89–170. Reprinted in C. A. Wilson (1989). *Astronomy from Kepler to Newton.* Variorum Reprints.

(1973). The inner planets and the Keplerian revolution. *Centaurus* **17**, 205–48. Reprinted in C. A. Wilson (1989). *Astronomy from Kepler to Newton.* Variorum Reprints.

(1978). Horrocks, harmonies, and the exactitude of Kepler's third law. *Studia Copernicana* **16**, 235–59. Reprinted in C. A. Wilson (1989). *Astronomy from Kepler to Newton.* Variorum Reprints.

(1980). Perturbations and solar tables from Lacaille to Delambre. *Archive for History of Exact Sciences* **22**, 53–304.

(1984). The sources of Ptolemy's parameters. *Journal for the History of Astronomy* **15**, 37–47. Review in R. R. Newton (1982). *The Origins of Ptolemy's Astronomical Parameters.*

(1985). The great inequality of Jupiter and Saturn: from Kepler to Laplace. *Archive for History of Exact Sciences* **33**, 15–290.

(1987a). D'Alembert *versus* Euler on the precession of the equinoxes and the mechanics of rigid bodies. *Archive for History of Exact Sciences* **37**, 233–73.

(1987b). On the origin of Horrock's lunar theory. *Journal for the History of Astronomy* **18**, 77–94. Reprinted in C. A. Wilson (1989). *Astronomy from Kepler to Newton.* Variorum Reprints.

(1989a). The Newtonian achievement in astronomy. In R. Taton and C. Wilson (eds), *Planetary Astronomy from the Renaissance to the Rise of Astrophysics. Part A: Tycho Brahe to Newton. The General History of Astronomy.* Volume 2. Cambridge University Press.

(1989b). Predictive astronomy in the century after Kepler. In R. Taton and C. Wilson (eds), *Planetary Astronomy from the Renaissance to the Rise of Astrophysics. Part A: Tycho Brahe to Newton. The General History of Astronomy.* Volume 2. Cambridge University Press.

(1993). Clairaut's calculation of the eighteenth-century return of Halley's comet. *Journal for the History of Astronomy* **24**, 1–15.

(1995a). The problem of perturbation analytically treated: Euler, Clairaut, d'Alembert. In R. Taton and C. Wilson (eds), *Planetary Astronomy from the Renaissance to the Rise of Astrophysics. Part B: The Eighteenth and Nineteenth Centuries. The General History of Astronomy*. Volume 2. Cambridge University Press.

(1995b). The work of Lagrange in celestial mechanics. In R. Taton and C. Wilson (eds), *Planetary Astronomy from the Renaissance to the Rise of Astrophysics. Part B: The Eighteenth and Nineteenth Centuries. The General History of Astronomy*. Volume 2. Cambridge University Press.

Wisdom, J. (1987). Chaotic dynamics in the solar system. *Icarus* **72**, 241–75.

Wisdom, J., S. J. Peale, and F. Mignard (1984). The chaotic rotation of Hyperion. *Icarus* **58**, 137–52.

Woolf, H. (1959). *The Transits of Venus. A Study of Eighteenth-Century Science*. Princeton University Press.

Wright, L. (1973). The astronomy of Eudoxus: Geometry or physics. *Studies in History and Philosophy of Science* **4**, 165–72.

Xia, Z. (1992). The existence of noncollision singularities in the N-body problem. *Annals of Mathematics* **135**, 411–68.

Yavetz, I. (1998). On the homocentric spheres of Eudoxus. *Archive for History of Exact Sciences* **52**, 221–78.

(2001). A new role for the hippopede of Eudoxus. *Archive for History of Exact Sciences* **56**, 69–93.

Yeomans, D. K. (1977). The origin of North American astronomy—seventeenth century. *Isis* **68**, 414–25.

Yoke, H. P. (1983). Tycho Brahe (1546–1601) and China. In *Hong Kong – Denmark lectures on science and humanities*. Hong Kong University Press.

Zinsser, J. P. (2001). Translating Newton's *Principia*: the Marquise du Châtelet's revisions and additions for a French audience. *Notes and Records of the Royal Society* **55**, 227–45.

Index

503